T0297687

Transformations of Lamarckism

Vienna Series in Theoretical Biology
Gerd B. Müller, Günter P. Wagner, and Werner Callebaut, editors

The Evolution of Cognition, edited by Cecilia Heyes and Ludwig Huber, 2000

Origination of Organismal Form: Beyond the Gene in Development and Evolutionary Biology, edited by Gerd B. Müller and Stuart A. Newman, 2003

Environment, Development, and Evolution: Toward a Synthesis, edited by Brian K. Hall, Roy D. Pearson, and Gerd B. Müller, 2004

Evolution of Communication Systems: A Comparative Approach, edited by D. Kimbrough Oller and Ulrike Griebel, 2004

Modularity: Understanding the Development and Evolution of Natural Complex Systems, edited by Werner Callebaut and Diego Rasskin-Gutman, 2005

Compositional Evolution: The Impact of Sex, Symbiosis, and Modularity on the Gradualist Framework of Evolution, by Richard A. Watson, 2006

Biological Emergences: Evolution by Natural Experiment, by Robert G. B. Reid, 2007

Modeling Biology: Structure, Behaviors, Evolution, edited by Manfred D. Laubichler and Gerd B. Müller, 2007

Evolution of Communicative Flexibility: Complexity, Creativity, and Adaptability in Human and Animal Communication, edited by Kimbrough D. Oller and Ulrike Griebel, 2008

Functions in Biological and Artificial Worlds: Comparative Philosophical Perspectives, edited by Ulrich Krohs and Peter Kroes, 2009

Cognitive Biology: Evolutionary and Developmental Perspectives on Mind, Brain, and Behavior, edited by Luca Tommasi, Mary A. Peterson, and Lynn Nadel, 2009

Innovation in Cultural Systems: Contributions from Evolutionary Anthropology, edited by Michael J. O'Brien and Stephen J. Shennan, 2010

The Major Transitions in Evolution Revisited, edited by Brett Calcott and Kim Sterelny, 2011

Transformations of Lamarckism: From Subtle Fluids to Molecular Biology, edited by Snait B. Gissis, and Eva Jablonka, 2011

Transformations of Lamarckism

From Subtle Fluids to Molecular Biology

edited by Snait B. Gissis and Eva Jablonka

with illustrations by Anna Zeligowski

The MIT Press
Cambridge, Massachusetts
London, England

First MIT Press paperback edition, 2015

This book was set in Times Roman by Toppan Best-set Premedia Limited.

Library of Congress Cataloging-in-Publication Data

Transformations of Lamarckism : from subtle fluids to molecular biology / edited by Snait B. Gissis and Eva Jablonka ; with illustrations by Anna Zeligowski.
p. cm. – (Vienna series in theoretical biology)
Includes bibliographical references and index.
ISBN 978-0-262-01514-1 (hardcover : alk. paper)–978-0-262-52750-7 (pb. : alk. paper)
1. Evolution (Biology)–History. 2. Natural selection. 3. Adaptation (Biology) 4. Evolutionary genetics. 5. Lamarck, Jean Baptiste Pierre Antoine de Monet de, 1744-1829. 6. Darwin, Charles, 1809-1882. I. Gissis, Snait, 1945- II. Jablonka, Eva. III. Title.

QH361.T73 2011
576.8′27–dc22

2010031344

The MIT Press is pleased to keep this title available in print by manufacturing single copies, on demand, via digital printing technology.

Contents

Series Foreword ix
Preface xi
Acknowledgments xv

INTRODUCTORY ESSAYS 1

1 **Lamarck, Darwin, and the Contemporary Debate about Levels of Selection 3**
Gabriel Motzkin

2 **Jean-Baptiste Lamarck: From Myth to History 9**
Pietro Corsi

I **HISTORY 19**

3 **Introduction: Lamarckian Problematics in Historical Perspective 21**
Snait B. Gissis

4 **Lamarck, Cuvier, and Darwin on Animal Behavior and Acquired Characters 33**
Richard W. Burkhardt, Jr.

5 **The Golden Age of Lamarckism, 1866–1926 45**
Sander Gliboff

6 **Germinal Selection: A Weismannian Solution to Lamarckian Problematics 57**
Charlotte Weissman

7 **The Notions of Plasticity and Heredity among French Neo-Lamarckians (1880–1940): From Complementarity to Incompatibility 67**
Laurent Loison

8 **Lamarckism and Lysenkoism Revisited 77**
Nils Roll-Hansen

9 **Lamarckism and the Constitution of Sociology 89**
Snait B. Gissis

II **THE MODERN SYNTHESIS 101**

10 **Introduction: The Exclusion of Soft ("Lamarckian") Inheritance from the Modern Synthesis 103**
Snait B. Gissis and Eva Jablonka

11 **Attitudes to Soft Inheritance in Great Britain, 1930s–1970s 109**
Marion J. Lamb

12 **The Decline of Soft Inheritance 121**
Scott Gilbert

13 **Why Did the Modern Synthesis Give Short Shrift to "Soft Inheritance"? 127**
Adam Wilkins

14 **The Modern Synthesis: Discussion 133**

III **BIOLOGY 143**

15 **Introduction: Lamarckian Problematics in Biology 145**
Eva Jablonka

16 **Lamarck's Dangerous Idea 157**
Stuart A. Newman and Ramray Bhat

17 **Behavior, Stress, and Evolution in Light of the Novosibirsk Selection Experiments 171**
Arkady L. Markel and Lyudmila N. Trut

18 **The Role of Cellular Plasticity in the Evolution of Regulatory Novelty 181**
Erez Braun and Lior David

19 **Evolutionary Implications of Individual Plasticity 193**
Sonia E. Sultan

20 **Epigenetic Variability in a Predator-Prey System 205**
Sivan Pearl, Amos Oppenheim, and Nathalie Q. Balaban

21 **Cellular Epigenetic Inheritance in the Twenty-First Century 215**
Eva Jablonka

22 An Evolutionary Role for RNA-Mediated Epigenetic Variation? 227
Minoo Rassoulzadegan

23 Maternal and Transgenerational Influences on Human Health 237
Peter D. Gluckman, Mark A. Hanson, and Tatjana Buklijas

24 Plants: Individuals or Epigenetic Cell Populations? 251
Marcello Buiatti

25 Instantaneous Genetic and Epigenetic Alterations in the Wheat Genome Caused by Allopolyploidization 261
Moshe Feldman and Avraham A. Levy

26 Lamarckian Leaps in the Microbial World 271
Jan Sapp

27 Symbionts as an Epigenetic Source of Heritable Variation 283
Scott F. Gilbert

IV PHILOSOPHY 295

28 Introduction: Lamarckian Problematics in the Philosophy of Biology 297
Snait B. Gissis and Eva Jablonka

29 Mind the Gaps: Why Are Niche Construction Models So Rarely Used? 307
Ayelet Shavit and James Griesemer

30 Our Plastic Nature 319
Paul Griffiths

31 The Relative Significance of Epigenetic Inheritance in Evolution: Some Philosophical Considerations 331
James Griesemer

32 The Metastable Genome: A Lamarckian Organ in a Darwinian World? 345
Ehud Lamm

33 Self-Organization, Self-Assembly, and the Inherent Activity of Matter 357
Evelyn F. Keller

V RAMIFICATIONS AND FUTURE DIRECTIONS 365

34 Introduction: Ramifications and Future Directions 367
Snait B. Gissis and Eva Jablonka

35 Lamarck on the Nervous System: Past Insights and Future Perspectives 369
Simona Ginsburg

36 Lamarck's "Pouvoir de la Nature" Demystified: A Thermodynamic Foundation to Lamarck's Concept of Progressive Evolution 373
Francis Dov Por

37 Prokaryotic Epigenetic Inheritance and Its Role in Evolutionary Genetics 377
Luisa Hirschbein

38 Evolution as Progressing Complexity 381
Raphael Falk

39 Epigenetics and the "New Biology": Enlisting in the Assault on Reductionism 385
Alfred I. Tauber

40 Epigenetic Inheritance: Where Does the Field Stand Today? What Do We Still Need to Know? 389
Adam Wilkins

41 Final Discussion 395

Appendix A: Mandelstam's Poem "Lamarck" 411
Appendix B: Mechanisms of Cell Heredity 413
Glossary 423
Contributors 433
Index 437

Series Foreword

Biology is becoming the leading science in this century. As in all other sciences, progress in biology depends on interactions between empirical research, theory building, and modeling. However, whereas the techniques and methods of descriptive and experimental biology have evolved dramatically in recent years, generating a flood of highly detailed empirical data, the integration of these results into useful theoretical frameworks has lagged behind. Driven largely by pragmatic and technical considerations, research in biology continues to be less guided by theory than seems indicated. By promoting the formulation and discussion of new theoretical concepts in the biosciences, this series is intended to help fill the gaps in our understanding of some of the major open questions of biology, such as the origin and organization of organismal form, the relationship between development and evolution, and the biological bases of cognition and mind.

Theoretical biology has important roots in the experimental biology movement of early-twentieth-century Vienna. Paul Weiss and Ludwig von Bertalanffy were among the first to use the term *theoretical biology* in a modern scientific context. In their understanding the subject was not limited to mathematical formalization, as is often the case today, but extended to the conceptual problems and foundations of biology. It is this commitment to a comprehensive, cross-disciplinary integration of theoretical concepts that the present series intends to emphasize. Today, theoretical biology has genetic, developmental, and evolutionary components, the central connective themes in modern biology, but also includes relevant aspects of computational biology, semiotics, and cognition research and extends to the naturalistic philosophy of sciences.

The “Vienna Series” grew out of theory-oriented workshops, organized by the Konrad Lorenz Institute for Evolution and Cognition Research (KLI), an international center for advanced study closely associated with the University of Vienna. The KLI fosters research projects, workshops, archives, book projects, and the journal *Biological Theory*, all devoted to aspects of theoretical biology, with an emphasis on integrating the developmental, evolutionary, and cognitive sciences. The series editors welcome suggestions for book projects in these fields.

Gerd B. Müller, University of Vienna and KLI
Günter P. Wagner, Yale University and KLI
Werner Callebaut, Hasselt University and KLI

Preface

The year 2009 was a memorable one for evolutionary biology as the year in which the whole world celebrated the 200th anniversary of Darwin's birth and the 150th anniversary of the publication of *The Origin of Species*. But 2009 was also the anniversary of another biological achievement that, although less noted, may have been equally monumental: namely, the publication of Lamarck's *Philosophie zoologique*. In a coincidence of history, the year of Darwin's birth was also the year in which the first comprehensive and systematic theory of biological evolution appeared.

This volume has its origin in a workshop held in Jerusalem, in June 2009, to celebrate the 200th anniversary of the publication of Lamarck's magnum opus. The aim of this workshop was twofold: first, to redress a critical lapse in historical memory (to our knowledge, ours was the only full-scale international meeting devoted to commemorating the publication of *Philosophie zoologique*), and second, to bring into sharper focus new developments in biology that make Lamarck's ideas relevant not only to modern empirical and theoretical research, but also to problems in the philosophy of biology and to the writing of the history and sociology of evolutionary theory.

We wish to make clear at the outset that it is not our intention to contribute to a debate about Lamarckism versus Darwinism. Such a dichotomy is both theoretically unjustifiable and historically misleading. The theories that Lamarck and Darwin originally formulated were radically transformed as biological research advanced since the early nineteenth century. Lamarckism in particular has been evolving or, to use Lamarckian terminology, *transforming*. There is little similarity between Lamarck's descriptions of the subtle fluids that lead to the transformations of organisms during the vast epochs of time and modern ideas and research on "Lamarckian" evolution, which are based on the inheritance of epigenetic variations and the evolution of self-organizing structures, and make use of sophisticated molecular and computational tools. There is, however, a certain logic, or stance, which differentiates the Lamarckian and Darwinian approaches. Selection is central to Darwinian thought, whereas the generation of developmental variations is central to Lamarckian thought. In the early twentieth century Delage and Goldsmith were already explaining what the Lamarckian stance is by contrasting it with neo-Darwinism, formulated as a challenge to the logic and relevance of Lamarckism:

> Neo-Darwinism, which has found its most complete expression in Weismann's writings, constitutes a well-harmonised system of conceptions relative to the structure of living matter, ontogenesis, heredity, evolution of species, etc. Lamarckism on the other hand is not so much a system as a point of view, an attitude towards the main biological questions.
> Whatever theory emphasises the influence of the environment and the direct adaptation of individuals to their environment, whatever theory given to actual factors the precedence over predetermination can be designated as Lamarckian. (Delage and Goldsmith [1909]; trans. Tridon 1912:244–245)

The two approaches, the one which stresses developmental variation, and the other which stresses selection, are complementary rather than mutually exclusive. Yet their different emphases have important implications for the kinds of questions biologists ask and for the type of research they conduct. We refer to the former stance as the "Lamarckian problematics," and we believe the time is ripe for an examination of the crystallization of this problematics in the modern era. This, in fact, is what the present volume is about. The volume is based on the papers presented and the discussions that took place in our workshop, and it describes work from a wide range of disciplines. Because of this we have included a glossary of some of the specialist terms that may be unfamiliar to those not directly involved in the various research areas.

The book begins with an introductory essay by Gabriel Motzkin, who considers how an engagement with Lamarckian problematics forces us to consider multiple levels of organization and analysis. As Motzkin argues, interactions between principles of causality that can be distinguished not only by the level at which they operate, but also by their different modes of operation, open up new approaches and new questions. Pietro Corsi then gives sketches of Lamarck's life and of his main ideas that dispel some common myths about the man and his thought, and provide a far more realistic, sophisticated, and interesting view of Lamarck's life and way of thinking.

Part I of the book deals with the historical transformations of Lamarckism from the 1820s until the 1940s. It illustrates how scientific, cultural, and national contexts and styles formed and shaped modes of understanding, thus producing different images of Lamarck and Lamarckism. At some periods in the nineteenth century, Lamarckism seemed compatible with Darwinism. In the late nineteenth century, however, neo-Darwinism emerged as an exclusively selectionist, anti–Lamarckist view of evolution. The historical section also includes a consideration of the impact of Lamarckian ideas on social thought.

The Modern Synthesis of evolution is the subject of part II. The fusion of ideas from various branches of biology that took place in the 1930s and 1940s and gave Mendelian genetics a central place in evolutionary thinking forms a bridge between the historical and biological parts of this volume. The workshop participants agreed that, in spite of the different ways biology developed in Great Britain and the United States, by the 1960s there was a clear eclipse of Lamarckism in both countries. With the establishment of the Modern Synthesis in the 1960s and its persistence until the early 1990s, the exclusion of "soft inheritance"—the term adopted to embrace all alternatives to Weismannism, including "the inheritance of acquired characters"—seemed final and irreversible. Many biologists regarded Lamarckism in much the same way they regarded the phlogiston theory in chemistry. It was seen as being of historical interest only, though perhaps having some heuristic value in the teaching of Darwinism. However, as the lectures and discussion that are reported in this part of the book show, there are still many unanswered questions about how and why the development-oriented approach to evolution was replaced by the Modern Synthesis view, and how evolutionary thinking became increasingly dominated by genes and by selection.

The essays in part III present theoretical and experimental work that focuses (directly or indirectly) on Lamarckian problematics. *Plasticity*—the capacity of organisms to change in response to varying conditions—is the theme of several contributions. It is a large topic, but, just as Lamarck anticipated, an understanding of plasticity is now recognized as being fundamental to an understanding of evolution. Another theme that is explored by some of the contributors is that of *soft inheritance*. Experimental work now shows that, contrary to the dogmatic assertions of many mid-twentieth-century biologists that it could not occur, even a form of "inheritance of acquired characters" does occur and might even be said to be ubiquitous. In particular, new variations induced by stress are sometimes inherited. The molecular mechanisms that underlie such inheritance—the epigenetic inheritance systems—are now partially understood, and are outlined in appendix B. As some of the contributors suggest, the existence of various types of soft inheritance affects how we have to see adaptive evolution and speciation. It also has implications for human health.

Part III ends with two essays on biological *individuality*, a theme that is related both to the nature of hereditary continuity and to the inheritance of acquired variations. It is now recognized that the acquisition of new genes and genomes from non-parents, including individuals belonging to other species, has been of major importance in microbial evolution. It is also clear that symbiotic interactions among organisms have led to the construction and evolution of new types of developmental systems, and new types of more or less cohesive superorganisms or consortiums—that is, to new types of biological individuals. These realizations create problems for conventional theories that assume linear descent with modification and treelike evolutionary lineages; they also provoke questions about the speed of evolutionary change and the nature of species.

Part IV of this volume is about the importance of the Lamarckian problematics—that is, of a developmental approach to evolution—in the philosophy of biology. Following the biologists, most earlier philosophers of biology also stayed away from Lamarckism. Nevertheless, since the 1990s, philosophers of biology have spearheaded the formulation of new ideas about the relationship among heredity, development, and evolution; in short, they have begun to formulate, as James Griesemer succinctly puts it, a philosophy *for* biology. One finds in this literature an empirically oriented approach to philosophical questions, as well as suggestions for new theoretical and research heuristics—heuristics that may even be of help to biologists in formulating relevant biological questions.

The volume ends (part V) with a series of short contributions and a summary of the discussion in the final session of the workshop. They illuminate some biological, philosophical, sociological, and historical issues raised by the Lamarckian problematics that open up future perspectives—issues that either had not been discussed previously or that had only been hinted at, but that we felt deserved our attention. In the final discussion, Adam Wilkins was asked to play the devil's advocate—to highlight problems, prod, and challenge previous presentations. The animated discussion and the different approaches expounded thereafter are indicative of the stimulating and generative character of his comments.

We hope that the volume conveys some of the excitement and intellectual fun that we had during the meeting. We realize that not all of the people who will be reading it will agree with the interpretations given to the findings described, and they may not share the feeling of some of us that we are at a turning point in our understanding of evolutionary biology. Nevertheless, we believe that even critics and skeptics will agree that the Lamarckian problematics, the developmental-variation-focused view, is returning to the center of evolutionary theorizing, complementing the Darwinian focus on gene selection.

Acknowledgments

This volume was produced following the Twenty-third Annual International Workshop on the History and Philosophy of Science, which was supported by three academic institutes in Israel: the Cohn Institute for the History and Philosophy of Science and Ideas at Tel-Aviv University (TAU); the Edelstein Center for the History and Philosophy of Science, Technology and Medicine at the Hebrew University of Jerusalem (HUJ), and the Van Leer Jerusalem Institute.

We are very grateful to the individuals and institutions that contributed so very generously toward the realization of the workshop on which the volume is based: to Bert and Barbara Cohn, who not only made this workshop possible but also made possible the founding of the Cohn Institute in 1983, and have generously supported it ever since; to Professor Gabriel Motzkin, the director of the Van Leer Jerusalem Institute, without whose support the whole enterprise could not have taken off; to Professor Leo Corry, who, in his capacity as chair of the Cohn Institute, helped in things big and small; to Professor Yemima Ben Menahem, the head of the Edelstein Center, HUJ, who was very helpful and supportive all the way through. We also want to express special thanks to the Van Leer team, Ms. Shulamit Laron, the director of public events, and Mr. Igor Sankin, coordinator of public events, who were responsible for much of the administrative organization of the workshop, and to Ms. Anat Zion, administrative executive of Cohn Institute TAU, for their invaluable efficiency, common sense, and cheerfulness (even when things were getting, as they always do on such occasions, a bit desperate).

We thank the external grant-giving bodies, which were not only helpful in financial terms but also expressed enthusiasm and support: the Israel Academy of Sciences and Humanities, which supported the Bat-Sheva Workshop on the Current Status of Soft Inheritance in Biology (the papers from which form part III of this volume), and Doctor Yossi Segal, who helped to make it come true; the British Council and Oxford University, which, thanks to the hard work of Professor Pietro Corsi and Ms. Sonia Feldman, the local Science and Education officer in Israel, supported our workshop as part of the British Council's *Darwin Now* project;

the Israel Science Foundation and Dr. Ohad Nahctomy, and especially Dr. Tamar Mittvoch-Yaffe of the ISF, who gave us much needed support in difficult moments. The volume is as good as the papers that were contributed to it, and we would like to thank all the contributors to the workshop and the volume, some of whom participated against their better political judgments.

First, we want to thank Marion Lamb, who not only contributed a chapter to the volume but also helped with the scientific and language editing. Had she not refused, she would have been listed as a collaborator on the volume. The book would not have seen the light of day without her inestimable and thorough help. We are also very grateful to Evelyn F. Keller for her editorial help when we were in a tight spot.

We are deeply grateful to Anna Zeligowksi, who put together an exhibition, Webs of Life, that accompanied the workshop, and whose pictures adorn this volume. We are also indebted to Yigal Liverant, who generously allowed us to use his new translation of Mandelstam's poem "Lamarck."

Some invited speakers could not come. Goulvent Laurant, an eminent French historian of Lamarckism, died in the autumn of 2008, and we would like to use this opportunity to honor his contribution to the study of Lamarckism. Two American historians of science had to cancel their participation because of illness: Richard Burkhardt and William Provine. However, Richard Burkhardt was able to contribute an essay for both the workshop and the book, and we are deeply grateful for his effort. We thank Adam Wilkins and Scott Gilbert for kindly agreeing to discuss the topic of Will Provine's lecture. We deeply regret that Minoo Rassoulzadegan, an eminent French biologist who studies the RNAi system, was prevented from coming to the workshop by the Israeli Ministry of Interior and the security services. They would not grant her an entrance visa because, although she has lived in France since 1970 and holds a French passport, she also holds an Iranian one. Although the officials at Van Leer Jerusalem Institute and Tel-Aviv University did everything that is humanly possible to try to overcome the Israeli bureaucratic security walls in order to have Professor Rassoulzadegan join us, they did not succeed. Minoo nevertheless agreed to contribute to this volume because, as she put it, "Science is more important than politics." We are most grateful to her. Another contributor to the volume who did not come for political reasons—in this case because of his strong objections to Israeli government actions—is Stuart Newman, and we thank him for his willingness to contribute a paper to the volume and take part in this intellectual adventure.

It has been historically shown that science and technology can flourish under various political regimes, but scientific cooperation does need a great measure of academic freedom. We want to express the hope that academic and research institutions in Israel will cherish this freedom within the country, and will also see it as part of their responsibility to help enable such freedom for the parallel Palestinian institutions.

INTRODUCTORY ESSAYS

1 Lamarck, Darwin, and the Contemporary Debate about Levels of Selection

Gabriel Motzkin

Like all of you, I took an undergraduate course (in my case, some forty-six years ago) in which I learned that Lamarck was bad and Darwin was good. It was the first course ever given on evolution as part of the program of general education at Harvard, and it was its first year. The excitement of the course turned on two themes. The first was the breakthrough in genetics that signified the ability to understand how biological material copied itself. I remember the excitement we experienced learning about DNA and RNA, concepts that were being taught to freshmen in general education courses in 1963. The second theme was the evolution of man. At that time, the theory that humans had evolved separately in different places was still being taken seriously, and so we had to compare and evaluate the single-origin hypothesis against the multiple- origin hypothesis (Coon 1962).

The neo-Darwinian model you all know, loosely associated with the name of Dobzhansky, was the reigning orthodoxy (Dobzhansky 1962). The book we read about the nineteenth century, *Apes, Angels, and Victorians*, introduced us into the heart of the development of the theory of evolution (Irvine [1956] 1963). We admired Lyell and despised Lamarck. We associated Lamarck vaguely with Lysenko, that commie pseudo geneticist. We despised Lamarck because it was obvious to all that acquired characteristics could not be inherited, that is, that there was no connection between, for instance, learning experience and genetic transmission. We did not stop to reflect on how, for example, learning might affect sexual selection, on how a living being can make decisions with genetic implications. The distance between DNA and the environment seemed to us to be very wide.

There was another reason, one that we did not then understand well, that affected our treatment of evolution. This was the legacy of Nazism. We all understood why the victory of a theory of separate origins for different races would reinforce racism, which at the time was the central question in American cultural politics. Feminism was only beginning to be visible. *The Feminine Mystique* was published in 1963 (Friedan 1963); birth control became an acute issue only some time later, after the Pill was introduced in 1964–1965. Race, the significant cultural question, was somehow related to the Holocaust, although no one discussed this connection. It was clear to all, however, that decisions about biological truths were also political decisions.

This accusation of racism had two dimensions: one question was whether races actually exist, that is, what is the genetic diversity of humankind, and can these races be distinguished according to degree of fitness or intelligence or whatever? This question has not, and will not, vanish from the world, for it is the question of whether subgroups in a species have traits that distinguish them from other subgroups, and whether these traits are indicators of an emergent speciation event. The second question was the degree to which Darwin was responsible for Hitler. At the time, this question was actually even more relevant than the question of separate human races because Hitler was still a recent memory for people who were not old. This question also hid a fundamental question, one that is very important for the future of mankind, and one on which we have made no progress: Suppose that a science is true, but that its truths have devastating consequences for our ethics. Should we choose our science or our ethics? There are countless attempts to bridge this gap between the true and the good, or to reconcile them, but if scientific truth is not absolutely good in itself, so long as there is science, this question will remain, and it will have to be answered differently in different situations, since there is no one overarching answer to the relation between the true and the good. What, then, does it do to our ethics to find out that we are animals and that our future depends on mechanisms of fitness, selection, and adaptation, some of which can be controlled intentionally and some of which cannot, and where the strategies to maximize individual and group robustness or evolution, whatever they are, may not be those of our ethics or our civilization?

The problem for us humans is not even the question of the inheritance of acquired characteristics; it is, rather, the question of whether our intentions, as individuals and as groups, have any evolutionary significance, that is, any future significance whatever. The theory of evolution, and certainly any neo-Lamarckism, requires some conception of intentionality, because sexual and natural selection do not just mean the interaction of genetic drift with environmental selective constraints but, rather, the interaction between choices made by organisms in relation to selective constraints. It is this element of contingency that makes Massimo Pigliucci think that evolution can be told only as a history, that is, that there is no iron link between statistics and necessity (Pigliucci and Kaplan 2006).

I now want to turn to the development of philosophy in the twentieth century to highlight a phenomenon that is in the process of disappearing but is still robust in Israel: namely, the love of logic and physics, and the consequent infatuation with reductionism. We are torn between the quest for a unidirectional causal link between levels of analysis and our sense that causality in evolution can operate in different directions between the levels of selection.

Recently, in an interview for the journal *Odyssey*, I argued that there are levels of evolution, and that biology is not physics. I was forewarned that I would be attacked in the next issue, in which the physicist Hanoch Gutfreund would argue that self-organization is evident in organic molecules, that is, that self-organization is a prerequisite for life rather than just a consequence; hence, by implication the distinction between the inorganic and the organic levels has less consequence than one might think (Gutfreund's article has not yet appeared). The line of defense is obvious: life depends on reproduction, and reproduction is not merely a question of errors in DNA replication but is a process that leads to functional specialization. Of itself, functional specialization, I think, is different from the organization of matter because it can occur when the bodies in question have largely the same material organization but do different things. It should be evident to all of you that the kind of science we do, and therewith our ultimate evaluation of the function of culture, depends on the question of whether all functions can be explained entirely in terms of interactions between particles.

I will now engage in a digression. This digression addresses the question of the source of our love for imitating physics as the basic model for the sciences, and why this imitation of physics has recently run into trouble. The love for physics is obviously a consequence of the fundamental breakthroughs in physics both in the seventeenth century and in the late nineteenth and early twentieth centuries. It found its reflection in culture: for example, the development and attraction of analytic philosophy was related to this love for physics, even though one could argue that analytic philosophy made some errors in its physics, notably the well-known reluctance to accept quantum physics as a repository of invisible entities.

The great divide between what we loosely call analytic philosophy and Continental philosophy hid common ground: namely, both rejected late nineteenth-century psychologism and both adopted reductionist positions. I can document this reductionism for my tradition of Continental philosophy: both Husserl's formalism and Heidegger's axiomatic analysis of existential questions are reductionist. For Husserl, the formal rules of consciousness are true for any consciousness whatsoever, whether animal, human, or Divine. For Heidegger, the axioms that govern human experience also structure the world; the world of nature is a subset of human experience. On the other side of the fence, the basis for the development of analytic philosophy was the logicism that stems from Frege, and it was just as reductionist, just as prone to assuming that there is only one level of reality. Both philosophical traditions presupposed that any division of reality into different levels of being with different operative principles of causality was redolent of theological implications. The difference between the two traditions was the characterization of this level of reality. The one tradition identified logic and science, and deployed logic to justify science, and science to justify logic, while the other substituted an ultimately anthropological conception of the one-level reality, respectively either consciousness or human life. And both traditions hence were attracted to the model of physics, Continental philosophy no less than analytic philosophy, because physics seemed to be that science for which reality is composed of elementary particles and their motions, and hence all superstructures must be completely expressed on this basic level.

While this tradition, as noted, is robust among physicists, and even computational neuroscientists sometimes think that their science is one in which all observations can be defined in terms of the interactions between elementary particles of different kinds (in this case neurons and electrical charges), the development of both cognitive science and biology in the last generation shows that the idea that the same forces are at work on different levels is unstable.

It is quite surprising that the development of a cognitive science suffused with computational (i.e., computer) metaphors would actually serve as an engine for eroding the attraction of physics. The obvious social explanation is the apparent lack of dramatic progress in physics since the 1950s. But there is an inherent structural reason for this development: namely, the development of cognitive science has reawakened the ancient issue of the relation between mind and brain. Early twentieth-century philosophies of mind could safely ignore the brain since they were wed to formal explanations of mental operations for which the question of brain states was largely irrelevant. It is perhaps paradoxical that this formalism correlated with physics-oriented ideas about science, but perhaps not: in both cases metaphors derived from the dramatically expanded scope of mathematical analysis–informed conceptions of both science and mind. In contrast, cognitive science reawakens the question of whether or not mind-level explanations are reducible to brain-level descriptions. Is there a coherent conception of mental causality which is different from the conception of physical causality? Philosophies of mind are in dissensus about this issue. The psychologist Stephen Kosslyn requires many intermediate levels between mind and brain in order to explain his description of mental events (Kosslyn 1994). Other philosophers, such as the Churchlands, require a characterization of mind–brain as a network, that is, one in which both mind and brain need to be redefined so that the argument can be advanced that they are identical (Churchland PM 1995; Churchland PS 2002). My point is that the reawakening of this discussion has led to a rebirth of late nineteenth-century psychologism, and hence to an erosion of mental formalism. Moreover, while one could conceive of mental events in a reductionist fashion, the entire debate about the mind is one about levels of events, which is not so different from the debate about "levels of selection," however one might distinguish between the operative principles in selection and the operative logic of mental events.

I do not need to inform this public about the debates in evolutionary biology. I merely wish to point out that any debate about neo-Lamarckism is also a debate about levels of selection. Now the debate about neo-Lamarckism can easily tend to a debate about what higher levels might be. Thus, when I read Jablonka and Lamb, I find myself very interested in the analysis of epigenetics and less sure about the characterizations of human conduct (Jablonka and Lamb 2005). But it is clear—just look at Pigliucci and Kaplan—that the purely genetic analysis of biological development is no longer tenable, that the invocation of genetic drift is just as prone to mysticism as the idea of genes as subject to complete control by phenotypical developments, which, I take it, was the reason why Lamarck was so vociferously rejected. In other words, what we now know about cellular development makes it clear that not all biological processes are reducible to each other, that some processes are consequences of functional specialization, and that consequently some processes do not function on all levels of organization in the same way. Now this is much messier as science, but it is a consequence of a perhaps hidden but nonetheless necessary assumption that not all locally anti-entropic organization, which is what life is, is organized in the same way. It is this variation in the organization of resistance to entropy that implies that not all processes are strictly reducible to the consequences of physical interactions. The new science and the new philosophy which we will obtain will not deny physics, but will assume that the rules of physical science permit a scope of possible variation within which different structures of organization are possible. In turn, those structures can be organized in such a way that they have different levels, and those different levels will allow for causal interactions between different levels that are not unidirectional. That is why I think that the engagement with neo-Lamarckism will not only be possibly fruitful; it will also be interesting as a description of how events on one level affect events on another. However, for such a path to be rewarding, we shall need to suspend our philosophical and cultural preconceptions about what those levels might be, such as mind or social groups. Rather, we will need to describe the levels in terms of the causal patterns that we observe. Thus, for example, what we define as a group may not be what nature defines as a group. The current confusion about the definition of what is a species is a good example of this kind of confusion: it assumes a coherent conception of species for causal analysis that is derived from observation. If a multilevel description of nature is to be convincing, it will, rather, need to map different patterns of causality, and then specify the levels in terms of these patterns.

There is a larger question, one that challenges the concept of science even while it may be the discovery of its limits: even if one kind of causality operates at all levels, different types of causality may supervene at successive levels of development. The issue then would not be the interaction between principles of causality that can be distinguished by the level at which they operate, but the interaction between principles of causality that operate in different ways, ways that can be conceptually defined irrespective of the level at which they operate. That is the challenge posed by the apparent debate between Lamarck and Darwin, and it is a challenge that we have not yet met.

References

Churchland PM. *The Engine of Reason, the Seat of the Soul. A Philosophical Journey into the Brain.* Cambridge, MA: MIT Press; 1995.

Churchland PS. *Brain-Wise. Studies in Neurophilosophy.* Cambridge, MA: MIT Press; 2002.

Coon C. *The Origin of Races.* New York: Knopf; 1962.

Dobzhansky T. *Mankind Evolving: The Evolution of the Human Species.* New Haven, CT: Yale University Press; 1962.

Friedan B. *The Feminine Mystique.* New York: Norton; 1963.

Irvine W. *Apes, Angels, and Victorians: The Story of Darwin, Huxley, and Evolution.* New York: McGraw Hill; 1963 (orig. pub. London: Weidenfeld and Nicolson, 1956).

Jablonka E, Lamb MJ. *Evolution in Four Dimensions: Genetic, Epigenetic, Behavioral and Symbolic Variation in the History of Life.* Cambridge, MA: MIT Press; 2005.

Kosslyn, SM. *Image and Brain. The Resolution of the Imagery Debate.* Cambridge, MA: MIT Press; 1994.

Pigliucci M, Kaplan J. *Making Sense of Evolution. The Conceptual Foundations of Evolutionary Biology.* Chicago: University of Chicago Press; 2006.

2 Jean-Baptiste Lamarck: From Myth to History

Pietro Corsi

Since the 1860s, biologists have repeatedly referred to Jean-Baptiste Lamarck (figure 2.1) as one of the founding fathers of modern evolutionary doctrines, a claim hotly denied by most of their colleagues. Lamarck's name has often been evoked to contrast his insistence on acquired individual variations with the mainstream interpretation of the Darwinian doctrine, which stresses the constant presence within a population of repertoires of variation that are subjected to the action of selection. As a consequence, reference to Lamarck has often been made with polemical intent, and has rarely been based on firsthand acquaintance with his works and the biological, philosophical, and even political debates of which he was part.

Although, as his taxonomical work abundantly proved, he was prepared to follow the path of specialization, accuracy, and rigor that many of his colleagues advocated, Lamarck insisted throughout his life that specialization constituted a betrayal of the Newtonian ideal of searching for the handful of ultimate principles on which all nature rested. Thus, he opposed Lavoisier and his school's claim to have proven the existence of a few score primary chemical elements, and agreed with Buffon that the Linnaean system of classification was introducing artificial partitions into the endless repertoires of variation that living nature offered to the philosophical observer.

As a rather typical eighteenth-century materialist, Lamarck believed that the universe and the Earth were as eternal as matter. For him, two principles were the basis of all natural phenomena: Newtonian attraction and life. During the first half of his career Lamarck did not lose hope that one day even life could be shown to be the result of strict physicochemical causes. In the very early years of the nineteenth century, Lamarck matured the conviction that this was indeed the case, though the demonstration he provided did not meet with universal acceptance. Until 1800, Lamarck believed that whereas universal attraction guaranteed that basic forms of matter derived from the four primary elements (earth, water, air, fire) coalesced into planetary and stellar bodies, only life could be responsible for the

Figure 2.1
Jean-Baptiste de Monet de Lamarck. Lithograph by L. Boilly (1821). Copyright Bibliothèque Centrale du Muséum National d'Histoire Naturelle.Used with permission.

building of complex compounds. As he repeatedly pointed out, organic chemistry was much more complex than inorganic.

Since only life was an active principle capable of generating complex structures, spontaneous generation was unthinkable, and species had always existed as they are today. Lamarck called living organisms "true chemical laboratories" constantly fabricating complex chemical and mineral substances. The very existence of the planet Earth depended on the ability of living things to cumulate mountains of shells and thick layers of coal or petroleum, thereby providing a continuous supply of material that filled in after the constant, relentless work of destruction that natural elements (rain, wind, ice, tides) were carrying out at the surface of the Earth. Eternal cycles of life and death provided material for the building of mountains destined to be leveled by erosion. Furthermore, and more important within the economy of Lamarck's philosophy of nature, once life stopped keeping together the particles making up compounds (the flesh, blood, and bones of organisms), the particles inexorably returned to their elemental state. So, without life there would be no complex chemical compounds and mineral substances: there would be no Earth. Without life, the Earth would be a spheroid of inert material elements surrounded by water.

The Revolutionary decade 1789–1799 opened up new opportunities for Lamarck, though he soon discovered that the ruling class of the old Academy of Sciences proved remarkably resilient in the long term. They still surrounded Lamarck's philosophical speculations with icy silence and ostentatious and at times offensive disregard. At the same time, substantial sections of the reading public showed increasing interest in his proposals, something that was also helped by the fact that after 1800 Lamarck gave up his open attacks on the new chemistry, and concentrated on his taxonomical work and yearly meteorological publications (1799–1810). The latter proved very successful with the reading public, though they deeply irritated Georges Cuvier and Pierre-Simon Laplace, and this led to the famous rebuke from Napoleon himself in 1809. Lamarck never abandoned his physicochemical convictions and, after some initial reluctance, followed fashion by joining in the popular debates on the origin and transformations of the Earth and of life.

In the long run, although not to his liking, being given the chair of the zoology of invertebrates in the newly established Muséum d'Histoire Naturelle (June 1793) provided Lamarck with unexpected advantages and philosophical rewards. Initially he limited himself to giving the required forty-hours-per-year course, mainly relying on the anatomical investigations of his younger rival Cuvier, who had moved to Paris in 1795. Until 1799 Lamarck stubbornly pursued his attempt to gain a reputation as a philosophical chemist, against the opposition of prominent colleagues at the Academy of Sciences. His eternalism and his philosophy of nature were also challenged by increasingly popular geological and evolutionary doctrines put forward by naturalists who, although they are ignored by historians, enjoyed European reputations during the very last years of the eighteenth and the first decade of the nineteenth centuries. Lamarck embarked upon a critical assessment of current "directionalist" cosmologies and life histories, which were centered on descriptions of the beginning of the Earth either by fire or by water; of the appearance of life due to the spontaneous generation of simple organisms or prototypes of all major animal and plant orders; and of the slow progress toward the appearance of man. Lamarck's original answer to trends he considered unfavorable to his theoretical stands matured during the years 1799–1802, and culminated in 1802 with the publication, in January, of his *Hydrogéologie* and, in July, of the *Recherches sur l'organisation des corps vivans*, which was the first draft, so to speak, of his better-known 1809 *Philosophie zoologique*. In 1801 he had published his *Système des animaux sans vertèbres*, a work containing his first ambitious taxonomical proposals, which circulated widely in Europe and the Americas.

In his *Hydrogeology*, Lamarck announced the project for a "terrestrial physics" grounded on three main disciplines: meteorology, the fluid dynamics responsible for all the phenomena of the atmosphere; hydrogeology, the fluid dynamics constantly and eternally shaping the surface of the Earth; and biology, the dynamics of the fluids present within living organisms and responsible for all their structural articulations and transformations. After 1800, "life" lost the status of "a principle for ever unknown to man," and was seen by Lamarck as the product of physical laws regulating the behavior of fluids constantly flowing within (blood, nervous fluids, lymph) and through living organisms (caloric, electricity, or magnetism). Spontaneous generations were now possible, since a tenuous membrane could accidentally surround a molecule of a gas or of a fluid subjected to expansions or contractions due to external thermal conditions. Spontaneous generations occurred every day, whenever local circumstances allowed, and were destined to live the space of a few moments and be forever invisible to man. Lamarck defined spontaneous generations as "inductions": a necessity of scientific reasoning and of his theory of life, a theoretical conclusion rather than a datum of observation. He never tired of repeating that even the simplest animal or plant forms known to naturalists were true animals or true plants, not spontaneous generations.

Exceptionally, a few spontaneous generations had been able to nourish and reproduce themselves, thereby giving rise to simple forms of life up to the structural level of radiates. More complex structures, showing serial segments and eventually bilateral symmetry, developed out of spontaneous generations occurring within already existing, simple organisms. Lamarck never believed in a single hierarchically articulated life plan, but in at least one plan for plants and at least two for animals. Moreover, higher levels of organization were not reached because of an innate tendency toward complexity, but were always due to the action of "extraordinary sets of circumstances." Thus, for instance, infusorians never developed into vertebrates, a fate reserved to parasitic wormlike forms that must necessarily have appeared later in the history of life. Simple spontaneous generations occurring in lukewarm marshy waters never ascended as far as wormlike parasites produced within already existing organisms.

Although known as the first and major theorizer on the principle of the inheritance of acquired characteristics, Lamarck never expressed himself in such terms, and clearly—and rightly—considered himself as one of the many naturalists convinced that the development of an organ during the lifetime of an organism, or the appearance of however slight behavioral propensities ("habits"), could be passed on to the next generation if a mating occurred between individuals that had experienced the same change. Yet, characters were not acquired or transmitted: only biological processes were. To Lamarck, organic fluid dynamics naturally gained strength within the parts more exposed to changing environmental circumstances, thereby contributing to their reinforcement and extremely gradual modification; equally, decreasing levels of stimulation would decrease the flow of nutritional, nervous, or other fluids to the relevant parts. Thus, it was not "characters" that were acquired during the lifetime of the organism, but only a higher or lower degree of organic fluid flow or, in general, a small difference in fluid distribution patterns. The subtle fluid contained in the sperm of animals undergoing sexual reproduction—a fluid akin to electricity or caloric, a specification of the very active element "fire"—was equally affected, and would organize the eggs of the next generation accordingly, thereby "transmitting" the slightly modified pattern of fluid distribution. Species, that is, the groups of individuals capable of reproducing among themselves, remained stable in stable conditions: taxonomy, in other words, was not a useless exercise; it did suggest filiations and provided the only reliable evidence of the changes and developments that life had undergone. Lamarck never relied on paleontological data, convinced as he was that all fossils, with a few exceptions (essentially the remains of animals destroyed by man), were still alive somewhere on Earth or at the bottom of the seas. For Lamarck, as for a number of his followers, the beautiful fossil ammonites found embedded in rocks were still thriving in the oceans. He also refused to accept contemporary claims that the phases of development of the embryos displayed strong analogies with the history of life on Earth, and expressed his opposition to the doctrine of the unity of plan put forward by Étienne Geoffroy Saint-Hilaire and others, since he did not believe that all animals were built from the same basic functional and structural elements (*Philosophie zoologique* 1809, I:375).

Lamarck is not an easy author to read: he rarely quoted contemporary naturalists and never gave his sources, even when they were colleagues who shared many of his convictions. Unlike Darwin, he almost never gave voice to the difficulties he was experiencing or the issues that were still in need of clarification. Thus, for instance, he never addressed the temporal dimension of the tree of life in explicit and sustained terms, as several of his contemporaries were doing. As noted earlier, his Earth needed all the life it could sustain, and not just a multitude of spontaneous generations striving to survive, in order to be an Earth. Moreover, where could a beginning in the eternal succession of astronomical and geological cycles be located? The fact that Lamarck classified fossil invertebrates together with the living ones reflected his deep uneasiness with the idea of a history of life on Earth, a history his theory of life and his taxonomical work strongly posited, but his view of an eternal universe and of an Earth of indefinite age found difficult to accommodate. As he had declared in 1794 (*Recherches sur les causes des principaux faits physiques*, 1:11), there could be no science of beginnings, and naturalists should rely only on what they see in the present. As was the case with spontaneous generations, the history of life on Earth could be deduced from the trees of life but could not be described in detailed historical terms. It is certainly significant that Lamarck never expressed himself with sufficient clarity on this point.

Equally difficult to understand is Lamarck's concept of "the power of life," usually taken to indicate a force inherent in life, pushing toward higher and higher levels of organic complexity. There is no doubt that Lamarck toyed with several concepts that were not always reconcilable with one another. In the 1802 *Recherches*, Lamarck appeared to indicate that the more complex life becomes, the more the biological fluid dynamics increases its efficacy and its ability to further specialize organs and functions. In other words, the longer life lasts, the more complex it will become. In one, albeit only one, case in the *Philosophie zoologique* (1809, 1:133) he also stated that if marine animals developed within absolutely stable conditions, they would display a perfect gradation of forms. Yet, talking of the transition from respiration through gills to respiration through lungs, he emphatically declared that this required a particularly important set of environmental circumstances: he did not, in other words, refer to the fact that sooner or later the "power of life" would have taken care of the transition—a point several contemporaries stressed, at times against Lamarck (*Philosophie zoologique* 1809, 1:108). Indeed, to a naturalist such as Antoine-Jacques-Louis Jourdan (the translator of Gottfried Reinhold Treviranus, Karl Friedrich Burdach, and Johann Friedrich Meckel into French), Lamarck had a good explanation for species production and perhaps the production of new genera, but could not explain major structural advances. Only the analogy between fetal development and the development of organic complexity throughout the animal series could do this—a view voiced in almost all the natural history publications of the 1800s, the 1810s, and the 1820s, and one that Lamarck never commented upon nor even mentioned. Finally, in the *Histoire naturelle des animaux sans vertèbres*, in which Lamarck formulated the law often associated with his name, that of a "power of life" pushing toward complexity (*Histoire naturelle* 1815–1822, 1:133, 181–182), he also stated that human beings have a tendency to see "tendency" when they contemplate the cumulative effect of the endless repetition of the same, simple physicochemical laws, and he explicitly used the growth in complexity of the animal series as a telling example (*Histoire naturelle* 1815–1822, 1:184) of the way in which a "physical necessity" was mistaken for a "finality."

In spite of the scholarly effort bestowed on Lamarck since the 1970s, the well-entrenched myths surrounding his life and career are difficult to overcome. This is surprising, since his theories and contributions to evolutionary thought are constantly referred to (and often misrepresented) in textbooks, histories of biology, and current debates. It could be argued that the stereotypes of Lamarck have been of service to evolutionary debates for so long that the point has been reached where there is little interest in what he really said, the difficulties he faced, and the contradictions he never resolved, or in his actual role in the scientific and cultural scene of his time.

One further difficulty is the almost inevitable anachronistic bias that still characterizes much of the history of science in all its methodological articulations. Ever since it has been assumed that a second scientific revolution stormed the physical and natural sciences of the last decade of the eighteenth century and the early years of the nineteenth, introducing—slowly but securely—increasing levels of professionalization based on higher levels of epistemological sophistication and institutional control, Lamarck's alleged uneasy fit with the requirements of the new scientific age has oriented the reading of his work. In fact, it is only by ignoring most of the natural history practitioners and publications of the time that we can support the view of a Lamarck barely tolerated by his contemporaries, the easy target of the "justified" strictures by Cuvier and his associates.

Systematic perusal of scientific and general culture periodicals, or of the scores of dictionaries, encyclopedias, and medical textbooks France exported throughout the world during the first three decades of the nineteenth century, reveals the complexity of contemporary scientific debates and the popularity of Lamarck. His *Histoire naturelle des animaux sans vertèbres* (1815–1822, 7 vols.) became a classic text for zoologists and, more important, for the growing population of practitioners of geology, a discipline that enjoyed enormous popularity throughout Europe during the first half of the nineteenth century. From the late 1810s, commentators sympathetic to Lamarck attempted to graft elements of his transformist doctrines onto theoretical systems not necessarily favorable to the philosophy of nature propounded by the French naturalist. Thus, for instance, in France, England, Germany, and Italy, comparative anatomists influenced by German authors such as Friedrich Tiedemann and Johann Friedrich Meckel, enlisted the Lamarckian mechanism of individual change to explain the production of varieties and of species within a given genus, although they relied on embryological considerations to explain the development of more complex anatomical and functional structures. In France, authors very sympathetic to Lamarck, such as Jean-Baptiste Bory de Saint-Vincent and Geoffroy Saint-Hilaire, selectively translated elements of the theories put forward by their colleague into the evolutionary or developmental views of life they had independently elaborated.

By the time of his death on December 18, 1829, Lamarck enjoyed in France and elsewhere a considerable scientific and political reputation. His courage and determination, the fact that he had become blind in June 1819, and his relative personal isolation due to his age and ill health all contributed to the creation of the myth of the naturalist–philosopher who pursued his research against all odds and in spite of the opposition from the powerful Cuvier. Indeed, the growing opposition to Cuvier, a leading political figure in the reactionary Restoration establishment of the 1820s, expressed itself in aggressive rhetorical contrapositions between the heroic Lamarck and the opportunistic Cuvier. It became common for the press to call Lamarck "the French Linnaeus" or "the Nestor" of French natural history, and to portray Cuvier as someone who had achieved power thanks to science, only to turn against his colleagues and oppose the advancement of research. Historians and commentators have failed to appreciate the fact that the myth of the poor, isolated, and blind Lamarck was created and diffused by his supporters at the height of his success with the reading public and the intellectual elites: there is no doubt that during the 1820s, Lamarck reached the peak of his popularity in France and in the rest of Europe.

As Geoffroy Saint-Hilaire remarked at the funeral of his colleague, ill health had plagued the last years of Lamarck's life, but the naturalist found consolation in the respect and admiration which surrounded him. Cuvier's more famous and better-known commemoration of Lamarck caused uproar and indignation, and was presented to the Academy of Sciences only in 1832. It is unfortunate that it is usually read as being a reliable portrait of Lamarck's supposed failure to attract any meaningful attention.

During the 1820s, in Edinburgh and London, Brussels and Turin, Naples and Boston, Lamarck's works received considerable attention, and provoked anxiety in moderate and conservative intellectual and scientific circles. It is significant that Charles Lyell felt compelled to devote eleven chapters of the second volume of his *Principles of Geology* (1832) to subjecting Lamarck to a respectful albeit thorough criticism, in order to distance himself from any suspicion that his geological doctrines could be construed as providing indirect support for transformism. Many of Lamarck's colleagues in the British Isles were indeed worried that the sympathy with which his doctrines met in radical circles would turn into an aggressive, materialist, and antiestablishment rallying point, as had been the case in France during the 1820s. Yet, it would be wrong to conclude that Lamarck's ideas found supporters only among radical antiestablishment doctors, journalists, and popular writers. Throughout Europe, even theologically minded naturalists and commentators started to explore ways to incorporate transformism (appropriately modified and expurgated of its materialistic overtones) into a renewed Christian apologetics capable of enrolling contemporary science in its ranks. This was due not only to Lamarck but also to a host of authors who endorsed a variety of explanations for the succession of life on the surface of the Earth and throughout its history. Indeed, it is deeply reductive to see pre–Darwinian natural sciences, and the debates they aroused, as locked into a confrontation between a majority of creationists (or in any case upholders of the fixity of species) and an absolute minority of authors, essentially Lamarck, whose views are represented as exemplifying the nefarious power of materialistic and atheistic prejudice. In fact, the perception of Lamarck as an isolated figure who never managed to attract the attention of his contemporaries is not supported by a thorough engagement with the complex scientific, intellectual, and political articulations of French and other European societies of the first half of the nineteenth century.

The real point historians need to investigate further is why, of the many alternative and at times opposed views of the changes life can undergo that were put forward during the first half of the nineteenth century in several European countries and in the United States, it was the views of Lamarck that continued to attract the attention of commentators throughout the nineteenth and twentieth centuries. The fact that today only a handful of scholars know even the name of naturalists and commentators such as Bory de Saint-Vincent and Jean-Baptiste-Julien d'Omalius d'Halloy, Frédéric Gérard and Pierre Boitard, Achille-Pierre Requin and Sir Richard Vyvyan, or any of the few score of colleagues whom Darwin (rightly, and at times wrongly) acknowledged had preceded him in believing in evolution, defies easy explanations: explanations have to be found case by case, different in different contexts, countries, and periods. Thus, as hinted earlier, in France in the 1820s the debate about Lamarck's ideas was part of a wider philosophical, scientific, and political debate involving the right-wing policies of the Restoration, the personal and scientific rivalry between Cuvier and Geoffroy Saint-Hilaire, and the wide coverage of the history and transformation of life found in large-circulation medical, natural history, and general culture periodicals and encyclopedias. After 1830, the public debate in France appeared to wane, although encyclopedias and dictionaries, including leading Catholic ones, continued to summarize, and at times distort, Lamarck's views from the 1830s to the 1850s. The reception and discussion of Lamarck's ideas in large-circulation dictionaries and encyclopedias has never been surveyed, though several leading naturalists, including Geoffroy Saint-Hilaire's son Isidore, took an active part in the debate. It is significant—to limit our remarks to a single eminent case—that the famous English writer Charles Kingsley became interested in evolution through reading the *Encyclopédie nouvelle*, edited from 1836 to 1843 by prominent French Saint-Simonians, in which the ideas of Lamarck were subjected to spiritualistic interpretations with which the French naturalist would have deeply disagreed. In the United States and Germany, awareness of evolutionary-like theories, and of Lamarck, remained high, with a few favorable pronouncements. In the British Isles, as already noted, the scientific authority of Lamarck refuted the commonly held view among conservative naturalists and commentators that only illiterate radicals could maintain such a theory, and some respected and far from radical naturalists came to believe that although the details of Lamarck's solutions might be wrong, the basic idea deserved attention and further research. Yet, even in the British Isles, where the discussion of transformism was probably the most popular and ideologically the most meaningful, the pre–Darwinian scene was highly diversified. Indeed, one of the most popular evolutionary books of the 1840s, Robert Chambers's anonymously published *Vestiges of the Natural History of Creation*, dismissed Lamarck and had little in common with the transformist tradition represented by the French naturalist.

It could be argued that the reputation Lamarck gained during the 1820s ensured that his name became representative of views that were often very different from the ones he had elaborated. As would be the case with "Darwinian" during the second half of the nineteenth century and in the twentieth, the term, and at times the epithet, "Lamarckian" came to indicate a large spectrum of doctrines having in common the recognition that life had a history and was subject to change, and the plurality of mechanisms put forward to account for this. Throughout the period and until today, the term "Lamarckism" also, and more specifically, has described the evolutionary role of individual variations that emerged during the life of an organism in response to environmental stress. "Lamarckian" elements were also appropriated by a variety of philosophical, teleological, and at times theologically oriented views of evolution, which insisted on concepts such as the power of life or the tendency for inexorable, predetermined growth. Religious, national, political, and broad philosophical considerations have played important parts in defining the role of Lamarck in the history of evolutionary debates since the nineteenth century. Yet, the Lamarckism of current and past mythical reconstructions of the history of evolution bears little or only occasional resemblance to what the French naturalist thought, taught, and wrote.

Further Reading

Burckhardt RW. *The Spirit of System: Lamarck and Evolutionary Biology*. Cambridge, MA: Harvard University Press; 1995.

Corsi P. *The Age of Lamarck: Evolutionary Theories in France, 1790–1830*. Berkeley: University of California Press; 1988. New ed. with the list of pupils attending Lamarck's lectures, *Lamarck. Genèse et enjeux du transformisme 1770–1830*. Paris: Éditions du CNRS; 2001.

Corsi P. Biologie. In P. Corsi et al. *Lamarck, Philosophe de la nature*. Paris: Presses Universitaires de France; 2006: 37–64.

Corsi P. For all the texts of Lamarck quoted: <http://www.lamarck.cnrs.fr/?lang=en>.

For the complete prosopographic list of the 973 pupils attending Lamarck's lectures from 1795 through 1823, see <http://www.lamarck.cnrs.fr/auditeurs/index.php?lang=en>.

I HISTORY

3 Introduction: Lamarckian Problematics in Historical Perspective

Snait B. Gissis

This historical section is a preliminary attempt to outline a general history of Lamarckian problematics. At present this history has many partial chapters, scattered in various books and articles. The section is a gesture toward a future integrated historical narrative of Lamarckism and Darwinism which will present them as intertwined throughout the nineteenth century, as separated and opposed during parts of the twentieth century, and once again in great need of being amalgamated and woven together.

What did Lamarck actually say? What was Lamarckism? And in what ways were the two related? At first glance these questions seem to be the most urgent. The historical portrait of Lamarck drawn by Pietro Corsi in chapter 2 dispels myths and gives us a portrait of the actual historical figure. It differs from the usual accounts of the ways in which what Lamarck wrote was read, interpreted, disseminated, and made use of throughout the nineteenth and early twentieth centuries in fields as widely divergent as paleontology, medicine, and the emerging social sciences of psychology, anthropology, sociology, economics, geography, and politics. Scientific, cultural, and national contexts and styles formed and shaped these modes of understanding, thus producing different "Lamarcks" and diverse "Lamarckisms." Nonetheless, one can argue that historically there were certain features that served as a common foundation for that diversity. These features can be identified and defined through some core questions:

- Was evolution perceived as internally directed, was it progressive?
- Were any assumptions made that accounted for an increase in complexity of organisms?
- Was there any role for teleological efficacy, or any role for a cohering of functions?
- Was there a necessary relation between an evolutionary view of nature and materialism?

- Was any constitutive role attributed to self-organization?
- What was the role of development in the description and analysis of organisms?
- Was the environment perceived as an active causal agency in adaptive change? In the process of evolution?
- Were organisms perceived as passive receptacles, as active or interactive agents, or as a combination of both?
- What was the role of chance in the ordering of the organic world?
- What was considered as a biological individual?
- Was there any room for the concept of population perceived as aggregated and diversified individuals?
- Were any mechanisms proffered for preserving changes acquired within the life span of a developing organism? And if there were, what were they?
- Were acquired changes transmitted transgenerationally?
- Was natural selection a principal "creative" mechanism, or a secondary eliminative one?
- Was any role attributed to competition, conflict, struggle?
- Was there any recapitulation mechanism—any ontogenetic sequence of processes in an individual that simulates the phylogeny of the lineage? And if there was, what was its explanatory purpose?

The nature of Lamarckian problematics becomes clear when the answers to these questions emphasize the causal role of "environment" in the formation and development of organisms rather than its selecting role; the role of development and its relation to mechanisms of heredity; the activities and interactions within organisms (including self-organization), among organisms, and between organisms and their external conditions; and the role of "behavior" in evolution.

The chapters in this section suggest that these questions were formulated and answered differently before and after Darwin, before and after Weismann, and before and after the rise of Mendelism. Furthermore, they show the ways in which adopting a certain position on biological issues was connected with specific explanatory mechanisms and narrative styles, with views of science at large and its methods. All the chapters depart—some strongly, others moderately—from the mainstream historiography of their topics. They do so because they chose a "Lamarckian perspective" on the history of relations between Lamarckisms and Darwinisms during the nineteenth and early twentieth centuries.

The historiography of Lamarck as well as that of the various pre-1930s Lamarckisms has increased and diversified since the late 1970s. It has benefited from the gradual construction of a continuous narrative of "evolution" instead of a focus on precursors and anticipators, as had usually been the case in the practice of the history of evolution. For example, this new history has incorporated many features of the social, political, economic, and cultural developments of the eighteenth century, and has addressed issues at the core of natural history theorizing and practice of that period. Thus Lamarck, too, has been integrated into the narrative rather than posited as an odd, lone figure. Historical context now includes the interactions among practices, learning institutions, and state machinery, as well as interactions between the various fields that were then considered as sciences.

The more recent historical writings on Lamarckism can be divided roughly into two types of narrative. The first argues that a conceptual gap existed between Darwin and the Lamarckism of that time, thereby implying that the two were perceived as incompatible, whereas the second suggests that complementarity or coexistence characterized the two conceptual schemes.

The main assumptions of the first approach are the following:

1. That Darwin's principal innovation was the mechanism of natural selection, which explained how populations of living forms adapt to changes in their environment by assuming that natural selection operated on chance variations. This resulted in what Darwin called "descent with modification," which was perceived by him and his contemporaries to be slow and gradual.
2. That most of Darwin's contemporaries did not or could not accept Darwin's mechanism—even as expounded in the later editions of *The Origin.*

Consequently, it was concluded by the historians advocating this outlook that Darwin had been interpreted through other contemporaneous and more generally accepted notions of progressive evolution.

The proponents of the second approach argued that Darwin's own position was more complex, having progressivist and recapitulationist as well as "Lamarckian" elements, and that his views underwent significant changes. Furthermore, they suggested that until sometime in the late 1880s there was enough room for positions that would be perceived as "hybrid" (by today's historians), on the one hand, as well as for exclusively "pure" neo-Lamarckian and neo-Darwinian positions, on the other. These polarized varieties, it is argued, emerged in the last decade of the nineteenth century and persisted through the first two thirds of the twentieth century.

The inheritance of acquired characters by use inheritance was perceived by historians as a common key feature of late nineteenth-century Lamarckisms, and tended to color the presentation of Lamarck's own position. Indeed, use and disuse formed an important mechanism in Lamarck's discussion of transmutation of life-forms. It was formulated within a framework in which the organism *inter*acted with the environment, and was not just acted upon by it. This was put in the context of animal behavior and constituted an important facet of evolutionary theory.

The theory of use inheritance was formulated rather late in the eighteenth century. Lamarck, as well as Pierre J.G. Cabanis and Erasmus Darwin, deployed the notion of "habit." Lamarck refined this cornerstone concept of eighteenth-century empiricism, and had it serve as the intermediary between changes in the "circumstances" (i.e., the environment) and the heritable modifications in the organism. Within the general eighteenth-century debate on the choice of bodily structures and mechanisms to explain sensations in general, there was quite a broad choice of models. Lamarck, in line with his geological and physical views, chose a hydraulic model, with various active fluids as transferring agents and with body structures that would go with it—pipes, tubes, canals, and other equivalent spaces. These structures functioned as linkages between inside and outside, as well as connections between various parts of the inside. More generally, they enabled the "rapports" between the environment and the body. This occurred by being translated into "behavior" (i.e., modes of experiencing and acting).

In his opening comments, Richard Burkhardt in chapter 4 appropriately calls for a richer and more multilayered discussion of Lamarck, implicitly suggesting that the relation between the narratives of Lamarckism and that of Lamarck should be reexamined. Burkhardt compares the ways in which Lamarck, Frédéric Cuvier (the brother of Georges Cuvier), and Darwin used behavior in explaining inheritance. Lamarck held that modifications of functions, of habits (use and disuse), preceded and caused change in structures, and these changes would be inherited, with cumulative inherited changes bringing about the transformation of species. However, his notions of animal behavior were not drawn from direct observation of living animals, but rather from natural history cabinets. Burkhardt describes the crystallization, deployment, and delimitation of the notion of "hereditary modifications" by Frédéric Cuvier and by Darwin, both of whom observed wild and domesticated animals. Darwin studied wild animals during his *Beagle* trip, and later at the London Zoo, whereas Cuvier observed animals only in the zoo of the Muséum d'Histoire Naturelle. Both discussed the formation of instincts and their relation to habit. However, while Lamarck and Darwin related their work to evolutionary issues, and looked upon use inheritance as a significant evolutionary mechanism, whether principal or auxiliary, Frédéric Cuvier avoided any discussion of species transformation. Furthermore, Burkhardt points out that from an early stage of his theorizing, Darwin was interested in a mechanism combining "habit" and "inheritance," and suggests that this could serve as a gauge of his varying attitude toward Lamarck's Lamarckism.

In chapter 2, Corsi notes that from the 1820s until the 1850s, numerous versions of transformist models were being advanced and put to various scientific and political uses, both by individuals and by groups, particularly in France, Germany, Great Britain, and Italy. Lamarck's theories were expounded in his central writings, especially in the *Philosophie zoologique* and in *Histoire naturelle des animaux sans vertèbres*. They came to represent for some naturalists (zoologists, physiologists, and anatomists), for some medical practitioners, and for some geologists and anthropologists the evolutionary account and the evolutionary mechanisms of the succession of life-forms and the explanation of the emergence of new species.

A number of rough generalizations can be used to distinguish deployments of Lamarck during the period up to 1859 from later ones. Rather than adopting a closely knit and cohering corpus, the later appropriations of Lamarck used particular elements, such as a specific recapitulation model, that were then deployed in conjunction with other explanatory components to produce new theories that in the historical literature have been discussed under the general heading of Lamarckism. As the foci of the natural sciences gradually moved from Paris/France to Germany and somewhat later also to Great Britain, scientific styles, as well as the cultural and political frameworks, became factors in the choice of problems and in the formulation of the solutions that were offered.

After the publication of *The Origin of Species* and later works by Darwin, the status of whatever was considered the relevant Lamarckian corpus and explanatory model changed—at first gradually and later more dramatically. Thus, toward the latter part of the century the "progress" and "growth in complexity" components that were earlier assumed to be factors in evolutionary change, gradually receded to the background. In his *History of Biology—A Survey*, Eric Nordenskiöld stated:

> ... Both Darwin and Haeckel based their doctrines of descent partly on the theory of variability and natural selection brought about by the struggle for existence among the variations, and partly on the assumption of the direct influence of environment upon the individual, and the inheritance of the changes thus brought about—that is, a Lamarckian conception.... one could with prejudice emphasize the idea of selection, or with equal prejudice maintain the influence of environment. (Nordenskiöld 1928:562–563)

I would like to suggest that one should view the history of evolutionary biology during the period discussed as always containing a spectrum of views at whose center would be found positions which exemplify—without a sharp demarcation and within one framework—an intertwining or at least a coexistence of some Lamarckian and Darwinian components; toward both poles would be found positions which in differing degrees deny such a possibility, and adopt a position of either exclusive self-defined Darwinism or exclusive self-defined Lamarckism. The chapters in the historical section illustrate these different positions, and thus implicitly refer to the general historiographical issues mentioned earlier, namely, the importance of the local and the contextual: What Lamarckism, which Darwinism, was employed by whom, when, where, and why?

A significant missing piece in this section is the contributions of the American neo-Lamarckists, primarily Edward Drinker Cope, Alpheus Hyatt, Alpheus S.

Packard, and John Ryder. On the spectrum between self-defined Darwinian positions and self-defined Lamarckian ones, they would be located as adherents of mostly nonselectionist, partially Lamarckian, evolutionary mechanisms. Packard, who also wrote a biography of Lamarck (1901), coined the name "neo-Lamarckianism" (1885). Historiographically these American neo-Lamarckists have been considered a rather more cohesive group, with a formulated agenda, a common journal (*The American Naturalist)*, and a common reliance on somewhat similar disciplinary sources. This is not surprising, since two of the more prominent members, Cope and Hyatt, were paleontologists, and all had come under the influence of Louis Agassiz at some stage of their professional development. Indeed, in spite of various distinctions and differences, common to them, at least in significant stages of their careers, was the assumption that the environment effected heritable adaptive reactions that brought about change of structures. These changes, the acquired characters, were viewed as stages that could be evolutionarily progressive, adding new stages to the ontogeny of the individual organism within a recapitulation model, or alternatively could be degenerative and thus retrogressive. Consequently, one can argue that their commonality with other European Lamarckisms of the last decades of the nineteenth century, particularly in response to Weismann's challenge, was the deployment of the inheritance of acquired characters as the primary mechanism to explain the causes of variation. In that sense, their views harbored the same tension between plasticity and rigidity that is pointed to by Loison in his discussion of the French neo-Lamarckists in chapter 7. Within their recapitulation model, they emphasized the regularity of development during the embryonic stages, and the regularizing mechanisms were later perceived as internalized, the variation systematically directed and thus orthogenetic.

All the chapters in this section exhibit the wealth of the life sciences sources that were used by Lamarckists, drawing on field and experimental data from disciplines as varied as geology, entomology, paleontology, embryology, and botany. This diversity partially accounts for the different emphases within loose groups of Lamarckists. Another common feature of all the chapters is the noting of the conspicuous shift of emphasis toward the last decades of the nineteenth century, before Weismann, with the "inheritance of acquired characters" becoming the hallmark of "Lamarckism." Perhaps this shift should be analyzed more thoroughly historically and thematically, and not be seen so much as a continuation of Lamarck's core interests. The issue would then be in what ways the use-inheritance mechanism, considered as a cause-and-origin of variations, can explain the contemporaneous emphasis on the development of organisms. Until the end of the nineteenth century, mechanistic conceptions were principally favored, and for the most part whole organisms were studied. But a diversity of interpretations was attributed to these conceptions among varieties of Lamarackism. Interpretations were materialist, utilitarian, and adaptionist, with "will" brought in to bear on the use-inheritance mechanism, wavering between mediated and direct response to changing environmental conditions. They emphasized external stimuli or internal direction/motivation, and some even stressed programmed changes that resulted in either a predetermined multiplicity of evolutionary paths or a single one (orthogenesis).

It is interesting to note that the late nineteenth-century physiologists, such as Hermann von Helmholtz and Michael Foster, did not feel it incumbent upon them to consider evolutionary matters. It was assumed that all organic systems were similar, not constrained by structures or boundaries of species. Their physiological experimentations were related to practical interests (e.g., medicine), and to a reductionist, materialist program. That program regarded physics and chemistry as model disciplines. It saw the mid-nineteenth-century developments in physics, in particular the establishment of the first and second laws of thermodynamics, as giving physics a universal character; its drive toward atomic and microscopic interpretations and explanations was seen as freeing it as much as possible from context dependence. The successful procedures of the physical and chemical sciences, the possibility of reduction from the macroscopic to the microscopic, enhanced the physiologists' tendency to investigate chemical and physical processes (e.g., organic chemical reactions), where the constraints were ahistoric and local, and had a universal character. The direction of their reduction—from the whole to the parts—was also that of bottom-to-top causal explanations. That meant that solutions were sought lower and lower in the hierarchy of levels, with smaller and simpler entities. Consequently this whole field of the life sciences seemed to be unaffected, at least directly, by the evolution controversies of the late nineteenth century. Yet, this reductionist approach pointed the way for evolutionary biologists to look for solutions at a lower level of analysis.

After Weismann, the discussion revolved principally around what were defined as the "material units" of heredity in a population of organisms, on the one hand, and a conception of "force" as imposing the character on the individual organism, on the other. Note, though, that under the pressure of changing discourse, some Lamarckists did suggest a discussion of heredity in terms of material units. Their difficulties were focused on supplying an account of the structure and causal agency of such units within their Lamarckian framework. It is mostly, though not solely, within this theoretical and experimental context that an analogy between memory and heredity was brought up as a potential resolution of the tension between open-ended plasticity and predetermined, rigid stability. The analogy likened the acquisition of a character to learning, and inheritance to memory, which was enhanced by repetition. Furthermore, a second level of analogy between mental memory and physiological heredity was suggested, whereby repetition that produced habit was analogous to memorization that produced instinct. Given the similarity between the orderly procession of development and recall, the analogy was applied to the recapitulation model. Ontogenetically acquired characters were assumed to be transmitted across generations and then "recalled" during the ontogenesis of descendants, as the memory of the organism's past history. Haeckel, Cope, Hyatt, Henry P. Orr, and Samuel Butler all made use of this analogy, and so did Richard Semon, who in the early twentieth century developed a wholly original, materialistic, and systematic heredity-as-memory view. The analogy between heredity and memory helped Lamarckists to interpret atavisms, reversion, lags in progress in the biological sphere, and, by implication, in the social sphere.

There were also diverse positions on the tempo of evolutionary change. Both Lamarck and Darwin were perceived as gradualists, but some of the late nineteenth-century Lamarckists suggested that the rate of evolutionary change can be rapid, with change occurring within a few generations, depending on the developmental rate of the acquired character. This perception of rate was associated with the recapitulation model, which played a predominant role among those whose emphasis was Lamarckian. The acquired character was thought to be imposed on the adult organism, which thus provided a stable state. New characters were expected to be added to the developing organism, and in order to make room for them, there was either a deletion of earlier stages of development or, as found more often among Lamarckists, a condensation of earlier stages, which thus allowed for an acceleration of the rate of development.

Given the above diversity, a "thick narrative" is called for, which will reconfigure the intertwining, coexisting, and separating of Lamarckisms and Darwinisms as their contemporaries experienced, conceptualized, and institutionalized them.

Presenting the Spectrum of Lamarckisms

Sander Gliboff (chapter 5) analyzes the German-speaking Lamarckisms in the period from 1863 until 1926, that is, Haeckel and his generation, and the generation after them. Gliboff suggests a view of the evolutionary history of that period as one in which Lamarckian and Darwinian components could be seen as compatible within a single theoretical framework. He argues against the Modern Synthesis demonization and lumping together of the various Lamarckisms, and for a renewed appreciation of the significant contribution that these "Lamarckian" theories made to the study and conceptualization of evolution in the twentieth century. The principal figure in Gliboff's historical narrative is Ernst Haeckel (1834–1919), whose evolutionary work was carried out primarily in morphology. Haeckel is depicted as advancing a view of selection that applied to both "inborn" and "acquired" variations. Though a proponent of "progress," Haeckel objected to both teleological and vitalist explanations, and presented selectable variations as mechanically induced by the direct effect of the environment. The inclusive position of Haeckel serves, according to Gliboff, as a common ground, which is then contrasted with other contemporary varieties of Lamarckism, such as the views of Karl Wilhelm von Nägeli, Julius von Sachs, Theodor Eimer, and August Pauly, who, in reaction to Weismann, polarized their positions and gave selection a merely negative role in evolution. The second generation Haeckelians—Semon, Paul Kammerer, and Ludwig Plate—had to cope with the rise of genetics, the growing importance of experimentation on living organisms, and the increased pressure to "insulate heredity from the environment." Gliboff posits that the perceived failure of Kammerer's experiments is both the crucial event and the excuse for excluding Lamarckian-type explanations from genetics in particular, and from evolutionary theory in general.

In chapter 6, Charlotte Weissman offers an internalist analysis of August Weismann's theory of germinal selection. She argues that this theory was Weismann's alternative solution to the contemporaneous Lamarckian challenge that focused on observations that seemed impossible to explain within the bounds of natural selection. Weismann agreed that a selectionist answer could not be given at the level of the individual organism, since selection could not account for all of the changes that occurred during the organism's evolution. However, he proposed that selection was still the explanation for these observations, albeit at a lower level of biological organization, namely, that of the determinants within the germplasm of the cell. Weissman shows why Weismann's solution was not that of "giving in to Lamarckian objections," as has been argued by mainstream historiography, but rather a fine-tuning of his basic position on the segregation of the germplasm and the central role of selection. By descending to the molecular level of the heredity material and thus making a case for multilevel selection, Weismann reaffirmed the all-sufficiency of selection.

This analysis of Weismann's work on heredity and selection enables one to understand the challenge Weismannism posed to Lamackists of all persuasions on the theoretical level, and how it served as a stimulus toward experimental investigations. Weismann's work sparked an intense and intellectually challenging debate between Herbert Spencer and Weismann, as well as a multiplicty of minor debates on this issue. Some historians have argued that the "Weismann effect" was a significant enabling condition for the acceptance of Mendelism and for the later emergence of genetics, as attention shifted to the germ and its structure. The "Weismann effect" also began to undermine the deployment of recapitulation models, which were partially abandoned when Mendelism became predominant. Others argue that Weismann seems all-important only if the narrative focuses on natural selection and adaptation as the principal tenets of Darwinism, while in fact his views were incompatible with "orthodox" Darwinism as well as with the varieties of Lamarkism. If one turns one's attention to certain commonalities between Lamarckism and

Darwinism, such as a naturalistic explanation, gradualism, and rejection of teleology, then Weismann is not perceived as such a significant polarizer. However, a long-term perspective may see him as a pioneer in a long line of theoreticians who argued for a one-way flow of biological information.

In chapter 7, Laurent Loison revises the narrative of French Lamarckists from the last decades of the nineteenth century until the late 1920s, arguing that theirs was a Lamarckism struggling to keep its rejection of Darwinian components in the face of fast-changing theorizing in the evolutionary field. Loison presents an internalist description of the work of several important French scientists. In contrast to some earlier work on these biologists, Loison suggests that they were indeed Lamarckists, but that they lacked a clear common agenda and a concerted synthetic corpus of positions. What united them was a certain style of doing science—positivistic, inclined toward physiological experimentation, and with a great measure of isolation from outside scientific theorizing. He suggests that the history of French neo-Lamarckism can be analyzed and structured through two principal notions: plasticity and heredity. The answer of these biologists to the question "What were the origins and the causes of hereditary variation?" was through "soft inheritance" operating in individual, indivisible organisms. This position became predominant in a variety of biological subdisciplines between the 1880s and the 1920s. Consequently, the effects of the environment at every level were the focus of their experimental work. Loison argues that within their Lamarckian conceptual framework, a notion of stability and rigidity was perceived to be a precondition to heredity, while soft inheritance seemed to allow for the possibility of change in the form of an adaptive response. Consequently, plasticity and heredity turned out to be incompatible within a framework devoid of Darwinian components, and this became a source of irreducible tension within French neo-Lamarckism. Loison ends by claiming that a theoretical impasse had been reached within French neo-Lamarckism in the first decades of the twentieth century, which turned experimental practice into a metaphysical dogma.

Chapter 8 looks at Trofim Lysenko and Lysenkoist science from the 1930s until the early 1950s. Nils Roll-Hansen has chosen to discuss Lysenkoist scientific tenets and practices from the perspective of their being perceived as Lamarckian. This perspective is located in the context of Soviet science, on the one hand, and the predominant evolutionary theory and its practice in the West, on the other, first with the increasing role of genetics and then with the convergence toward the Modern Synthesis. From that perspective Lysenkoist science was closely related not just to the political powers in the Soviet Union, the facet usually discussed, but also to a philosophical Marxist worldview, with its emphasis on future-oriented malleability. Furthermore, it was founded on a scientific approach that gradually became subsidiary, and was not taken seriously or was even ignored by evolutionary scientists. That approach emphasized interaction with the environment and a version of the inheritance of acquired characters. Roll-Hansen insists on its presence in both the Soviet Union and in Europe at least up to the establishment of the Modern Synthesis, and even after that, that is, from the 1920s to the 1950s (see Marion Lamb and others in the section on the Modern Synthesis). It may be easy to see the effect of ideology in constraining science in the Soviet Union, but Roll-Hansen asks readers to turn their attention to the systematic exclusion of non–Modern Synthesis views in the West, and the ways in which such views were implicitly or explicitly related to Lysenkoism during the Cold War. In appendix A, we have included a new English translation of a poem dedicated to Lamarck, written by the well-known Russian poet Osip Mandelstam, which illustrates the widely differing uses of Lamarckism within Russian–Soviet culture just before the rise of Lysenkoist science.

Throughout the nineteenth century, Lamarckism was an important framework for conceptualizing the development of human societies. In chapter 9 I look at evolutionary biology between the 1850s and the 1890s as a resource for the emerging social science of sociology. I argue that the transfer of models, metaphors, and analogies from evolutionary theory was a significant enabling condition for the rise of sociology as a discipline. I point out that, contrary to some received views, versions of Lamarckian evolutionism, rather than the Darwinian one, were at least as significant in this development, particularly in Great Britain and France. Thus, the transfer into the sociological realm of Lamarckian notions of the inheritance of acquired characters, the mechanism of use inheritance, and the role of "habit" are analyzed within the theoretical systems of two central founding figures of sociology—Herbert Spencer and Émile Durkheim. I suggest viewing the relationships between individuals and collectivities as stemming from the interactions between social thinking and Lamarckian evolutionary theories. Thus, Spencer is interpreted as a Lamarckian evolutionary collectivist in disguise, and the notions of "environment" and of social plasticity are shown to have played an important role in both Durkheim's emerging sociology and the Spencerian system.

The narratives in the historical section stretch from the first decades of the nineteenth century until the period after World War II, tracing actual relationships among practitioners and conceptual connections among theories. They delineate the space where natural selection was not the sole, and perhaps not even the predominant, mechanism of evolution. The possibility of such a space was temporarily discarded. The papers in this section retell some of the histories contained in that space from a perspective of Lamarckian problematics. This, we hope, may bring about the reopening and broadening within historical writing of what Provine (1988) called "the evolutionary constriction."

Further Reading

Allen G. *Life Science in the Twentieth Century*. Cambridge: Cambridge University Press; 1978.

Bowler PJ. *The Eclipse of Darwinism*. Baltimore: Johns Hopkins University Press; 1992.

Burian R, Gayon J, Zallen D. 1988. The singular fate of genetics in the history of French biology, 1900–1940. J Hist Biol. 21:357–402.

Corsi P, gen. ed. Lamarck Web site: <http://www.lamarck.cnrs.fr/?lang=en>.

Gould SJ. *Ontogeny and Phylogeny*. Cambridge, MA: Belknap Press of Harvard University Press; 1977.

Gayon J. *Darwinism's Struggle for Survival: Heredity and the Hypothesis of Natural Selection*. Cobb M, trans. Cambridge: Cambridge University Press; 1998.

Lewontin R, Levins R. The problem of Lysenkoism. In: *The Radicalisation of Science: Ideology of/in the Natural Sciences*. Rose H, Rose SPR, eds. London: Macmillan; 1976:32–64.

Nordenskiöld E. *The History of Biology: A Survey*. New York: Alfred A. Knopf; 1928.

Packard AS. *Lamarck, The Founder of Evolution: His Life and Work*. New York: Longmans, Green, and Co., 1901.

Provine W. Progress in evolution and meaning in life. In: *Evolutionary Progress*. Nitecki MH, ed. Chicago: University of Chicago Press; 1988:49–74.

Richards RJ. *The Tragic Sense of Life: Ernst Haeckel and the Struggle over Evolutionary Thought*. Chicago: University of Chicago Press; 2008.

Roger J., ed. 1979. Les Néo-lamarckiens français. Rev Syn. 95/96:279–469.

4 Lamarck, Cuvier, and Darwin on Animal Behavior and Acquired Characters

Richard W. Burkhardt, Jr.

In the history of evolutionary biology, the name of Jean-Baptiste Lamarck is closely associated with the idea of behavior as an agent of organic change. In 1800, in his earliest presentation of his ideas on organic mutability, he told his students that he "could prove that it is neither the form of the body nor of its parts that gives rise to the habits, to the way of life of animals, but that, to the contrary, it is the habits, the way of life, and all the influencing circumstances that have over time constituted the form of the body and of the parts of animals. With new forms, new faculties have been acquired, and little by little nature has reached the state where we now see it" (Lamarck 1801:15).[1]

Lamarck's ideas on the role of habit in species change are easily summarized. He maintained that as animals moved into new environments, or as the environments in which they lived changed (an inevitability, according to his theory of the Earth), animals experienced new needs and contracted new habits. This led them to use some of their organs more, and some of their organs less, than previously. The changes acquired as the result of the use or disuse of organs were passed on to succeeding generations. The result, over time, was the transformation of species.

To illustrate his views, Lamarck offered a variety of examples. The most famous of these is no doubt that of the giraffe, of which Lamarck wrote in 1809:

> In regard to habits, it is interesting to observe a product of them in the particular form and height of the giraffe (*Camelo-pardalis*). This animal, the largest of the mammals, is known to live in the interior of Africa in places where the earth is nearly always arid and without pasturage, obliging it to browse on the leaves of trees and to continually strive to reach up to them. It has resulted from this habit, maintained for a long time by all the individuals of the race, that the forelegs have become longer than the hind legs and its neck has so lengthened itself that the giraffe, without standing on its hind legs, raises its head and reaches a height of six meters (nearly twenty feet). (Lamarck 1809, 1:256–257)

The giraffe, however, was neither the only example Lamarck offered nor the first. His first examples were birds, including shorebirds, about which he wrote:

> One may perceive that the bird of the shore, which does not at all like to swim, and which however needs to draw near to the water to find its prey, will be continually exposed to sinking in the mud. Wishing [*voulant*] to avoid immersing its body in the liquid, it acquires the habit of stretching and elongating its legs. The result of this for the generations of birds that continue to live in this manner is that the individuals will find themselves elevated as on stilts, on long naked legs. (Lamarck 1801:14)

This example proved unfortunate, not only because it was easy to caricature but also because Lamarck seemed to be suggesting that wishing (or willing) was an important part of the evolutionary process. That, however, was not his position. This becomes clear when we consider his views on the correlation between organic complexity and animal faculties—one of the main themes of his biological theorizing.

At the beginning of the nineteenth century, animals were typically defined as organic beings endowed with the faculties of sensation (*sensibilité*) and voluntary movement. Lamarck himself resorted to this definition on occasion in his early teaching (e.g., Lamarck 1801:5), but he later came to insist that voluntary movement was not a property of all animals. Only a small set of the creatures in the animal kingdom were capable of "wishing" or "willing" of any kind (Lamarck 1802:186). He explained this to his students in 1806 with a simple set of observations: "Put a hydra in a glass of water, and when it is fixed on one side of the glass, turn the glass in such a way that the light strikes it on the opposite side. You will always see the hydra go slowly to place itself in the place that the light strikes, and to remain there as long as you do not change this place" (Lamarck 1806, in Giard 1907:135). This was no different, Lamarck said, from the way that the parts of a plant turned involuntarily toward the light. Equally involuntary, he claimed, was the way that the hydra brought to its mouth whatever corpuscle its tentacles encountered. The corpuscle might be digested in whole or in part (or completely rejected), but there was never any volition involved, never any choice that would lead to any variation in the animal's actions. From this Lamarck was happy to conclude: "No, it is not true, what has always been said, that the faculty of sensation and that of voluntary motion are general and common to all animals" (Lamarck 1806, in Giard 1907:135).

Lamarck elaborated on this point in his *Philosophie zoologique*. There he explained that the least perfect animals (the lower invertebrates) are incapable of voluntary action; they move only in response to external stimuli. The more highly organized invertebrates have nervous systems that endow them with *feeling* and with *instincts*, but they are still incapable of *thinking* and voluntary action, because they lack a sufficiently developed brain. Only in the vertebrates, and primarily in the birds and mammals, does one find the capacity for thought and voluntary action, but even there—and even among humans—most actions occur without thinking or volition being involved (Lamarck 1809, 2:336–338).

Lamarck did not explain the emergence of new faculties in the same way that he explained change at the species level. Instead he invoked what he called "the power of life" to explain the general trend toward increased complexity in the organ systems of the animal classes. This phrase sounds vague and vitalistic, but the process Lamarck had in mind was wholly materialistic, amounting to nothing other than the interplay between fluids and solids. He thus ended up offering a two-factor theory of organic evolution in which (1) the power of life explained the progressive, increased complexity of the organ systems of the different animal classes and (2) environmentally induced changes in habits explained change at the species level (and why, unlike the animal classes, the species could not be arranged in one simple series). It deserves to be noted, however, that this clear contrast between "the power of life" and the influence of environmental circumstances breaks down upon closer scrutiny, given that, according to Lamarck's account of the very beginnings of life, environmental factors in the form of subtle fluids (caloric and electricity) are critical for bringing form and movement to the simplest life-forms (Burkhardt [1977] 1995:147, 151–157).

Lamarck typically portrayed his evolutionary views as readily comprehensible. On at least one occasion, though, he admitted that a certain class of phenomena might seem impossible for his theory to explain. As he put it to his students in 1803:

> . . . would not one be stopped . . . by the sole consideration of the admirable diversity one notices in the *instinct* of the different animals, and by that of the marvels of all sorts that their diverse sorts of *industry* present? Would one dare carry the spirit of system building [*l'esprit de système*] so far as to say that it is nature that has, by herself, created this astonishing diversity of means, of ruses, of skills, of precautions, of patience, of which animal industry offers us so many examples! What we observe in this regard, in the class of insects alone, is it not a thousand times beyond what is necessary to make us feel that the limits of nature's power in no way permit her to produce by herself so many marvels! And to force the most obstinate philosopher to recognize that here the author of all things has been necessary, and has alone sufficed to bring into existence so many admirable things?

Lamarck for his own part was undaunted. Instincts, no less than animal structures, he went on to explain, could be explained as the result of habits developed and carried out for long periods of time in different environmental settings. He acknowledged that one would be out of his senses "to claim to assign limits to the power of the primary author of all things." He allowed, however, that his own admiration of the "primary cause of all things" would not be diminished if it turned out that, instead of attending to all the various details of creating, developing, and changing everything in existence, the "primary cause" had delegated such matters to nature. Nature, he believed, was responsible for all animal actions, from the very simplest to those involving instinct and industry, and finally to those involving reasoning and the exercise of will (Lamarck an XI [1803] in Giard 1907:100–101).

Lamarck's stance in this regard was reminiscent of that of his onetime mentor, Georges-Louis Leclerc, Comte de Buffon. Half a century earlier Buffon had scoffed at René-Antoine Ferchault de Réaumur's detailed studies of insect life and the way Réaumur had attempted to portray insect behavior as evidence of the Creator's wise design. As Buffon had put it: "Who, in fact, has the greatest idea of the Supreme Being, he who envisions him creating the universe, setting in order the forms of life, founding nature on invariable and perpetual laws; or he who looks for him and wants to find him intent upon managing a republic of flies, and very much occupied with the way in which a beetle must fold its wing?" (Buffon 1753, 4:95).

One might have expected Lamarck, as professor of "the insects, worms, and microscopic animals" at the Museum of Natural History, to have been more concerned than Buffon had been about the details of insect behavior. Though that may be so, the fact remains that Lamarck's daily practices as a scientist involved studying the remains of dead invertebrate animals, not the behavior of live ones (Burkhardt 1997). It seems emblematic of his whole approach to accounting for behavior in evolutionary terms that he had never seen a live giraffe, only a stuffed one, when he offered his explanation of how the giraffe got its long neck. For all his claims about habits preceding structures, Lamarck sought his insights into the behavior of animals by examining their structures, not by observing their actions.

Interestingly, while Lamarck's contemporaries objected to his general "system building" and likewise to his specific examples of change at the species level, they do not seem to have criticized him for supporting the idea of the inheritance of acquired characters. He had encapsulated the idea in a pair of laws laid out in his *Philosophie zoologique*:

> *First Law*: In every animal that has not reached the end of its development, the more frequent and sustained use of any organ will strengthen this organ little by little, develop it, enlarge it, and give to it a power proportionate to the duration of its use; while the constant disuse of such an organ will insensibly weaken it, deteriorate it, progressively diminish its faculties, and finally cause it to disappear.
> *Second Law*: All that nature has caused individuals to gain or lose by the influence of the circumstances to which their race has been exposed for a long time, and, consequently, by the influence of a predominant use or constant disuse of an organ or part, is conserved through generation in the new individuals descending from them, provided that these acquired changes are common to the two sexes or to those which have produced these new individuals. (Lamarck 1809, 1:235)

The primary reason Lamarck was not criticized for the second of these laws was no doubt that in his day the idea of the inheritance of acquired characters was widely accepted. He himself treated the idea as self-evident. As he put it in 1815: "The law of nature by which new individuals receive all that has been acquired in organization during the lifetime of their parents is so true, so striking, so much attested by the facts, that there is no observer who has been unable to convince himself of its reality" (Lamarck 1815:200). What was novel in Lamarck's use of the idea was his assertion that the cumulative effect of the use or disuse of organs was not limited, but could instead proceed indefinitely, bringing new species into being. Thus, while he did not claim credit for the idea of the inheritance of acquired characters per se, he did claim to have been the first to show how the effects of use and disuse of organs in the individual illuminated "the causes that have brought about the astonishing diversity of animals" (Lamarck 1815:191).

We can get some sense of what was distinctive and what was commonplace about Lamarck's thinking by comparing his ideas with those of one of his contemporaries, Frédéric Cuvier (on whom see Richards 1987 and Burkhardt 2001). The younger brother of Georges Cuvier, Frédéric Cuvier seems to have been the first zoologist to use the words "hereditary" and "heredity" in their now familiar biological senses (see Gayon 2006).[2] Significantly, he used both words in the context of promoting the idea of the inheritance of acquired characters. In a paper presented in 1812 he went so far as to represent the idea of the inheritance of acquired characters as a "rule" that he had "for a long time" believed he could establish (Cuvier 1812).

What historical sense can we make of Frédéric Cuvier's ideas? More than Lamarck's ideas about behavior, Cuvier's ideas about behavior were closely connected to his zoological researches. Initially called to Paris in 1797 by his brother, Georges, to help with the comparative anatomy collection Georges was developing, Frédéric benefited from institutional circumstances that led him to study living animals as well as dead ones. In 1803 he was appointed to a new position at the Paris Museum of Natural History, that of *garde* (superintendent) of its menagerie. Among his official responsibilities were "to attempt the acclimatization and naturalization of foreign species that would be of some benefit for the rural or domestic economy" and to study the habits [*moeurs*] of animals, "[recording] with care the times which begin and end rutting, gestation, laying, molting, and generally all the changes of state that happen to animals each time of the year" (Archives Nationales de France, AJ.15.591, 29 Frimaire an XII).

Over the next thirty-five years, Frédéric proceeded to make the study of animal behavior his specialty. He paid particular attention to such topics as the nature of animal instincts, the features that made species amenable to domestication (or not), and the nature and extent of the changes that domesticated species undergo. He set for himself the general goal of establishing a new science based on the study of animal actions.

Cuvier's first paper on animal behavior appeared in 1807. In an examination of the rutting periods of different mammals, he first used the word "hereditary" (*héréditaire*). He employed it to refer to acquired characters. Writing about how domestication had changed the rutting periods of the males of certain species, he raised the possibility that "qualities that are at first only accidental, can finally become hereditary, if the causes that produce them exercise their influence over a certain number of generations" (Cuvier 1807:129–130).

The next time Cuvier used "hereditary" was again in the context of modifications undergone by domesticated animals. In 1808 he offered "some reflections on the moral faculties of animals" as a preface to a description of the race of dogs associated with the Australian aborigines. He presented as a presumptive truth the idea "that some of the qualities that are regarded as belonging to instinct in animals are subject to the same laws as those that depend on education." Qualities acquired through education, he said, "become finally instinctive or hereditary as soon as they have been exercised over a series of sufficient generations." By the same token, "they become obliterated or wear away more or less, after their exercise ceases to fortify or sustain them" (Cuvier 1808: 462). Cuvier credited this idea to the naturalist Charles Leroy, the former keeper of the king's game at Versailles, a man recognized for his "long experience and profound sagacity." Cuvier made no mention of Lamarck.

Cuvier next used the word "hereditary" in 1811, once again with respect to modifications produced in domesticated animals. He did so in discussing the physical traits of the different races of domestic dogs. This time, however, he was talking about the inheritance of features that had been developed not through use or disuse but instead through a process of selection (though he himself did not use the word "selection"). Having noted that dogs display a wide range of variations in their coloration and other characters, he wrote: "When one takes care to reunite the individuals of the same color, the race ordinarily perpetuates itself, and it is the same for the majority of the characters that we have already examined: new proofs that accidental modifications end up always by becoming hereditary." It was "the care that one has generally taken to mate in each race only the individuals of the same color" that explained the characteristic colors and coat patterns of the different dog races (Cuvier 1811:349).

In January 1812 Cuvier delivered to the Société Philomathique a paper on the intellectual faculties of brutes. In it, according to the report that appeared in the Society's journal, Cuvier presented "the rule, which he for a long time has believed able to establish, that acquired faculties propagate themselves by generation and become hereditary." Cuvier used the rule to show "the cause of the existence of races and what they owe to this heredity," and he furthermore suggested how the knowledge of this "law" might be applied to "the management of animals in general" (Cuvier 1812:217–218). This appears to have been the first occasion on which the word "heredity" (*l'hérédité*) was used in a biological context to refer to the transmission of acquired traits.

Cuvier thus clearly endorsed the idea of the inheritance of acquired characters, calling upon it to explain the formation of the different races of domesticated animals. However, he explicitly refrained from endorsing the idea of species mutability. He believed that the changes exhibited by domestic animals (or animals in the wild) were ultimately limited. Despite the great diversity of the different races of dogs, they all belonged to a single species. Their differences, as he put it in 1811, were an indication of "all the modifications of which these animals are susceptible *for us*." They could not, however, be legitimately cited "in support of the systems by which one has wanted to deduce the different forms of animals from the diverse circumstances which have been able to influence them." The authors of such systems, Cuvier said, had been misled by "the lights of a false analogy" (Cuvier 1811:353). Lamarck was surely one of the writers Cuvier had in mind. Once again, though, he did not mention Lamarck's name.

In the course of his career, Frédéric Cuvier had much to say about animal instincts. He continued to endorse the idea that instincts developed through the inheritance of acquired characters. At the same time, he continued to steer clear of Lamarck's theory of evolution. The species question, Cuvier believed, was scientifically inaccessible. To ask about the cause of the differences between species was to ask in vain, because "the force that gave birth [to species] has ceased to act, or hides itself in depths where our sight cannot reach." One could study the creation of varieties—one could see this happening with domestic animals—but not the creation of species (Cuvier 1819).

Like his brother, Georges, Frédéric Cuvier seems to have thought it best to give contemporary evolutionary theorists as little publicity as possible. Nonetheless, in 1834, two years after Georges Cuvier's death (and five years after Lamarck's), Frédéric felt compelled to defend his brother against new advocates of the idea of species transformation. In his introductory remarks to the fourth edition of his brother's *Recherches sur les ossemens fossils* (Researches on Fossil Bones), Frédéric attacked the idea of evolution. He assured his readers that neither that "system" which Buffon had "developed with so much lucidity and eloquence," nor "that to which Lamarck brought so many arbitrary suppositions," had any observational or experimental basis. The same was true, he said, of more recent transformist theories. As Frédéric put it: "Ah! If there existed the feeblest proof, I would say not even of the transformation, but of the possibility of the transformation of one species into another species, how would it be possible for an anatomist, a physiologist, or a naturalist to be able to direct his attention thereafter to any other phenomenon?" But Cuvier did not believe species transformation was possible. For it to happen would be "miraculous." He did allow, however, that if the idea of species transformation were ever recognized as possible, then that "very instant" there would be "a fundamental revolution in all the sciences that have animals as their subject" (Cuvier 1834, 1:xxi–xxii).

We have thus far identified a number of ironies with respect to the now traditional association of Lamarck's name with the idea of the inheritance of acquired characters. In the first place, we have seen that Lamarck treated the idea as commonplace and did not claim any special credit for the idea itself—though he did claim credit for having shown how the long-term results of the use and disuse of organs would lead to species transformation. In the second place, we have seen that Frédéric Cuvier, while endorsing the idea of the inheritance of acquired characters—and using the words "heredity" and "hereditary" in their now familiar biological sense for the first time—adamantly rejected the idea of species change. A third irony regarding the common linking of Lamarck's name with the inheritance of acquired characters appears when we consider the work of Charles Darwin. Darwin himself proves to have been a firm believer that the effects of use and disuse could be inherited. Though in the years after his death, Darwinian natural selection came to be contrasted with the "Lamarckian" idea of the inheritance of acquired characters, Darwin himself put great stock in use inheritance as an auxiliary mechanism in the evolutionary process. Indeed, in 1880, responding in the journal *Nature* to some misguided criticism from Sir Wyville Thomson, Darwin wrote: "Can Sir Wyville Thomson name any one who has said that the evolution of species depends only on Natural Selection? As far as concerns myself, I believe that no one has brought forward so many observations on the effects of use and disuse of parts, as I have done in my *Variation of Animals and Plants Under Domestication*; and these observations were made for this special object" (Darwin 1880:32).[3]

With that historical irony to ponder, let me now offer a few remarks about Darwin's thoughts on animal behavior and evolution. Paying attention to the behavior of living animals was part of his multifaceted scientific practice. He observed animals both in the wild and in captivity. Early in his career, observations from his *Beagle* voyage informed his thinking about species mutability, as did his visits to the London Zoo. Behavioral considerations came to figure prominently in a number of his writings, most notably in his chapter on instinct in *On the Origin of Species* (1859); in his extended discussions of human evolution and sexual selection in his book *The Descent of Man, and Selection in Relation to Sex* (1871); and in the whole of his book *The Expression of the Emotions in Animals and Man* (1872). Late in his career, he studied the habits of earthworms and wrote an entire volume devoted to their role in the development of the Earth's soil (Darwin 1881).

Space limitations here do not allow for a detailed survey of Darwin's behavioral interests (see Burkhardt 1983, 1985; Richards 1987). I will limit myself to a brief discussion of the role of habits in Darwin's earliest theorizing and to a survey of what Darwin had to say about Lamarck.

Darwin scholars have shown that it was not until the winter or spring of 1837—after he had returned from the *Beagle* voyage—that Darwin took up the idea of species transmutation (e.g., Herbert 1974). One can follow his early thoughts on the subject by observing what he wrote in his manuscript notebooks of 1837 and 1838. In this period he was strongly attracted to the idea of habit preceding structure. During the *Beagle* voyage, he had been struck by the behavior of various animal species, in particular that of a woodpecker from the pampas that seemed to feed exclusively on the ground rather than on trees. In his C Notebook of 1838 he wrote: "The circumstances of ground woodpeckers,—birds that cannot fly &c. &c. seem clearly to indicate those very changes which at first it might be doubted were possible,—and it has been asked how did the otter live before it had its web-feet. All nature answers to the possibility" (Barrett et al. 1987:257). Here, in effect, Darwin was answering one of Georges Cuvier's objections to Lamarck. Later in his C Notebook, Darwin gave Lamarck the highest praise he would ever give him, calling him "the Hutton of Geology" (one imagines Darwin meant to write "the Hutton of Zoology"). Darwin allowed at this point that Lamarck "had few clear facts" but was "so bold" in his thinking, with "such profound judgment," that he might be said to have been endowed with "the prophetic spirit in science," "the highest endowment of lofty genius" (Barrett et al. 1987:275).

Such praise of Lamarck by Darwin would not be typical. Darwin was not inclined to see himself as following in Lamarck's footsteps, even when the similarities between their views were strong. In his N Notebook, commenting on Lamarck's *Philosophie zoologique*, Darwin wrote: "Habits becoming hereditary form the instincts of animals.—almost identical with my theory." Tellingly, he did not put it the other way around, that is, "my theory is almost identical with Lamarck's." He went on to note of the same book: "no facts, mingled with much hypothesis" (Barrett et al. 1987:589).

After reading Malthus and coming to the idea of natural selection, Darwin gave habits less of a role in his theorizing about species change. Natural selection became his primary agent of change, and habits (and their inherited results) became material for natural selection to work upon. In 1842, writing out his first extended sketch of his theory, he remarked:

> It must I think be admitted that habits whether congenital or acquired by practice often become inherited; instincts, influence, equally with structure, the preservation of animals; therefore selection must, with changing conditions tend to modify the inherited habits of animals. If this be admitted it will be found *possible* that many of the strangest instincts may be thus acquired. (F. Darwin 1909:18)

Darwin would pursue this same line of reasoning in 1859 in *On the Origin of Species*.

Darwin evidenced his desire to distance himself from Lamarck on a number of occasions. In January 1844, when first telling J.D. Hooker of his thoughts on species mutability, Darwin wrote: "Heaven forfend me from Lamarck nonsense of a 'tendency to progression,' 'adaptations from the slow willing of animals,' etc." He went on to admit, however, that "the conclusions I am led to are not widely different from his; though the means of change are wholly so" (F. Darwin 1887, 2:24). Later, in his chapter on instincts in *On the Origin of Species*, Darwin was pleased to note that while the instincts and structures of neuter insects could be explained by natural selection, they could not be explained by "the well known doctrine of Lamarck" (Darwin 1859:242).

Darwin's desire to distance himself from Lamarck also manifested itself in a letter he wrote to Charles Lyell in 1860. In it he told Lyell, "you refer repeatedly to my view as a modification of Lamarck's doctrine of development and progression. If this is your deliberate opinion there is nothing to be said, but it does not seem so to me. Plato, Buffon, my grandfather before Lamarck, and others, propounded the *obvious* view that if species were not created separately they must have descended from other species, and I can see nothing else in common between the 'Origin' and Lamarck" (F. Darwin 1887, 3:14). In contrast, when Darwin added a "historical sketch" to the third edition of *On the Origin* of *Species* (1861), his public comments about Lamarck were much more generous.

All around the world in 2009 biologists celebrated the 200th anniversary of Darwin's birth. They tended to pay less attention to the fact that 2009 was also the bicentenary of the publication of Lamarck's *Philosophie zoologique*. But Lamarck and his bold, pathbreaking theory of organic evolution also deserve proper recognition. To pay him his due, he should be remembered for much more than his endorsement of the idea of the inheritance of acquired characters. When one compares his case with those of Frédéric Cuvier and Charles Darwin, one is reminded that the inheritance of acquired characters should not be seen as a distinctively "Lamarckian" idea. One would make better historical sense of Lamarck, and one might even find certain inspiration for the future, if one looked at the whole range of Lamarck's biological theorizing. That theorizing featured his recognition that nature has been responsible for successively producing all the different forms of life on Earth, from the simplest to the most complex, and that the living animal should be seen simultaneously as an agent and as a consequence of biological evolution.

Notes

1. All translations in this chapter are by the author.

2. In the present chapter I identify earlier uses of the words *héréditaire* (1807) and *hérédité* (1812) than those identified by Gayon. I find Frédéric Cuvier first using the word *héréditaire* in 1807, whereas Gayon cites Cuvier's paper of 1811, and I find Cuvier using *hérédité* in 1812 to refer to the transmission of acquired traits, whereas Gayon points to a much later review of Cuvier's writings by Pierre Flourens (1841) for the first use of *hérédité* in this sense.

3. Darwin went on to say of his book: "I have likewise there adduced a considerable body of facts, showing the direct action of external conditions on organisms" (Darwin 1880: 32).

References

Archives Nationales de France. AJ.15.577–678. *Minutes des procès verbaux des assemblées des professeurs, et pièces annexes.*

Barrett PH, Gautrey PJ, Herbert S, Kohn D, Smith S, eds. *Charles Darwin's Notebooks, 1836–1844.* Ithaca, NY: Cornell University Press; 1987.

Buffon GLL, Comte de. Discours sur la nature des animaux. In: Buffon, *Histoire naturelle, générale et particulière.* Vol. 4. Paris: Imprimerie Royale; 1753:3–110.

Burkhardt RW, Jr. *The Spirit of System: Lamarck and Evolution Biology.* Cambridge, MA: Harvard University Press; [1977] 1995.

Burkhardt RW, Jr. Lamarck's understanding of animal behaviour. In: *Lamarck et son temps, Lamarck et notre temps.* NN, ed. Paris: Vrin; 1981:11–28.

Burkhardt RW, Jr. The development of an evolutionary ethology. In: *Evolution from Molecules to Men.* Bendall DS, ed. Cambridge: Cambridge University Press; 1983:429–444.

Burkhardt RW, Jr. Darwin on animal behavior and evolution. In: *The Darwinian Heritage.* Kohn D, ed. Princeton, NJ: Princeton University Press; 1985:327–365.

Burkhardt RW, Jr. Unpacking Baudin: Models of scientific practice in the age of Lamarck. In: *Jean-Baptiste Lamarck, 1744–1829.* Laurent G, ed. Paris: Éditions du CTHS; 1997:497–514.

Burkhardt RW, Jr. 2001. Frédéric Cuvier and the study of animal behavior. Bull hist épistémol sci vie 8:75–98.

Cuvier F. 1807. Du rut. Ann Mus Hist Nat. 9:118–130.

Cuvier F. 1808. Observations sur le chien des habitans de la Nouvelle-Hollande, précédés de quelques réflexions sur les facultés morales des animaux. Ann Mus Hist Nat. 11:458–476.

Cuvier F. 1811. Recherches sur les caractères ostéologiques qui distinguent les principales races du chien domestique. Ann Mus Hist Nat. 18:333–353.

Cuvier F. 1812. Essais sur les facultés intellectuelles des brutes. Nouv Bull Sci Soc Philomat. 3:217–218.

Cuvier F. Le bouc de la Haute-Égypte. In: Geoffroy Saint-Hilaire E, Cuvier F. *Histoire naturelle des mammifères.* Vol. 1. Paris: C. de Lasteyrie; 1819. All entries in this work are separately paginated.

Cuvier F. Observations préliminaires. In: *Recherches sur les ossemens fossils.* 4th ed. Vol. 1. Paris: E. d'Ocagne; 1834:i–xxiv.

Darwin C. *On the Origin of Species by Means of Natural Selection, or the Preservation of Favoured Races in the Struggle for Life.* London: John Murray; 1859.

Darwin C. *The Descent of Man, and Selection in Relation to Sex.* 2 vols. London: John Murray; 1871.

Darwin C. *The Expression of the Emotions in Man and Animals.* London: John Murray; 1872.

Darwin C. 1880. Sir Wyville Thomson and natural selection. Nature 11, November: 32.

Darwin C. *The Formation of Vegetable Mould, Through the Action of Worms.* London: John Murray; 1881.

Darwin F, ed. *The Life and Letters of Charles Darwin.* 3 vols. London: John Murray; 1887.

Darwin F, ed. *The Foundations of the Origin of Species. Two Essays Written in 1842 and 1844 by Charles Darwin.* Cambridge: Cambridge University Press; 1909.

Flourens P. *Résumé analytique des observations de Frédéric Cuvier sur l'instinct et l'intelligence des animaux.* Paris: C. Pitois; 1841.

Gayon J. Hérédité des caractères acquis. In: Corsi P, Gayon J, Gohau G, Tirard S, eds. *Lamarck, philosophe de la nature.* Corsi P, Gayon J, Gohau G, Tirard S, eds. Paris: Presses Universitaires de France; 2006:105–163.

Herbert S. 1974. The place of man in the development of Darwin's theory of transmutation. Part I. To July 1837. J Hist Biol. 7:217–258.

Lamarck J-B. *Système des animaux sans vertèbres.* Paris: Déterville; 1801.

Lamarck J-B. *Recherches sur l'organisation des corps vivans.* Paris: Maillard; 1802.

Lamarck J-B. *Philosophie zoologique.* 2 vols. Paris: Dentu; 1809.

Lamarck J-B. *Histoire naturelle des animaux sans vertèbres.* Vol. 1. Paris: Déterville; 1815.

Lamarck J-B. Discours d'ouverture d'un cours de zoologie, prononcé en prairial an XI. In: *Discours d'ouverture des cours de zoologie donnés dans le Muséum d'Histoire Naturelle (an VIII, an X, an XI et 1806) par J.-B.* Lamarck. Giard A, ed. Paris; 1907a.

Lamarck J-B. Discours d'ouverture du cours des animaux sans vertèbres, prononcé dans le Muséum d'Histoire Naturelle en mai 1806. In: Giard A, ed. *Discours d'ouverture des cours de zoologie donnés dans le Muséum d'Histoire Naturelle (an VIII, an X, an XI et 1806) par J.-B. Lamarck.* Giard A., ed. Paris; 1907b.

Richards RJ. *Darwin and the Emergence of Evolutionary Theories of Mind and Behavior.* Chicago: University of Chicago Press; 1987.

5 The Golden Age of Lamarckism, 1866–1926

Sander Gliboff

After two hundred years, the theories that now pass as "Lamarckism" would hardly be recognizable to its original author, Jean-Baptiste de Lamarck (1744–1829) (Burkhardt 1977) or to his early supporters (Corsi 1988, 2005; and chapter 2 in this volume). It started out as a dynamic system of physics, mineralogy, and biology unified by the transformative effects of energetic particles and subtle fluids, and was once considered an important milestone in the history of evolutionary thought (Glass, Temkin, and Straus 1959). After 1859, "Lamarckian" notions about heredity were sometimes considered essential to Darwinism, and the two pioneering transformationists could be treated as allies in support of gradual, progressive, but non-teleological change. But by the time of the Modern Synthesis of the 1930s and 1940s, Lamarckism had come to be seen as an ill-supported theory of heredity and an illegitimate alternative to Darwinism.

In Synthesis-era depictions, "Lamarckism" was hardly even a coherent "-ism" anymore, but rather a collection of obsolescent ideas and mechanisms, including the transformative effects of the organism's own striving, will, or perceived needs; the communication of acquired bodily changes to the hereditary material; a teleological drive toward morphological perfection, going straight up an idealized scale of nature; and multiple origins of life, as opposed to universal common descent. Proponents of the Modern Synthesis, such as Ernst Mayr, viewed such "Lamarckian" notions (along with some others not necessarily attributed to Lamarck) as obstacles to be overcome (Mayr 1982), and by some interpretations, the main achievement of the Synthesis was to rule them out (Provine 1992).

Today, when proponents of epigenetics or evo-devo challenge the Synthetic theory, they are often tempted to muster Lamarck or later "Lamarckians" as antecedents and rhetorical allies, and to reposition and redefine Lamarckism yet again (Jablonka and Lamb 1995; Vargas 2009). An old pattern is thus repeating itself.

Ever since 1859, Lamarck has served alternately as a foil and an ally to Darwin. Darwin's supporters have tried in various ways either to appropriate Lamarck as a precursor and partner, or else to define Lamarckism narrowly and to dismiss it as naive and obsolete. Contrarily, Darwin's rivals have tried to redefine Lamarck as a figurehead for their own theories and approaches. These Lamarcks, and their rhetorical and biological importance, were especially creative, diverse, and hotly debated in the six decades between 1866 and 1926, the period I have chosen to discuss here and to call a "golden age."

My emphasis on the diversity of interpretations, motivations, and relationships between factions will contrast with much of the existing secondary literature, in which there is a tendency to lump together everything and everyone that was called "Lamarckian" or invoked environmental effects. Standard representations dwell mainly on the competition between Lamarckism and Darwinism, in which Lamarckism is the loser, but I think that many more of these nineteenth-century theories made positive contributions to the later Evolutionary Synthesis and even to post–Synthesis developments than is commonly acknowledged.

Haeckel's Lamarck: A Precursor and a Partner to Darwin

I begin a bit earlier than 1866, with a short prelude to the golden age. In September 1863, Ernst Haeckel (1834–1919) made his debut as a spokesman for Darwin at the big meeting of the Gesellschaft deutscher Naturforscher und Ärzte (German Society of Naturalists and Physicians). His talk in the plenary session struck a masterful balance between the revolutionary and the traditional, proclaiming the dawn of a new era while also building bridges to the old establishment and situating Darwin within a familiar and unthreatening intellectual context that included Lamarck. He argued that species transformation and common descent were not new with Darwin but had long been recognized, and that a pantheon of great naturalists was on the Darwinian side, including not only Lamarck but also Johann Wolfgang von Goethe, Lorenz Oken, Étienne Geoffroy Saint-Hillaire, and even Lamarck's opponent, the antievolutionary Georges Cuvier (Haeckel 1863).

It is easy to say that in making even a staunch antievolutionist like Cuvier into a forerunner of Darwin, Haeckel was mangling the history of science. Indeed he was, but there was method in his mangling. The historical figures he invoked represented a wide variety of approaches and assumptions, and he united them all, rhetorically, behind Darwin, leaving no good figureheads for the antievolutionists to rally around.

Of course, Haeckel presented his pantheon of naturalists in "Darwinized" form. His Lamarck, for example, never said anything about subtle fluids, an inherent drive to perfection, or any role for the needs and strivings of the organism. The purely mechanistic, heritable effects of use and disuse or environmental factors were Lamarck's contribution to the present theory. They complemented Darwin's theory by supplying a mechanism of variation.

Some variations, Haeckel explained, "... are in part already present in the germ of the individual, in the egg" (in other words, inborn), but "In part, the individual characteristics are only acquired during the life of the individual, through adaptation to the external conditions of life, especially through the interrelationships that every organism has with all the others around it" (Haeckel 1863:22). In other words, some variations are acquired in what we now think of as the "Lamarckian" manner. Natural selection would act upon both sorts of characteristics, the inborn and the acquired. Thus, again, Lamarck was integrated into the Darwinian mechanism of change, as a source of variation. (Darwin himself had made use of the same mechanisms of variation, but did not make the association with Lamarck.)

In his 1866 *Generelle Morphologie* (General Morphology), Haeckel went over the mechanisms of change in greater detail, with greater emphasis on ridding biology of teleology and every kind of predetermined, providential, or self-directed process. *Generelle Morphologie* was intended above all to be a "mechanical morphology," and to this end, Darwinian natural selection was essential:

> *We see in Darwin's discovery of natural selection in the struggle for existence the most striking evidence for the exclusive validity of mechanically operating causes in the entire field of biology. We see in it the definitive death of all teleological and vitalistic interpretations of organisms.* (Haeckel 1866, 1:100 [emphasis in original])

Nothing was more important to Haeckel than to purge biology of mind and purpose: "We do not think we can emphasize this extremely important point enough. It is the unassailable citadel of scientific biology" (Haeckel 1866, 1:150). For Haeckel's purposes, Darwin had not yet been explicit enough in his rejection of supernatural agency. Haeckel was trying to bring evolutionary biology into line with what he considered to be good scientific practice and good monistic philosophy, which ruled out the existence of any independent realm of mind or spirit. It also allowed Haeckel to draw on Darwinism in his polemics against special creation and organized religion, and to argue that the religious viewpoint had been ruled out (Gliboff 2008; on Haeckel's personal and ideological motives for rejecting mind, purpose, and religion, see also Richards 2008).

But in order for Darwinian natural selection to play this role in biology, it had to be supplied with *mechanically* induced variations upon which it could act. Haeckel required that variations be caused by unpredictable stimuli from the environment. Over and over he emphasized that the process of acquiring a new characteristic was a purely mechanistic response to either physical conditions or biological interactions, and that such mechanical causes sufficed to explain all evolutionary change.

Consequently, any adaptive changes had to be initiated by the environment, not by the organism itself, because the latter might again open the door to teleology, will, and striving. How could the internal workings of the organism produce adaptations all on their own? That would mean anticipating environmental change instead of reacting to it, and people might argue that some providential or purposeful agency was directing the changes.

It must be noted, however, that even though environmental influences had no special biological purpose for Haeckel, that did not mean that their effects on the organism had to be random in the modern sense. In Haeckel's system they clearly were not. They usually went in an "adaptive" or "progressive" direction, but not always, and never to an extent that would account for evolution and adaptation without additional mechanisms.

According to Haeckel, environmental conditions were so many and varied that no two individuals would experience them and react to them in precisely the same way: "*No organic individual remains absolutely the same as the others*," he wrote (Haeckel 1866, 2:192 [emphasis in original]). *Variation* was the principal effect of the environment, but it was only part of a larger process that transformed species. Natural selection was still needed to mold this variation into adaptations and give evolution its progressive direction and mechanistic, nonteleological character. Thus, in Haeckel's depiction, Darwin's natural selection *added* to Lamarck's theory but did not supplant it.

The relegation of environmental effects to the role of producing variation distinguished Haeckel's theory from the "Lamarckism" that is usually attributed to him, and it is the key to his synthesis of Lamarck and Darwin. That synthesis, as it stood in 1866 in *Generelle Morphologie,* went further than the loose coalition building of his 1863 address. No longer content with unifying biology behind the general ideas of species transformation and common descent, he had now specified that a nonteleological theory required both Darwin and Lamarck: Darwin for struggle and selection, Lamarck for the strictly mechanistic causes of variation. For the rest of his long career, Haeckel opposed all new accounts of variation, from August Weismann's germplasm theories to the various mutation concepts of the early twentieth century. That was because they were either not as well founded empirically as the direct effects of the environment, or not mechanistic in the precise sense that he required.

Haeckel's Rivals: Refuting or Reconceiving Lamarck

Haeckel's synthesis did not go unchallenged, of course, and obvious points to attack were the appropriation of Lamarck and his subordination to Darwin. Several nineteenth-century rivals consequently tried to win Lamarck away from Haeckel's coalition. There were several ways of doing this, each contributing to the proliferation of modern Lamarckisms. Not all of them were designed to be viable alternatives to natural selection. Some were straw men to be knocked down and removed from the realm of possible sources of variation.

August Weismann (1834–1914) was most effective with this latter strategy. He defined Lamarckism very narrowly—and not too differently from Haeckel's version—as a mechanism of inheritance of environmental effects. As is well known, Weismann argued mainly on theoretical grounds, but also on the basis of some experimentation, that no such mechanism could exist.

Weismann's own mechanisms of generating variation were designed to take the place that environmental effects had occupied within Haeckel's (and, for that matter, Darwin's) system. He tried out several different mechanisms, sometimes allowing only mixing and remixing of unchanging germplasm elements, sometimes allowing the environment to modify the germplasm directly, sometimes isolating the germplasm completely, and, late in his career, also allowing the internal environment to modify the germplasm (Winther 2001; Weissman, chapter 6 of this volume).

Other rival evolutionists employed similar strategies: construing Lamarck as supplying the mechanism of variation, then undermining and replacing that mechanism with something else. For example, the botanist Carl von Nägeli (1817–1891), plant physiologist Julius Sachs (1832–1897), and many others sought to replace it with systems of internally generated variation. They argued that life processes were orderly and operated according to discoverable laws, and that it was the job of the physiologist to find the laws of variation and evolution. Unfortunately, from the physiologists' point of view, all sorts of zoologists, comparative morphologists, paleontologists, and anthropologists kept intruding into the discussion and overstepping the limits of their disciplines (Nägeli 1884:3–4; see also Sachs 1894).

For a competent physiologist, it could never suffice, according to Nägeli, to say, generally, that the environment caused variation. One had to investigate, specifically, what sorts of environmental stimuli caused what sorts of changes. When one did, it became apparent that such factors as climate or nutrition hardly ever effected lasting changes in important characteristics (Nägeli 1884:103ff.). Environmental effects, even if they occurred, could not explain much of evolution.

Nägeli aimed to replace environmental effects with what he called a "physiological" mechanism that would generate variation according to internal laws of growth and development: "The germ developed *in accordance with its own organization*, independently of and unperturbed by the dissolved (nonorganized) nutrients taken in by the mother" (Nägeli 1884:109; emphasis added). As an effect of preexisting internal organization and laws of growth and development, variation could neither be random nor go in many different directions, and so could not provide meaningful choices for natural selection to make. Natural selection was therefore largely superfluous, according to Nägeli. It might have a role in changing unimportant characteristics or honing adaptations once they arose, but on the whole, it could not redirect the lawful process of variation.

Recovering the "Real" Lamarck

In the middle of my spectrum of Lamarckisms are the ones that functioned as complete accounts of evolutionary change, either as one of several mechanisms of change in a pluralistic system or as full-fledged alternatives to Darwin. An example of the full-fledged Lamarckian alternative is provided by Theodor Eimer (1843–1898). He is usually counted as an orthogenecist, and indeed he seems to have coined the term "orthogenesis," but his orthogenetic drive was just one component in a larger scheme. Lamarck, according to Eimer, recognized more causes of species transformation than just the direct effects of the environment and use and disuse. In lecture notes, he enumerated four Lamarckian causes:

> The direct effects of water, air and light, and climate;
>
> Hybridization;
>
> An inner developmental drive [*Entwicklungstrieb*] (similar to Schopenhauer's will as the highest principle);
>
> and habit (or disuse) (Eimer n.d.)

Eimer had little or no role for natural selection, and he pointedly titled his three-volume book, in an allusion to Darwin's, *The Origin of Species, by Means of Inheritance of Acquired Characteristics, According to the Laws of Organic Growth* (Eimer 1888–1901). He attacked Weismann bitterly but seems to have been friendly with Haeckel.

The most pluralistic of all were the American neo-Lamarckians such as Edward Drinker Cope (1840–1897), Alpheus Hyatt (1838–1902), and Henry Fairfield Osborn (1857–1935), whose theories allowed for a plethora of mechanisms, including the effects of use and disuse of organs, direct environmental effects, natural selection, internally generated variations such as Nägeli would approve of, and special laws or energies that ensured progress or orthogenesis, as well as the will or consciousness of the organism itself (Rainger 1981).

Not all of these mechanisms were attributed explicitly to Lamarck, but the association was strongly implied. In some cases, their use of Lamarck could be very similar to Haeckel's. Cope, for example, had Lamarck supply variation by means of environmental effects and use and disuse of organs, but he denied Haeckel any originality in recognizing this, and he was less definite than Haeckel about the need for natural selection to sort out these variations (Cope 1896:2–9).

Most of the nineteenth-century Lamarckians interpreted Lamarck in mechanistic terms, but in the early twentieth century, a group of "psycho–Lamarckians" or, as they sometimes also called themselves, "Eulamarckians," argued instead that Lamarck's most important insights concerned the role of the mind in evolution. The mind, according to their Lamarck, recognized and assessed the organism's (or even the single cell's) needs and initiated the appropriate morphological change and adaptation.

The leading figure in this group was August Pauly (1850–1914), a professor of forestry at Munich, who described his approach as a "psycho-physical teleology." The idea was that the psyche was tied to the material of the body and nervous system, so it was not transcendental, vitalistic, or mystical (Pauly 1905). To Haeckel's school, however, that seemed insincere. If a single cell had the same capacity for evaluating and responding to evolutionary needs as an animal with a complex brain and nervous system, that implied that the physical basis of psyche could not be very important, and that the whole approach was not respectable science (Plate 1909).

Early Twentieth-Century Renewal

Toward the end of his long career, Haeckel wearied of debating with rival evolutionists. He wrote to Thomas H. Huxley in 1893 that he would give up biology and devote himself to his philosophy of monism and the fight against religious dogma (Uschmann and Jahn 1959/1960:27). But just as Haeckel went off to bash the clerics, the biological sciences were changing rapidly. A "revolt against morphology" led away from Haeckel-inspired reconstructions of evolutionary history, and toward experimentation on living organisms (Coleman 1977; Allen 1978, 1979; but see also Maienschein, Rainger, and Benson 1981 and response by Allen 1981). The burgeoning of genetics after 1900 challenged evolutionists of all stripes to reexamine their assumptions about heredity. The eugenics movement raised expectations that biologists would address social and political problems. And soon World War I would shake the faith of nineteenth-century evolutionists in progress and in the benefits of struggle and selection.

A younger generation of evolutionists took over from Haeckel and rose to these challenges, while also trying to preserve what they took to be essential to their Darwinism (and Haeckel's and Darwin's own): namely, the role of the environment in initiating variation. The leading figures of this generation of Haeckelians were Richard Semon (1859–1918), Ludwig Plate (1862–1937), and Paul Kammerer (1880–1926). Plate referred to himself, Semon, Kammerer, and a few others as the *Altdarwinisten*, or, as I shall call them, the "old-school Darwinians"—in contradistinction to the *Neudarwinisten* or neo-Darwinians, who followed Weismann in insulating heredity from the environment and rejecting the inheritance of acquired characteristics.

Semon made his scientific reputation as an explorer–collector in Africa and Australia, and as a comparative morphologist in the Haeckelian mold (Semon 1893–1913), but he turned to theoretical work after a scandalous love affair closed most laboratory and university doors to him (Anon. 1920; Kammerer 1920). His Mneme theory treated heredity as an analogue of memory and proved to be a seminal work for psychologists, even if not for biologists. Semon's theory was designed to explain away the experimental results of genetics and *Entwicklungsmechanik* without admitting the existence of hereditary particles such as genes or Weismannian determinants. It was one of the last bastions of Haeckel's system of environmentally induced variation (Semon 1904, 1908; Schatzmann 1968; Schacter 1982).

Plate was Haeckel's powerful successor to the chair of zoology at Jena, and an influential reviewer, journal editor, and commentator on developments in genetics, eugenics, and evolution. His extreme nationalism and anti–Semitism, and, late in his life, his ties to National Socialism, make him a pivotal figure in both the scientific and the social interpretation of evolution.

Under Plate's coeditorship, the *Archiv für Rassen- und Gesellschafts-Biologie* (Archive for Racial and Social Biology) was a very important forum for interdisciplinary work in the biological and social sciences and for discussing general problems in heredity and evolution, in addition to human applications. Plate's editorial voice was very strong, as he wrote frequent short comments, literature reviews, critiques, and reports on foreign publications. These writings, together with a series of books on the state of evolutionary theory, made him an important arbiter of what was good and new in evolution and heredity. In contrast to Semon's rejection of the Mendelian mechanisms of heredity, Plate's strategy was to compromise, accepting genetics and the chromosome theory to a large extent, but denying that they sufficed to explain all of heredity (Robinson 1975; Harwood 1993; Penzlin 1994; Levit and Hossfeld 2006).

Kammerer, a Viennese zoologist, went the furthest with the new experimentalism and genetics, which he tried to assimilate to Haeckel's old synthesis of Lamarck and Darwin. His goal was to show that environmental effects on morphology and behavior could be communicated to the chromosomes and cause heritable genetic changes there—a goal that was not widely shared among the various Lamarckian factions (Gliboff 2006).

Moreover, Kammerer's whole system was designed to serve political as well as scientific purposes. In evolution as in human affairs, Kammerer upheld the importance of improving the unfit, as opposed to selecting the fit, and he argued for the evolutionary efficacy of cooperation in addition to competition and struggle. After World War I, he promoted an alternative eugenics, based on his own version of Darwinism, that would use surgery, hormone treatments, testicle transplants, and other innovative methods to create heritable improvements in deficient individuals (Kammerer 1924; Gliboff 2005).

Kammerer committed suicide in 1926, following accusations that he faked his experimental results. His example was cited (as late as Mayr 1999) as evidence of how desperate and depraved the "Lamarckians" had become by the eve of the Evolutionary Synthesis, but Kammerer was just one of many kinds of Lamarckians. Very few of the others thought that Kammerer spoke for them or that his experiments had any bearing on their versions of Lamarck. If anything was discredited by the Kammerer scandal, perhaps it was the middle ground and the case for a synthesis of Lamarck with Darwin.

Epilogue and Conclusion

After 1926, it is hard to find anyone still defending the middle ground of old-school Darwinism. Haeckel, Semon, and Kammerer were dead, and Plate was becoming irrelevant. If ever the field was polarized between Lamarckians and neo-Darwinians, it was after this time. Genetics and the Evolutionary Synthesis moved ahead without meaningful input from the remaining Lamarckian factions, who seemed to the geneticists and the purer selectionists to stand for obsolete assumptions about environmental effects, internal drives, energies, and forces of progress.

Perhaps as a result of the decline of the old school, Synthesis-era histories tend to see only a single amorphous, oppositional "Lamarckism" rather than the complex of theories and factions that I have depicted. As we commemorate the Lamarck bicentennial here, I wonder whose version of Lamarck most of us have in mind and whether we are celebrating a precursor and a partner to Darwin, or the originator of a rival theory.

References

Allen G. *Life Science in the Twentieth Century.* Cambridge: Cambridge University Press; 1978.

Allen G. 1979. Naturalists and experimentalists: The genotype and the phenotype. Stud Hist Biol. 3: 179–209.

Allen G.1981. Morphology and twentieth-century biology: A response. J Hist Biol. 14: 159–176.

Anon. 1920. The death of Richard Semon. J Hered. 11: 78–79.

Burkhardt R. *The Spirit of System: Lamarck and Evolutionary Biology.* Cambridge, MA: Harvard University Press; 1977.

Coleman W. *Biology in the Nineteenth Century: Problems of Form, Function, and Transformation.* Cambridge: Cambridge University Press; 1977.

Cope E. *The Primary Factors of Organic Evolution.* Chicago: Open Court; 1896.

Corsi P. *The Age of Lamarck: Evolutionary Theories in France, 1790–1830.* Berkeley: University of California Press; 1988.

Corsi P. 2005. Before Darwin: Transformist concepts in European natural history. J Hist Biol. 38: 67–83.

Eimer T. Entstehung der Arten [undated, unpublished notebook]. Theodor Eimer Papers, Universitätsarchiv Tübingen, item 13.

Eimer T. *Die Entstehung der Arten auf Grund von Vererben erworbener Eigenschaften, nach den Gesetzen organischen Wachsens*. 3 vols. Jena: Gustav Fischer; 1888–1901.

Glass B, Temkin O, Straus W, Jr, eds. *Forerunners of Darwin: 1745–1859.* Baltimore: Johns Hopkins University Press; 1959.

Gliboff S. 2005. "Protoplasm . . . is soft wax in our hands": Paul Kammerer and the art of biological transformation. Endeavour 29,4: 162–167.

Gliboff S. 2006. The case of Paul Kammerer: Evolution and experimentation in the early twentieth century. J Hist Biol. 39,3: 525–563.

Gliboff S. *H. G. Bronn, Ernst Haeckel, and the Origins of German Darwinism: A Study in Translation and Transformation.* Cambridge, MA: MIT Press; 2008.

Haeckel E. 1863. Über die Entwicklungstheorie Darwins. Gesell deutsch Naturforsch Ärzte, Amtlich Ber. 38:17–30.

Haeckel E. *Generelle Morphologie der Organismen: Allgemeine Grundzüge der organischen Formen-Wissenschaft, mechanisch begründet durch die von Charles Darwin reformierte Descendenz-Theorie.* 2 Vols. Berlin: Georg Reimer; 1866.

Harwood J. *Styles of Scientific Thought: The German Genetics Community, 1900–1933.* Chicago: University of Chicago Press; 1993.

Jablonka E, Lamb M. *Epigenetic Inheritance and Evolution: The Lamarckian Dimension.* Oxford: Oxford University Press; 1995.

Kammerer P. 1920. Richard Semon: Zur Wiederkehr seines Todestages. Der Abend (Vienna), December 27.

Kammerer P. *The Inheritance of Acquired Characteristics.* New York: Boni & Liveright; 1924.

Levit G, Hossfeld U. 2006. The forgotten "Old-Darwinian" synthesis: The evolutionary theory of Ludwig H. Plate 1862–1937. NTM 14: 9–25.

Maienschein J, Rainger R, Benson K. 1981. Were American morphologists in revolt? J Hist Biol. 14,1: 83–87.

Mayr E. *The Growth of Biological Thought: Diversity, Evolution, and Inheritance.* Cambridge, MA: Belknap Press of Harvard University Press; 1982.

Mayr E. Informal lecture at the summer seminar "Why Ernst Haeckel?" Marine Biological Laboratory, Woods Hole, MA: June 5, 1999.

Nägeli C. *Mechanisch-physiologische Theorie der Abstammungslehre.* Munich: R. Oldenbourg; 1884.

Pauly A. *Darwinismus und Lamarckismus: Entwurf einer psychophysischen Teleologie.* Munich: Ernst Reinhardt; 1905.

Penzlin H, ed. *Geschichte der Zoologie in Jena nach Haeckel 1909–1974.* Jena: Gustav Fischer; 1994.

Plate L. 1909. Gegen den Psychovitalismus: Nachwort zu dem vorstehenden Aufsatze von O. Prochnow, "Mein Psychovitalismus." Arch Rassen-Gesell-Biolog. 6:237–239.

Provine W. Progress in evolution and meaning in life. In: *Julian Huxley: Biologist and Statesman of Science*. Van Helden A, Waters C, eds. Houston, TX: Rice University Press; 1992:165–180.

Rainger R. 1981. The continuation of the morphological tradition: American paleontology, 1880–1910. J Hist Biol. 14: 129–158.

Richards R. *The Tragic Sense of Life: Ernst Haeckel and the Struggle over Evolutionary Thought.* Chicago: University of Chicago Press; 2008.

Robinson G. Plate, Ludwig Hermann. In: *Dictionary of Scientific Biography*. Gillispie C, ed. New York: Scribner's; 1975: sub nomen.

Sachs J. 1894. Physiologische Notizen, VIII: Mechanomorphosen und Phylogenie. Flora 78: 215–243.

Schacter D. *Stranger Behind the Engram: Theories of Memory and the Psychology of Science.* Hillsdale, NJ: Lawrence Erlbaum Associates; 1982.

Schatzmann J. *Richard Semon (1859–1918) und seine Mnemetheorie.* Zurich: Juris-Verlag; 1968.

Semon R. *Zoologische Forschungsreisen in Australien und dem malayischen Archipel.* 5 vols. Jena: Gustav Fischer; 1893–1913.

Semon R. *Die Mneme als erhaltendes Prinzip im Wechsel des organischen Geschehens.* Leipzig: Wilhelm Engelmann; 1904.

Semon R. *Die Mneme als erhaltendes Prinzip im Wechsel des organischen Geschehens.* 2nd ed. Leipzig: Wilhelm Engelmann; 1908.

Uschmann G, Jahn I, eds. 1959/1960. Der Briefwechsel zwischen Thomas Henry Huxley und Ernst Haeckel: Ein Beitrag zum Darwin-Jahr. Wissenschaft Zeitschr Friedrich-Schiller-Universität Jena, math-naturwissenschaft Reihe. 9,1/2: 7–33.

Vargas A. 2009. Did Paul Kammerer discover epigenetic inheritance? A modern look at the controversial midwife toad experiments. J Exp Zool. 312B,7: 667–678.

Winther R. 2001. August Weismann on germ-plasm variation. J Hist Biol. 34,3: 517–555.

6 Germinal Selection: A Weismannian Solution to Lamarckian Problematics

Charlotte Weissman

Although, as I shall argue in this chapter, Weismann never abandoned his selectionist agenda, two of the arguments used by the Lamarckians led him to recognize that natural selection, acting at the level of the individual, could not account for all the changes organisms have undergone during their evolution. He accepted that traits that are selectively neutral at the individual level can nevertheless accumulate during evolutionary time, and that the environment can affect heritable traits directly. To account for these observations while retaining his selectionist agenda, Weismann added to classical Darwinian individual selection a lower level of selection: competition and selection among the hereditary units *within* the germplasm, a process that he called *germinal selection* (Weismann 1896).

Germinal selection allowed Weismann to explain both adaptive and nonadaptive cumulative changes. Adaptive traits were selected at the personal level, which meant that the corresponding germplasms came to dominate in the population. In this case, selection acted indirectly on the composition of the germplasm through its effect on individual survival and reproduction. In contrast, nonadaptive variations were the result of direct, intragermplasm selection—the result of altered intracellular competitive conditions, which were neither advantageous nor disadvantageous at the organism level. Selection within the germplasm offered an explanation for nonadaptive trends and for environmentally influenced changes. Unlike most other interpreters of Weismann's ideas, I maintain that Weismann made no concessions to the Lamarckians, nor did he add elements which conflicted with his early model of the germplasm. Rather, the debate gave him an opportunity to extend the scope of selection and further develop his germplasm theory.

Weismann's Germplasm Theory

In order to understand the development of Weismann's ideas and his response to the Lamarckians' challenge, a few words about his background and his heredity theory are necessary. Weismann qualified in medicine at Göttingen and worked briefly as an assistant in a chemical laboratory in Rostock. After military service he carried out embryological studies under Rudolph Leuckart (1822–1898) at Giessen, and became docent in zoology and comparative anatomy at Freiburg im Breisgau, presenting a habilitation on the metamorphosis of insects in 1863; he then had a lifelong career as professor of zoology at Freiburg. Weismann was a Darwinian and a committed evolutionist, and believed from the outset of his career that natural selection was the main mechanism of evolutionary change. In the 1880s, he developed a theory of heredity, based on the continuity of the germplasm, which integrated the latest discoveries about cells and embryological data with the theory of evolution. It was a unique synthesis. He argued that the germ line was not affected by the changes undergone by the somatic cells during the organism's lifetime, so somatic adaptations could not be transmitted to the germplasm. This recognition led him, from 1883 onward, to a polemical refutation of the inheritance of acquired characteristics, a position to which he adhered throughout his life.

In *Das Keimplasma: Eine Theorie der Vererbung* (The Germ Plasm), published in 1892, Weismann described the structure and dynamics of the hereditary material in the germplasm of the cell nucleus. He assumed that a quantity of the germplasm of the fertilized egg, containing the full information for an individual's development, is segregated from the rest of the material during development, and is kept unchanged in the germ cells that give rise to the next generation. In somatic cells the full complement of the germplasm is *not* preserved. Different portions of it are used by the different somatic cell lines for the development of the individual's various tissues and systems. The functions of heredity and development are thus steered along separate tracks, although they are served by the same kind of building blocks.

Weismann hypothesized that the germplasm consists of a hierarchical architecture of nonhomogeneous particles capable of assimilation, growth, and self-replication. The basic units were termed *biophors* (the bearers of life), which combined to form *determinants*, which in turn made up *ids*, which made up *idants* (equivalent to chromosomes). The determinants controlled the "hereditary character" of the body, a *determinate*. Particular determinants (different in different cell types) broke up into their constituent *biophors*, which then migrated through the nuclear membrane into the cytoplasm and controlled separate parts of a cell. During embryonic development, Weismann suggested, cell division in the soma is unequal, and each cell gets a different combination of determinants. Ontogeny, the multiplication of cells and their diversification, was depicted as an ongoing process of control and determination effected by the determinants. Only germ cells (and some other special cells) retained a whole set of preserved idants. This segregated and preserved germplasm was transmitted to the next generation via the germ cells, and provided the continuity between generations. Weismann maintained that the germ line was not affected by the changes undergone by the somatic cells during the life of the organism, and that heritable, adaptive evolutionary changes resulted from the operation of natural selection on individuals, a form of selection he called "personal selection."

The Neo-Lamarckians' Challenge

Before I present the neo-Lamarckians' challenge to Weismann, it is important to highlight a common assumption of both neo-Lamarckians and neo-Darwinians: both believed that all variations have small effects, and that evolution is therefore very gradual. The question was whether minute changes in traits are adaptive, and therefore whether their progressive accumulation can be explained as the consequence of evolution by natural selection. George Romanes (1893), Theodor Eimer (1888), Oskar Hertwig (1894), and particularly Herbert Spencer (1893a, 1893b, 1894, 1895), whose polemics encompassed most of the arguments against Weismann's views, all raised objections to Weismann's belief in the omnipotence of natural selection (Weismann 1893).[1] The objections focused on the evolutionary explanation of changes that could not be explained by natural selection, for example, characteristics that could not have an adaptive significance, such as the extraordinary tactile perception of the tongue. Very small alterations leading to evolutionary trends were another problem for a Weismannian worldview. For example, the progressive and very gradual reduction in the size of the hind limbs during whale evolution seemed impossible to explain as a consequence of natural selection, since each minute reduction in the size of the limbs was deemed to be selectively neutral. Moreover, the neo-Lamarckians argued that even the evolution of an adaptive and useful characteristic had to pass through a stage in which it had not yet reached selective value. So how could such a transition occur?

Another problem was that of developmental trends, such as those seen in the color of butterflies' wings, which seemed to be the direct result of changes in the environment, or of changes in plant structures that resulted from changed conditions. These cases seemed to be best explained as a consequence of the inheritance of characters acquired during the lifetime of the individual. The neo-Lamarckian view was that the effects of environment, such as nutrition and climate, affected the organism directly, were directly inherited, and that such inherited effects accumulated. Experiments in which organisms were exposed to environmental changes such as temperature shifts in animals, or different climatic regimes in plants, produced variations that could also be found in natural populations, where they were often heritable. These phenomena suggested that the similarity between natural and experimentally induced variations was not purely contingent, but reflected the Lamarckian principle (inheritance of acquired characters) at work. The bottom line of the neo-Lamarckian position was that progressive, yet nonadaptive, trends could not be explained within the framework of personal natural selection.

Rising to the Challenge: The Theory of Germinal Selection

Weismann accepted the validity of the Lamarckians' observations, but not their explanations. He conceded that personal selection was not sufficient to account for all evolutionary changes in organisms, and suggested internal, within-cell selection of determinants: "Everywhere equivalent parts are contending one with another, and everywhere it is the best that prevails" (Weismann 1894:12). A year later he refined his proposition, accepting the reality of some directed hereditary variation, and called the process leading to such effects "germinal selection." Weismann argued that useful variation—at some level of organization—always exists.

Weismann's germinal selection hypothesis was based on Roux's principle of "the struggle of the parts," in which processes of selection take place among every kind of unit within the organism—not only in cells and tissues, but also in the smallest conceivably living parts (Roux 1881). Germinal selection transferred this internal struggle to the constituents of the germplasm, the determinants and biophors: the determinants could compete with each other and hence could be subject to internal selection within the cell. Weismann suggested that each trait was represented in the germplasm by groups of plus and minus determinants, which fluctuated around a theoretical average value. A trait's development was underlain by a diversity of determinants which, through the unequal growth, multiplication, and assimilation caused by an ever-changing stream of nutrients, determined the eventual strength or weakness of the corresponding characteristic. Weismann suggested that one reason for the fluctuations in the stream of nutrition used by the determinants might be the influence of environmental changes on the cell. The mechanism that led to variations was supposed to be the result of a shift in the equilibrium among determinants, resulting in a change of their quantitative relationships. This change was not reflected at the whole-organism level, and was not initially affected by personal selection. It depended on selection among determinants within the cell, and this selection may have been invisible at the level of the whole organism. However, personal selection still retained a major role: "Personal selection is [thus] in no way rendered superfluous by germinal selection, only it does not produce the augmentation of the distinguishing characters, but is chiefly instrumental in fixing them in the germ-plasm" (Weismann 1904, 2:131). Weismann accepted that there may be rare situations of "a conflict between germinal and natural selection" in which personal natural selection "cannot control" germinal selection. He interpreted the demise of the Irish elk, which was supposed to have been driven to extinction by the overdevelopment of its antlers, in these terms. Personal selection therefore could not always reverse excessive growth driven by internal selection, so the species would become extinct.

Let us take a closer look at the two problems that challenged Weismann's original focus on personal selection, and see how he interpreted them in terms of his germinal selection theory. The first problem was the explanation of the disappearance of an unused organ, for example, the rudimentary eyes of the cave fish, or the complete disappearance of eyes of crustaceans that live in the depths of the ocean, or the extreme diminution of the hind limbs of whales. In these cases, the change in the environment leads to a change in use, but the environment does not directly alter developmental conditions (as changes in temperature or nutrition may do). Weismann recognized that what he called "panmixia"—the relaxation of natural selection—was not a sufficient explanation of the complete degeneration of parts which became superfluous. Such parts, which are not present in the mature animal, appear in the embryonic primordium and disappear completely in the course of ontogeny. The question was why they are completely absent, with no remnant remaining. This was difficult to explain by natural selection at the organism level, since a very small vestige was assumed to make no selective difference. Weismann suggested that when there is no selection at the level of the individual, the process of internal selection going on between the determinants within the cell is responsible for the observed degeneration. When no natural selection at the individual level operates, the determinants that cause the part to be weaker (the (–) determinants) are no longer eliminated. The (+) determinants become fewer, because they are not supported by natural selection anymore, and hence the (–) determinants get (relatively) more nutrition. A process of positive feedback will occur among the (–) elements of the germ, and in the competition between (–) and (+) determinants, the former will predominate, leading in the course of generations to degeneration. Weismann argued that this struggle takes place over "an inclined plane" (Weismann 1896:25), a point of no return, since when the determinants for a weak organ are no longer eliminated, the other kind will continually be "robbed" of nutrition, the (–) determinants will prevail, and the specific part will degenerate. From this point onward, an ascent is no longer possible, and personal selection does not intervene to reverse the process.

Weismann's second problem was to explain the direct and heritable effects of external conditions, such as temperature, on organismal traits (for example, on butterfly wing color), which were interpreted as evidence for the operation of a neo-Lamarckian principle. Examples of climatic and nutritional effects in natural populations of animals and plants, which were identical to experimentally induced morphological changes in the laboratory, could be explained by the inheritance of developmental changes that occurred during the life of the individual and their accumulation over many generations. As in the case of degeneration, Weismann came to the conclusion that his original explanation, in terms of personal selection, was not sufficient, and he made use of his new germinal selection model to explain the apparently nonadaptive, yet heritable and cumulative, environmentally induced changes.

Weismann argued that an external condition, such as climate, might influence internal conditions in the cells, changing the nutrition stream in the germplasm. The change in nutritive conditions favors one type of determinants over an alternative one, so there is a change in the numerical composition of determinants, which is reflected in changes at the level of visible traits (e.g., the color of a butterfly's wing). Personal selection would (initially) play no role in such cases. The gradual accumulation of such variations in the course of generations could lead to the transformation of a species, such as is seen in cases of seasonal dimorphism. Weismann thus believed that there could be a direct external influence on the numerical composition of the determinants in the germplasm: "We must assume the occurrence of these [the possibility of variations of the germplasm through direct external influences] *on a priori* grounds, if we refer—as we have done—individual hereditary variations to fluctuations in the nutrition of the individual determinants of the germ plasm" (Weismann 1904, 2:272).

One of the things that convinced Weismann that external conditions can have direct nonadaptive effects on visible traits was his experimental work on the little red-gold fire butterfly *Polyommatus plaeas*. (Today this species has the genus name *Lycaena*, but in many of his works Weismann referred to the genus as *Chrysophanus* as well; see Churchill 1999.) This butterfly has two forms: one lives in the north of Europe and in Germany, and has a red-gold upper wing surface with a narrow black outer margin. The second form is found in the south of Europe, and the red-gold color has been replaced by black. When pupae that developed from eggs taken from the southern part of Europe were reared in Germany at low temperatures, the butterflies that resulted were not as dark as the southern types. And when the pupae of the German type were exposed to high temperatures, the butterflies were darker than the ordinary northern types. Such experiments pointed to the direct effect of external conditions on the soma. The question was whether the somatic changes were transferred to the germplasm, as the neo-Lamarckians suggested. Cases in which noninduced progeny of such experimentally induced parents showed the same change proved that the germplasm had been affected, and this supported the Lamarckians' contention that the somatic effects of temperature were hereditary. But Weismann suggested that such cases could be explained as instances of "parallel induction," in which both the soma and the germ were affected in the same way (but not with the same intensity) by the same external influences. Although it seems as if the change in wing color was directly inherited via the soma and accumulated in the course of generations, Weismann argued that "... in reality however it is not the somatic change itself which is transmitted but the corresponding variation evoked by the same external influence in the relevant determinants of the germ plasm within the germ-cells, in other words, in the determinants of the following generation" (Weismann 1904, 2:273). Moreover, in such cases, germinal selection provided an explanation in which the environment could affect the germplasm even when there was no useful function at the individual level. Weismann argued that in natural populations the coloration is the result of a long process of inheritance and of the accumulation of those determinants which were favored in the specific climatic conditions. The germplasms of the northern and southern forms varied in different directions because in one climatic condition, one type of determinant was favored in the germplasm, whereas in the other climatic condition, another type of determinant was favored. In both cases, however, it was internal selection within the germplasm which influenced the direction taken, and not personal selection.

Weismann supported his theory of germinal selection as the means by which environmental factors could influence the germplasm with another experiment. The pupae of various species of *Vanessa* butterflies were exposed to low temperatures; some of the resulting butterflies possessed colorations that he considered as "aberrations" (Weismann 1904). Such individuals were occasionally found in natural populations as well, and were thought to be reversions to earlier phyletic stages. Weismann concluded that the development of phyletically older markings implied that the germplasm was affected during the experiments. Another inference was that the germplasm of the modern *Vanessa* contains some ancestral determinants that were favored in the experimental conditions, which were similar to the ancestral conditions. Moreover, these results suggested to Weismann that for many such "aberrations" in Lepidoptera, cold conditions favor some preexisting variations but do not create anything new. He argued that sudden variations, such as reversions, should not be compared with the varieties known as sports, which are the mutations of plants.

Weismann did not believe that direct effects could produce elaborate adaptations. Such adaptations depend on classical natural selection, which in turn depends on the fluctuations of the elements in the germplasm. It was not clear how important the direct influence of climate and food was in the transmutation of species, but Weismann believed that in most cases it was uncertain whether the germplasm itself was directly affected at all. However, he accepted that fluctuations in determinants and internal selection within the germplasm might play a role as long as they are not harmful. Personal selection steps in at the moment when such direct effects become harmful, and if it cannot reverse the trend taken, the species will go extinct.

Germinal selection was for Weismann a theoretical cellular mechanism which could generate cumulative heritable variations in a trait. The concession made in accepting a direct influence of external conditions on the germplasm, as suggested with "parallel induction," in which personal selection played no role, has provoked modern historians to interpret Weismann as opening a door to neo-Lamarckian principles. For example, Peter Bowler (1979) argued that Weismann's germinal selection theory shows that he was forced to accept some form of an internal directing force, although Bowler stresses that for Weismann "there was no structure within the germ which could impose a predetermined limit on variation" (Bowler 1979: 58). The second concession was that "it is clear that he [Weismann] no longer saw utility as the sole driving force of evolution" (Bowler 1979:58).

Novak, too, interprets Weismann's germinal selection as a concession to the neo-Lamarckian explanation of inheritance caused by direct external influence: "Not only could the composition of germ-plasm therefore change, but it did so through an environmental influence, since in Weismann's view the intracellular environment often ultimately depended on the environment in which the organism lived. In short, germ-plasm could change in response to an environmental stimulus, which made germinal selection essentially a neo-Lamarckian mechanism, albeit with a selectionist twist" (Novak 2008:211). Therefore Weismann's work on seasonal dimorphism "contributed what seemed like a solid case of Lamarckian inheritance" (Novak 2008:280).

Ernst Mayr took a similar view. The effects of differential nutrition on the determinants caused Weismann's theory to be "no longer a selection theory," and Mayr argued that "it approached Geoffroyism" (Mayr 1991:126). As Mayr saw it, through germinal selection the constancy of the genetic material was abandoned. The historian Rasmus Winther took an even more radical position with regard to Weismann's theory of germinal selection being a concession to the Lamarckians. He saw in Weismann's germinal selection a basic deviation from his earlier views, and claimed that Weismann was not a Weismannian because he adopted an externalist view of the cause of hereditary variation, and therefore accepted the inheritance of acquired germplasm variations (Winther 2001). Due to the direct influence of the environment, Winther argued that "Selection, even at the germinal level, as Weismann pointed out subsequently, does not produce variation—it requires variation on which to act" (Winther 2001:538).

The position taken by Frederick Churchill (1999) is very different. He argues that Weismann's germinal selection "did not imply a smuggling of Lamarckian inheritance through the back door," but he sees in germinal selection "an ad hoc supplement" which "dulled the crisp separation of the processes of transmission and development that had been at the focus of the original germ-plasm concept." Churchill's view of Weismann's theory of germinal selection suggests that Weismann "failed to make clear whether the germinal competition took place simply in the germplasm, and so influenced not the bearer but the next generation; or in the somatoplasm, and so became a feature in the development of the bearer; or in both" (Churchill 1999:764). He sees in parallel induction in the soma and germplasm an invalidation of the independent status of the germplasm.

In my view, germinal selection is an extension and fine-tuning of Weismann's germplasm theory of heredity. His recognition that personal selection could not account for the development of nonadaptive variations led him to descend to the molecular level of the hereditary material itself and to add a level of selection. In doing so, Weismann did not abandon the principle of the segregation of the germplasm. The constitution of the germplasm was affected by internal selection among the units of heredity, but only variation in the germ line could be passed on. For Weismann, the variation occurring in the germplasm was the basis for the developmental processes of the organism, and was constrained by the constitution of the organism. When occurring in the germ line, this variation determined the direction of evolution. Only when discussing parallel induction did Weismann admit an ad hoc change in the germplasm as well as in the soma, but there was no direct communication from the soma to the germ, as suggested by the Lamarckians. I therefore agree with Churchill that Weismann did not "smuggle Lamarckian inheritance in through the back door," because he remained firm in his previous beliefs that variations occur through the (direct or indirect) operation of selection in the germ, and that changes of the soma are not transferred to the germ line. Weismann never denied that the germplasm could be affected directly by the environment, and in 1893, in a letter to Mrs. Victoria W. Martin, he explained that unequal nutrition of the components of the germplasm is the root of individual variations (Churchill and Risler 1999). Although he used the term "induction," his notion of induction referred not to the *expression* of latent determinants activated by a stimulus, but to *selection* between various homologous and heterologous determinants. For Weismann everything was accomplished through selection, but selection acted at different levels on quantitative germplasm variations. Selection was always the major organizing principle in Weismann's evolutionary theory. Although he accepted the validity of the neo-Lamarckians' observations, his solution remained consistent and selectionist.

Note

1. Although Spencer coined the term "the survival of the fittest," he did not believe that natural selection was an explanatory principle sufficient to account for evolutionary changes, especially in human society.

References

Bowler PJ. 1979. Theodor Eimer and orthogenesis: Evolution by definitely directed variation. J Hist Med Allied Sci. 34: 40–73.

Bowler PJ. *The Eclipse of Darwinism: Anti-Darwinian Evolution Theories in the Decades Around 1900.* Baltimore: Johns Hopkins University Press; 1983.

Churchill FB. August Weismann: A developmental evolutionist. In: *August Weismann: Ausgewählte Briefe und Dokumente/Selected Letters and Documents.* Churchill FB, Risler H, eds. Vol. 2. Freiburg im Breisgau: Universitätsbibliothek Freiburg im Breisgau/Bärbel Shubel; 1999:749.

Churchill FB, Risler H, eds. *August Weismann:Ausgewählte Briefe und Dokumente/Selected Letters and Documents.* 2 Vols. Freiburg im Breisgau: Universitätsbibliothek Freiburg im Breisgau/Bärbel Shubel; 1999.

Eimer T. *Die Entstehung der Arten auf Grund von Vererben erworbener Eigenschaften nach den Gesetzen organischen Wachsens. Ein Beitrag zur Auffassung der Lebewelt.* 3 vols. Jena and Leipzig: Gustav Fischer; 1888–1901.

Hertwig O. *Zeit und Streitfragen der Biologie.* Heft I: *Präformation oder Epigenese Grundzüge einer Entwicklungstheorie der Organismen.* Jena: Gustav Fischer; 1894.

Mayr E. *One Long Argument. Charles Darwin and the Genesis of Modern Evolutionary Thought.* Cambridge, MA: Harvard University Press; 1991.

Novak J. *Alfred Russel Wallace's and August Weismann's Evolution: A Story Written on Butterfly Wings.* Ph.D. dissertation. Princeton University; 2008.

Romanes E,. ed. *Life and Letters of George John Romanes.* London: Longman's, Green; 1896.

Romanes GJ. *An Examination of Weismannism.* Chicago: Open Court; 1893.

Roux W. *Der Kampf der Teile im Organismus.* Leipzig: Engelmann; 1881.

Spencer H. 1893a. The inadequacy of "natural selection." Contemp Rev. 63: 152–166, 439–456.

Spencer H. 1893b. A rejoinder to Professor Weismann. Contemp Rev. 64: 893–912.

Spencer H. 1894. Weismannism once more. Contemp Rev. 66: 592–608.

Spencer H. 1895. Heredity once more. Contemp Rev. 68: 608.

Weismann A. *Das Keimplasma. Eine Theorie der Vererbung.* Jena: Gustav Fischer; 1892.

Weismann A. 1893. The all-sufficiency of natural selection. A reply to Herbert Spencer. Contemp Rev. 64: 309–338, 596–610.

Weismann A. *The Effect of External Influences upon Development.* Wilson G, trans. [Romanes Lecture 1894]. London: Henry Frowde; 1894.

Weismann A. *On Germinal Selection as a Source of Definite Variation*, trans. TJ McCormack. Chicago: Open Court, 1896.

Weismann A. *The Evolution Theory.* Thomson JA, Thomson M, trans. 2 vols. London: Edward Arnold;1904. [This is a translation of the second German edition.]

Winther GR. 2001. August Weismann on germ-plasm variation. J Hist Biol. 34: 517–555.

7 The Notions of Plasticity and Heredity among French Neo-Lamarckians (1880–1940): From Complementarity to Incompatibility

Laurent Loison

Although it was Lamarck's homeland, France was one of the last Western countries to accept the hypothesis of a progressive transformation of living things. Indeed, it was necessary to wait until the 1880s to see transformism finally emerging as the general framework for interpreting biology. It is well known that the type of evolutionism which was developing at that time was sharply non-neo-Darwinian, and that this inclination was going to remain a peculiarity of French biological thought (Bowler 1992). Yet, however powerful it was, this neo-Lamarckian tradition was never embodied in a synthetic work that would had been a doctrinal reference. If Lamarck had his *Philosophie zoologique* (1809) and Darwin his *Origin of Species* (1859), French neo-Lamarckism was a transformism that was scattered in the contents of the publications of its main representatives. Thus the historian of science faces a curious object which, although it is powerfully rooted in a given space-time, is at the same time scattered, and not structured by a precise text or by a specific scientist. For those who wish to understand the history of this transformism, the only possibility is to try to reconstruct the work of the various evolutionists. Clarifying the internal logic of this French evolutionism becomes necessary for understanding its history. It seems to me that this logic rested almost exclusively on two notions: that of plasticity and that of heredity. The history of this transformism can therefore be understood as the ceaseless dialogue between these two notions, in a complex relationship of complementarity/incompatibility.

In this chapter, I argue that French neo-Lamarckism was initially structured by the notion of plasticity. Implicitly or explicitly, this notion was used to make the results of experimental transformism (1880–1910) understandable. Since phyletic evolution was seen as the large-scale consequence of individual changes, French neo-Lamarckian biologists had to accept soft inheritance. This kind of conception, however, caused many theoretical problems. In the third part of this chapter I focus on the problem of the articulation of these two notions, plasticity and heredity, in order to understand the explanatory impotence of this transformism during the 1920s and the 1930s.

The Notion of Plasticity: Explaining Variation

As Burian, Gayon, and Zallen (1988) noted, "the country of Lamarck was also the country of Claude Bernard and Louis Pasteur," a statement that highlights the fact that biological work in France was greatly influenced by experimental physiology and microbiology. This orientation was very strong during the period 1860–1910, the time of the reception of Darwin's theory. That is why the French community was disappointed by Darwin's book: it did not present a single (experimental) *fact* of species transformation. In order to convince French biologists, the first partisans of the transformist hypothesis had to show that it was indeed a scientific theory—that it was susceptible to experimental tests. Hence, from the beginning of the 1880s, French biologists started to envisage a project of experimental, physiologically oriented transformism.

This project developed at the same time in several branches of biology. In microbiology, many experiments performed by Pasteur and his colleagues showed that some bacteria could be transformed by varying culture conditions. These experiments, at least at the beginning, were not performed to address evolutionary issues, but to deal with practical problems (to protect livestock from the very dangerous bacterium *Bacillus anthracis*). Nevertheless, the results obtained during the 1880s seemed to show that an individual organism—a bacterium—was able to react to the environment and to adjust to it. These results were interpreted in terms of individual plasticity (see, for example, Duclaux 1898).

In teratology (the study of developmental abnormalities), Étienne Rabaud (1868–1956), who was a student of Camille Dareste (1822–1899), tried to advance the experimental research program of his master. He wanted to show that embryological development was controlled by the environment, and that phyletic evolution followed exactly the same rules as individual embryonic development (Rabaud 1914). That is why, at the end of the nineteenth century, he carried out many experiments with bird embryos. According to Rabaud, it was necessary to accept that embryonic development was not driven from the inside by heredity, but mainly from the outside by environmental changes. Thus, the organism had to be plastic.

In zoology, Frédéric Houssay (1860–1920) wanted to explain, using mechanical considerations, the present-day morphology of organisms. In particular, he was very interested in the hydrodynamic shape of fish. His idea was that the shape of most extant species of fish was the result of long-term effects of water pressure on a plastic body. Because of these stresses, ancestral organisms had been slowly transformed into present-day fish (Houssay 1912).

However, the most spectacular and clear results were obtained in botany. This research program was developed under the supervision of Gaston Bonnier (1853–1922), professor at the Sorbonne, and Julien Costantin (1857–1936), professor at the

National Museum of Natural History of Paris. They performed numerous experiments in order to establish that the morphology, anatomy, and physiology of plants were dominated by abiotic parameters such as luminosity, temperature, and humidity. Bonnier, for example, started a huge program of experimentation at the beginning of the 1880s. Cuttings of the same seedling were planted at stations in the Alps (at altitudes of 1060 to 2030 meters) and in his laboratory near Paris (Bonnier 1895). Many characteristics of the plants were rapidly affected by the new environment: their size, color, general shape, for example, were changed. The results showed clearly that by changing growing conditions, it was possible to directly (i.e., without the need for natural selection) transform living plants. As far as Costantin was concerned, the laboratory work he carried out at the same time was supposed to clarify Bonnier's results. He realized that it was possible to transform one organ into another by imposing drastic changes in growing conditions. Indeed, by cultivating a stem under a mass of thick soil, one could observe transformations which slowly made the stem look like a root (Costantin 1883). He also obtained interesting results by pushing land plants into the water during their growth, which led to the disappearance of stomata (Costantin 1886).

Whatever the trait, at every possible level and on every scale, it seemed that living organisms were capable of conforming to the requirements of their environment. All these results strengthened the idea of the transformability of life, and were widely discussed in France at the end of the nineteenth century. The data were rationalized by the notion of plasticity, that is, by the capacity of the living organism to conform to the physicochemical characteristics of its environment. This notion had physiological roots, and was used to reinforce the idea that, just as Claude Bernard had emphasized, the organism was a self-adjusting system. Individual plasticity was therefore necessary in order to maintain the integrity of the "milieu intérieur," the internal environment, a key concept in Bernard's view of homeostasis.

The mechanism of adjustment was never clearly articulated by these scientists. In general, they believed that the intimate relationship of an organism with its environment (biotic and abiotic) was a sufficient cause to explain the transformation. The physiological working of the body was the way through which the environment affected morphology.

In this theoretical framework, the adaptations of living things were explained as individual physiological acclimations. This means that phyletic evolution was totally reduced to individual changes, and thus the organism was the only relevant level for studying the operations of evolutionary mechanisms. Alfred Giard (1846–1908), holder of the first chair of "évolution des êtres organisés" at the Sorbonne, emphasized this point many times (see, for example, Giard 1904). But to have any evolutionary significance, these individual transformations had to be at least partially hereditary. That is why all of these scientists assumed that the inheritance of acquired characters was a fact.

The Notion of Heredity: Explaining Continuity

Natural selection was not well understood in France at the end of the nineteenth century (Conry 1974). Most biologists accepted it, but always by reducing its evolutionary role to almost nothing. Natural selection was seen as being responsible for destroying the unfit, but certainly was not responsible for the creation of the fittest. Because of this common, negative interpretation of natural selection, adaptation was explained using physiological arguments. It was necessary to accept soft inheritance, because otherwise phyletic adaptation could not be explained. Such acceptance was natural for these biologists, because the phenomenon of reproduction was identified with simple budding. There was strict protoplasmic continuity between organisms from generation to generation, and thus an inevitable morphological re-formation. Reproduction was indeed a *re-production*, because of the continuity of the protoplasmic material (Perrier 1881).

This view was reinforced by merotomy experiments: when a unicellular organism was cut into pieces, those pieces that contained the nucleus were able to develop and re-form the initial organism with its specific shape. It seemed that the chemical composition of protoplasm was the only factor that was necessary to bring about morphological features. There was no need to imagine hypothetical particles like pangenes or gemmules. Félix Le Dantec (1869–1917), a famous disciple of Giard, played a central role in constructing a developmental notion of heredity in France. After 1895, when he stopped working as an experimental biologist, he constructed a theoretical system to explain life as a mechanical process, and the evolution of such a system as a neo-Lamarckian phenomenon. In one of his major books, published in 1907, he argued that there is a necessary link between protoplasmic chemistry and morphology, suggesting it as a new theorem of general biology (Le Dantec 1907).

When considering unicellular organisms such as bacteria, this notion of heredity as phenotypic continuity was quite convincing, but major problems arose when this notion was applied to multicellular organisms. These problems were clearly recognized by August Weismann (1834–1914), and led him to reject the inheritance of acquired characters (1883, translated into French 1892; see Weismann 1892). According to Weismann, it was impossible to think of a mechanism through which modifications of the soma could be incorporated into the germ line. This was one of the *theoretical reasons* that led Weismann to develop his germplasm theory during the 1880s and the 1890s. However, French biologists did not pay much attention to

Weismann's conception. For a long time his critique was considered irrelevant, because his explanatory system seemed to be an excessively metaphysical, speculative construction. The theory of the germplasm was indeed very different from the positivist expectations of the French scientists of the second half of the nineteenth century (Canguilhem, 2002).

Because of their belief in a universal and complete determinism (France was the country of Pierre-Simon Laplace, and his influence was evident then and even more so later in the century), and because of their conception of the organism as an indivisible totality, French biologists were convinced that, in one way or another, environmentally induced phenotypic modifications could reach the germ cells and become inherited. The Bernardian concept of "milieu intérieur" seemed to be a sufficient foundation for this notion of heredity. The precise mechanisms, once again, were never a matter of interest for these biologists.

Moreover, evolutionary mechanics was identified as a simple process which could be understood and articulated by using two seemingly simple concepts: plasticity and heredity. At least until the first years of the twentieth century, an evolutionary explanation based on these concepts seemed to be sound because, unlike the notion of natural selection, it did not appear to raise major theoretical problems. At the same time, it seemed that, unlike the germplasm theory, it was easy to experimentally test the proposed evolutionary mechanisms.

The Impossible Link between the Notions of Plasticity and Heredity

A conception of the evolutionary process based on the notion of heritable, plastic responses has to deal with at least three theoretical problems. The first is the well-known problem of adaptability. How can one explain that individuals have the ability to respond adaptively to environmental changes? Adaptability has to be explained, and cannot be taken for granted: a complete theory of evolution has to make the origins and evolution of adaptability understandable. This point was clearly stated by Weismann (1892:350). The answers of the French neo-Lamarckians were quite complicated, and differed depending on the scientist one focuses on (for detailed discussion of different views, see Loison 2008a). But for most of them, this property of the living was just a physiological characteristic inherent in life processes (see, for example, Costantin 1901); there was no need for a specific explanation of plasticity. Furthermore, the question of the *origin* of a natural phenomenon (in this case, natural adaptiveness or adaptability) was seen as going beyond the limits of scientific inquiry (Costantin 1898).

The second problem is to explain the inheritance of acquired characters in multicellular organisms. As I described earlier, this, too, was taken for granted because of general physiological considerations. Indeed, it was impossible for these biologists to construct a precise hypothesis of heredity without renouncing their metaphysical conceptions. Thus, this problem was never clearly articulated in France during these years (Loison 2008a).

The third problem, on which I focus here, is the link between the notion of plasticity and that of heredity. These two notions were at first seen as complementary, but this simple articulation hid a very complicated theoretical issue. This was pointed out by the British zoologist Edwin Ray Lankester in 1894. In a short paper he emphasized the self-contradiction inherent in the statement "the inheritance of acquired characters." He wrote:

> And it seems to me, that in considering this we are led to the conclusion that *the second law of Lamarck is a contradiction of the first.* [. . .] What Lamarck next asks us to accept, as his "second law," seems not only to lack the support of experimental proof, but to be inconsistent with what has just preceded it. [. . .] Since the old character (length, breadth, weight) had not become fixed and congenital after many thousands of successive generations of individuals had developed it in response to environment, but gave place to a new character when new conditions operated on an individual (Lamarck's first law), why should we suppose that the new character is likely to become fixed after a much shorter time of responsive existence, or to escape the operation of the first law? Clearly there is no reason (so far as Lamarck's statement goes) for any such supposition, and *the two so-called laws of Lamarck are at variance with one another.* (Lankester 1894:102, italics added)

The "two so-called laws of Lamarck" were the basis of every Lamarckian theory at the end of the nineteenth century, both in France and in England. These "laws" were published by Lamarck in 1809 to explain the deviations from the progressive sequence of evolution. The lateral branches of the chain of being were the consequences of an adaptation process resulting from environmental changes. Lamarck stipulated:

> *First Law*: In every animal which has not passed the limit of its development, a more frequent and continuous use of any organ gradually strengthens, develops and enlarges that organ, and gives it a power proportional to the length of time it has been so used; while the permanent disuse of any organ imperceptibly weakens and deteriorates it, and progressively diminishes its functional capacity, until it finally disappears.
> *Second Law*: All the acquisitions or losses wrought by nature on individuals, through the influence of the environment in which their race has long been placed, and hence through the influence of the predominant use or permanent disuse of any organ; all these are preserved by reproduction to the new individuals which arise, provided that the acquired modifications are common to both sexes or at least to the individuals which produce the young. (Lamarck [1809] 1963:113)

These two laws show the same theoretical structure as the one I discussed for neo-Lamarckians' evolutionism: the first law implies plasticity; the second, heredity. But for Lamarck these processes were of secondary importance, because evolution was mostly driven by a progressive internal, mechanical force. For most of the French neo-Lamarckians, this march of progress did not exist, and evolution was only about environmental adaptation. Because of this reframing of Lamarckian evolutionism, they had to face the problem of the contradictory relationship between the notions of plasticity and heredity. "Inheritance" presupposed the stability of a character in heredity in spite of environmental changes. If that were so, heredity was "stronger" than plasticity. On the other hand, the notion of "acquired" presupposed phenotypic change as a result of environmental actions. If that were so, plasticity was "stronger" than heredity. A neo-Lamarckian organism should have been able to stabilize its form and, *at the same time*, should have been capable of adaptive response. This theoretical problem was not explicitly formulated in France until 1911 (Cuénot 1911). But even though the problematic character of the Lamarckian statement was not explicitly recognized, the *tension* between the notions of plasticity and heredity led to various attempts of reconciliation.

Although in botany the program of the experimental transformism showed some real success, some plants could not be experimentally modified. Costantin recognized this as a fact. If the period of time was long enough, then plasticity seemed to fade in favor of the almighty "ancestral heredity" (the accumulated effect of the environment on organisms [Costantin 1899:120]). A new character could evolve only if a drastic change in abiotic conditions happened, because it was necessary to break the old and powerful balance between the organism and its environment. Hence, for some of the French botanists, evolution became a saltationist phenomenon: most of the morphological transformations occurred only during the rare periods of organic plasticity, under extreme, stressful conditions. This theoretical tension started the phase of theoretical schizophrenia which had been intensifying from the 1910s to the late 1930s.

In zoology, Le Dantec developed the idea that plasticity could be a vital property which decreased as the complexity of the organism increased. Biological evolution was then identified with the universe's thermodynamic transformation: plasticity should follow the same laws as entropy, but in the opposite direction (Le Dantec 1910). Only the simplest organisms could be transformed by the environment; for others, heredity had become stronger than plasticity. The general idea that evolution seemed to be a process which was about to stop was further developed by one of Le Dantec's colleagues at the Sorbonne, the zoologist Maurice Caullery (1868–1958) (Loison 2008b). As time went on, Caullery became more and more skeptical about the possibility that an acquired character could be effectively inherited. Since he did not understand that natural selection may play a creative role in evolution, he believed that the inheritance of acquired characters had to be the main evolutionary mechanism, but that it had operated only in the past (Caullery 1931). On the one hand, fossils showed that organisms were already very complex during the early eras of life. For Caullery that meant that a progressive evolution had occurred in the *past* because of the efficiency of the inheritance of acquired characters. And, on the other hand, experiments failed to demonstrate that *present* evolution was driven by the physiological effects of the environment. That is why he believed that Lamarckian processes were the main factors of evolution only during the first stages of life's transformation.

These examples show that the tension between plasticity and heredity was not a simple one. At first, French biologists were interested in life's transformations: they focused on phenotypic plasticity. Plasticity seemed to be the best explanation of the variation of individual organisms. However, around 1900, as genetics was starting to develop, they had to pay much more attention to the possibility that variations could be inherited. That is why they became more and more focused on explaining the persistence of induced characters. But the respective weights of plasticity and conservative heredity remained an unsolved question which had been first underestimated by these biologists. This tension was theorized at that time by the very vague notion of balance (*équilibre*): organisms were torn between past causes (because of their heredity) and present causes (because of their plasticity). This state of theoretical schizophrenia finally led French neo-Lamarckism to an explanatory impotence: after the 1930s, adaptation was no longer a phenomenon that could be explained by the present laws of nature. Hence, for some biologists, adaptation belonged to mechanisms that acted in the past (Caullery), while for others adaptation simply did not exist (Rabaud 1922).

Conclusion

To conclude, I would like to emphasize three points. The first is a historical point. During their later careers, Costantin, Le Dantec, Caullery, and others reached a theoretical impasse which prevented transformism from becoming an experimental theory of evolution. According to them, because of the progressive intensification of the power of heredity, the important steps of evolution happened in the past, and could not be reproduced in the present. Thus, experimental science had nothing to tell us about (past) evolutionary mechanisms. This important renunciation transformed French neo-Lamarckism from an experimental practice (experimental transformism, 1880–1910) to a metaphysical and dogmatic position (1910–1940). Subsequently, because of its explanatory impotence, this type of evolutionism slowly disappeared.

The second point is a more philosophical one. The failure of this version of Lamarckism seems to show that it is impossible to construct a consistent theory of evolution that is founded *only* on the two notions of plasticity and heredity. Note that this was not the case for Lamarck's own theory, since he used additional assumptions and notions, nor is it true for the current epigenetic notion. Epigenetic explanations take place (or have to take place) within the framework of the Darwinian synthesis.

This leads us to the third point, which is a more scientific one. Epigenetic inheritance is a phenomenon which does exist (Jablonka and Raz 2009). Its importance *in natural conditions* is a scientific question that will be discussed by scientists in the next few years. As I see it, however, the main challenge to epigenetic inheritance is theoretical: it is to find an articulation within the classical assumptions of Darwinian evolutionism. Like other kinds of inheritance (ecological, for instance), it is important to understand how and how much these epigenetic phenomena could affect fitness, and thus have an evolutionary role.

Acknowledgments

I would like to thank Eva Jablonka and Snait Gissis for their very kind invitation to the international workshop commemorating the publication of the *Philosophie zoologique* (1809). It was for me a unique opportunity to meet very interesting people and to have fruitful discussions.

References

Bonnier G. Recherches expérimentales sur l'adaptation des plantes au climat alpin. *Ann sci nat, botan.* 1895; 7th ser. 7, 20:217–360.

Bowler P. *The Eclipse of Darwinism.* London: Johns Hopkins University Press; 1992:107–117.

Burian R, Gayon J, Zallen D. The singular fate of genetics in the history of French biology, 1900–1940. J Hist Biol. 1988;21: 357–402.

Canguilhem G. La philosophie biologique d'Auguste Comte et son influence en France au XIXe siècle. In *Études d'histoire et de philosophie des sciences concernant les vivants et la vie.* Paris: Vrin; [1968] 2002: 61–74.

Caullery C. *Le problème de l'évolution.* Paris: Payot; 1931.

Conry Y. *L'Iintroduction du darwinisme en France au XIXe siècle.* Paris: Vrin; 1974.

Costantin J. Étude sur les feuilles des plantes aquatiques. *Ann Sci Nat.* 1886;3:94–162.

Costantin J. *L'hérédité acquise.* Paris: C. Naud; 1901:51–52.

Costantin J. *Les végétaux et les milieux cosmiques.* Paris: Alcan; 1898:88–89.

Costantin J. *La nature tropicale*. Paris: Alcan; 1899.

Costantin J. *Étude comparée des tiges aériennes et souterraines des Dicotylédones.* Paris: Masson; 1883.

Cuénot L. *La genèse des espèces animales.* Paris: Alcan; 1911: 188–189.

Darwin C. *On the Origin of Species*. London: John Murray; 1859.

Duclaux E. *Traité de microbiologie*. Vol. I, *Microbiologie générale.* Paris: Masson; 1898:605.

Giard A. Les facteurs de l'évolution. In *Controverses transformistes.* Paris: C. Naud; 1904:109–134.

Houssay F. *Forme, puissance et stabilité des poissons.* Paris: Hermann; 1912:7.

Jablonka E, Raz G. Transgenerational epigenetic inheritance: Prevalence, mechanisms, and implications for the study of heredity and evolution. Q Rev Biol. 2009; 84, 2: 131–176.

Lamarck J-B. *Zoological Philosophy*. H Elliott, trans. New York: Hafner; [1809] 1963: 113.

Lankester ER. 1894. Acquired characters. Nature 51: 102–103.

Le Dantec F. *Éléments de philosophie biologique.* Paris: Alcan; 1907:194.

Le Dantec F. *La stabilité de la vie. Étude énergétique de l'évolution des espèces.* Paris: Alcan; 1910.

Loison L. *Les notions de plasticité et d'hérédité chez les néolamarckiens français (1879–1946). Éléments pour une histoire du transformisme en France*. PhD dissertation, University of Nantes 2008a:111–135.

Loison L. La question de l'hérédité de l'acquis dans la conception transformiste de Maurice Caullery. In: Morange M, Perru O, eds. *Embryologie et évolution (1880–1950). Histoire générale et figures lyonnaises*. Paris: Vrin and Lyon: IIEE; 2008b):99–127.

Perrier E. *Les colonies animales et la formation des organismes.* Paris: Masson; 1881:142–143.

Rabaud E. *L'adaptation et l'évolution.* Paris: Chiron; 1922.

Rabaud E. *La tératogenèse. Étude des variations de l'organisme.* Paris: Octave Doin; 1914.

Weismann A. *Essais sur l'hérédité et la sélection naturelle*. H. de Varigny trans. Paris: C. Reinwald; 1892.

8 Lamarckism and Lysenkoism Revisited

Nils Roll-Hansen

How important was Lamarckism in the Lysenko affair? Was belief in the inheritance of acquired characters a major reason for the suppression of genetics in the Soviet Union between 1935 and 1965? The answer is disputed. Biologists saw lingering Lamarckian ideas as a main factor, in tandem with a Marxist science policy. A direct formative role for the environment was attractive to Soviet social utopianism, and the combination stimulated wishful science (Cook 1949; Zirkle 1949, 1959; Huxley 1949). Historians of science, on the other hand, have mostly considered scientific issues and Marxist philosophy to be peripheral and unimportant.

In standard history of science, Lysenkoism is depicted as the product of brutal and ignorant political intervention into strictly scientific matters. According to this account, it had little to do either with serious scientific issues or with Marxist theories about science. "Lysenkoism rebelled against science altogether," and Marxist theory played a minor role, claimed David Joravsky (1970: ix, 228). Lysenkoism "was a chapter in the history of pseudoscience rather than science," and Soviet Marxist philosophy of science, so-called dialectical materialism, often had a positive effect on scientific creativity, argued Loren Graham (1972: 195, 430–440). More recent works by Valerii Soyfer (1994) and Nikolai Krementsov (1997) follow Joravsky's and Graham's classical and authoritative claims that Lysenkoism requires primarily a sociopolitical explanation (Roll-Hansen 2008).

The numerous wildly pretentious claims about breeding and genetics made by the poorly educated leader of the movement, Trofim D. Lysenko, fit this interpretation. He lacked a broad understanding of biological science as well as the intellectual qualifications to understand Marxist theory.

However, a closer look at the historical origins of Lysenkoism reveals two major weaknesses in the standard history. First, Lysenkoism, including Lysenko's personal career, was the product of a science policy based on Marxist theory epitomized in the slogan "unity of theory and practice." Second, Lysenko's scientific reputation derived from his work in developmental physiology of plants. Only in the mid-1930s did Lysenko confront classical genetics. This suggests that the standard historical account does not fit the origin of Lysenkoism, and that neo-Darwinian neglect of ontogenesis was an important factor in the rise of Lysenko. This chapter will focus on the role played by specific scientific research questions in the development of Lysenkoism. The role of Marxist theory in science policy has been discussed elsewhere (Roll-Hansen 2005, 2008).

Hard Inheritance

Darwin was well aware that the source and the nature of hereditary variation constitute a fundamental problem for evolutionary theory. But the solution that he proposed in *The Variation of Plants and Animals Under Domestication* (Darwin 1868), his theory of pangenesis, was little more than a repetition of Hippocrates' ancient speculations. Only by the end of the nineteenth century, as cytology had laid the foundations of a deeper understanding of growth and reproduction, were conditions ripe for the new discipline of genetics. It was born from a nexus of questions in cytology, embryology, heredity, taxonomy, evolution, and practical plant breeding.

Continuous variation in heredity was the usual assumption common to most evolutionary theories around 1900. But with the discovery of stable and reproducing microscopic structures within the cells, discontinuity became a clear and tangible possibility. "Hard inheritance" (Mayr 1982:681–726) meant that basic hereditary factors are highly stable and mostly do not change from one generation to the next. Hereditary variation is produced by occasional, accidental, and stepwise change in individual factors ("mutations") and by their recombination in sexual reproduction. This was contrary both to Lamarckian and to orthodox Darwinian theories of evolution.

The German zoologist August Weismann prepared the ground with his strong and tenacious criticism of the inheritance of acquired characters. But a clear conception of hard inheritance, supported by precise experiment, was presented by the Danish plant physiologist Wilhelm Johannsen (1903, 1909). His distinction between genotype and phenotype was quickly applied to numerous species of both plants and animals. The international breakthrough of Johannsen's genotype theory was marked by a special symposium at the December 1910 annual meeting of the American Society of Naturalists.[1]

Johannsen introduced the term "gene" (*Gen*) to designate a disposition (*Anlage*) for specific properties in the developed organism (phenotype) (Johannsen 1909:124). He explained the genotype as "the sum total of all the 'genes' in a gamete or in a zygote," and presented it as "in the least possible degree speculative." In contrast to Weismann's germplasm (*Keimplasma*), his genotype did not involve any "correspondence to special organs" (Johannsen 1911:131–133), that is, no preformation.

The Mendelian chromosome theory of heredity was launched by Thomas Hunt Morgan and his pupils in *The Mechanism of Mendelian Heredity* (1915). It soon became the paradigm of classical genetics and the neo-Darwinian theory of evolution (Amundson 2005:149–153). Morgan and his pupils presented Johannsen's bean selection experiment as "convincing" evidence for stability of hereditary factors, that is, as evidence against a continuous hereditary change (Morgan et al. 1915:204). But Johannsen never accepted their atomistic view of heredity as reducible to the factors that they mapped on the chromosomes. He praised the chromosome theory as an epoch-making step but continued to insist that it was only a partial answer to the fundamental questions of heredity and evolution (Johannsen 1923).

The historiography of genetics and evolution has tended to pit Lamarckism against neo-Darwinism and to neglect those who found both theories inadequate. This third view was common, for instance, in Germany, in the opposition to the "nuclear monopoly" of the Morganists (Harwood 1993). It was also strong in the Soviet Union, where the importance of embryology and interaction with environment was stressed, and where skepticism of strict Mendelian and neo-Darwinian theories was common.

The Lysenko affair stimulated a polarization in which Mendelian neo-Darwinism and Lamarckism appeared as the only alternatives. Even in the West, where neo-Darwinism prevailed, it was sometimes hard to get support and credit for work that appeared open to Lamarckian interpretations.

Thus the American geneticist and protozoologist Tracy Sonneborn devoted a presidential address (1949) to the American Society of Naturalists to a survey of the borderland between Lysenkoism and classical genetics. He explained how phenomena that had been claimed to support Lysenkoist theories could be explained by an extension of Mendelian theory. For instance, the transfer of heredity in graft hybrids could be explained by transport of viruses, and his own experiments on a killer trait (death factor) in paramecia could be explained by particles (plasmagenes) that were located outside the nucleus and did not behave according to the Mendelian rules of hybridization. Such phenomena were well worth exploring but provided no good reason to prefer Lysenkoist theories over classical genetics (Sonneborn 1949).

Darwinism Reborn

The theory of evolution by natural selection was in a serious slump from the late nineteenth century well into the twentieth. As late as the 1920s and 1930s many, perhaps most, biologists took a bleak view of its future (Bowler 1983:5). Revival came as the neo-Darwinian "synthesis" of Mendelian genetics and natural selection was promoted by population geneticists in the 1920s and 1930s and achieved a general breakthrough around 1940 (see part II, on the Modern Synthesis, in this volume). However, it was only after DNA had been pinned down in 1953 as the basic molecule of heredity that neo-Darwinism reached its maximum influence. Early molecular genetics gave rise to the "central dogma" holding that information could go only from DNA to proteins and never in the opposite direction.

Criticism of neo-Darwinism came from experience in practical breeding as well as from theoretical work in systematics, ecology, paleontology, and other biological disciplines. Critics held that the genes of the chromosome theory were quite insufficient to explain the evolution of new species, so-called macroevolution (Harwood 1993). There was also continuing criticism from the perspective of embryological development by Richard Goldschmidt, Ivan Schmalhausen, Conrad Waddington, and others, leading eventually to present-day evo-devo (Amundson 2005:198).

The inheritance of acquired characters remained an attractive hypothesis in spite of numerous experimental failures. The possibility that Lamarckian mechanisms could play a significant role in the evolution of species was constantly present and discussed. This possibility appeared plausible to many biologists and other scientists (such as Niels Bohr and other physicists interested in biology), and was perhaps the common attitude among the majority of biologists as late as the 1950s. (Roll-Hansen 2000: 433–435; see also chapter 11 in this volume). Cytoplasmic inheritance and *Dauermodifikationen* (lingering modifications) were phenomena that stimulated speculation on mechanisms for inheritance of acquired characters. Most of these phenomena could, in principle, be explained by Mendelian mechanisms (Harwood 1993), but there was a continuing accumulation of facts, mainly in plants and microorganisms, which indicated that some kind of Lamarckian inheritance could be part of the evolutionary story. In plants, graft hybridization was a particularly contested area.

Soviet Biology in the 1920s and 1930s

While enthusiastic young Soviet geneticists were inspired by the new chromosome theory, the skepticism of established biologists working on systematics, ecology, physiology, and paleontology was strong. Neo-Darwinism did not appear convincing as the sole or main explanation of evolution. There was genuine scientific openness to the possibility that the inheritance of acquired characters could be an important factor in the evolution of species. The enthusiastic neo-Darwinian geneticists appeared as young turks with a strong but exaggerated message.

Soviet biology differed from Western biology by its tie to a philosophical worldview that was also the ideological foundation of the state. "Creative Darwinism" was part of the official world picture, a substitute for traditional religious beliefs.

Ideological pressure on scientists grew as the Stalinist political system tightened its grip from the late 1920s on. Lamarckism was attractive to Marxist ideology because it promised mutual reinforcement between improvement of the social environment and improvement of the "gene fund" of the population. New progressive scientific institutions established in the 1920s were from the start inclined toward Lamarckism. The standard argument against Mendelism was that its concept of stable genes contradicted evolutionary change and progress, biological as well as social. But toward the end of the 1920s new work on mutations helped geneticists temporarily turn the tables on Lamarckism.

Aleksander Serebrovskii was the leader of the geneticist camp. In a 1927 article in *Pravda*, the Communist Party newspaper, he popularized the controlled production of mutations with X-rays by Morgan's student Hermann Müller. The title, "Four Pages That Shook the Scientific World," played on John Reed's famous book about the October Revolution, *Ten Days That Shook the World*. Serebrovskii argued that Müller's results had demonstrated how genes were changeable and by no means were in conflict with Marxist dialectics (Gaissinovich 1980:41–46).

In 1929 the Communist Party Central Committee decided that it was time to resolve the disagreement between Lamarckians and geneticists. A national conference was organized "in order to draw results from the protracted discussions between mechanists and dialecticians on the fundamental questions of philosophy and science." With the help of favorable staging, the geneticists ("dialecticians") prevailed over the Lamarckians ("mechanists") at this first in a series of fateful conferences. The geneticists' uncompromising attitude and extensive use of philosophical arguments set a dangerous precedent of polarization in science (Gaissinovitch 1980:46–51).

The period of cultural and agricultural revolution around 1930 was marked by strong practical demands on science, guided by the ideological principle of "unity of theory and practice." The Lenin Academy of Agricultural Science was formed in 1929 to coordinate the numerous agricultural research institutes that had grown up during the 1920s. Its first president was the plant scientist Nikolai Vavilov, who had been the main entrepreneur in agricultural science since the early 1920s. He underlined the central role of the Lenin Academy in collectivization. It was to be "the academy of the general staff of the agricultural revolution," the general staff being the Ministry of Agriculture (Roll-Hansen 2005:89–92). Thus, the Lenin Academy had to take some of the responsibility for the tragic failures of forced collectivization, and the government reacted by tightening political control and planned steering in 1935. Food production remained a headache and a key issue in Soviet politics for decades. Up to the 1950s the Lenin Academy was a central scientific institution with great clout in biology, rivaling the influence of the traditional "big" Academy of Sciences.

The next important crossroads was the December 1936 national conference on "the two directions in genetics" organized by the Lenin Academy. By this time Lamarckian ideas were gaining an upper hand in the eyes of public opinion under strong guidance from the Communist Party. Genetics carried a heavy burden of involvement with eugenics and an impractical "ivory tower" approach to agricultural science. This growing ideological polarization made the situation uncomfortable for those who were skeptical of both neo-Darwinism and Lysenkoism. At the 1936 conference many biologists still tried to balance the two. They pointed to the practical shortcomings of neo-Darwinism as well as to the primitive speculation and lack of evidence in Lysenko's Lamarckian theorizing. But such sober criticism drowned in the intellectual chaos of Stalinist terror.

The official summing up of results from the December 1936 conference was carried out in a spirit of compromise and cooperation by the plant breeder and geneticist G.K. Meister. He was a loyal Communist Party member who had written extensively on the Marxist theory of science, and was vice president of the Lenin Academy. When the president, the old-time Bolshevik Aleksander Muralov, was arrested, Meister succeeded him. But by early 1938 he, as well as other Communist Party members in the leadership of the Lenin Academy, had been arrested, and in February 1938 Lysenko became president (Soyfer 1994:110–112).

Thus Stalinist terror paved the final steps for Lysenko. Decision at the top political level, probably Stalin's own, put Lysenko into the most influential position of Soviet biology. His social background as a son of a peasant with little formal education was a valuable asset in the turbulent 1930s. Nevertheless, his quick rise would not have been possible without widespread sympathy for Lamarckism and skepticism toward neo-Darwinism among leading Soviet plant scientists throughout the 1920s and 1930s.

A few examples of notable Soviet biologists illustrate the extent of such attitudes.

Vladimir Liubimenko (1873–1937) was the Nestor of plant physiology and a pioneer in ecological physiology, Lysenko's special field. In the 1920s he admitted that Johannsen's experiments had set strict limits to inheritance of acquired characters, but insisted that it still played a role. He argued that neo-Darwinism was far from a full explanation of evolution (Liubimenko 1923:882–885).

The botanist Vladimir Komarov (1869–1945) served as vice president of the Academy of Sciences from 1930 to 1936, and as president from 1936 to 1945. He was thus closely involved in the squeezing out of classical genetics from the institutes of the Academy between Lysenko's ascent to the presidency of the Lenin Academy in 1938 and the events around Vavilov's arrest in 1940. In Komarov's 1944 monograph on the nature of species, neo-Darwinism is characterized as inadequate as a general theory of evolution. In the spirit of *Dauermodifikationen,* Komarov thought that permanent mutations are likely to arise from environmental pressure over generations (Komarov 1944:105, 167).

Nikolai Vavilov (1887–1943) is most famous as the martyr who defended genuine genetic science against Lysenko. He was arrested in 1940, charged with espionage or collaboration with foreign powers, and sentenced to death; he perished in prison in 1943. But Vavilov also promoted Lysenko's early career (Popovskii 1984). Although he was highly critical of Lysenko's genetic theories and ideas on breeding, Vavilov found his methods to control plant development appealing. In the manual of plant breeding, *Theoretical Foundations of Plant Selection*, edited by Vavilov (1935), both classical genetics and Lysenko's physiological methods are well represented. As late as July 1935 Vavilov was adamantly defending Lysenko in scientific discussions (Roll-Hansen 2005:172–173).

In the 1920s Vavilov drew public attention to the amateur fruit breeder Ivan Michurin (1855–1935), praising him as a Russian counterpart of Luther Burbank. Michurin held nineteenth-century ideas about heredity, including inheritance of acquired characters, and regarded graft hybridization as a main breeding method. His meager practical contributions were hyped to make him an icon of progressive gardening and agriculture (Joravsky 1970:39–54). After Michurin's death in 1935, Lysenko promoted himself as the true heir. And as late as the 1950s even the sharpest geneticist critics of Lysenko continued to declare themselves "Michurinists."

The Content of Lysenko's Biology

Lysenko first achieved scientific fame through his experiments treating germinating seeds with low temperatures to speed up flowering in cereals. This was by no means an unknown phenomenon, but Lysenko's approach appeared to be a possible solution to serious problems in the cultivation of cereals. The influence of temperature and light on plant development was at the time a new and exciting field with rapid theoretical development as well as promises of important practical application. The Imperial Bureau of Plant Breeding in Great Britain, established in 1929 to serve as an international scientific information center, soon picked up Lysenko's work. His Russian term *iarovizatsiia,* meaning "spring-making," was latinized into "vernalization," which soon became a standard scientific term (Roll-Hansen 2005:143–148). Today vernalization is a hot topic in plant science, at the intersection of genetics and developmental physiology. Although the effects of vernalization are not inherited between generations, as Lysenko had claimed, the "mystery of vernalization-induced flowering" is "only partly worked out," concludes an up-to-date review of the state of the art (Dennis and Peacock 2009).

Lysenko had an overblown belief in the practical usefulness of general theorizing. His ideas about vernalization soon grew into a general theory of stages in plant development which was widely discussed, both in the Soviet Union and internationally. However, specialists soon found that Lysenko's work lacked contact with the rapidly advancing knowledge of physiology, in particular, plant hormone research. In the early 1930s Vavilov, who was hard pressed for practical results from his grand world collection of cultivated plants, became interested in Lysenko's vernalization techniques as an instrument in plant breeding. This stimulated Lysenko's engagement in breeding and developing systems of seed production, and his inclination for ambitious general theory quickly expanded his theory of stages into a comprehensive account of development and heredity (Roll-Hansen 2005:155–169).

In stressing heredity as a property of the whole organism and its close interaction with the environment, Lysenko was in line with the common contemporary criticism of neo-Darwinism. But his Lamarckian preference for malleability and continuous variation of heredity was incompatible with classical genetics.

In Lysenko's lecture at the December 1936 conference, Johannsen's genotype theory was still his primary target. Under the label of "pure line theory" it had been a guiding principle in Soviet plant breeding and seed husbandry. But according to Lysenko it contradicted true "creative" Darwinism. It was a fundamental misunderstanding to deny a "creative role for selection in the evolutionary process." The chromosome theory of the American geneticist T.H. Morgan made the same mistake, said Lysenko, referring to the main lectures from the geneticist camp, which were given by Herman Müller and Aleksander Serebrovskii (Roll-Hansen 2005:198–199). After this first sharp confrontation between the "two directions in genetics," Morgan's chromosome theory became the main target of Lysenko's theoretical attacks.

It is significant that in December 1936 many breeders were quite critical of neo-Darwinism. Vavilov kept a low profile at the conference, but one of his close coworkers argued that Lysenko's theory of stages might well be a key to the questions of phylogeny and directed change in heredity (Sinskaia 1937).

Graft Hybridization Rehabilitated

In addition to environmental "training" to produce adaptive hereditary change, graft hybridization was a favorite technique of Lysenkoism. The production of new varieties of plants by transmission of hereditary properties between scion and stock is an ancient idea that was never completely abandoned by plant breeders even in the heyday of Mendelism. Darwin, for instance, saw it as support for his theory of pangenesis. Observations suggesting transfer of hereditary factors have been a continuous challenge to classical Mendelian genetics. In a 1946 critical survey, titled *The New Genetics in the Soviet Union,* the Imperial Bureau in Cambridge found primarily one area where solid empirical evidence contrary to neo-Darwinism might have been produced:graft hybridization (Hudson and Richens 1946:45–51). Similarly, when the famous British geneticist and Communist J.B.S. Haldane was pressed in the late 1940s for some minimal empirical support of Lysenkoism, he pointed to graft hybridization (Haldane 1947).

With the development of molecular genetics from the 1950s on, more specific chemical mechanisms for the transfer of genetic materials in the spirit of graft hybridization became conceivable. By the early twenty-first century it would be surprising if transfer of heritable variation through graft hybridization did not occur under suitable circumstances.

In 2006 the reputable journal *Advances in Genetics* published a review paper presenting evidence for graft hybridization in relation to a possible molecular mechanism. The author, Yongsheng Liu, is a Chinese plant breeder. He presents recent results as support for a revival of Darwin's theory of pangenesis as well as a rehabilitation of some of the work done by Lysenko's followers. A substantial part of the paper describes research done by IE. Glushchenko and other Lysenkoists from the 1940s to the 1970s, including a group of French "Michurinists." Liu concludes that further progress in fruit tree breeding is dependent on better understanding of graft hybridization and "requires that we reconsider Michurin's principles and methods" (Liu 2006:125). In later papers Liu has developed the link to Darwin's pangenesis (Liu 2007, 2008). He argues that the new understanding of graft hybridization shows that "observations on the inheritance of acquired characteristics are increasingly compatible with current concepts in molecular biology." But he adds, prudently, that this does not "in any way revive the theories of Lamarck or Lysenko" (Liu 2007:802).

Historically the link between Darwin's pangenesis and modern molecular genetics is a long shot. Liu's rehabilitation of work by Lysenko's followers in the "Michurin tradition" is also questionable. It may be fair to say that his review of biological data is well documented and interesting, but the historical interpretation is inadequate and misleading. Nevertheless, Liu's publications indicate how ideology has constrained the freedom of scientific research in the West. During the early decades of the Cold War, condemnation of Lysenkoism made it hard to get support or credit for legitimate scientific work that tasted of the inheritance of acquired characters. However, by the early twenty-first century both Lysenkoism and the Cold War are long past, and graft hybridization has again become a legitimate research topic, albeit now based on molecular biology methods that did not exist in Lysenko's time.

In an obituary notice for the prominent British molecular embryologist Dame Anne McLaren, Liu thanks her for help and support leading up to the paper in *Advances in Genetics* (2006). She was a student of J.B.S. Haldane as well as a persevering peace activist and active Communist Party member. As a member of the Royal Society of London she became its international secretary.[2] McLaren was always open-minded about the possibility of soft inheritance and Lamarckian-type mechanisms.

Conclusions

In this chapter I have argued that the neo-Darwinism promoted by geneticists in the controversy with Lysenko was more controversial and unsettled than standard history of science assumes. These standard accounts also neglected the role of developmental physiology of plants in the rise of Lysenkoism.

The origin of Lysenko's ideas in developmental physiology is significant because this was an area of biology in continuing opposition to neo-Darwinism throughout the middle decades of the twentieth century. Recognition of the historical importance of this countercurrent to neo-Darwinism has grown in parallel with the impact of so-called evo-devo theorizing on the understanding of evolution. There has even been some rehabilitation of phenomena and ideas previously denigrated due to their association with "Lysenkoism," such as graft hybridization.

Preoccupation with the "long series of social, political and economic events" (Graham 1972:6–7) that framed the rise of Lysenkoism has led standard history of science accounts (Joravsky 1970; Graham 1972; Soyfer 1994; Krementsov 1997) to neglect the scientific issues. Without some minimal scientific plausibility there would have been no Lysenkoism. From a broad sociopolitical point of view, Lysenkoism was a part of the history of science and not merely of pseudoscience. That inheritance of acquired characters had a role in the evolution of species still appeared plausible and attractive to many leading scientists as Lysenkoism reached its maximal influence around 1950. Throughout the middle decades of the twentieth century such ideas represented an important countercurrent to dominant neo-Darwinian views.

The ideological dichotomy of the Cold War and the scientific dominance of neo-Darwinism have now receded into the past. But questions about autonomy and the limits to legitimate political governance of science are once more highly important topics of political debates about science all over the world. Against this background Lysenkoism appears as an interesting historical "experiment" that deserves reexamination—with special attention to the interaction of scientific and science policy issues (Roll-Hansen 2005, 2008).

Notes

1. For a detailed account of Johannsen's genotype theory, see Roll Hansen 2009.
2. <http://www.gurdon.cam.ac.uk/anne-mclaren.html>. Accessed April 30, 2009.

References

Amundson R. *The Changing Role of the Embryo in Evolutionary Thought. Roots of Evo-Devo.* Cambridge: Cambridge University Press; 2005.

Bowler P. *The Eclipse of Darwinism.* Baltimore: Johns Hopkins University Press; 1983.

Cook RC. 1949. Lysenko's Marxist genetics. Science or religion? J Hered. 40: 169–202.

Darwin C. *The Variation of Plants and Animals Under Domestication.* 2 vols. London: John Murray; 1868.

Dennis ES, Peacock WJ. 2009. Vernalization in cereals. J Biol. 8: paper 57.

Gaissinovich AE. 1980. The origins of Soviet genetics and the struggle with Lamarckism, 1922–1929. J Hist Biol. 3: 1–52.

Graham L. *Science and Philosophy in the Soviet Union.* New York: Alfred A. Knopf; 1972.

Haldane JBS. *Science Advances.* London: Allen & Unwin; 1947.

Harwood J. *Styles of Scientific Thought: The German Genetics Community 1900–1933.* Chicago: University of Chicago Press; 1993.

Hudson PS, Richens RH. *The New Genetics in the Soviet Union.* Cambridge: English School of Agriculture, Imperial Bureau of Plant Breeding and Genetics; 1946.

Huxley J. *Soviet Genetics and World Science. Lysenko and the Meaning of Heredity.* London: Chatto and Windus; 1949.

Johannsen W. *Ueber Erblichkeit in Populationen und reinen Linien. Eine Beitrag zur Beleuchtung schwebender Selektionsfragen.* Jena: Gustav Fischer; 1903.

Johannsen W. *Elemente der exakten Erblichkeitslehre.* Jena: Gustav Fischer; 1909. (Thoroughly revised 2nd and 3rd eds. published 1913 and 1926).

Johannsen, W. 1911. The genotype conception of heredity. Am Nat 45: 129–159.

Johannsen W. 1923. Some remarks about units in heredity. Hereditas 4: 133–141.

Joravsky D. *The Lysenko Affair.* Cambridge, MA: Harvard University Press; 1970.

Komarov V. *Uchenie o vide u rastenii* (Theory About Species in Plants). Moscow and Leningrad: Izdatel'stvo Akademiia Nauk SSSR; 1944.

Krementsov N. *Stalinist Science.* Princeton, NJ: Princeton University Press; 1997.

Liu Y. 2006. Historical and modern genetics of plant graft hybridization. Adv Genet. 56: 101–129.

Liu Y. 2007. Like father like son. A fresh review of the inheritance of acquired characteristics. EMBO Rep. 8,9: 798–803.

Liu Y. 2008. A new perspective on Darwin's pangenesis. Biol Rev 83: 141–149.

Liubimenko V. *Kurs obshchei botaniki* (Course in General Botany). Berlin: R.S.F.S.R. Gosudarstvennoe Izdatel'stvo; 1923.

Mayr E. *The Growth of Biological Thought. Diversity, Evolution, and Inheritance.* Cambridge, MA: Belknap Press of Harvard University Press; 1982.

Morgan TH, Sturtevant AH, Muller HJ, Bridges CB. *The Mechanism of Mendelian Heredity.* New York: Henry Holt; 1915.

Popovskii M. *The Vavilov Affair.* Ann Arbor, MI: Hermitage; 1984.

Roll-Hansen N. 2000. The application of complementarity to biology: From Niels Bohr to Max Delbrück. Hist Stud Phys Biol Sci. 30: 417–442.

Roll-Hansen N. *The Lysenko Effect. The Politics of Science.* Amherst, NY: Humanity Books; 2005.

Roll-Hansen N. 2008. Wishful science: The persistence of T.D. Lysenko's agrobiology in the politics of science. Osiris 23,1: 166–188.

Roll-Hansen N. 2009. Sources of Wilhelm Johannsen's genotype theory. J Hist Biol. 42,3: 457–493.

Rossianov K. 1993. Editing nature. Joseph Stalin and the "new" Soviet biology. Isis 84: 728–745.

Sinskaia EN. (Contribution to Discussion). Spornye voprosy genetiki i selektsii. Raboty IV sessii VASKh-NILa 19–27 dek. 1936g. (Disputed Issues in Genetics and Breeding, Works of IV Session of VASKhNIL, December 19–27 1936). Moscow and Leningrad: Lenin Academy of Agricultural Science; 1937: 252–255.

Sonneborn TM. 1949. Heredity, environment, and politics. Science 111, 19 May: 529–539.

Soyfer V. *Lysenko and the Tragedy of Soviet Science.* New Brunswick, NJ: Rutgers University Press; 1994.

Vavilov N, ed. *Teoreticheskie osnovy sleksktsii rastenii* (Theoretical foundations of plant breeding). Moscow-Leningrad: Sel'khozgiz; 1935.

Zirkle C. *Evolution, Marxian Biology, and the Social Scene.* Philadelphia: University of Pennsylvania Press; 1959.

Zirkle C, ed. *Death of a Science in Russia.* Philadelphia: University of Pennsylvania Press; 1949.

9 Lamarckism and the Constitution of Sociology

Snait B. Gissis

This chapter deals with the emergence of sociology as a discipline in Great Britain and France during the second half of the nineteenth century, and does so by considering the transfer of concepts, models, metaphors, and analogies from contemporaneous evolutionary biology. Sociology emerged in continued interaction with this biology. Moreover, this evolutionary biology had a marked Lamarckian/neo-Lamarckian perspective and emphasis both in France and in Great Britain. Insights into the relationships between individuals and collectivities, as conceptualized in the formulation of sociology in these countries, are obtained by analyzing the interactions and transfers between social thought and Lamarckian evolutionary theories.

The decades from the 1850s onward witnessed the beginning, and thereafter the triumphal march, of evolutionism as a metanarrative. Evolutionism appeared in most fields of knowledge, and became absorbed into the progressivist mode of thought. It subsumed the organic world under a law of necessary advancement, within which human progress had specific traits, and left its marks in terms such as the evolution of civilization, of culture, of mind. After the publication of *The Origin of Species*, the diffusion of Darwin's work was neither "pure and simple" nor a straightforward matter. Rather, as Peter Bowler has indicated (1983, 1988), it was culturally integrated into a progressivist conceptual framework which had been there previously, partially formed from Lamarckian transformist–transmutationist components (see chapters 2 and 4 in this volume). Darwinian components, including natural selection, could be and were perceived by numerous biologists as compatible with Lamarckian mechanisms, within one theoretical framework (see chapters 3 and 5 in this volume). The resulting discourse often had "progress," "development," and/or "the inheritance of acquired characters" as its principal explanatory terms, rather than natural selection and variations, even though these evolutionary mechanisms were often present.[1] Furthermore, some of the then-extant models of recapitulation formed a component of this discourse (see chapter 3 in this volume).

The ongoing debate about the meaning and impact of the processes of industrialization and urbanization was part and parcel of the concerns of the emergent sociology in the second half of the nineteenth century. The explanation of two sets of issues was deemed crucial in order for sociology to acquire an autonomous status: first, how to account for the emerging features of modern society and its novel institutional, political, and governmental machinery and, by extrapolation, of new orders in general; and second, how to frame this within a historical narrative. Sociologists had to address four key issues:

1. that of change and its tempo
2. the novel perception of the gaps between classes
3. the apparent stability of the new order and its cohesion
4. the possible use of a "social science-oriented" policy of reform, which would resonate with current values, symbolic systems, institutions, and laws, that is, current social imaginaries.

The new social sciences—sociology, anthropology, social psychology, criminology, economics—that emerged during the second half of the nineteenth century were molded by the natural sciences that were perceived as models, with these sciences supplying presuppositions, methods, and boundaries. Biology began to assume that role for social theorizing at midcentury, even though physics was still paradigmatic for all the sciences. Transfer between fields of knowledge creates rather than points to already existing similarities. This stipulation of similarities depends on their being perceived as plausible by the professional practitioners in those fields of knowledge as well as by their audiences. The transfer of models, metaphors, and analogies from biology to sociology could take place only within a cultural context which allowed for the assumption that there was a fundamental correspondence/similarity/analogy between organic nature and social life, between mechanisms of biological and social development, and between types of regularities observed in both fields. The success of the transfer implied that it enabled sociologists to conceive more than they could otherwise do.

It is within this context that I shall discuss certain aspects of the work of two social theorists who were profoundly influential internationally and in their respective countries, Great Britain and France: Herbert Spencer (1820–1903) and Émile Durkheim (1858–1917). In important respects their work was part of a general framework of Lamarckian modes of thought that became significant around the mid-nineteenth century, and whose impact lasted until the end of that century. However, it should be recalled that during that period the signification, influence, and institutional localizations of both "Lamarckism" and "sociology" were in the process of being formed, changed, and re-formed.

Herbert Spencer

Spencer's use of the models, metaphors, and analogies he took from contemporary biological thought—to which he was a contributor—enabled him to claim psychology and sociology as scientific fields by evolutionizing them.

The 1855 edition of Spencer's *The Principles of Psychology,* together with some of his pre-1858 essays, conveys his views of the processes and mechanisms of evolution and development. The latter were one and the same thing for him, and it was a top-down phenomenological explanation. His approach did not alter when he added, somewhat later, a more generalized physical framework, which he hoped would allow for a universalized, reductionist, bottom-up explanation applicable to all scientific disciplines (Peel 1971; Jones, Peel 2004; Francis 2007).

For Spencer the "course of evolution" was based on Karl Ernst von Baer's law of individual development seen as divergence from an archetype, and thus as an alternative to linear recapitulation. But just as significantly, it was looked upon as a process of differentiation and specification, drawing on Henri Milne-Edwards's conception of the division of physiological labor in the economy of the organism. Thus, more complex organisms had more successfully adapted to their external milieu and had achieved a new and better equilibrium in their internal milieu. The cohesion of the components of the organism was seen as a model of biological cooperation and solidarity. Spencer assumed the mechanism of evolution to be primarily Lamarckian (Corsi P, Gayon J, Gohau G, Tirard 2005).The organic world was seen as a graded continuity, the evolutionary process was seen as gradual, and whatever directionality and irreversibility there was in evolutionary change, it tended toward greater complexification by way of adaptation. In his later work, though, Spencer did consider the possibility of reversible courses. Spencer's evolutionary sociology adhered to that view of evolution in *The Study of Sociology* (1873), *Descriptive Sociology* (1873–1881), and *The Principles of Sociology* (1876–1896).

For Lamarck, "external conditions," "force of circumstances," and similar terms were posited as the principal cause of change, and they were conceived as solely physical. Already in 1855, in *The Principles of Psychology*, Spencer substituted for them the unified concept of "environment," which can be looked upon as an equivalent of August Comte's "milieu" (as can be inferred from Harriet Martineau's English adaptation of Comte). This conceptualization was richer than Lamarck's in that it included the effect of other living organisms and their spatial distribution (as already suggested by Lyell), and was extended to include social conditions. Thus, the environment, whether external, internal, physical, organic, or social, effected changes directly through adaptation. A predominant role was assigned to the environment by virtue of "posing problems and difficulties" to the actively reacting organisms.

Thus, the basic methodological unit was the relationship between environments and organisms, which Spencer termed "correspondence" in *The Principles of Psychology* of 1855 (see Gissis 2005). The mechanisms in the process of adaptation were those of use/disuse and habit and habituation, that is, they were behavioral in the broad sense of the term.

The results of the process of the individual organisms' adaptation—their "functional modifications" (adaptive acquired traits)—were transmitted transgenerationally. This meant that a trait acquired socially or psychologically would be inherited biologically, just as whatever was acquired biologically could serve as a foundation for social or psychological processes in a back-and-forth movement between the social and the biological. The gradual change in individuals of a species occurred in such a fashion that all members of the group, ignoring individual differences, turned out to incorporate the same effects. In human societies, experience and behavior became manifestations of collective hereditary transmission, equally affecting all possible generations of individuals of the relevant group. What was hereditarily transmitted was not solely a psychological–cultural pattern, but rather a changed biological (i.e., a neural) pattern, a pattern which provided the members of the collectivity with modified competences through which their experiences would be organized, molding the physical and psychological behavior of the individual. Thus Spencer was not a simple, straightforward individualist, neither in his biology nor in his social sciences, but rather an ambivalent collectivist-in-disguise. Moreover, as James Elwick (2007) has shown, Spencer's model organisms[2]—common and simple invertebrates, such as chitons, annelids, hydrozoans—were considered by him to be compounds of individual units, and thus, I would argue, could be looked upon as various kinds/levels of collectivities. Topics discussed in biological terms, such as the relations of parts to wholes, dependencies and interdependencies, individuality and collectivity, were all applicable to the *superorganism*, that is, to the social collectivity and its interdependent institutions.

The utopian dimension of Spencer's sociological work—the changed ethical quality of life—is of relevance here. The social enabling conditions for it were very high degrees of mutual interaction and cooperation, which served as indices of social evolution. Although in Spencer's narrative the changes would seemingly happen to individuals and concerned their happiness, they could become socially meaningful only when the collectivity at large was affected through a biological–cultural back-and-forth civilizing process. What made possible the alternation between the cultural and the biological in Spencer's sociology, and made room for psychology as a unique science of the individual, was the role assumed by the collectivity: Spencerian individuals contained the collective within themselves as their mode of adapting.

After the publication of *The Origin of Species,* Spencer included natural selection as yet another branching evolutionary path, albeit as a secondary evolutionary mechanism. For Spencer, natural selection attributed causality and agency to the environment and it operated merely as a constraint. Thus, for him it explained the survival of existing populations but not the appearance of novelty. Spencer qualified natural selection's explanatory power further by stating that it could perhaps be instrumental in explaining change in lower forms, but certainly not in humans. Thus he was positioning himself closer to Alfred Russel Wallace's position in the controversy over the explanation of human cognitive evolution. But, contrary to Wallace, he did not invoke a nonnaturalist explanation for that. Spencer held this position on "the survival of the fittest" more or less consistently throughout his life. It became more explicit in the exchange of essays with August Weismann in the 1890s, in *The Factors of Organic Evolution* (1887), and in the revised edition of *The Principles of Biology* (1898).

Émile Durkheim

During the last third of the nineteenth century, evolutionary biology, including Lamarckian transformism, played a vital role, particularly for sociologists in France, Belgium, and Italy, in molding programs, actual practices, cultural rhetoric, and their self-image as scientists (see discussion in Conry 1974). In France, from the 1860s onward, Lamarckian transformism was a significant presence, with Spencer as an important resource. It was informed by three core tenets: the central role of the environment in adaptation; evolution understood as a development from less to more complex; and the inheritance of adaptive acquired traits (see chapter 7 in this volume). Social thinkers, perhaps like some of the biologists, failed to inquire what sort of patterns and traits were distinct enough to be passed on biologically. When discussing Durkheim I am focusing on his work from the mid-1880s until the late 1890s, the period when biology served as a principal source of his models, metaphors, and analogies (Lukes 1973; Alexander and Smith 2005).

Durkheim's innovative claim "to explain the social by the social" was made possible to a significant extent by the sophisticated manner in which he transposed fundamental biological tenets and terms to the social field (see Durkheim [1893] 1984, [1895] 1982, [1900] 1960). He did this in two ways: one strategic—in order to legitimize; the other theoretical—in order to construct. Society was to be explained by using social categories, but the conceptual tools were to be borrowed from another science, evolutionary biology. Yet Durkheim repeatedly referred to the distinction between the two sciences, the existing biology and the emerging sociology, and uniquely delimited the latter. Furthermore, he demanded that the role of the entities transposed into sociology—the biological concepts, models, metaphors—be apposite in the two disciplines. Durkheim invested great efforts, both in his rhetoric and in his formulation of theory and methodology, to distinguish between his views and those of others who regarded society straightforwardly as an organism. He thereby hoped to constitute sociology as an autonomous discipline.

In the social as well as in the political discourse of the time, "the organism" was applied to a variety of entities. Durkheim was not a Spencerian-style organicist. Spencer thought that each individual could be considered as an actual organism, and society was to be considered as a real "superorganism." Durkheim used the "organism" as a necessary component in his argument about the *relations* between individuals and their collectivities. These relationships—but not the actual individuals, nor the actual collectivities—could be viewed, in *some* of their aspects, as analogous to parts of the organism in relation to its totality. He insisted on a very partial application of this analogy to individuals, and admonished against any literal understanding of it. For most of his contemporaries, individuals were conceived to be epistemologically the real entities, observable ones. The reality of the collectivity was often envisaged as secondary, deriving from the reality of the individual and thus unobservable on its own. Durkheim's usage of the analogy purported to make the collectivity epistemologically real, to construct its "visualization" at a time when the visibility of "Nature" had become a significant scientific issue. He thought thus to have provided a semblance of observability to the collectivity. Evolutionary biology was perceived as relying on a partially observable past, which could be reconstructed by supplementary hypotheses, but no causal mechanism of the present could be exemplified. To a large extent, the force of the gradually emerging Durkheimian group and its claim to academic and cultural legitimation resided, for Durkheim's contemporaries, in the claim that it was a science which relied on the emulation of a culturally paradigmatic science, contemporary biology, while it concurrently provided sociophilosophical underpinning to an emerging hegemonic ideology of solidarité-ism.

The particularity of Durkheim's use of "the biological" was that he was a Lamarckian evolutionist but not an organicist. Like Spencer, he accepted the cluster of notions related to the role of the environment, the economy of the organism, the physiological division of labor, coordination, and so on. He transposed the differentiation in the economy of the organism into the social differentiation resulting from the division of labor in society. Social plasticity meant that when a primary process of adaptation to the environment occurred, either biological or social, it resulted in a certain practice, a social pattern. This pattern would then be transmitted as a habit and custom, that is, through processes of socialization and acculturation. Adaptation of individuals could be understood only within their culture. This was considered by Durkheim to be the sociological equivalent of the inheritance of acquired traits. The division of labor as a mechanism of change could induce alterations in the moral, ideational, and civilizing practices and beliefs of societies, and thereby transform those of individuals. For Durkheim, in spite of admitting variability in populations, changes and transformations applied to all members of a group, for they were considered to be identical as far as the relevant social trait was concerned. Durkheim's transfer from biology can also be exemplified by the reasoning which informed his view of modern society as having a degree of differentiation and specification in the division of labor. This resulted in a complex social stratification and a sophisticated economy. Socially, it produced complex, nonoverlapping networks of relations of interdependence and multilevel coordinations among individuals, resulting in complex "organic" solidarity.

I would argue that the consequences of the social translation of the "environment" which were semi-implicit in Spencer became foundational assumptions for Durkheim. His thought can best be understood as an evolving, at times problematical and contradictory, construction of classificatory social continuities—from individuals to collective formations. This mode of thought can be interpreted as illustrative of the deep impact of Lamarckian evolutionism. "Habit," whose meaning stretched between reflective, conscious acts that were informed by intentions, motives, and reasoning, and those that were the result of nonreflective, habitual behavior exemplifies that. Note that Durkheim's position with respect to habit paralleled that of Spencer. A version of French neo-Lamarckism (e.g., that of Edmond Perrier) translated into the Durkheimian idiom served a constitutive role in Durkheim's sociology, but it disappeared from the forefront of the works and the programmatic declarations of the Durkheimians after the Dreyfus affair. Thus, at the height of the battle between neo-Lamarckians and neo-Darwinians in France (see chapter 7 in this volume), the Durkheimians were moving elsewhere.

Evolutionary Mechanisms: A Comparative Discussion

Let me very briefly weave together and compare a number of features which mark the notions of environment, heredity, social plasticity, and social complexity in the early and middle work of Spencer and Durkheim, exemplifying the particular enmeshing of evolutionary mechanisms. Within the specific sociocultural and political context of the Third Republic, this enmeshing was instrumental in making the Durkheimian group into "French sociology." Spencer's work had appeared almost a generation earlier, and this implied that it could not, within the specificities of British social and political practices, directly effect the establishment of sociology there at that time. But such an enmeshing was a significant enabling condition for its institutionalization in the next generation—that of Leonard Hobhouse at the London School of Economics.

Both Spencer and Durkheim believed in social plasticity. Both professed some notion of human progress and assumed an evolving social complexity. Thus, both looked upon sociality at large, as well as upon specific social formations, as emergent. Yet both also wanted to have a notion of laws governing the social, though Spencerian laws did not leave any place for contingency, while one could argue that Durkheim was groping toward a notion of probabilistic laws. Both argued that one could neither plan nor change society by rationality alone (and in that sense were overtly non–Enlightenment). This had implications for their notion of progress as it related to social plasticity (and to evolution). Both strove to keep a balance between individuals and collectivities within their explanatory framework, in particular in the context of plasticity. At issue was the question whether individuals or collectivities were to be the fundamental explanatory unit. Spencer focused on how to understand individuals, and Durkheim, collectivities. For both, the evolutionary biology of their time (and how they understood it) supplied crucial elements in their formulation of problematics and in arriving at answers in the social field. Both used the findings of cell theory of their respective times. Both used some notion of animal social grouping related to work done by their contemporaneous biologists to illustrate notions of individuals and of collectivities. Both struggled with the question of how to accommodate Darwinian explanatory mechanisms within some version of Lamarckism. For both, the inheritance of adaptive acquired traits was central for the description and explanation of the movement from past states to future ones. Furthermore, both deployed biological models, metaphors, and analogies, thereby claiming the social fields to be scientific by evolutionizing them. But concurrently they also pointed to modes of distinguishing and severing sociology from biology in order to regard the former as autonomous. The different deployment of evolutionary biology sources by Spencer and Durkheim can be at least partially attributed to the following:

a. the periods during which each of them worked out his mode of transfer—Spencer primarily from the 1850s until the mid-1870s, with some changes in his later writings, and Durkheim in the 1880s and 1890s

b. the differing practices of investigating the social in Great Britain and France

c. whether there had already been professional or scholarly frameworks (institutional or voluntary) for discussing social theory

d. the nature of the cultural field at large—the roles of the ideal of science, of the scientific method, and of scientifically based social reform.

As elaborated in previous chapters, the differences and discrepancies among versions of Lamarckism and between those versions and the Darwinism ones were not clear-cut. This is all the more so when one deals with the practitioners of social theory of that time and their transfers from the contemporaneous evolutionary and nonevolutionary biology. Spencer was perceived by many of his contemporaries to be a Darwinian and, moreover, as a principal agent of the diffusion of Darwinism. But Spencer was active for over fifty-five years, and at the end of the nineteenth century he was perceived by neo-Darwinians as a proper Lamarckian. Durkheim was perceived by one of the major groupings of sociologists at the time as a hybrid of Lamarckism and whatever they called organicism. Some recent historians of sociology have dubbed him a Darwinian (e.g., Limoges 1994). I, on the other hand, believe and argue that he was closest to some version of Lamarckian tenets held by his contemporaries (Gissis 2002).

To further clarify why this is so, let me elaborate on social plasticity within late nineteenth-century evolutionism. Any evolutionary framework transferred to social thought which posited adaptive relations between organisms and environments—social, organic, physical—could be, and was, interpreted in one of two ways. One was as a facet of a hard-core biological determinism that would trump whatever effects social changes of environment could have. The capabilities of the individual would then be conceived as merely reflections of the conditions of existence of the species, and thus would be bindingly hereditary. Biological determinists could, and indeed did, apply this viewpoint to concepts of biological purity such as "race" and "class." This more rigid conception of heredity played a significant role in the constitution and boundary formation of the criminology and psychology of the period as medicalized–biologized disciplines, and in the split among anthropology practitioners in France after Paul Broca's death. That split was between those who later came to be known as cultural anthropologists and the physical anthropologists who remained within the framework of medical schools. The second way was to regard the role of the environment as central in the formation and the transmission of social and cultural functionally adaptational patterns such as habits, customs, and traditions. This approach could therefore highlight "progress" as an open-ended endeavor. To a large extent it could be identified with "progress" when applied to humans (including "race" distinctions).

Lamarckism offered a double perspective on ethics and on society at large. It emphasized the overall importance of the milieu/environment in shaping present and future generations through the inheritance of acquired characters, with "use-repetition-habit-habituation" as its principal mechanism. This implied the possibility of shaping the future: present changes could be bequeathed as prospective biological or cultural traits to be further elaborated and complexified in the future. The seeming determination of the present by the past was one consequence of such thinking, the other being the molding of the present in light of a projected future. This "future perspective," which was associated with Lamarckism in the last decades of the nineteenth century, was of the utmost importance to those social scientists who considered the emerging discipline to be interwoven with social reform, assumed a continuous narrative of humanity/society, and formulated their analysis of modern society in terms of a progressively complexified division of labor. However, the implicit possibility of anchoring such a perspective in biology was at least partially undermined by the Weismannian arguments in answer to Spencer in the early 1890s (see chapter 6 in this volume), for they were understood as severing social inheritance from biological heredity.

Concluding Remarks

The evolutionary metanarrative supplied the emerging field of sociology with a framework in which to formulate theories of the constitution of modern society and of the genealogy and evolution of its various institutions and their functioning, that is, of mechanisms of change and of stabilization. Though "modern society" was virtually the sole object of investigation in sociology, the evolutionary framework enabled practitioners to deploy materials relating to past and present societies (Western and non–Western) in order to construct a single continuous narrative. Note that degeneration, which was perceived at the time as the opposite of evolutionary progress, arose within this narrative and not outside it. The mechanisms advanced were claimed to be scientific, for they were presented as variations or continuations of those operating in evolutionary biology.

Although the emerging social sciences have been viewed for a long time as having been molded and regulated by Darwinian evolutionism, I believe that versions of Lamarckian evolutionism were at least equally significant, particularly in Great Britain and France. The differences in the choice of an evolutionary framework stemmed from the concrete societies and national cultures in which the social scientists lived and worked, and were the result of varying interpretations given to these evolutionary frameworks in the cultural–political discourses of those societies. I believe that whatever was considered Lamarckism in the late nineteenth century was deemed congenial to constituting emerging sociology in Great Britain and France, and therefore to be a significant source. The existence of a variety of interpretations of what "evolution" consisted of, and the fact that these were meaningfully different in their "socially translated" presuppositions and implications, can be of help in explicating the complex relationship between the sociopolitical worldview and the adoption of a specific scientific stance. It can also provide a useful historical perspective on the checkered history of such transfers.

Acknowledgment

I am grateful to Yemima Ben Menahem and Sam Schweber for their useful critical comments.

Notes

1. This downward scaling of Darwin's mechanism to a secondary role was also translated into a socio-political competition and/or a "struggle for life" among various types of collectivities, such as states, nations, or races, rather than among individuals.

2. One can call them model organisms because they were selected as exemplars of widely observed features of life, were accessible and commercially obtainable, and were perceived as typical, serving as an index to the group of instances (Ankeny 2001).

References

Alexander JC, Smith P, eds. *The Cambridge Companion to Durkheim.* Cambridge: Cambridge University Press; 2005.

Ankeny RA. 2001. Model organisms as models: Understanding the "lingua franca" of the human genome project. Philos Sci. 68D,3(suppl.): S251–S261.

Bowler PJ. *Evolution: The History of an Idea.* Berkeley: University of California Press; 1983.

Bowler PJ. *The Non-Darwinian Revolution.* Baltimore: John Hopkins University Press; 1988.

Conry YL. *L'introduction du Darwinisme en France.* Paris: Vrin; 1974.

Corsi P, Gayon J, Gohau G, Tirard S, eds. *Lamarck, philosophe de la nature.* Paris: Presses Universitaires de France; 2005.

Durkheim E. *De la division du travail social.* –The Division of Labor in Society:trans. WD Halls. New York: Free Press; [1893] 1984.

Durkheim E. *Les règles de la méthode sociologique.* The Rules of Sociological Method.trans. WD Halls. New York:Free Press; [1895] 1984.

Durkehim E. Sociology and its Scientific Domain, K Wolff, trans. In Wolff KH ed. *Émile Durkheim 1858–1917.* Columbus: Ohio State University Press; 1960 [1900]:354–375.

Elwick J. *Styles of Reasoning in the British Life Sciences: Shared Assumptions, 1820–1858.* London: Pickering & Chatto; 2007.

Francis M. *Herbert Spencer and the Invention of Modern Life.* Chesham, UK: Acumen; 2007.

Gissis SB. 2002. Late nineteenth century Lamarckism and French sociology. Perspect Sci. 10,1: 69–122.

Gissis SB. 2005. Biological Heredity and Cultural Inheritance in the two editions of Herbert Spencer's The Principles of Psychology: 1855 and 1870/72. In: *A Cultural History of Heredity.* Vol. 3. Berlin: Max Planck Institute; preprint 296: 137–152.

Jones GR, Peel RA, eds. *Herbert Spencer: The Intellectual Legacy.* London: Galton Institute; 2004.

Limoges C. Milne-Edwards, Darwin, Durkheim and the division of labour: A case study in reciprocal conceptual exchange between the social and the natural sciences. In: *The Natural Sciences and the Social Sciences: Some Critical and Historical Perspectives.* Cohen IB, ed. Dordrecht, Netherlands: Kluwer; 1994:317–344.

Lukes S. *Emile Durkheim: His Life and Work.* London: Allen Lane; 1973.

Peel JDY. *Herbert Spencer—The Evolution of a Sociologist.* London: Heinemann; 1971.

Spencer H. *The Principles of Psychology.* London: Longman, Brown, Green, and Longmans; 1855.

Spencer H. *Descriptive Sociology, or Groups of Sociological Facts.* Classified and arranged by Herbert Spencer; compiled and abstracted by David Duncan, Richard Scheppig, and James Collier. New York: Appleton; 1873–1881.

Spencer H. *The Study of Sociology.*London:Henry S. King; 1873.

Spencer H. *The Factors of Organic Evolution*. Reprinted, with additions, from *The Nineteenth Century.* New York: Appleton; 1887.

Spencer H. *The Principles of Sociology.*London:Wlliams & Norgate; 1876–1896.

Spencer H. *The Principles of Biology.* Revised and enlarged ed. New York: Appleton; 1898.

II THE MODERN SYNTHESIS

10 Introduction: The Exclusion of Soft ("Lamarckian") Inheritance from the Modern Synthesis

Snait B. Gissis and Eva Jablonka

During the construction and consolidation of what became known as the Modern Synthesis of evolution, Lamarckism seemed to have come to an end. As William Provine explained when we were discussing the contribution he hoped to make to the workshop, by the end of the Synthesis period not only had the views of Lamarck and Lysenko been totally rejected in the Anglo–American world, but so, too, had the neo-Lamarckian views of Darwin. Will intended to look at the Lamarckism of Lysenko, particularly the vernalization of wheat and other crops, and try to separate Lysenkoism as a scientific theory from the terrible consequences of Lysenko as a scientific administrator of Soviet heredity. Unfortunately, illness prevented Will from attending the workshop, so we decided to change the format of the session, which we viewed as a bridge between the historical and the biological aspects of Lamarckism, to enable a general discussion of how and why, by the middle of the twentieth century, most evolutionists had excluded soft (Lamarckian) inheritance from their theorizing.

In two short talks, Scott Gilbert and Adam Wilkins looked at different aspects of early twentieth-century American evolutionary thinking. Scott discussed the split between development and heredity, and Adam considered some of the reasons why soft inheritance was rejected. Marion Lamb then described the changing attitudes toward soft inheritance in Britain. These talks were followed by a discussion open to all workshop participants. The whole session, which was chaired by Alfred Tauber, was recorded and transcribed. What is presented here are abridged versions of these talks, which have been edited by their authors, and a summary of the discussion which was prepared by Snait Gissis and Eva Jablonka and approved by those who participated.

The "Modern Synthesis," a term borrowed from the title of a book by Julian Huxley (1942), refers to the wide-ranging consensus about the nature and dynamics of evolutionary change that emerged among biologists between the 1920s and 1950s. The Synthesis included the classical Mendelian genetics studied by experimentalists in the lab; the population dynamics of Mendelian genes as worked out by theoreticians; studies of natural populations made by geneticists and by systematists interested in speciation; and the analysis of the fossil record by paleontologists. Fundamental to the Synthesis was the belief that the frequency of Mendelian genes carried by nuclear chromosomes can change in predictable and quantifiable ways as a result of natural selection, migration, random drift, mutation pressure, and altered mating patterns (Fisher 1930; Haldane 1932; Wright 1932). The Mendelian gene was assumed to be a well-behaved, stable entity which is not altered by the behavior of the organism (by use and disuse), or by the direct effect of the environment, or by internal guiding processes.

Several major books, published between 1937 and 1950, are regarded as the bedrock of the Synthesis. In the first of these, by the geneticist Theodosius Dobzhansky (1937), it was argued that the only direction-giving process in evolution is natural selection of nondirected genetic variations. Natural selection can account for cumulative, gradual, adaptive evolution in populations because many genes, each with a small effect, underlie most characters. Five years later, Ernst Mayr, a systematist, described how, with some auxiliary assumptions such as geographic isolation, new species could arise (Mayr 1942). Natural selection of small, nondirected, genetic variations was also shown to apply to many aspects of plant evolution (Stebbins 1950), and to be compatible with the observed patterns of macroevolution—with evolution above the species level (Simpson 1944). Patterns of macroevolution were explained in the same terms as the microevolution occurring within populations. In all these major works, "Lamarckian" modes of inheritance, orthogenetic directional processes, and saltational changes were excluded (Mayr and Provine 1980; Mayr 1982; Smocovitis 1996).

The starting point for this evolutionary synthesis, and the way it took place, were different in different countries; they also occurred at different rates in different disciplines (Mayr and Provine 1980). While in the United States the Synthesis involved the resolution of the conflict between the approaches of laboratory-based geneticists and the systematists who studied natural populations and processes of speciation (Allen 1979; Mayr and Provine 1980; Mayr 1982), in chapter 11 of this volume Marion Lamb suggests that there was very little conflict in England. Other scholars outside the United States also argue that evolutionary biology developed without much conflict: according to Adams (1980), in the Soviet Union there was no conflict at all, and Reif, Junker, and Hössfeld (2000) maintain that the Synthesis was a long-term project carried out between 1930 and 1950 by a large number of biologists in several countries. In all countries, however, it took longer for paleontological data—for the modes and rates of macroevolutionary change found in the fossil record—to be accommodated within the emerging view of gene-based, gradual evolution by natural selection. Developmental biology was not integrated, and in fact was explicitly excluded by some of the early founders of the Synthesis. In chapter 12 of this volume, Scott Gilbert shows how the idea of the morphogenetic field, which is reconstructed during every ontogeny and which was seen by embryologists as a unit of both development and heredity, was left out of the geneticists' study of heredity. Although Waddington in the United Kingdom and Schmalhausen in the Soviet Union explored the evolutionary implications of developmental–genetic variations that lead to plasticity and canalization, their ideas did not become part of mainstream Modern Synthesis thinking. As we discuss in chapter 15, these ideas became an important part of evolutionary biology only in the 1990s.

In the fifty years following the 1947 Princeton Conference, which can be seen as a self-conscious celebration of the consensus about the mechanism of evolution that had been reached in the United States, the Synthesis itself "evolved" and incorporated new data and new ideas. However, its core tenets remained intact. The tenet which was of special interest to the workshop participants was the strict adherence to neo-Darwinism. Neo-Darwinism was a term coined by Romanes (1893) to describe Weismann's version of Darwinism: it is Darwinism (gradual evolution through natural selection of small, individual heritable variations) without any kind of "Lamarckian inheritance" of acquired characters or internal guidance. The assumption that acquired characters are not inherited is therefore one of the defining features of the neo-Darwinian version of Darwinism.

The inheritance of acquired characters is seen as an aspect of soft inheritance,[1] a term which Mayr (Mayr and Provine1980:15) characterized in the following way: "I use this term [soft inheritance] to designate the belief in a gradual change of the genetic material itself, either by use or disuse, or by some internal progressive tendencies, or through the direct effect of the environment." Mayr thus used "soft inheritance" as a general term encompassing not only the inheritance of acquired characters but also other processes which neo-Lamarckians and orthogeneticists had suggested could alter heredity in a directional manner. He saw the belief in soft inheritance as an obstacle to the building up of a population-based, synthetic, neo-Darwinian interpretation of evolution, and stated that "It was perhaps the greatest contribution of the young science of genetics, to show that soft inheritance does not exist" (Mayr and Provine1980:17). This is a surprising statement: the young science of genetics, which was based on Mendelian genes, did not prove that soft inheritance does not exist; it showed that Mendelian inheritance is usually stable, but it did not rule out occasional directional effects of the environment on the gene, and it certainly did not prove that the Mendelian gene is the *sole* hereditary factor. As Sapp (1987) has documented, many biologists thought that there were other hereditary factors in the cytoplasm. It is clear that Mayr's assertion is in line with his own conviction that the "hard" Mendelian gene is the exclusive material basis of heredity, and with the general marginalization of cytoplasmic inheritance and other non–

Mendelian phenomena in the Modern Synthesis. Twenty-two years later, Mayr had not changed his mind, although by then the more general concept of the DNA sequence had replaced the Mendelian gene: in an interview with Adam Wilkins, Mayr claimed that one of the four major contributions of molecular biology to issues in evolution "is Francis Crick's Central Dogma, in that information can go only from nucleic acids to proteins but never from proteins to nucleic acids and that principle was of course the final nail in the coffin of the inheritance of acquired characteristics" (Wilkins 2002:965). For Mayr, a consensus about there being a single type of inheritance system and a single type of variation was crucial for the construction of a unified view of evolution. Adam Wilkins's workshop paper (chapter 13 in this volume) points to three additional important reasons why evolutionary biologists rejected "soft inheritance."

In spite of the enduring beliefs of Mayr and other contributors to the Modern Synthesis, the status of the Mendelian gene as the sole unit of heredity was repeatedly challenged. During the 1950s and 1960s, in both the United Kingdom and the United States, there were many ideas about additional factors and processes that were involved in cellular heredity. Not only was it discovered that chloroplasts and mitochondria contain DNA, and that many of the known cases of cytoplasmic inheritance could be interpreted in terms of organelle heredity, but other types of non–Mendelian inheritance, both nuclear and cytoplasmic, including some remarkable cases of soft inheritance, were also reported (Sapp 1987; Lamb, chapter 11 in this volume). However, these observations had little effect, and Marion Lamb suggests several reasons for their neglect. One of the reasons was the powerful advocacies of the "founding fathers" (there were no "mothers") of the Synthesis, who convinced biologists that alternative interpretations were unnecessary. In chapter 12 of this volume, Scott Gilbert argues that the scientific and historical narratives written by the "founding fathers" were very influential in disseminating their views. The institutionalization of evolutionary biology in the United States by the architects of the Modern Synthesis was another important factor in the formation of "one voice." The founding of the Society for the Study of Evolution (SSE) by the Modern Synthesis's evolutionary biologists and the establishment of the journal *Evolution*, whose first editor was Ernst Mayr and which was the vehicle for disseminating and consolidating the views of the Modern Synthesis adherents, were important factors in the strength and domination of the Modern Synthesis in the United States (Smocovitis 1994). In 1998, in a new preface to the second edition of the 1980 volume he and Provine had edited, Mayr claimed that all the controversies in evolutionary biology "are within the synthesis theory" (Mayr and Provine 1998:xiii). Although, as the concluding discussion in that volume makes clear, some of the authors strongly disagreed with Mayr's position, there is no doubt that the formidable persistence of the Modern Synthesis is still the context within which new ideas about heredity and evolution are evaluated.

Yet, in spite of the persistence of the Modern Synthesis view, the chapters in part III clearly show that the concept of heredity has expanded beyond the gene, and that the generation of new hereditary variations, at the genomic and epigenomic level, is no longer seen as a "blind" process. It seems that a new consensus is being formed, in which heredity is seen as an aspect of development, and evolution includes far more than changes in gene frequencies.

Note

1. The terms "soft inheritance" and "hard inheritance" were used in Darlington CD. *Darwin's Place in History*. Oxford: Blackwell; 1959. The first mention of them is on pages 14–15.

Further Reading

Adams MB. Soviet Union. In: *The Evolutionary Synthesis: Perspectives in the Unification of Biology*. Mayr E, Provine WB, eds. Cambridge, MA: Harvard University Press; 1980:242–278.

Allen G. 1979. Naturalists and experimentalists: The genotype and the phenotype. Stud Hist Biol. 3:179–209.

Dobzhansky T. *Genetics and the Origin of Species*. New York: Columbia University Press; 1937.

Fisher RA. *The Genetical Theory of Natural Selection*. Oxford: Clarendon Press; 1930.

Haldane JBS. *The Causes of Evolution*. London: Longman, Green; 1932.

Huxley JS. *Evolution: The Modern Synthesis*. London: Allen & Unwin; 1942.

Mayr E. *Systematics and the Origin of Species*. New York: Columbia University Press; 1942.

Mayr E. *The Growth of Biological Thought: Diversity, Evolution, and Inheritance*. Cambridge, MA: Belknap Press of Harvard University Press; 1982.

Mayr E, Provine W., eds. *The Evolutionary Synthesis: Perspectives on the Unification of Biology*. Cambridge, MA: Harvard University Press; 1980; 2nd ed. 1998.

Reif W-E, Junker T, Hössfeld U. 2000. The synthetic theory of evolution: General problems and the German contribution to the synthesis. Theory Biosci. 119,1: 41–91.

Romanes G. *An Examination of Weismannism*. London: Longmans, Green; 1893.

Sapp J. *Beyond the Gene*. New York: Oxford University Press; 1987.

Simpson GG. *Tempo and Mode in Evolution*. New York: Columbia University Press; 1944.

Smocovitis VB. 1994. Organizing evolution: Founding the Society for the Study of Evolution (1939–1950). J Hist Biol. 27,1:241–309.

Smocovitis VB. *Unifying Biology: The Evolutionary Synthesis and Evolutionary Biology*. Princeton, NJ: Princeton University Press; 1996.

Stebbins GL. Variation and Evolution in Plants. New York: Columbia University Press; 1950.

Wilkins A. 2002. Interview with Ernst Mayr. BioEssays 24:960–973.

Wright S. The roles of mutation, inbreeding, crossbreeding and selection in evolution. In: Proceedings of the 6th International Congress of Genetics. Brooklyn, New York: Brooklyn Botanic Garden 1932; 1:356–366.

11 Attitudes to Soft Inheritance in Great Britain, 1930s–1970s

Marion J. Lamb

Biologists frequently use expressions like "before the Modern (or Evolutionary) Synthesis," or "as a result of the Modern Synthesis," without giving much thought to when the Modern Evolutionary Synthesis actually took place or exactly what it entailed. The view of one biologist-turned-historian, Ernst Mayr, was that for the first third of the twentieth century there were two camps of evolutionists, the experimental geneticists and the naturalists–systematists. Neither camp knew or appreciated what the other camp was doing. Then, "a meeting of the minds came quite suddenly and completely in a period of about a dozen years, from 1936 to 1947" (Mayr 1982:567). Both camps accepted that almost all evolution, including macroevolution and speciation, occurs through the gradual, cumulative effects of natural selection acting on small genetic differences between individuals. Earlier beliefs in "Lamarckian" mechanisms, saltationism, and orthogenesis were abandoned.

A very different account of the Synthesis has been given by another biologist-turned-historian, Stephen Jay Gould (2002: chap. 7). He put its beginning much earlier, citing Fisher's 1918 integration of Mendelian genetics and biometry as one of "the first rumblings of synthesis." Gould also stressed how the leading American contributors of the 1930s and 1940s moved away from their initial positions, adopting a narrower, more restrictive version of the Synthesis in the 1950s and 1960s.

Given such diverse views about its history and nature, one begins to wonder whether the term "the Modern Synthesis" is worth retaining. It seems to have about the same descriptive value as "the Dark Ages." However, like the latter term, its appropriateness may be different in different places, because the history of evolutionary thinking is not the same in all countries (Mayr and Provine 1980). In what follows, I will argue that for Great Britain there is no evidence of a great gap between the evolutionary thinking of geneticists and of other biologists in the 1930s. Evolutionary theory then, and subsequently, was based on a mixture of ideas from different branches of biology, and was constantly being modified as new facts and concepts emerged and the importance of different institutions, fields, and individuals changed. I will focus mainly on attitudes toward the "Lamarckian" aspect of evolution that became known as "soft inheritance." This term is usually attributed to Mayr, who was the first to formally define it (Mayr 1980:15), but it was used by Cyril Darlington at least as early as 1959. I use a modification of Mayr's (1982:959) definition: soft inheritance is "inheritance during which the hereditary [genetic] material is not constant from generation to generation but may be modified by the effects of the environment, by use or disuse, or other factors." "Hereditary" is substituted for Mayr's "genetic," because present usage makes "genetic" too restrictive.

From 1930 to the Second World War

Although in the 1920s and 1930s a few people were still supporting Lamarckism, the evolutionary question that interested most British biologists was whether evolution was gradual, as Darwin suggested, or occurred through big jumps, as some of the early Mendelians believed. Two books published in the early 1930s—*The Genetical Theory of Natural Selection* (Fisher 1930) and *The Causes of Evolution* (Haldane 1932a)—helped to return British evolutionary thinking to Darwin's gradualism. Fisher and Haldane, along with Sewall Wright in America, demonstrated mathematically how the interactions between gene mutation, migration, chance, and natural selection could produce gradual, cumulative, evolutionary changes. They showed that Darwin's natural selection could work; they renewed confidence in it.

The two books are very different. Fisher's prose, as well as the mathematics, make his book difficult to read. I doubt whether many biologists had a firsthand knowledge of Fisher's arguments, even though the book included topics such as mimicry, the evolution of dominance, sexual selection, and eugenics. Fisher was a strict neo-Darwinian, and had no room for any form of Lamarckism. He presented an absurd caricature of Lamarck's view of the origins of hereditary novelties: "It may be supposed, as by Lamarck in the case of animals, that the mental state, and especially the desires of the organism, possess the power of producing mutations of such a kind, that these desires may be more readily gratified in the descendants" (Fisher 1930 [1958]:12). Naturally, it was easy for Fisher to dismiss this "Lamarckian" explanation of new variations.

Haldane's book, which was based on a lecture series, is an easy read. Wisely, he left the mathematics to an appendix. In the preface (Haldane 1932a:vi), he wrote that he particularly wanted "to prove that mutation, Lamarckian transformation, and so on, cannot prevail against natural selection of even moderate intensity," and he went on to show how evolution can be explained by the interplay between gene mutation, migration, and natural selection. However, Haldane didn't rule out everything else. He certainly had no enthusiasm for soft inheritance, describing, for example, experiments with *Drosophila* which showed that flies' eye size and tendency to move toward light were unaffected by being kept in darkness for sixty-nine generations. But he left the door open: he couldn't find fault with MacDougall's experiments with rats, which seemed to show the inheritance of induced behavioral change; he noted that Weismann's germ line–soma argument against the inheritance of acquired characters did not apply to plants, which had no segregated soma; he discussed non-Mendelian, cytoplasmic inheritance in plants; and he did not dismiss work suggesting that melanism might be induced by lead and magnesium salts. He also recognized that changes in chromosome structure and number had been significant in evolution.

In spite of the differences in style and focus of their books, Haldane and Fisher had things in common. Both believed in the importance of mathematics. Fisher hammered this home in his preface: genetics, he claimed, had its origin in the work of "a young mathematician, Gregor Mendel" (Fisher 1930 [1958]:viii), and all biologists would benefit from being trained in mathematical thinking. Haldane, too, put emphasis on mathematics, stating at the end of the appendix that if the history of science is an adequate guide, the permeation of biology by mathematics would continue (Haldane 1932a:215). He was right: other branches of biology, such as ecology, were also gradually adopting a more mathematical approach (see Elton 1938; Thorpe 1940). There seems to have been a feeling that mathematics made biology more "scientific," more like chemistry and physics, and, as quantification became more important over the next few decades, respect for descriptive biology decreased. In this atmosphere, there was little profit in worrying about the fuzzy findings and one-off results that were likely outcomes of soft inheritance.

Something else that Fisher and Haldane had in common was that neither had a degree in biology. Nevertheless, both had biological interests that went well beyond theoretical population genetics. Fisher was a statistician at Rothamsted Experimental Station, but from the mid-1920s he collaborated closely with E.B. Ford, a traditional English naturalist interested in butterflies. *Mendelism and Evolution* (Ford 1931) was one of the first books to bring the mathematical genetics of Fisher and others to a wide audience. Fisher continued to interact with Ford and his students for many years, and in the two decades following the Second World War these scientists made ecological genetics one of the strongest disciplines in British evolutionary biology.

Haldane's interests were different. Until 1932 he was reader in biochemistry at Cambridge. Besides *The Causes of Evolution*, his publications around this time included another book, *Enzymes* (Haldane 1930), and a lengthy paper titled "The Time of Action of Genes, and Its Bearing on Some Evolutionary Problems" (Haldane 1932b). This paper describes how genes can act at any stage in the life cycle—in the gametes, the embryo, the immature organism, and even the next generation (through maternal effects) or previous generation (through effects of the fetus on the mother).

Consequently, if the time of action of genes is altered by mutation, it could have profound effects on evolution, resulting, for example, in neoteny or paedogenesis. Haldane suggested how certain paleontological data could be explained in these terms.

Given the breadth of interests evident in Fisher's and Haldane's publications, I see no reason for thinking that there was a rift between the evolutionary thinking of geneticists and of other biologists in Great Britain in the early 1930s. Nor do I see any evidence that embryology and development were sidelined. The contents of *Evolution: The Modern Synthesis* (Huxley 1942) reinforce this view. Before 1942, Huxley had authored or coauthored books and papers about behavior, genetics, experimental embryology, sexual selection, relative growth, and systematics. His *Modern Synthesis* incorporates all these topics, and he also describes possible evolutionary consequences of mutations affecting gene-controlled development in some detail. As would be expected, he considered the evidence for soft inheritance, but he dismissed almost all of it. For Huxley, natural selection was sufficient to explain adaptive evolution.

Although a strong adaptionist, Huxley gave a lot of space to Cyril Darlington's work. Darlington had made his mark early in life with two books dealing with chromosomes and evolution (Darlington 1932, 1939). Using a mass of observations on chromosome numbers, structures, and behaviors (mainly in plants), he unraveled many aspects of chromosome mechanics. He also showed that the genetic–reproductive system has itself evolved. There is not one system of heredity, but many: asexual systems; sexual systems that lead to inbreeding; systems that force outcrossing; systems with a high, or low, rate of crossing-over; haplodiploid systems; and so on. This has evolutionary consequences. It affects, among other things, the likely occurrence and survival of chromosome aberrations and duplications, hybridization and polyploidization—the types of sudden genetic changes that can lead to rapid speciation. Darlington pointed out that changes in chromosomes and reproductive systems survive not because they benefit the individual in which they arise, but because they benefit posterity, sometimes very remote posterity. Consequently, they cannot be the result of any Lamarckian, direct adaptive response to the environment.

Although Darlington's cytological work was frequently cited, for reasons that are not clear to me his ideas about genetic systems were never fully integrated into mainstream evolutionary biology, and interest in them waned in postwar England. However, showing that there was more to evolution than slow, gradual adaptation through natural selection was important. There was a quite widespread feeling in Great Britain in the 1930s that there were things missing from the Mendelian–Darwinian version of evolutionary theory. It can be seen in *The Variation of Animals in Nature* (Robson and Richardson 1936). In this widely read book, the authors, both taxonomists, evaluated the evidence for selection in natural populations. They concluded it was very meager. They also thought that some of the assumptions made in Fisher's and Haldane's mathematical treatments of Darwinism (e.g., that mutation frequencies were low) were unwarranted. But they were equally critical of other evolutionary explanations, including soft inheritance and orthogenesis. What particularly worried them was that in the vast majority of cases, the characters that distinguish species had no known adaptive significance. Clearly, they concluded, there was much more to be discovered about the mechanisms of evolution.

One of the people who suggested an additional mechanism was Conrad Waddington, an embryologist and geneticist whose early work had been in paleontology. In "Canalization of Development and the Inheritance of Acquired Characters," Waddington (1942) proposed a mechanism, later called "genetic assimilation," through which induced adaptive characters can become inherited characters. He illustrated it with an example from Robson and Richards (1936)—the sternal and alar callosities of the ostrich. These skin thickenings are undoubtedly beneficial, preventing abrasive damage when the animal is squatting. They are present in the embryo, so they are inherited, but if they were not there, almost certainly they would form in response to abrasion. So, has an acquired developmental response become an inherited character? Or are the inherited callosities merely the result of selection of chance mutations that happened to cause skin thickening in the right place? Waddington's explanation was more subtle. He suggested that inherited callosities are the result of selection for the genetically controlled developmental capacity to respond to rubbing: those individuals that produced the most appropriate calluses in the right place survived best. Gradually, the response became "canalized"—through selection it became adjusted so that the skin thickened after even light abrasion, and eventually no stimulus at all was required for it to form.

Later, in the 1950s, Waddington illustrated this idea with his famous pictures of "epigenetic landscapes," a notion he had first introduced in his influential book *Organisers and Genes* (Waddington 1940). Development is represented as a ball rolling down a tilted landscape with branching valleys (figure 11.1). The steepness of a valley indicates how much the system can vary—how well buffered it is—and this depends upon the genes underlying it. Through natural selection, the steepness of the valleys, and hence the outcome of development, can be altered.

The Postwar Period: From 1945 to the 1960s

During the Second World War, most British biologists were directly or indirectly involved in war work, but they continued to meet and to write. Almost all were believers in Darwinian gradualism but acknowledged the importance of sudden chromosomal changes. Soft inheritance was generally ruled out.

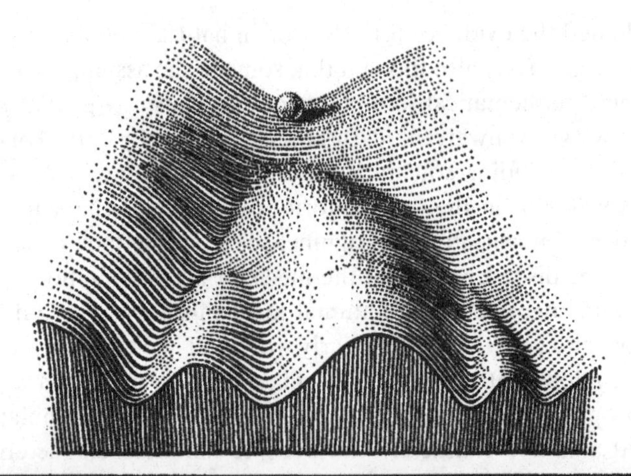

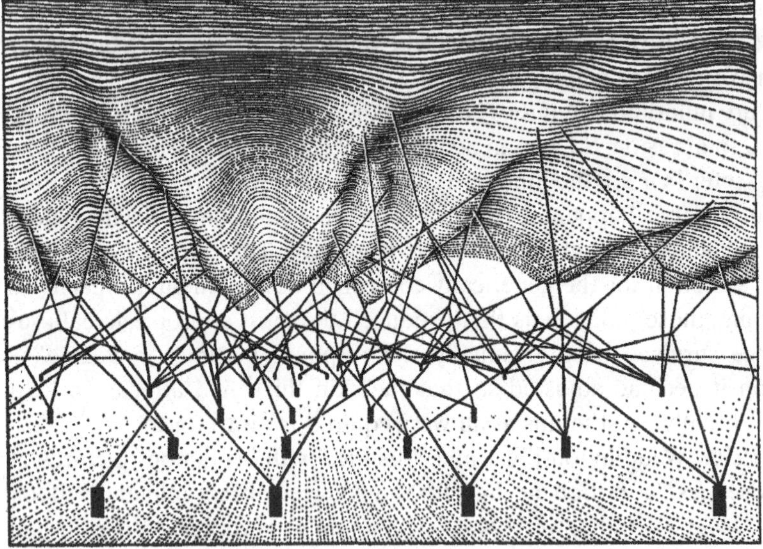

Figure 11.1
Waddington's pictures of epigenetic landscapes. In the upper picture, the branching valleys represent alternative developmental pathways descending from a plateau (the initial stage in the fertilized egg) to particular end states, such as part of the skin, brain, or heart. The lower picture shows the processes and interactions underlying this developmental landscape. The pegs represent genes, and the guy ropes represent the products of genes; these interact to determine the shape of the epigenetic landscape, the steepness of the valleys, and the final phenotype. Reproduced with permission of the publishers from C.H. Waddington, *The Strategy of the Genes*. London: Allen & Unwin; 1957:29, 36.

When the war ended, positive attempts were made in Great Britain to renew scientific contacts with the rest of the world. For example, the Genetical Society invited people from continental Europe to meet and review what had been happening in genetics research in each country (Lewis 1969). The British genetics scene began to change: people who had been refugees from Europe in the 1930s became established and influential figures; scientists from the physical sciences moved into biology; and, as a result of social and educational reforms, more working- and lower-middle-class people moved into academic posts. Genetics research in the universities expanded rapidly, but the biggest postwar expansion was in government-funded research institutes such as the Molecular Biology Unit at Cambridge, the Radiobiology Unit at Harwell, and the Animal Breeding and Genetics Research Organisation at Edinburgh (Lewis 1969; Fincham 1993). Biometry and ecological genetics remained strong, but new organisms, such as the fungus *Aspergillus* and the ciliate *Paramecium*, which have genetic systems that are very different from those of the plants and animals used in classical Mendelian genetics, began to have a substantial role in genetical research. Significantly, there were no British societies, journals, or departments devoted solely to evolutionary biology.

Perhaps because of the influence of the new people and new organisms, during the 1950s and early 1960s there seems to have been a greater interest in and willingness to accept "soft inheritance" and pseudo–Lamarckian phenomena than there had been before the war. At meetings of the Genetical Society many papers were presented describing cytoplasmic inheritance and various types of "weird" genetics involving unstable genes or environmental effects on heredity. The Society for Experimental Biology symposium on evolution held in 1952 shows the diversity of approaches to inheritance and evolution. I will highlight three papers. The first is by Waddington (1953), who claimed that mathematics had contributed nothing new to evolutionary thinking (something that Haldane, in the foreword to the symposium volume, hotly denied). Waddington maintained that current theory did not explain adaptation, did not explain macroevolution, and did not explain paleontological trends, whereas his developmental approach did. The second paper, by Hinshelwood (1953), is on bacterial adaptation, a process which, he claimed, was not always the result of selection of chance mutations; it could also occur through processes that modify chemical equilibria within most cells in the population (i.e., are "Lamarckian"). In the third paper, written as a postscript to the symposium volume, Danielli (1953) speculated that the internal environment of the cell might produce gradual quantitative changes in the chemical composition of the gene, which eventually would cause a qualitative change in function that simulates sudden mutation. Danielli also suggested that genetic material might flow from one species to another.

In his foreword to the symposium volume, Haldane summed up the state of evolutionary theory in Great Britain at the time:

> To sum up, then, a number of workers are groping from their own different standpoints towards a new synthesis, while producing facts which do not fit too well into the currently accepted synthesis. The current instar of the evolution theory may be defined by such books as those of Huxley, Simpson, Dobzhansky, Mayr and Stebbins. We are certainly not ready for a new moult, but signs of new organs are perhaps visible. The papers here collected represent points of view too diverse to be capable of coadaptation at present. They certainly achieved less unity than did the symposiasts of Princeton. This is not a bad sign. It points forward to a broader synthesis in the future. (Haldane 1953:xviii–xix)

The range of opinions in British evolutionary biology in the 1950s can also be seen in *A Century of Darwin* (Barnett 1958), a volume which celebrated various aspects of Darwin's work—speciation, corals, the descent of man, sexual selection, and so on. Two of the fifteen contributors were American; the rest were British. One of the American contributors, Theodosius Dobzhansky, is worth quoting briefly because of the sign he gave us of things to come:

> Heredity is, in the last analysis, self-reproduction. The units of heredity, and hence of self-reproduction, are corpuscles of macromolecular dimensions, called genes. The chief, if not the only, function of every gene is to build a copy of itself out of the food materials; the organism, in a sense, is a by-product of this process of gene self-synthesis. (Dobzhansky 1958:21)

One of the most interesting papers in the volume is by Donald Michie, who argued that Darwin's ideas on heredity were "one hundred and ten years ahead of their time" (Michie 1958:56). Darwin had a two-way theory of heredity, wrote Michie: as well as the germplasm determining the form of the body, pangenesis allowed environmental influences on the body to be transmitted to the germplasm. Weismann and those who followed him, including the founding fathers of genetics, had a one-way view: the germplasm determines the body form but is not affected by it, so environmental influences on the body are not transmitted to future generations.

Michie suggested that genetics had gone through two stages. The first was classical chromosomal genetics. It was based on organisms that could be studied using the techniques of Mendelian genetics, ignored asexual reproduction, and had little to say about how genes produced phenotypes. In the second stage, which Michie identified with the postwar period, some of the findings that didn't fit the classical Mendelian model were accommodated. The cytoplasm and cytoplasmic particles were brought in to explain the inheritance of differentiated cell types; the asexual propagation of variant clones of protozoa, fungi, and bacteria; and various atypical phenomena in sexual crosses. There were thus two systems of heredity—the classical, one-way, chromosomal system in the nucleus, and a two-way, cytoplasmic system, which is responsive to the environment. Through the latter, environmentally induced traits (acquired characters) can be stably inherited through many asexual generations.

Now, Michie proclaimed, it was time for the third stage in genetics: recognition that the germplasm can be influenced by the soma. The evidence he cited included work showing (1) that when the nuclei from differentiated cells of frog embryos are transplanted into enucleated eggs, they are progressively less able to support development; (2) that constructing *Amoeba* pseudohybrids, which have the nucleus of one species and the cytoplasm of another, causes heritable changes in both nucleus and cytoplasm; (3) that genetic material can be moved from one bacterial cell to another by phages; and (4) that when scions of one variety of tomato are grafted onto the rootstock of another, some of the rootstock's characters appear among the progeny of the graft. Such evidence convinced Michie that a third stage in genetics, one that acknowledged the two-way relationship between the soma and the germ line, was just beginning. He was wrong, of course, even though throughout the 1950s and 1960s experimental data that neither Mendelian genetics nor cytoplasmic inheritance could explain mounted. For example, work with peas (Highkin 1958), with flax (Durrant 1962), and with tobacco (Hill 1965) showed the inheritance of environmentally induced variations, and the work of McClintock (1951) demonstrated that genes can move around in the genome.

What, then, hindered further progress toward the third stage in genetics? I believe that there were several reasons:

1. The idea that inheritance was soft was associated with Lysenko and the Soviet Union. Although Great Britain didn't have McCarthyism, the Cold War affected British biologists, and it would have been unwise, particularly for young people, to give Lamarckian interpretations to their data.

2. The problems that many geneticists were studying were unlikely to lead to investigations of soft inheritance. For example, I was one of a number of *Drosophila* geneticists studying radiation-induced mutation rates. The nature of our experiments was such that mutations had to be clear-cut—in my lab, in order to be counted, a new mutation had to show itself cleanly in two successive generations. "Iffy" ones were not counted or investigated further. This was true of much work in genetics—only cleanly segregating genes were studied. Bateson (1908) had told us, "Treasure your exceptions," but most of us of didn't.

3. Animal geneticists were studying a very limited group of organisms, mainly insects and mammals, which had distinct and relatively early-segregating germ lines. Most evidence for non–Mendelian inheritance came from microorganisms, which have no germ line, and plants, where some somatic cells can become germ cells. The separation and lack of interaction between disciplines in British universities meant that zoologists, who had the greatest influence on evolutionary thinking, were often unaware of botanists' and microbiologists' findings.

4. By the mid-1960s, the advocacy of the world's leading evolutionary biologists had convinced people that no new interpretations were needed. Selection and competition experiments in the lab, and studies of polymorphisms in natural populations of snails, butterflies, moths, and *Drosophila*, were showing that selection, drift, and other processes worked more or less as the algebra of Fisher, Haldane, or Wright said they should. The Mendelian–Darwinian version of evolutionary theory was deemed to be sufficient.

5. There were no mechanisms to explain soft inheritance.

Inheritance Hardens in Evolutionary Biology as It Softens Elsewhere

During the 1960s and 1970s, two lines of research affected evolutionists' views about soft inheritance. First, theoretical work on the evolution of social behavior and altruism, traits that were difficult to explain in conventional Darwinian terms, led increasingly to the gene, rather than the individual, being seen as the focus of selection. Such a view of evolution, exemplified by *The Selfish Gene* (Dawkins 1976), makes sense only if inheritance is assumed to be hard. Second, molecular biology influenced evolutionary ideas. One consequence was that although Waddington embraced molecular biology with enthusiasm, his own work on genetics and development, which was neither quantitative nor molecular, began to be seen as woolly and old-fashioned. Biologists lost interest in it. DNA, the genetic code, transcription, and translation were far more exciting.

How molecular genetics affected evolutionists' ideas about soft inheritance can be seen by comparing the first and second editions of *The Theory of Evolution* (Maynard Smith 1958, 1966). In the first edition, Maynard Smith discusses and illustrates unorthodox, non-Mendelian inheritance in a chapter titled "Heredity." In the second edition, which incorporates molecular genetics, the illustration is dropped, and the material is covered briefly in a new chapter, "Weismann, Lamarck, and the Central Dogma." "Central dogma" is the term Crick (1958) used for his hypothesis that information can flow from nucleic acids to proteins, but not from proteins to nucleic acids. Maynard Smith shows the parallels between the central dogma and Weismann's germ line–soma segregation concept in a figure, and declares, "The greatest virtue of the central dogma is that it makes it clear what a Lamarckist must do—he must disprove the dogma" (1966:66). In other words, to prove that soft inheritance takes place, it has to be shown that it involves changes in DNA sequences. Yet, five pages later, when discussing the inheritance of determined states in cell lineages, Maynard Smith acknowledges that there is more to heredity than DNA. He writes, "The view generally taken by geneticists of differentiation, when it is not simply forgotten, is that the changes involved are too unstable to be dignified by the name 'genetic', or to be regarded as important in evolution. I tend to share this view, although I find it difficult to justify" (1966:71).

In spite of Maynard Smith's reluctance to see the inheritance of differentiated states as truly genetic, studies of mammalian cells in culture, serially transplanted *Drosophila* imaginal discs, X-inactivation, position effects, paramutation, and other developmental phenomena had already convinced many geneticists that phenotypic differences between cells with identical genes could be stably transmitted, and they were already speculating about the mechanisms behind this epigenetic, soft inheritance (e.g., see Ephrussi 1958; Mather 1961). The suggestion, made in 1975, that DNA methylation played a role in the control and inheritance of differentiated cell states kick-started new lines of research (see chapter 21 in this volume), and cell heredity became an integral part of developmental and molecular genetics. However, Maynard Smith was right about evolutionists' attitude: it was not until much later—the late 1980s—that soft (epigenetic) inheritance began to creep back into evolutionary thinking.

Acknowledgment

I would like to thank the staff of the John Innes Centre Library and Archive Service for their help with my studies of the archives of the Genetical Society and other material.

References

Barnett SA, ed. *A Century of Darwin.* London: Heinemann; 1958.

Bateson W. *The Methods and Scope of Genetics: An Inaugural Lecture Delivered 23 October 1908.* Cambridge: Cambridge University Press; 1908.

Crick FHC. On protein synthesis. In: *The Biological Replication of Macromolecules.* Symposia of the Society of Experimental Biology. Cambridge: Cambridge University Press; 1958:138–163.

Danielli JF. Postscript. On some chemical and physical aspects of evolution. In: *Evolution.* Symposia of the Society of Experimental Biology. Cambridge: Cambridge University Press; 1953:440–448.

Darlington CD. *Recent Advances in Cytology.* London: Churchill; 1932.

Darlington CD. *The Evolution of Genetic Systems.* Cambridge: Cambridge University Press; 1939.

Darlington CD. *Darwin's Place in History.* Oxford: Blackwell; 1959.

Dawkins R. *The Selfish Gene.* Oxford: Oxford University Press; 1976.

Dobzhansky T. Species after Darwin. In: *A Century of Darwin.* Barnett SA, ed. London: Heinemann; 1958:19–55.

Durrant A. 1962. The environmental induction of heritable change in *Linum.* Heredity 17: 27–61.

Elton C. Animal numbers and adaptation. In: *Evolution: Essays on Aspects of Evolutionary Biology.* De Beer GR, ed. Oxford: Clarendon Press; 1938:127–137.

Ephrussi B. 1958. The cytoplasm and somatic cell variation. J Cell Comp Physiol. 52(suppl.): 35–53.

Fincham JRS. 1993. Genetics in the United Kingdom—the last half-century. Heredity 71: 111–118.

Fisher RA. *The Genetical Theory of Natural Selection*. Oxford: Clarendon Press; 1930. 2nd ed. New York: Dover; 1958.

Ford EB. *Mendelism and Evolution.* London: Methuen; 1931.

Gould SJ. *The Structure of Evolutionary Theory.* Cambridge, MA: Belknap Press of Harvard University Press; 2002.

Haldane JBS. *Enzymes*. London: Longmans, Green; 1930.

Haldane JBS. *The Causes of Evolution*. London: Longmans, Green; 1932a.

Haldane JBS. 1932b. The time of action of genes, and its bearing on some evolutionary problems. Am Nat. 66: 5–24.

Haldane JBS. 1953. Foreword. In: *Evolution*. Symposia of the Society of Experimental Biology. Cambridge: Cambridge University Press; 1953:ix–xix.

Highkin HR. 1958. Transmission of phenotypic variability within a pure line. Nature 182: 1460.

Hill J. 1965. Environmental induction of heritable changes in *Nicotiana rustica.* Nature 207: 732–734.

Hinshelwood CN. 1953. Adaptation in micro-organisms and its relation to evolution. In: *Evolution*. Symposia of the Society of Experimental Biology. Cambridge: Cambridge University Press; 1953: 31–42.

Huxley JS. *Evolution: The Modern Synthesis.* London: Allen & Unwin; 1942.

Lewis D. The Genetical Society—the first fifty years. In: *Fifty Years of Genetics*. Jinks J, ed. Edinburgh: Oliver and Boyd; 1969:1–7.

Mather K. 1961. Nuclear materials and nuclear change in differentiation. Nature 190: 404–406.

Maynard Smith J. *The Theory of Evolution*. Harmondsworth, UK: Penguin Books; 1958. 2nd ed. 1966.

Mayr E. Prologue: Some thoughts on the history of the evolutionary synthesis. In: *The Evolutionary Synthesis: Perspectives in the Unification of Biology*. Mayr E, Provine WB, eds. Cambridge, MA: Harvard. University Press; 1980:1–48.

Mayr E. *The Growth of Biological Thought.* Cambridge, MA: Belknap Press of Harvard. University Press; 1982.

Mayr E, Provine WB, eds. *The Evolutionary Synthesis: Perspectives in the Unification of Biology.* Cambridge, MA: Harvard University Press; 1980.

McClintock B. 1951. Chromosome organization and genic expression. Cold Spring Harb Symp Quant Biol. 16: 13–47.

Michie D. The third stage in genetics. In: *A Century of Darwin*. Barnett SA, ed. London: Heinemann; 1958:56–84.

Robson GC, Richards OW. *The Variation of Animals in Nature*. London: Longmans, Green; 1936.

Thorpe WH. Ecology and the future of systematics. In: *The New Systematics*. Huxley JS, ed. Oxford: Clarendon Press; 1940:341–364.

Waddington CH. *Organisers and Genes.* Cambridge: Cambridge University Press; 1940.

Waddington CH. 1942. Canalization of development and the inheritance of acquired characters. Nature 150: 563–564.

Waddington CH. 1953. Epigenetics and evolution. In: *Evolution*. Symposia of the Society of Experimental Biology. Cambridge: Cambridge University Press; 1953: 186–199.

12 The Decline of Soft Inheritance

Scott Gilbert

Soft inheritance sounds opposed to hard fact, a weak analogue kind of inheritance as opposed to the digital inheritance of the chromosomes. But there was, in fact, a "soft," developmental version of inheritance during the early twentieth century, and evolution was understood by many leading biologists in terms of the rules of development. In 1893, the evolutionary champion Thomas Huxley wrote, "Evolution is not a speculation but a fact; and it takes place by epigenesis." He didn't say that it takes place by natural selection: he said that it takes place by epigenesis, by development. He was looking at a level different from the struggle of variations within a population. Rather, he was looking at the origin of variation.

Before World War I, the fields of genetics and development were united in the science of heredity (Coleman 1971; Gilbert, Opitz, and Raff 1996). There were many modes of inheritance, and one of the most popular and most researched was the mode called the morphogenetic field. One of the major research programs was *Gestaltungsgesetze*, the attempt to discover the laws by which ordered form was established at each generation.

A morphogenetic field was assumed to have definite boundaries, and to be made up from a collection of cells that were specified in a way that told them that they were members of the field. The interactions of these cells led to the formation of a particular organ, and at each generation, fields were established for the creation of body parts. The model organism in which the inheritance of morphogenetic fields was studied was the flatworm *Planaria*. The flatworm splits by binary fission—either transverse or lengthwise—and each part regenerates the other part. A head will regenerate a tail, and a tail will regenerate a head. In *Planaria*, the inherited information was embodied in the gradient that enabled the organism to form a head at one end and a tail at the other. Upon splitting, each half inherited the ability to make a whole and properly organized animal. This is reproduction, inheritance, and development all wrapped in one (see Child 1915, 1941). This notion of heredity was based on cells, not chromosomes, as the focal units of inheritance. Cells were seen as organizing into morphogenetic fields, and the morphogenetic fields then formed organs (Gilbert, Opitz, and Raff 1996; Gilbert 2003).

The genetics program of biology was originally in direct opposition to the concept of morphogenetic fields, and the demise of the morphogenetic field approach in the United States was linked to the rise of genetics and to the research and writings of an eminent American embryologist-turned-geneticist, Thomas Hunt Morgan. At the turn of the last century, there were two main researchers studying morphogenetic fields in the United States: C.M. Childs at the University of Chicago and Thomas Hunt Morgan, then at Bryn Mawr College. At the beginning of his career, Morgan studied embryogenesis and regeneration, and he published several important articles and books on these subjects. However, in the early twentieth century Morgan abandoned the study of regeneration and lost belief in the ability of embryogenesis to serve as the scientific basis for the understanding of evolution. He started studying the genetics of *Drosophila,* where inheritance of variant traits (e.g., red versus white eye color) was shown to obey Mendel's laws and suggested the involvement of nuclear chromosomes. At first, he attempted to place his new mutations into a framework of development (see Falk and Schwartz 1993). Each of his mutant phenotypes was seen to represent a different developmental step on the way to the final phenotype. A mutation represented a frustrated developmental pathway. However, by 1913, Morgan realized that he, too, was frustrated in his attempts to make a unified genetics of transmission and development. In 1926 he claimed that genetics and the study of ontogeny have to be separated (Morgan 1926; Gilbert 1998). He then called the *Planaria* program unscientific and blocked the attempts of Child and his students to publish their findings. He considered such work old-fashioned and not good science (Mitman and Fausto-Sterling 1992). Indeed, Mitman and Fausto-Sterling sternly conclude that Morgan was so adamant in his ridiculing the field notion because, in the 1930s, the morphogenetic field was an alternative to the gene as the unit of ontogeny.

In order to see how the field notion and soft, developmentally constructed inheritance were removed from the evolutionary synthesis in the United States, we need to go to one of the founding narratives of genetics. "The Rise of Genetics" (Morgan 1932b) could have been subtitled "The Decline of Embryology," because in this article, as well as in *The Scientific Basis of Evolution* (Morgan 1932a), Morgan portrays embryology, and soft inheritance with it, as a failed research program. Genetics, he claims, is its victorious successor. Indeed, *The Scientific Basis of Evolution* was about what Morgan regarded as the *only* scientific basis for evolution—genetics. Everything else was the *unscientific* basis of evolution. Paleontology, morphology, and embryology were all considered to be "philosophical" old schools: "older speculative methods of treating evolution as a problem of history" (Morgan 1932a: 13). Genetics, however, "has brought that subject evolution an exact scientific method of procedure" (Morgan 1932b: 287).

Morgan tells a story, which would be amplified later by Dobzhansky and Mayr, about the rise of genetics and the fall of embryology and the developmental modes of inheritance. This story follows what philosophers of religion call the supersessionist paradigm, the narrative myth by which the early Christians claimed to have superseded the Jews, and the narrative by which Protestants later claimed to have evolved or developed past the Catholics. Morgan starts by redefining embryology as a science of gene expression (something that embryologists had not looked at), and having redefined it in such a way that it can't succeed, claims that genetics has superseded it. Moreover, all the things that embryology wanted to explain and had failed to do, genetics could do (Gilbert 1998).

When Morgan claimed in 1926 that heredity had to be split into genetics and embryology, he redefined both areas in terms of genes. Genetics was the science of gene transmission, and embryology was the science of gene expression. Then he claimed that the embryologists could not bridge the gap between the genes and the trait, which, in fact, *was not* what embryologists wanted to do. Embryologists were dealing with morphogenetic fields and cells, not with genes. (Indeed, many embryologists were distinctly anti-genetic and felt that the cytoplasm, not the genes, directed development.) Embryology, Morgan (1932a) said, ran a while after false gods and landed in a maze of ontological subtleties. Where embryology went into metaphysical subtleties, genetics went into concrete chemistry and math. It is noteworthy, however, that in the 1930s the gene was not any more "material" than the field. Neither "field" nor "gene" had been directly observed. Both were postulated on the basis of results of experimental data, and both sought to explain inheritance.

Mendelian genetics was very successful in explaining the inheritance of many traits, but there were hereditary phenomena that did not fit into its framework. It is fitting to recall that this year (2009) is the centenary of Wolterek's paper on phenotypic plasticity and the notion that what is inherited are potentials for development, and those potentials can be impacted by the environment. The example studied by Wolterek was the water flea *Daphnia* (Woltereck 1909). It develops in different ways in different environments: when the individual lives in a safe environment, it has a normal round head, but in an environment inhabited by predators it grows a protective helmet. And this developmental response is passed on to the next generation. It is not easy to explain this mode of inheritance within the classical framework of chromosomal inheritance. But these kinds of findings can be conveniently ignored if they do not fit into one's theoretical scheme. And they were often ignored. As Benkemoun and Saupe (2006) commented when discussing past work on the genetics of fungi, "exceptions" were routinely autoclaved. We do not treasure our exceptions, as William Bateson urged us to do. We autoclave them.

The Cold War and the rise of Lysenkoism in the Soviet Union provided much of the context in which genetics flourished and in which "soft inheritance" was marginalized or discredited (see Gilbert and Epel 2009). Due to the possible mutational aspects of radiation, the Atomic Energy Commission was responsible for funding many projects of population genetics in the United States, and in 1959 the American Genetics Society hired a public relations team to spread the good news of genetics: not to bad-mouth Lysenkoism, because the American geneticists feared that their colleagues in Russia would be hurt that way, but to tell that Mendel plus Darwin gives you evolution (Wolfe 2002; Gormley 2007.) Clearly, to stop a competing research program, you need not kill the scientists; killing their research funding will do.

And, of course, one can also write a historical narrative into which the competing data do not fit. This, as we saw, was one of the strategies of Morgan, soon followed by Dobzhansky and Mayr (Sapp 1987). Dobzhansky (1937) wrote a history of evolutionary biology, largely the history of experimental population genetics, saying that evolution is studied by looking at changes in gene frequency, and concluded by 1951 that "evolution is a change in the genetic composition of a population. The study of the mechanisms of evolution falls within the province of population genetics" (Dobzhansky 1951: 16). So we have a proper subset. Evolution is a proper subset of the mathematics of population genetics.

While Morgan and Dobzhansky wrote developmental mechanisms out of the history of evolution, Ernst Mayr wrote embryology and soft inheritance out of the philosophy of evolution (Mayr 1966, 1982). Embryology was seen to be mired—pardon the pun—in typological thinking rather than population thinking. It was, Mayr claimed, essentialist. But, as Polly Winsor (2006) has shown, in order to keep evolutionary biology focused on intrapopulation processes, Mayr invented a dichotomy that had actually been resolved well before the time of Darwin. Although Mayr did not think that evolution was merely a matter of changes in gene frequencies, his analysis led him to focus only on evolutionary/genetic changes within species. He maintained that evolution could be studied without paying any attention to the mechanisms by which genotypes generate phenotypes (Mayr 1982). And since evolution had no need for a theory of body construction, it had no need for the possibilities of soft inheritance.

We now can observe that inheritance can be affected by several means. DNA methylation can inactivate a gene as well as mutation, and gene methylation differences (epialleles) can be inherited from one generation to the next. Moreover, symbionts provide a parallel system of inheritance, and in many cases the symbionts can alter gene expression in their hosts or provide various proteins to their cells (see Jablonka and Raz 2009; Gilbert and Epel 2009; chapter 27 this volume). Here, variation can be provided by "soft inheritance." This analysis could not have been accomplished without restriction enzyme analysis, polymerase chain reaction, and high-throughput RNA analysis. It is therefore ironic that the most reductionist, the most analytical, molecular tools have actually ended up showing the necessity, and indeed the reality, of the soft inheritance that had been excluded from the study of evolution.

References

Benkemoun L, Saupe SJ. 2006. Prion proteins as genetic material in fungi. Fungal Genet Biol. 43,12: 789–803.

Child CM. *Individuality in Organisms.* Chicago: University of Chicago Press; 1915.

Child CM. *Patterns and Problems of Development.* Chicago: University of Chicago Press; 1941.

Coleman W. *Biology in the Nineteenth Century: Problems of Form, Function, and Transformation.* New York: Wiley; 1971.

Dobzhansky T. *Genetics and the Origin of Species*. New York: Columbia University Press; 1937.

Dobzhansky T. *Genetics and the Origin of Species.* 3rd ed. New York: Columbia University Press; 1951.

Falk R, Schwartz S. 1993. Morgan's hypothesis of the genetic control of development. Genetics 134: 671–674.

Gilbert SF. 1998. Bearing crosses: A historiography of genetics and embryology. Am J Med Genet. 76,2: 168–182.

Gilbert SF. *Developmental Biology,* 7th ed. Sunderland, MA: Sinauer Associates; 2003.

Gilbert SF, Epel D. *Ecological Developmental Biology.* Sunderland, MA: Sinauer Associates; 2009.

Gilbert SF, Opitz JM, Raff RA. 1996. Resynthesizing evolutionary and developmental biology. Dev Biol. 173,2: 357–372.

Gormley, M. L. C. Dunn and the Reception of Lysenkoism in the United States. PhD Thesis, University of Pennsylvania.

Jablonka E, Raz G. 2009. Transgenerational epigenetic inheritance: Prevalence, mechanisms, and implications for the study of heredity and evolution. Q Rev Biol. 84,2: 131–176.

Mayr E. *Animal Species and Evolution.* Cambridge, MA: Belknap Press of Harvard University Press; 1966.

Mayr E. *The Growth of Biological Thought.* Cambridge, MA: Belknap Press of Harvard University Press; 1982.

Mitman G, Fausto-Sterling A. Whatever happened to Planaria? C. M. Child and the physiology of inheritance. In: *The Right Tool for the Right Job: At Work in Twentieth-Century Life Sciences*. Clarke AE, Fujimura JH, eds. Princeton, NJ: Princeton University Press; 1992:172–197.

Morgan TH. *The Theory of the Gene.* New Haven, CT: Yale University Press; 1926.

Morgan TH. *The Scientific Basis of Evolution.* New York: W.H. Norton; 1932a.

Morgan TH. 1932b. The rise of genetics. Science 76: 261–288.

Sapp J. *Beyond the Gene: Cytoplasmic Inheritance and the Struggle for Authority in Genetics*. Oxford: Oxford University Press; 1987.

Winsor MP. 2006. The creation of the essentialism story: An exercise in metahistory. Hist Philos Life Sci. 28: 149–174.

Wolfe, A. 2002. Speaking for Nature and Nation: Biologists as Public Intellectuals in Cold War Culture. PhD Thesis. University of Pennsylvania.

Wolterek R. 1909. Weitere experimentelle Untersuchungen über Artveränderung, speziell über das Wesen quantitativer Artunterschiede bei Daphniden. Verhandl Tsch Zool Gesell. 110–172.

13 Why Did the Modern Synthesis Give Short Shrift to "Soft Inheritance"?

Adam Wilkins

I am not a historian by training but a geneticist. Nevertheless, I am going to present an interpretation of a matter of scientific history, one that concerns genetics. The matter is obscure because, unfortunately, there is a dearth of testimonials from the individuals involved. Nevertheless, it is important.

The question to be addressed is why the architects of the Modern Synthesis dismissed Lamarckism, or soft inheritance, so completely. The term "soft inheritance" was used by Ernst Mayr to describe a hypothetical type of unstable inheritance in which variations are malleable and can be produced initially by the effects of the environment, or by use or disuse, but are passed on to offspring in a less robust and predictable manner than "hard" Mendelian factors (Mayr 1982).

The Modern Synthesis was shaped by many people and took form over an extended period of about thirty years. Even to begin to answer the question posed above, it will be helpful to take a brief look at its history. I have broken this history into four main stages (Wilkins 2008) and will give a capsule description of each. The first stage involved the maturation of classical genetics and the development of population genetics. It was essentially a fourteen-year period, between 1918 and 1932, when the main ideas of classical transmission genetics, based on Mendelian inheritance, were put into a framework of population thinking to create the science of population genetics. Although studies of heredity in populations, biometrical studies, had existed since the late nineteenth century, these studies dealt with the distribution and inheritance of phenotypes, not genotypes (Provine 1971), and the new science of population genetics did not build on the work of the biometricians. The names of the three critical figures who established it will be familiar to all biologists: Ronald A. Fisher (1890–1962), J. B. S. Haldane (1892–1964), and Sewall Wright (1890–1989). They showed, using mathematical models, how the genetic structure of a population can change through selection, random genetic drift in small populations, migration, mating patterns, and mutation pressure. The models showed that natural selection acting on Mendelian genes could explain cumulative adaptive evolution. The interaction of natural selection with these other factors could, in theory, explain the specific patterns of adaptations observed in specific populations.

The second stage was the integration of the population-based way of thinking with the work of naturalists and systematists. The two key figures here were Theodosius Dobzhansky (1900–1975) and Ernst Mayr (1904–2005), and they contributed strikingly early in their careers, with two major books: *Genetics and the Origin of Species* (Dobzhansky 1937) and *Systematics and The Origin of Species from the Viewpoint of a Zoologist* (Mayr 1942). Their titles were, of course, inspired by Darwin's *The Origin of Species* (to give it its abridged title, by which it is best known). Dobzhansky integrated results of studies in natural populations with the insights from experimental Mendelian genetics and theoretical population genetics (Dobzhansky 1937), thus bringing into the picture the evolution of populations in nature. Mayr marshaled much data, for many different species, and developed ideas that showed how new species are formed when populations become geographically isolated and gene flow between the separated subpopulations ceases (Mayr 1942).

The third stage was very close chronologically, but it was nevertheless distinct, and involved one key individual: George Gaylord Simpson (1902–1984). He presented his thinking on evolution in his book *Tempo and Mode in Evolution* (Simpson 1944). In this book, Simpson integrated paleontological data with ideas from population genetics. He argued that macroevolution is an extension of microevolution, and therefore relatively short-term evolutionary changes in populations can be extrapolated to explain evolutionary processes above the species level, over much longer intervals.

By this time, the mid-1940s, much had been unified: genetics, natural populations and their genetics, systematics, and paleontology. George Ledyard Stebbins (1906–2000) added yet another tier with his book *Variation and Evolution in Plants* (Stebbins 1950), which brought plant biology into the Synthesis. I regard this as the fourth stage in the Modern Synthesis in the United States, although of course each of these stages had its own prehistory and involved many other contributors and contributions.

One might ask why, with so many participants, each of whom had his own distinct perspective on biology, soft inheritance got written out of the Synthesis. After all, as other chapters in this volume show, one can believe in both natural selection and soft inheritance. Indeed, Darwin himself believed in both, as did most nineteenth- and early twentieth-century biologists. There is nothing about believing in Darwinian natural selection that would lead one to automatically eliminate soft inheritance.

I think there are three main reasons for downplaying or excluding Lamarckian inheritance from the Modern Synthesis. The first is the legacy of belief in August Weismann's separation of germ line and soma. The second is the absence, from the

1920s through the 1940s, of any strong evidence for the inheritance of acquired characteristics, at least in animals (and animals were the focus of most of the people constructing modern neo-Darwinian theory in this period). The third reason, which usually is not given enough emphasis, is, in my opinion, of central importance. To have accepted Lamarckian soft inheritance would have meant undermining or weakening one of the key conceptual foundations of the Modern Synthesis. Let us look more closely at each of these three reasons.

First, there was the legacy of adhering to August Weismann's ideas about the discreteness of the germ line. Figure 13.1 presents a simplified view of the Weismannian belief about the separation of the germ line and the soma, which was adopted by the founders of the evolutionary synthesis. The basic assumption was that there is continuity between the germ lines of each generation, and that the germ line is

The view attributed to Weismann:

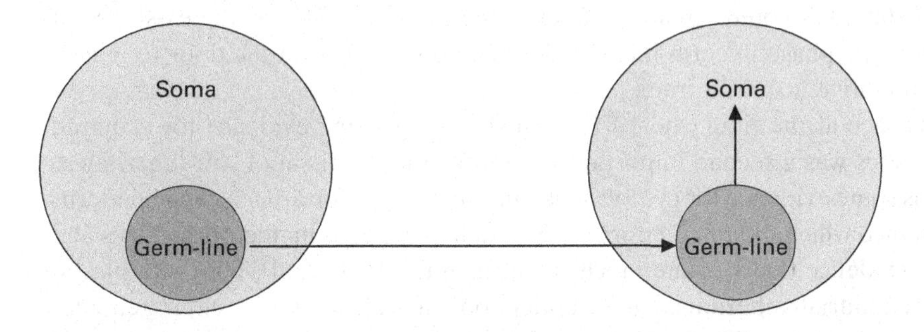

In reality, for most animal types,* the germ-line goes through a somatic phase

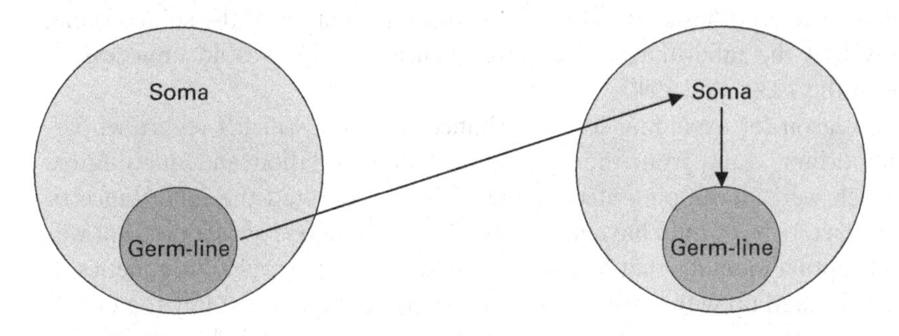

*see Extavour and Akam (2003)

Figure 13.1
Soma germ line relations. A simplified view of the Weismannian belief about the separation of the germ line and the soma (top), and a more realistic view that applies to most animals (bottom).

never derived from the soma. Hence, somatically acquired adaptations cannot be passed on to the next generation.

Today we know that in most animals, and of course in plants, the germ line actually does arise from somatic cells. In plants, the germ line is continuously generated from somatic cells. In most animals, too, as the detailed study by Extavour and Akam (2003) showed, the cells that will become the germ line in the embryo have a somatic phase. Every animal embryo starts with the fusion of two gametes, the sperm and the egg, and the zygote then starts undergoing cell divisions. Those cells that are initially formed are somatic cells; a portion of them becomes directed to becoming germ line cells. During the somatic phase the cells may adapt to new conditions, and these adaptations may pass to the germ line when these somatic cells become germ cells. This might seem to give an opening for Lamarckian inheritance, but we must be aware that in most animals, this happens only in the early embryo, when there are still relatively few cells. Thus, for such animals, it would be hard to imagine how adaptive alterations in somatic tissues of the adult could possibly influence the germ line. Hence, the existence of a short somatic phase in germ line development does not open the door to inheritance of adaptive acquired traits.

The absence at the time (the 1920s–1940s) of convincing evidence for acquired characteristics was a second important reason for the dismissal of soft inheritance. There was some evidence for cytoplasmic inheritance and for other strange patterns of inheritance which did not conform to Mendelian genetics in microorganisms and in plants, evidence that became much stronger in the 1950s and1960s (with plastid and mitochondrial inheritance now understood in terms of the small genomes carried by these organelles). However, there were no indisputable cases of inherited acquired characteristics in animals. So, with apologies to Laplace, who said that he did not need “this hypothesis” (God) to explain the formation of the solar system, one can say that the inheritance of acquired characteristics was an unnecessary hypothesis in the 1930s and 1940s.

The third reason for excluding soft inheritance, a reason which I regard as particularly important, came from the assumptions of population and quantitative genetics which were developing at the time. To have accepted the inheritance of acquired characteristics would have meant in effect turning one’s back on what was a major conceptual element that lay at the foundations of population genetics. It was the idea, backed up with a fair amount of experimental evidence by the end of the 1910s, that complex traits in animals and plants were underlain by the cumulative effects of multiple Mendelian factors. It sounds like a very simple point, but the nature of heredity was hugely controversial at the beginning of the twentieth century (Provine 1971).

It became clear early in the twentieth century that complex traits in animals and plants do not behave in heredity as Mendelian characters do, so one school of biologists, the biometricians (mentioned earlier in this account), argued against the importance of Mendelian genes in evolution. The great dispute between the Mendelian geneticists and the biometricians was resolved when it was realized that each character is the product of the activity of many Mendelian genes. This was the basis for the development of population genetics by Fisher, Haldane, and Wright. To have argued that the Lamarckian inheritance of acquired characteristics or any other form of soft inheritance is important in evolution would, in a sense, have undermined or made more trivial the very findings that were at the heart of population genetics: the crux of these findings was that heredity was not about the inheritance of *traits*, but about the inheritance of the numerous *genes*, each with small effect, that contributed to their construction. I think that this is a crucial point in understanding why Lamarckian inheritance never even got a look-in as the Modern Synthesis was being formulated.

The Modern Synthesis was, thus, essentially neo-Darwinian in the sense developed by Fisher, Haldane, and Wright. All inheritance was assumed to be hard and to involve changes in Mendelian genes. It also was assumed that mutation in genes and recombination between genomes occur independently of any adaptive value they might have, and that those events occur "at random," a somewhat ambiguous term. Chance and natural selection act on these genetic variations, and this determines which ones survive and spread in the population. Mutations that have deleterious effects will tend to be eliminated. Those that create some sort of adaptive advantage in particular niches will tend to be propagated and to spread in the population. From this point of view, evolution consists of the transformations in the frequencies of genotypes, with whole populations changing their genetic constitution over long periods of time.

It is interesting that many biologists in the 1930s and 1940s still adhered to soft inheritance, but once they realized that complex traits are underlain by multiple Mendelian factors, they usually dropped this belief. On two different occasions, I talked to two major figures in evolutionary biology, Ernst Mayr and John Maynard Smith, and they both told me that, yes, they were believers in Lamarckian inheritance until they saw the genetic findings. And for John Maynard Smith, this conversion was quite late historically (in the late 1940s). He was not trained as a biologist, and it was only after the Second World War, during which he had worked as an aircraft engineer, that he started thinking about biology. When he first came into biology, he assumed that the inheritance of acquired traits did occur. After examining the evidence, however, Maynard Smith decided that there was no evidence for the inheritance of acquired characteristics and dropped the idea. People often reinvent the past in ways that are subtly favorable to themselves, but since both Mayr and Maynard Smith were admitting to a belief that was—and largely is—regarded by evolutionists as heresy, there is no reason to doubt that they were accurately recounting this matter.

There may be an additional possible contributory factor in the exclusion of Lamarckism which is worth considering. If you are developing a new idea and you really want people to pay attention to it, it helps to have a counteridea, to be able to say "My idea, Y, is so much better than your idea, X." Lamarckian evolution was the strongest alternative scientific hypothesis to the twentieth-century neo-Darwinian view of evolution. So I think that, to some extent, Lamarckian inheritance has had the bashing that it has had because this strengthened the novelty and solidity of the Modern Synthesis interpretation. I think, however, that if we had then all the evidence that we have today, and that will be discussed in subsequent chapters, the neo-Darwinian synthesis would have had a somewhat different look in some of its basic propositions.

References

Dobzhansky T. *Genetics and the Origin of Species.* New York: Columbia University Press; 1937.

Extavour CE, Akam M. 2003. Mechanisms of germ cell specification across the metazoans: Epigenesis and preformation. Development 130,24: 5869–5884.

Mayr E. *Systematics and The Origin of Species from the Viewpoint of a Zoologist.* Cambridge MA: Harvard University Press; 1942.

Mayr E. *The Growth of Biological Thought: Diversity, Evolution, and Inheritance.* Cambridge MA: Belknap Press of Harvard University Press; 1982.

Provine W. *The Origins of Theoretical Population Genetics*. Chicago: University of Chicago Press; 1971.

Simpson GG. *Tempo and Mode in Evolution.* New York: Columbia University Press; 1944.

Stebbins GL. *Variation and Evolution in Plants.* New York: Columbia University Press; 1950.

Wilkins AS. *Neo-Darwinism.* In: *Icons of Evolution.* Regal B, ed. Vol. 2. Westport, CT: Greenwood Press; 2008: 491–515.

14 The Modern Synthesis: Discussion

The discussion that followed the three papers in this section dealt with the open questions the speakers raised. Although the discussion involved many fascinating issues, three main topics received special attention: (1) the neglect of data, ideas, and theories that did not fit the Modern Synthesis view; (2) the role of social/political factors, and in particular the effect of newly established institutional structures in the United States; (3) the influence of several experimental studies relevant to evolutionary theorizing, including the Luria-Delbrück experiments on the origin of bacterial mutations, the studies of melanism in the peppered moth, and the work on hemoglobin and sickle-cell anemia, which gave rise to the "globin paradigm" of evolution.

Neglected Ideas, Overlooked Theories: Darlington and the Plasmagene Theories

The Fading Influence of Cyril Darlington

Why were Darlington's ideas about evolution neglected? As Marion Lamb noted, his ideas were rarely discussed in the United Kingdom during the 1950s and 1960s, although his significant contributions to the chromosome theory were generally acknowledged. Indeed, when the history of the Modern Synthesis was being written by Mayr and Provine in the late 1970s, they found it very difficult to place the role of chromosome research within the evolutionary synthesis. They did think that it had played an important role, but it remained unclear precisely what that role had been. In the discussion, Oren Harman, author of *The Man Who Invented the Chromosome: A Life of Cyril Darlington*, suggested that this difficulty was indicative of the way Darlington was received. Harman suggested several factors that may have contributed to the neglect of Darlington's evolutionary views and the negligible impact his work had on the theorizing that was part of the Modern Synthesis:

1. Darlington put forward a complex model of evolution with different levels (or governments) of inheritance. If one compares Darlington's complicated diagrams, full of arrows indicating circular feedback relationships, with Fisher's elegant axioms and Haldane's plodding but accurate predictions, one gets an immediate feel for why Darlington was respected but not understood. Moreover, his models were difficult to test. Whereas Fisher's models could be tested by E. B. Ford and his students, the kinds of claims Darlington was making in *Recent Advances in Cytology* (1932) and *The Evolution of Genetics Systems* (1939) remained untestable.

2. By the mid-1940s, Darlington had left chromosome mechanics behind and become interested in cytoplasmic inheritance and the relationship of heritable cytoplasmic entities to nuclear determinants of heredity. According to Darlington, there were three levels of governance in the cell. The first was that of nuclear genes, the second was of genes belonging to entities such as chloroplasts (plastogenes), and the third was a "molecular system" of self-duplicating cytoplasmic particles (plasmagenes). For lack of a tradition and a mass of researchers on such topics in the United Kingdom, Darlington communicated primarily with French and German colleagues. There were few biologists in the United Kingdom who pursued these research directions; the British biological community was simply not interested.

3. Darlington did not fare well in the United States. When *Recent Advances in Cytology* came out in the United States in 1934, Stebbins, the country's leading evolutionary botanist, wrote a scathing review of the book, which he called "a masterpiece of mythogenesis," adding that it was "as informative on the content of Dr. Darlington's mind as it is uninformative on the subject of evolution." Darlington was shouted down at the International Congress of Genetics held in Ithaca, New York, in 1932. The American biologists thought little of his ideas, which were seen as speculative and unscientific.

4. By the time Watson and Crick came along, Darlington the chromosome man was thought of as obsolete. In the fall of 1953 he moved to Oxford to head the Botany Department, and the focus of his work became the writing of popular books on heredity, eugenics, and history from a biological–deterministic point of view.

5. Personal reasons contributed, too: as a young man, Darlington had no patron. He had worked almost entirely independently at the John Innes Institute, and had a falling out (for both professional and political reasons) with J.B.S. Haldane, the part-time head of genetical research, who was initially very friendly and appreciative of his work. Moreover, Darlington's well-known abrasiveness attracted few students who were interested in continuing his legacy.

In many ways, Harman concluded, one can say that by the mid-1950s Darlington had fallen off the map.

Hard and Soft Inheritance and the Demise of Plasmon and Plasmagene Theories

Hard inheritance and soft inheritance, the two terms that were originally coined by Darlington, tend to be identified with DNA and non–DNA inheritance, respectively. This is not, however, how these terms were always understood. As Eva Jablonka pointed out, Carl Lindegren (1966) argued that "hard inheritance" is any type of pattern-dependent, high-fidelity, hereditary reconstruction. He suggested that hard inheritance included, in addition to DNA replication, membrane reconstruction and the reconstruction of complex protein structures such as ribosomes, of basic non–DNA components of chromatin such as histones, and of certain cytoplasmic entities. Soft inheritance, Lindegren proposed, is associated with specific structures of chromatin—with "gene receptors." These "gene receptors," which he thought are coiled proteins, are specific to each gene, respond to environmental signals by changing their structure, and affect gene action. According to Lindegren, although they are very malleable, altered configurations of gene receptors could be inherited, and this might underlie the cases of soft inheritance described in the scientific literature. In some sense Lindegren anticipated the idea of the epigenetic inheritance of chromatin factors, such as patterns of protein marks.

Carl Lindegren was one of several prominent geneticists who, in spite of their scientific status, had very little lasting influence on the theory and practice of genetics and evolutionary biology. He was one of T. H. Morgan's students, and worked out the genetics of the bread mold *Neurospora*. In the course of this work Lindegren found that inheritance in *Neurospora* shows many non-Mendelian patterns, and this led him to suggest types of inheritance additional to the Mendelian nuclear one. He saw the general neglect of non-Mendelian inheritance in the United States as the result of anti-intellectual scientific indoctrination. In *The Cold War in Biology* (Lindegren 1966) he sharply criticized both American and Soviet indoctrination. He described in detail the ideas of scientists who were marginalized in the United States by the Morgan school, and highlighted the large sets of data that clearly pointed to the insufficiency of neo-Darwinian theories based on classical Mendelian genetics, as well as the need to address additional types of nuclear and cytoplasmic inheritance.

Another discussant, Jan Sapp, called attention to the fact that there were many non–Mendelian inheritance theories. Such theories were common and influential in Germany between the world wars. In 1926, Fritz von Wettstein in Germany applied the term *Plasmon* to the hereditary element of the plasm, and distinguished it from the *Genom*, the hereditary material in the chromosomes (a term suggested by Winkler in 1924). Wettstein created a moss hybrid (having maternal genome A and paternal genome B) and repeatedly backcrossed it to the paternal type (B), so the cytoplasm was maternal (A) and the nucleus almost entirely paternal (B). He showed that some of the characteristics of the mother, which donated only the cytoplasm, persisted for many generations, and hence that the cytoplasm must contain inheritable components. Similar experiments demonstrated the existence of cytoplasmic heritable components in many other plant species, in fungi, and in protists. The plasmon was divided by Michaelis, another important German botanist, into two components: the genes in the plastids (the chloroplasts), and other diffuse heritable elements in the cytoplasm. The experiments performed by Michaelis and his students were very complex and, unlike classical Mendelian genetic experiments, took many years to accomplish. Michaelis and his colleagues published their results on plant hybrids after twenty years of work.

Several plasmagene theories were developed in the United States and England after the Second World War. In the United States, Tracy Sonneborn, one of the country's most influential geneticists in the postwar period, described many types of non–Mendelian inheritance in the unicellular organism *Paramecium*, and suggested that some plasmagenes could mutate and interact with nuclear genes. As pointed out earlier, Darlington suggested a plasmagene theory, and Lindegren put forward a theory whereby heritable elements (cytogenes) moved between the cytoplasm and the nucleus and were able to replicate in the cytoplasm. Sol Spiegelman proposed that nuclear genes made replicas of themselves which were sent to the cytoplasm and, once there could replicate autonomously. These theories can be seen as a sort of precursor to the concept of messenger RNA, a kind of copy of the gene which gets to the cytoplasm and exerts its influence.

As Jan Sapp explained, these theories were associated with the elusive and long-standing problems of cellular differentiation in the face of nuclear equivalence. It was clear that Weismann was wrong on almost all accounts, and that his theory of devolvement was totally wrong—the genetic material in somatic cells did not disintegrate as differentiation proceeded. The genome was the same in all somatic cells: in skin cells, in liver cells, in kidney cells. Moreover, the genetically identical but phenotypically different somatic cells divided and retained their characteristics. One needed to understand not only cell differentiation but also cell heredity, and this was regarded as important by major and influential American biologists, for example, by Sewall Wright, one of the founders of population genetics. The question was: what regulates cellular differentiation and cell heredity. Where is the regulator? For many biologists this regulator was in the cytoplasm. So when evidence arose in Germany for cytoplasmic inheritance between the world wars, and in the United States and Europe after the Second World War, mainly as a result of experiments on microorganisms, this was seen as a theory of cellular differentiation. Of course the theory had implications for heredity in general, and therefore, inevitably, also for evolution.

Although most plasmagene theorists addressed the problem of differentiation, some tried to account for morphogenesis. To explain the regularity and species specificity of morphogenesis, plasmagene theorists proposed that there was a spatial property that was inherited and that guided the development of form, and they associated it with the cell cortex. The existence of a heritable, spatial organizing principle was experimentally demonstrated by Sonneborn and Beisson in 1965. They showed that an experimentally modified organization of cilia on the *Paramecium* cortex can be inherited through many generations. Such structural non–DNA inheritance was assumed to be important for multicellular organisms, too: the organization of the cortex of the egg was believed to direct the early stages of embryogenesis.

Sapp maintained that the plasmagene paradigm was wiped out by Jacob and Monod's operon theory. Regulation was explained as a nuclear phenomenon: genes could directly participate in cellular differentiation, and the model was integrated with other revolutionary discoveries in molecular biology. Monod wrote *Chance and Necessity* in 1971 to get rid of the dialectics in the cell: there were no dialectics in the cell—there was nuclear control. As a fierce Darwinian selectionist, Monod could explain not only heredity and development, but also all evolution, in terms of nuclear control.

There were additional reasons for the decline of plasmagene theories. It was found that some of the examples that were considered important evidence for plasmagenes—the Kappa elements that Sonneborn discovered in *Paramecium* (killer particles residing in the cytoplasm) and the Sigma factors of *Drosophila* studied by Charles-Louis L'Héritier in France—were infectious particles. They were therefore dismissed as parasites, examples of "infectious heredity," of little importance for development and evolution. The notion that "infectious heredity" is important is now returning to evolutionary thinking in a new context and with much supporting data, said Jan Sapp (see chapter 26 in this volume).

Reinforcing Institutional and Political Factors

Sam Schweber highlighted the importance of professionalization and institutional factors that shaped biological research and determined research priorities in the United States. There was a program to try to make biology into a basic science, as opposed to an applied one, for most of the geneticists in the United States were agricultural geneticists.

The Rockefeller Foundation played an important role in the transformation of genetics into a quantitatively oriented science. During the 1920s one of the things that the Foundation did in the United States was to support physics and mathematics, putting American physics and mathematics in the global spotlight. In 1932 the focus changed when Warren Weaver, a mathematician and engineer, became the head of the Natural Sciences Division of the Foundation and channeled a good deal of Rockefeller funds into biology and in particular into the Cold Spring Harbor Laboratory. In the summer of 1933 the first Cold Spring Harbor Symposia on Quantitative Biology were held there. According to the organizers of the summer events, "mathematicians, physicists, chemists and biologists, actively interested in a specific aspect of quantitative biology, or in methods and theories applicable to it, will be invited to carry on their work, to give lectures and to take part in symposia at the Laboratory." Weaver modeled these meetings on the Michigan Summer School, which had trained an entire generation of American physicists who played key roles in the Second World War.

After the Second World War, the Atomic Energy Commission supported genetics, not only the genetic work that was done in Hiroshima and Nagasaki, but also all the field and lab work of Dobzhansky, for example, including his experiments in South America. Cold War politics played a role in this process. The Genetics Society of America counteracted Lysenkoist propaganda by publishing *Genetics in the Twentieth Century* (1951), a project that was sponsored by the American Institute for Biological Sciences. The Rockefeller Foundation gave a big grant to the Genetics Society "to protect the freedom of science," and the money was used to disseminate information about genetics. Institutional factors, certainly in the United States, played a role in reorganizing and supporting new groups and new approaches.

Experimental Support for the Modern Synthesis View

The Modern Synthesis excluded certain approaches to heredity and evolution, but crucial to its spread and dominance were theoretical and experimental findings which supported its major claims. As indicated earlier, Jan Sapp talked about the importance of the operon model, and how it replaced the plasmagene theories of differentiation. In addition to this, three other studies that supported the Modern Synthesis view were discussed: the Luria and Delbrück experiments showing that mutations in the bacterium *E. coli* arose independently of environmental selecting conditions; Kettlewell's experiments on the peppered moth, which showed the effectiveness of selection in natural populations; and "the globin paradigm," as Scott Gilbert called the series of studies which explored the biology of sickle-cell anemia.

The Luria-Delbrück Experiments

Luria and Delbrück studied the occurrence of mutations that allowed bacteria to live in the presence of killer phages (bacterial viruses). Their goal was to determine whether resistant mutations were random with respect to the selecting agent (the phage) or, alternatively, were induced by the phage (and in this case could be seen as directed, "Lamarckian" mutations). The test they performed showed that the distribution of mutants on plates taken at different periods in the growth of the bacterial culture fit the hypothesis that the resistant bacteria were present before the exposure to the phage, and were not induced by their presence. This conclusion was confirmed by experiments performed later by Lederberg and Lederberg in the early 1950s, using a very different technique known as replica plating.

It is interesting that the original Luria and Delbrück experiments, performed in 1943, had very little effect on plasmagene theories. These experiments, as well as those of the Lederbergs, started to be widely cited as a conclusive disproof of environmentally induced mutations only when molecular studies became prominent about two decades later, by which time the Modern Synthesis had gained dominance in both the United States and Europe. As we now realize, the experiments were overinterpreted: the selective conditions were lethal, so there was no chance for the bacteria to adapt. Only preexisting mutations could be observed in these conditions. In experiments using nonlethal adverse conditions, Lindegren showed that when yeast cells that were unable to use a specific sugar were replica-plated onto selective media lacking this nutrient, after a few days they showed varied and nonidentical distributions of mutants, suggesting that some mutations arose after plating on the selective medium (summarized in Lindegren 1966). However, these experiments were generally ignored, and it was only in the late 1980s that similar experiments addressing the possibility of induced mutations were carried out (Jablonka and Lamb 1995).

Industrial Melanism and the Study of the Peppered Moth *Biston betularia*

Ehud Lamm argued that one should see the discussion of soft inheritance within the framework of other evolutionary debates, such as that surrounding the spread of the dark-colored form of the peppered moth, *Biston betularia*. The story of the evolutionary change in the moth is well known: the light-colored morph of the peppered moth was common in England until the industrial revolution, and the dark morph was very rare. During the industrial revolution, however, the dark morph became common and the light one dramatically decreased in frequency. Following the Clean Air Act in the United Kingdom in the 1960s, the light morph increased in frequency and the dark form declined. H. B. D. Kettlewell coordinated three surveys of morph frequencies at sites all over the United Kingdom, two in the 1950s and one in the 1960s. The hypothesis was that in areas where pollution had darkened the moths' resting sites, the darker moths were better camouflaged and less likely to be eaten by birds. Under less polluted conditions, the light-colored moths prevailed for similar reasons. Kettlewell showed that selective predation was indeed one reason for the frequency distribution of the moths. Earlier "Lamarckian" explanations, put forward in the 1920s by John S. Heslop-Harrison, had suggested that airborne pollutant particles had caused the mutation of genes for melanin production; however, these ideas were based on experiments that could not subsequently be reproduced, and by the 1950s his conclusions were rejected. Kettlewell's results were also significant in the debates about the relative importance of selection, mutation, migration, and drift in natural populations. Observations by Kettlewell himself and by others during the 1950s suggested that in addition to selective predation, other factors, such as drift and constraints on the moths' migration, must be added to explain the moths' distribution. Nevertheless, Kettlewell's work was generally considered to be a classic and nonproblematic demonstration of natural selection in action. It fit population genetics assumptions such as those of R. A. Fisher, and was used to argue for the general validity of neo-Darwinian evolution. As Scott Gilbert reminded us, in America one of the uses of Kettlewell's experiments was to counter creationism. Not accidentally, the review that Kettlewell wrote for the popular journal *Scientific American* was titled "Darwin's Missing Evidence" (Kettlewell 1959). It explicitly stated that genetics will show that Darwinism is correct. The peppered moth story was therefore important in the dissemination and popularization of the neo-Darwinian theory of evolution.

The Globin Paradigm

Scott Gilbert emphasized the contribution of the globin paradigm to the consolidation of the Modern Synthesis. Studies of the molecular basis and genetics of sickle-cell anemia were integrated with population studies explaining the frequency of the disease and of the genes causing it in populations.

Sickle-cell anemia is a severe genetic disease characterized by the sickle shape of the red blood cells, which causes them to obstruct capillaries, leading to anemia and extreme pain. It is an autosomal recessive trait, so only a person with two copies of the mutant gene (the homozygote) has the disease. In 1949, Linus Pauling had discovered that the reason for the disease is that the hemoglobin protein of people with sickle-cell anemia is abnormal. In the late 1950s, it was found that a single amino acid change—replacement of glutamic acid at position 6 in the β-globin chain by a valine—resulted in sickle-cell hemoglobin. Once the genetic code was discovered, it was possible to infer the DNA base change that led to sickle-cell disease, and this was later confirmed by DNA sequencing. Sickle-cell disease was the first genetic disorder whose molecular basis was known.

The reason for the high frequency of the disease in some regions of the world was at first a mystery, because the disease usually causes death. However, two scientists, J. B. S. Haldane and Anthony Allison, suggested that the heterozygous carriers (who have one normal and one abnormal globin allele) must have an advantage. They noticed that the disease was frequent in areas where malaria is common, and therefore suggested that the heterozygotes might resist malaria better than both normal non-carriers and people with sickle-cell anemia. If heterozygotes have the highest fitness, it would explain the prevalence of the sickle gene. It was indeed later confirmed that heterozygotes do have increased resistance to malaria, and that the geographic distribution and the frequency of the sickle allele can be explained by selection for heterozygotes in regions where malaria and the mosquito that transmits it are common.

Scott Gilbert argued that the studies that led to a better understanding of sickle-cell anemia integrated evolutionary genetics, transmission genetics, molecular genetics, human genetics, and developmental genetics. They brought together a huge group of things and showed how molecular biology is able to provide evidence for the Modern Synthesis. Left out were behavioral genetics and plant genetics, but other than that, the globin paradigm really made a coherent story which was of major importance for the consolidation of the Synthesis in the 1960s. It was easy to teach, and it provided a framework for thinking about evolution.

During the discussion, Everett Mendelsohn posed a series of questions that took us from the Modern Synthesis to the twenty-first-century studies that legitimize the developmental approach to heredity and evolution. Everett asked:

> When you have a series of people dealing with questions that we would label, in retrospect, soft inheritance, the question I want to ask is, why does it take so long until their discussion becomes part of a consensual discussion? What is it that is happening in a field that creates a consensus? Science has not been static, the consensus has moved. So how did it move? And of course the new consensus hasn't emerged yet, has it? We know in this room there are a number of the strong proponents for a new consensus, but how does it change? In a field like this, how do you move from the single figures to something which then generates the new research programs, generates the new publications, and in a sense defines what it is the field is looking at?

Further Reading

Darlington's Work

Darlington, CD. *Recent Advances in Cytology*. London: Churchill; 1932.

Darlington, CD. *The Evolution of Genetic Systems*. Cambridge: Cambridge University Press; 1939.

Harman OS. *The Man Who Invented the Chromosome: A Life of Cyril Darlington*. Cambridge, MA: Harvard University Press; 2004.

Plasmon and Plasmagene Theories and Monod's View

Lindegren CC. *The Cold War in Biology*. Ann Arbor, MI: Planarian Press; 1966.

Sapp J. *The New Foundations of Evolution: On the Tree of Life*. New York: Oxford University Press; 2009.

Monod J. *Chance and Necessity: An Essay on the Natural Philosophy of Modern Biology*. New York: Alfred A. Knopf; 1971.

Institutional and Political Factors Associated with Cold Spring Harbor

<http//:www.symposium.cshlp.org>.

Early Experiments on Mutation Induction

Jablonka E, Lamb MJ. *Epigenetic Inheritance and Evolution: The Lamarckian Dimension*. Oxford: Oxford University Press; 1995: 58–61.

Industrial Melanism

Kettlewell HBD. 1959. Darwin's missing evidence. Sci Am 200 (March): 48–53.

The Globin Paradigm

Gilbert SF, Epel D. *Ecological Developmental Biology: Integrating Epigenetics, Medicine, and Evolution*. Sunderland, MA: Sinauer Associates; 2009.

III BIOLOGY

15 Introduction: Lamarckian Problematics in Biology

Eva Jablonka

This section offers some answers to Everett Mendelsohn's query about the reasons for the current "move in the consensus" in evolutionary biology. Most of the answers are associated with the revival of an approach that gives explanatory primacy to development. Selection is still seen as crucial, but the nature, origins, construction, and inheritance of developmental variations are deemed to be just as important.

Chapters 16 through 27 provide some illustrations of the twenty-first-century incarnation of the "developmental–variation first" approach to evolutionary problems. It is difficult to encapsulate the views they present, because they touch upon many aspects of modern biology and a variety of issues in Lamarckian problematics. The topics discussed fall into three broad, overlapping categories: (1) the evolutionary significance of various facets of developmental plasticity; (2) the role of epigenetic inheritance and induced genome changes in evolution; (3) how changes in the notion of individuality affect evolutionary thinking.

Developmental Plasticity in Evolution

Since the 1990s, there has been an enormous expansion of research into the molecular processes that underlie development. Two topics that have been major subjects of theoretical and molecular studies are of particular significance for evolutionary biology. The first is developmental plasticity—the ability of a particular genotype to produce a variety of morphological, physiological, and behavioral features in response to developmental signals and environmental circumstances. The second, canalization, is in some ways the mirror image of plasticity: it is the active, developmental buffering that leads to an unchanged phenotype in the face of internal and external variations.

Molecular studies of plasticity and canalization have centered on cellular networks, cell memory, switching mechanisms, and signal transduction cascades. Usually the studies have been carried out without any evolutionary questions in mind, and initially the findings had little impact on mainstream evolutionary theorizing. More surprisingly, the analysis of developmental plasticity in populations living in different ecological conditions, which is clearly relevant to evolutionary questions, was also seen by many as peripheral to mainstream evolutionary biology, but this view is now changing.

Many of the arguments for adopting a development-first approach and making plasticity central in evolutionary thinking were synthesized in the influential book *Developmental Plasticity and Evolution* (West-Eberhard 2003). In it, West-Eberhard argues that it is phenotypic continuity across generations and the plasticity of the developing organism that are fundamental to evolution through natural selection. Her point of departure is therefore the developing phenotype rather than the gene: evolutionary change starts with a modification of development—usually a modification brought about by a change in the environment. If this is repeated over many generations, selection of genes that stabilize (or destabilize) the developmental pathways that produced the initial developmental adjustments will take place. As West-Eberhard aptly put it, "genes are followers in evolution": developmental responses to the environment are primary, and can be fine-tuned, stabilized, or ameliorated by subsequent genetic changes in populations. Although West-Eberhard avoids the word "Lamarckism" and distances herself from its connotations, in our terms she is exploring some major aspects of Lamarckian problematics. For Lamarck, the responsiveness of living organisms to environmental changes—many of which are the result of organisms' own activities—was a fundamental property, inherent in his characterization of life. Hence, processes of self-organization and adjustment are the basis of Lamarck's theory of adaptation. As Laurent Loisson (chapter 7 in this volume) noted, plasticity/responsiveness is the essence of Lamarck's famous first law.

Plasticity can take many forms, and there are different types of plastic responses to environmental changes. Responses can be reversible or irreversible, adaptive or nonadaptive, active or passive, continuous or discontinuous, programmed and with a limited range and repertoire or open-ended. In chapter 16, Stuart Newman and Ramray Baht discuss unprogrammed, generic developmental plasticity, which stems from the self-organizing properties inherent in the physics of soft, excitable, living matter. They see in Lamarck's developmental–physicalist approach a precursor of their own views on the centrality of inherent plasticity in evolution, and refer to it as "Lamarck's dangerous idea." The physics of biological matter, especially the physics of the developing embryo as it interacts with changing conditions, can generate, they argue, large morphological innovations leading to evolutionary saltations. The active organism, reorganizing its morphology and physiology, and constructing its own niche, is the agent of evolution.

Another example of the generative potential of unprogrammed complexity is discussed in Arkady Markel and Lyudmila Trut's chapter (17) about the great domestication experiment in Novosibirsk. Selection for tameness in silver foxes exerted a severe neurohormonal stress on the animals. It destabilized their whole developmental system, so that a lot of previously cryptic heritable variation became apparent. There was a burst of morphological, physiological, and behavioral variation, which after a few generations of selection culminated in the production of tame and docile, doglike foxes. Markel and Trut suggest that this special example points to something much more general: the creative role of stress in evolution, including the evolution of our own species, *Homo sapiens*.

Erez Braun and Lior David's research suggests a general principle for the way in which many of the adaptive solutions to new challenges arise (chapter 18). They show that when yeast cells are challenged, intracellular processes can generate many new biochemical networks; those networks enable the cell to grow and divide, and eventually become stabilized. The challenge they presented to the cells was a consequence of artificially rewiring some regulatory parts of the genome; this meant that if the cells were to grow and reproduce, they had to change the way they controlled some of their genes. Braun and David show that an adaptive regulatory response occurred in a very large proportion (50 percent) of the cells, and the response was heritable for many generations. They believe that such molecular trial-and-error "learning," which involves the random or semirandom generation of multiple biochemical responses followed by the selective stabilization and inheritance of those that are adaptive, is a general principle of adaptability, applicable to various levels of biological organization.

Responsiveness is a fundamental property of living organisms, but it, too, can evolve. The result may be programmed and limited plastic responses, such as the response to drought that Sonia Sultan has studied in the colonizing weed *Plygonum persicaria* (chapter 19), and the switching between dormant and active states seen in bacterial populations (chapter 20). Important evolutionary insights can be gained by taking such plasticity into account. As Sultan documents, selection for ranges and patterns of responsiveness to environmental inputs such as light and drought can explain the geographical distribution of plant species and populations. This has implications for studies of biodiversity, because it enables predictions to be made about a species' ability to cope with the drastic and destructive environmental changes caused by human activity.

In chapter 20, Sivan Pearl, Amos Oppenheim, and Nathalie Balaban explore the evolutionary significance of a special type of bacterial plasticity in which there are switches between a persistent dormant state that protects the bacteria against life-threatening conditions, and an active, dividing, but vulnerable state. They discuss two evolutionary aspects of switching when phages (viruses that are the bacteria's predators) are present: first, the possibility that switching between the two states is an evolved adaptive strategy that enables the bacteria to cope with fluctuating, unpredictable conditions; and second, the idea that switching between states decreases prey–predator oscillations, and hence the chances of extinction.

Although very different in subject matter, the five chapters just outlined show that, directly or indirectly, plasticity affects evolutionary dynamics by influencing the rate and direction of change. Both open-ended and programmed plasticity may increase evolvability, the capacity to generate heritable, selectable phenotypic variation, including the capacity to generate variation that leads to novel functions (Wagner 2005). As other chapters in this part and in part IV show, the evolution of evolvability is tightly linked to Lamarckian problematics.

Another aspect of Lamarckian problematics that is touched on by many contributors is how initially transient and often unstable developmental responses can become permanent, stable, heritable traits. For example, Newman and Bhat argue that the body plans of all animal phyla stem initially from the conditional, generic developmental plasticity of early metazoan organisms; only later were these body plans stabilized against environmental and genetic perturbations. In other words, during metazoan evolution they became more canalized. Similarly, in her discussion of the evolution of plant responses, Sultan also considers both how plasticity evolves and how conditional plasticity can become unconditional and genetically canalized. In fact, the question of how induced responses can become permanent, canalized traits is so significant in Lamarckian problematics that it is not surprising that the importance of *genetic assimilation*—the framework within which the process is now usually discussed—was a dominant topic in the discussion that took place in the final session of the workshop. Because it is so important, an outline of the background to the topic may be helpful.

At one time, the inheritance of the effects of use and disuse seemed to provide straightforward answers to many puzzling observations. For example, Darwin, as well as those more usually seen as Lamarckians, explained the presence of thick skin on the soles of newborn babies' feet as the outcome of many generations of pressure and abrasion having eventually become inherited. Similarly, the inborn fear of snakes in some birds evolved, according to Lamarckians (including Darwin), because learned fear became inherited after many generations of being acquired through learning. After doubts were cast on the assumption that there can be direct transfer of induced or learned information from one generation to the next, another kind of interpretation was put forward. Around 1900, two psychologists, James Mark Baldwin and Conway Lloyd Morgan, and a zoologist, Henry Fairfield Osborn, who were all interested in the evolution of animal behavior, independently came up with an idea that is now referred to as the Baldwin effect. They suggested that when animals are faced with a new challenge, they first adapt by learning; this allows the population to survive until a congruent hereditary change, which simulates the learning-based adaptation, occurs. Hence, a character that was originally learned becomes an inherited character.

In the mid-twentieth century, a Darwinian–Mendelian explanation for the inheritance of induced or learned responses was put forward by Conrad Waddington in Great Britain, and independently by Ivan Schmalhausen in the Soviet Union. Both proposed that selection for the genetic basis of the developmental capacity to respond adaptively to a new environmental stimulus leads to the construction of a genetic constitution that facilitates such an adaptation. Waddington termed the process *genetic assimilation*; Schmalhausen called it *stabilizing selection*.

Genetic assimilation occurs because individuals in a population vary in their responsiveness to changed conditions, some responding more rapidly or appropriately than others. Those individuals that respond to new inducing or learning conditions in an efficient manner have a selective advantage. Consequently, as a result of selection, genetic constitutions that stabilize an induced response (i.e., make it more reliable and precise) and ameliorate any detrimental side effects will become more common. Over time, individuals in the population will come to respond more and more appropriately and readily. Eventually, a threshold may be crossed and the trait that originally was induced or learned appears without any external induction or learning. When this happens, the trait is said to be fully genetically assimilated.

Waddington (1942) first described this idea in a short paper provocatively titled "Canalization of Development and the Inheritance of Acquired Characters." He later conducted a series of experiments that gave substance to the idea. Through selection for responsiveness, traits that were initially dependent on environmental induction became genetically assimilated—they developed in the absence of the inducing stimulus. In one experiment Waddington exposed early embryos of the fruit fly *Drosophila melanogaster* to ether vapor. A proportion of the exposed flies developed four wings instead of the two that flies normally have. By selecting four-winged flies as parents for twenty generations, Waddington produced a strain in which a high proportion of the flies developed four wings even in the absence of ether vapor. The explanation he gave for this and for similar experiments was that the new conditions decanalized (destabilized) development, thus exposing cryptic heritable variation. The ability to respond to the stimulus (ether) by developing the new phenotype (four wings) was, Waddington argued, the result of particular interacting combinations of alleles of several genes. After the sexual reshuffling of genes and selection for responsiveness, "responsive" alleles became more common, until eventually a regulatory network was constructed that made induction unnecessary. Assimilation experiments conducted forty years later supported Waddington's interpretation by showing that there are indeed several alleles in regulatory genes underlying the development of the inducible four-wing phenotype. Recent experiments that use sophisticated molecular techniques have uncovered the molecular basis for genetic assimilation of various induced traits in fruit flies, yeasts, and plants.

The concept of genetic assimilation can explain many aspects of adaptive evolution, most obviously the evolution of complex, fixed behavioral patterns (instincts), which are very difficult to explain in other ways. It provides a way through which developmentally acquired (induced or learned) characters can, by means of Darwinian selection for responsiveness, become inherited traits.

Epigenetic Inheritance and Genomic Responsiveness under Stress

The theoretical reevaluation and the empirical study of molecular processes of development is part of what is now called *epigenetics*, another term coined by Waddington. He suggested it ". . . as a suitable name for the branch of biology which studies the causal interactions between genes and their products which bring the phenotype into being" (cited in Waddington 1968:12). The "genetics" part of "epigenetics" highlights the major role of genes in development, and the "epi" part of the term, which means "upon" or "over," points to the need to study developmental processes that go "over" or "beyond" the gene. Today the term has become less general, and epigenetics refers to the study, in both prokaryotes and eukaryotes, of the processes that underlie developmental plasticity and canalization, and bring about *persistent* developmental effects. At the cellular level these are the processes involved in cell determination and differentiation, and at higher levels of biological organization they are the processes that lead to the physiological, morphological, and behavioral persistence of induced or developmentally "acquired" traits.

Epigenetics has become a buzzword in twenty-first-century biology. After a lag of nearly fifty years, there is now intense research activity in the field, something that is reflected not only in the publication of many reviews, books, and textbooks, but also in the mushrooming of a new terminology. Although epigenetics is sometimes used as a synonym for *epigenetic inheritance*, the latter is only part of the study of epigenetics, albeit a part that is getting increasing attention. Because epigenetic inheritance can lead to soft inheritance, it features strongly in the revival of Lamarckian problematics.

Epigenetic inheritance occurs when phenotypic variations that do not stem from variations in DNA base sequence, or from present environmental conditions, are transmitted to subsequent generations of cells or organisms. As the short sketch of the history of epigenetics makes clear (chapter 21), many of the discoveries about epigenetic inheritance *between* generations of organisms come from studies of cell heredity—of inheritance in somatic cell lineages *within* an organism. During development, a person's kidney stem cells and skin stem cells generally breed true, even though their DNA sequences are identical and the developmental stimuli that led to their different phenotypes are long gone. Heredity mechanisms similar to those that were discovered in such mitotically dividing somatic cells have been found in asexually reproducing single-celled organisms such as bacteria and protists, and in sexually reproducing organisms. In the latter, epigenetic variations that occur in the germ line, or are transferred from somatic cells to germ cells, can be transmitted from one generation of individuals to the next through the gametes. We describe the mechanisms underlying the transmission of cellular epigenetic variations in appendix B.

Although the definition of epigenetic inheritance is straightforward, the term is used in both a broad sense and a narrow sense (see Jablonka and Lamb 2007). *Epigenetic inheritance in the narrow sense* refers to the *cellular epigenetic inheritance* just described, in which the unit of transmission is the cell. *Epigenetic inheritance in the broad sense* includes *body-to-body (or soma-to-soma) information transfer between generations of individuals*. This can take place through developmental interactions between mother and offspring, through social learning, through symbolic communication, or through the transmission of symbionts and pieces of DNA between individuals.

All of the chapters in part III touch on epigenetic mechanisms and epigenetic inheritance, but they are the direct focus of chapters 21–25. In chapter 21 I give an overview of the field of cellular epigenetic inheritance, surveying its extent and range; I also discuss the continuity between past and present terms associated with soft inheritance. Chapters 22 and 23 describe some recent experimental studies, and the possibilities and problems they open up. On the basis of the evidence from her work on RNA-mediated inheritance in mice, Minoo Rassoulzadegan (chapter 22) suggests that in outbred populations, where there is a lot of heterozygosity and the difference between alleles can affect meiotic pairing, small heritable RNAs from one or both alleles are generated in the germ cells and used for modulating development in the progeny. RNA-mediated inheritance is, she argues, important for adaptive evolution within populations, as well as for speciation. In chapter 23 Peter Gluckman, Mark Hanson, and Tatjana Buklijas review evidence showing that transgenerational epigenetic inheritance has medical implications. They point out the difficulty of deciding whether the observed transgenerational effects are based on epigenetic inheritance through the germ cells, or on soma-to-soma transmission, or on both. As they emphasize, solving this problem and working out the molecular mechanisms involved are crucial for the further development of this important field.

Another topic that is currently getting increased attention is genomic plasticity—the ability of genomes to respond to challenges by reorganizing their structure. This is something that was central to Barbara McClintock's prophetic view of the genome more than half a century ago. Genomic reorganization involves interactions between genetic and epigenetic systems. It can occur as a response to environmental stresses, such as a heat shock or starvation, or in response to genomic challenges such as hybridization and polyploidization. Braun and David's work on the responses of yeast cells to novel regulatory challenges (chapter 18), and Markel and Trut's work on domestication (chapter 17), show that as well as extensive epigenetic changes, genetic changes occur, too.

Plants provide a wealth of evidence for genomic plasticity and repatterning in conditions of stress. Chapters 24 and 25 discuss this evidence and its evolutionary importance. Marcello Buiatti (chapter 24) reviews the rich data showing how the life strategies of plants facilitate the inheritance of somatic genetic and epigenetic variations. He discusses the different types of genomic reorganization mechanisms seen in both ontogeny and phylogeny, and points to the importance of selection for coordinating different levels of heredity. In chapter 25, Moshe Feldman and Avraham Levy describe their research on wheat hybrids, which shows that allopolyploidization (hybridization between two species followed by genome doubling) accelerates genome evolution by triggering an immediate and radical genomic revolution. They suggest that the nonrandom, reproducible changes that occur during the first generation(s) of newly formed allopolyploids might be natural defense mechanisms against the genetic shock of hybridization and genome doubling. These predictable and highly targeted genomic repatternings lead to heritable changes which have an important role in the adaptation of nascent allopolyploids to their environment. They are a major mechanism of speciation.

Global, stress-induced genomic changes are not specific to plants. An important subject which, unfortunately, is not represented in this volume, is the study of the mechanisms underlying global stress responses in bacteria. Patricia Foster (2007) reviewed data showing that in bacteria, especially in the most intensely studied bacterium species, *Escherichia coli*, stresses such as nutritional deprivation, DNA damage, temperature shifts, and exposure to antibiotics induce both wide-ranging changes in gene expression and an increase in mutability. The stress responses include up-regulation and activation of error-prone DNA polymerases, down-regulation of error-correcting enzymes, and movement of mobile genetic elements. All these result in the induction of genetic changes, which may be important for adaptive evolution. Although the specific genomic changes that are triggered by the environmental stresses are random, and therefore not necessarily adaptive, the global genomic stress response is adaptive.

Studies of genomic changes in response to challenge highlight the fact that the genome itself is a developmental system, a point that is developed by Ehud Lamm in part IV of this volume. It is something that blurs fast and easy dichotomies such as development and evolution, plasticity and evolvability. Developmental mechanisms are part of both proximate and ultimate causation—a point that is critical for Lamarckian problematics.

Rethinking Biological Individuality

Biological individuality is notoriously difficult to define, but functional coherence and being systematically a unit of selection and evolution are considered crucial criteria. In their influential book *The Major Transitions in Evolution*, Maynard Smith and Szathmáry (1995) argue convincingly that transitions to new levels of individuality require mechanisms that ensure and enforce cooperation between the constituent entities, and make the new individual resistant to the selfish interests of its components. However, the notion of individuality discussed in chapters 26 and 27 goes beyond this concept of the individual, because it recognizes the developmental interactions and exchanges between organisms of different species.

Jan Sapp documents in chapter 26 how the acquisition of new genes and genomes through horizontal gene transfers from non-parents, including from individuals belonging to other species, is a hallmark of microbial evolution. In addition, transfer of genetic material through viruses, and symbiosis with viruses, seem to be important in all kingdoms of life (Ryan 2009). Genomic acquisitions of all sorts lead to the formation of new composite biological individuals whose genealogies form complex nets rather than trees. As Sapp insists, evolution is characterized by genetic integration as well as genetic divergence.

New and surprising discoveries about the extent and the sophistication of gene acquisition keep appearing. Recently a defense system has been discovered in bacteria which is based on the acquisition of DNA from their natural enemies, the phages. Bacteria acquire parts of the genomes of the phages that attack them, and integrate them into their own genomes; then, when a phage with complementary DNA attacks, the bacterium transcribes the DNA it has previously acquired, the RNA binds (with the help of some dedicated proteins) to the attacking phage's DNA, and the complex is degraded (Koonin and Wolf 2009). In this case, a new character (phage DNA) is acquired from the environment and has a highly specific and heritable adaptive defense function. This, as Koonin and Wolf note, is a Lamarckian system par excellence.

In chapter 27, Scott Gilbert brings together the two great traditional "Lamarckian" issues of developmental variation and heredity, and that of individuality. He describes how symbiotic relations among organisms produce codevelopmental interactions that have long-term transgenerational effects. They lead to the construction and evolution of new types of more or less cohesive superorganisms or consortiums, with new types of developmental systems—in other words, new types of biological individuals. The coordinated developmental interactions seen in symbioses mean that ideas about individuality, heredity, and developmental variation need to be revisited: individuality is porous; there are multiple forms of hereditary transmission; and developmental variation is (almost) always codevelopmental. Evolution is coevolution.

The chapters in part III show that "Lamarckian problematics" in the twenty-first century include varied empirical and theoretical approaches. These share a common emphasis on development and developmental responsiveness. Today, endorsing "Lamarckian problematics" does not entail commitment to Lamarck's specific (and sometimes inconsistent) views, nor to the views of later Lamarckians. For example, *acquired* variations do not have to be *required* (adaptive) variations, and there is no commitment to gradualism, because contrary to Lamarck's gradualist view, saltation is both possible and likely. Nor is there any commitment to a specific mechanism of inheritance, and certainly there is no requirement that developmentally acquired variations in the phenotype should be directly transformed into corresponding changes in the genome. This may sometimes happen (although highly sophisticated and dedicated mechanisms are needed), but soft inheritance does not require such specific phenotype-to-genotype transfer when it involves epigenetic mechanisms. Most important, stressing the role of developmental variations does not negate the role of natural selection: Lamarckian and Darwinian problematics are complementary, not conflicting.

Further Reading

Foster PL. 2007. Stress-induced mutagenesis in bacteria. Crit Rev Biochem Mol Biol. 42,5: 373–397.

Gilbert SF, Epel D. *Ecological Developmental Biology: Integrating Epigenetics, Medicine, and Evolution*. Sunderland, MA: Sinauer Associates; 2009.

Jablonka E, Lamb MJ. *Evolution in Four Dimensions: Genetic, Epigenetic, Behavioral, and Symbolic Variation in the History of Life*. Cambridge, MA: MIT Press; 2005.

Jablonka E, Lamb MJ. 2007. Précis of Evolution in Four Dimensions. Behav Brain Sci. 30,4: 353–365.

Koonin EV, Wolf YI. 2009. Is evolution Darwinian and/or Lamarckian? Biol Direct. 4: 42. doi:10.1186/1745-6150-4-42.

Maynard Smith J, Szathmáry E. *The Major Transitions in Evolution*. Oxford: Freeman; 1995.

Ryan F. *Virolution*. London: Collins; 2009.

Sapp J. *The New Foundations of Evolution*. New York: Oxford University Press; 2009.

Schlichting CD, Pigliucci M. *Phenotypic Evolution: A Reaction Norm Perspective*. Sunderland, MA: Sinauer Associates; 1998.

Waddington CH. 1942. Canalization of development and the inheritance of acquired characters. Nature 150: 563–565.

Waddington CH. The basic ideas of biology. In: *Towards a Theoretical Biology*. Waddington CH, ed. Vol. 1: *Prolegomena*. Edinburgh: Edinburgh University Press; 1968:1–32.

Wagner A. *Robustness and Evolvability in Living Systems*. Princeton, NJ: Princeton University Press; 2005.

West-Eberhard MJ. *Developmental Plasticity and Evolution*. New York: Oxford University Press; 2003.

Youngson NA, Whitelaw E. 2008. Transgenerational epigenetic effects. An Rev Genom Hum Genet. 9:233–257.

16 Lamarck's Dangerous Idea

Stuart A. Newman and Ramray Bhat

The scientific ideas of Jean-Baptiste Lamarck (1744–1829) are reentering the discourse of biology after nearly a century of marginalization and opprobrium. Partly this is due to increased recognition of discordances between genotype and phenotype in a wide variety of organisms. In many cases organisms with given genotypes assume different phenotypes under different conditions. This presents a problem for the standard model of evolution, the Darwinian Modern Synthesis, if the variant phenotypes are found not to be manifestations of evolved alternative programs, but instead due to the inherent plasticity of any material system. Such phenomena are part of Lamarck's, but not Darwin's, theoretical framework.

The implausibility of gradualist trajectories for certain macroevolutionary episodes including, but not limited to, the origination of complex multicellular forms, has also drawn scrutiny of the standard model and created an opening for reconsideration of Lamarckian processes. Although Lamarck, like Darwin, favored incrementalist scenarios, his theory, unlike natural selection, did not require them. Lamarck's theory also included concepts such as "inner feeling" and "will," by which organisms purportedly evolved in preferred directions, which were related to what was later referred to as "orthogenesis." Modern versions of these concepts help account for empirically discerned evolutionary trends, but all are anathema to the Darwinian synthesis.

The main impediment to incorporation of Lamarck's ideas into a revised evolutionary theory, however, is the assertion of inheritance of characteristics acquired during the lifetimes of individual organisms. In this chapter we will describe the relationship of Lamarck's ideas to (and the discordance of Darwin's with) new knowledge concerning generation of biological forms during development by physical and physicochemical processes pertaining to complex materials. Since the laws and regularities of the physical world, with their conditional but stereotypical outcomes, are transmitted to developing organisms at least as certainly as their genes, the inheritance of acquired characteristics (as well as the increasing agency of organisms vis-à-vis their ecological niches) will be seen to be less of a problem for a naturalistic theory of evolution than for Darwin's version of it.

Environment Dependence of Phenotype

It is difficult to deny a role for the environment in specifying phenotypes if bacterial biofilms (Cho, Jönsson, Campbell, Melke, Williams, Jedynak, Stevens, et al. 2007), pine trees (Paiva, Garnier-Géré, Rodrigues, Alves, Santos, Graça, Le Provost, et al. 2008), or salmon (Gross 1991), for example, can, despite genotypic constancy, exhibit different morphotypes under different circumstances. The Darwinian view of the environmental induction of disparate phenotypes with constant genotypes is that alternative responses to environmental cues are encoded in the genome. Hagen and Hammerstein (2005:87), for instance, state that "we reject the view…that this sensitivity is well explained by a general plasticity. Instead we argue that flexibility is founded on a genetically encoded strategy, most or all of which is shared by the members of a population."

There is little doubt that certain dimorphic responses to environmental cues, such as sexual differentiation in alligators and turtles (Marshall Graves 2008), as well as many other purported Lamarckian phenomena (Torkelson, Harris, Lombardo, Nagendran, Thulin, and Rosenberg 1997), are products of evolutionary trajectories that resulted from extended episodes of selection. However, not all phenotypic plasticity is adaptive (see West-Eberhard 2003 for a review), and some initially nonadaptive plasticity—for example, in directional asymmetry (Palmer 2004) and sexual size dimorphism (Badyaev 2009)—contributes to evolutionary change. Indeed, we have proposed that unprogrammed phenotypic and developmental plasticity has been a source of major morphological innovations both at the origin of the multicellular animals and in the later evolution of body plans and organ forms (Newman and Müller 2000; Newman, Forgacs, and Müller 2006; Newman and Bhat 2008, 2009).

Our main focus in this chapter is the role played in the evolution of organismal form by physical forces involved in development (see Forgacs and Newman 2005 for a review). A key element of our argument is a rejection of the uniformitarianism inherent in Darwinian scenarios. That is, we propose that the dominant mechanisms of morphogenesis and pattern formation in ancient multicellular organisms and at the origination of novel morphological motifs in organs and appendages were different—more plastic and conditional—than the hierarchically regulated developmental programs seen in modern organisms.

It is worth noting, however, that the concept of inheritance, even for extant organisms, has been broadened in the last few years to encompass conditional processes that extend beyond the strict nucleic acid definition of the gene. Jablonka and Lamb, for example, have discussed several forms of epigenetic inheritance (Jablonka and Lamb 1995, 2005), and have proposed plausible hypothetical scenarios in which evolution may occur by employing such mechanisms in the absence of change in DNA (Jablonka and Lamb 2005). Gorelick and Laubichler (2008) have questioned the distinction between hard and soft inheritance (other than as extremes on a continuum), as well as the exclusive association of genes with DNA, and Griesemer (2000) and Salazar-Ciudad (2008), using the concepts of "reproducers" and "generative systems," have argued that focusing on the transmission and change of hereditary information that (like DNA) can be copied is too restrictive for understanding how such systems can pass on phenotypic change and evolve.

Our view is consonant with these expanded notions of inheritance, but also stresses that much evolution that followed the establishment of metazoan body plans, and novelties such as insect appendages and the vertebrate limbs, has had the effect of autonomizing the respective structures (i.e., stabilizing them against environmental and genetic change) and developmental noise and other intra-embryonic perturbations (Müller and Newman 1999). This means that the morphological characters of ancient counterparts of present-day metazoans, being under less stringent control, were likely to be more plastic than present-day forms (Newman 1994; Newman and Müller 2000). Could the large phenotypic variations resulting from "primitive plasticity" prior to this autonomization also be inherited?

Natural Selection and the Road Not Taken

In order to describe how conditionally acquired morphological characteristics might be propagated transgenerationally, it will be important to contrast the "physicalist" view of inheritance, for which we find precedent in Lamarck's writings, with the widely held notion of inheritance contained in the Modern Synthesis. As a prelude, we will summarize the key points of Darwin's theory:

(1) Organisms present themselves as "types," perpetuating themselves "each according to their kind"; (2) organismal types comprise actual individuals that are all somewhat different from one another; (3) part of this variability is passed on from one generation to the next, so offspring are not only recognizable members of their type but also carry on some of their parents' particularities; (4) as external circumstances change—for example, by depletion of a certain foodstuff or a rise in the ambient temperature—subpopulations with particular quirks, or differences from the norm, will survive or thrive to a better extent than average, contributing disproportionately higher numbers of descendants to succeeding populations; (5) the average properties of the selected subpopulations will thus be different at later

times from what they were at earlier times; (6) after enough generations have passed, the original type may no longer be recognizable in the selected subpopulations: a new type of organism will have emerged; (7) if no individuals of the new type can productively interbreed with any individuals of the originating population, *speciation*, the smallest step of phylogenetic significance, will have taken place; (8) the conditions and processes described in propositions 1–7 constitute the mechanism by which new biological forms arise over time: all large-scale differences, e.g., between plants and animals or between insects and mammals, were generated by a series of many small species-level diversification events.

The observations contained in propositions 1–5 were uncontroversial in the nineteenth-century European context in which Darwin (and his coformulator of the hypothesis, Alfred Russel Wallace) presented the idea. Only with items 6 and 7 does the mechanism of natural selection emerge, and even then there are few, even among present-day creationists, who would disagree with its implications. Proposition 8 is the central tenet of the Darwinian theory since it asserts that natural selection can account for large-scale (macro) evolution. The contest between the currently mainstream Modern Synthesis and newer concepts from the field of evolutionary developmental biology importantly depends on how successfully they account for macroevolution.

Darwin was part of a new breed of biologists with materialist aspirations (Lenoir 1982). The physical science of the time, however, embodied in Newtonian mechanics, was not up to the challenge of explaining the form and function of living systems. For Newton, matter was inert, changing its state in a continuous fashion, deviating from its track only when acted upon by external forces. Darwin and Wallace hewed closely to the Newtonian worldview, and forever gained the allegiance of proponents of "mechanistic materialism" (Lenin 1909 [1972]; Allen 1983).

In actuality, living matter is protean and abrupt in its transformations, and (in the terminology of the philosopher Immanuel Kant [1724–1804]), "self-organizing." A school of "teleomechanists" (e.g., J. F. Blumenbach and C. F. Kielmeyer; see Lenoir 1982), influenced by Kant's writings on teleology, sought to bring more dynamical physicalist ideas into biology. The "rational morphologists" (e.g., E. Geoffroy Saint-Hilaire, L. Oken), inspired in part by the ideas of Georges–Louis Leclerc, comte de Buffon (1707–1788) and Johann Wolfgang von Goethe (1749–1832), proposed that there were laws for the generation of biological form analogous to the laws of planetary motion. This differed from teleomechanism in providing an opening for preferred directions in the evolution of form (i.e., orthogenesis).

The teleomechanists are now forgotten, and the rational morphologists nearly so. In the century following Darwin, research agendas that sought deeper regularities of biological organization tended to be dismissed as vitalist or otherwise misguided. The geneticist William Bateson's oscillatory model for segmentation (Bateson 1894;

Bateson and Bateson 1928), for example, "simply retarded scientific progress," in the view of the influential evolutionist Ernst Mayr (1982:42), although it was later shown to be essentially accurate (reviewed in Newman 2007).

Darwin avoided similar criticisms by formulating in natural selection a theory of biological change that did not depend on any knowledge of the source of variation, inherited or noninherited, in organismal form and function. The theory seemed materialist, since it was a commonplace of contemporary manufacturing technologies that a material object will often have properties slightly different from those of other copies of the same item, and that small differences can occasionally be advantageous.[1] From there it was a small step (once the possibility had been pointed out) to conceive of continuous trajectories of change between present-day organisms and their presumed ancestral forms, based on the small differences that appear in each generation.[2] It took another century before the scientific concepts and tools were produced that could deal adequately with embryonic development and demonstrate that tissue morphogenesis is often a discontinuous and nonlinear process (see Forgacs and Newman 2005 for a review of such applications). As the American embryologist E. E. Just presciently stated, in the years just before these new methods emerged, what was needed was "a physics and chemistry in a new dimension . . . superimposed upon the now known physics and chemistry" (Just 1939:3).

Lamarck, a near contemporary of Kant's, studied both physical and biological questions, and when he eventually concluded, against his earlier beliefs, that biological forms can undergo transmutation, he proposed models based on purported physical principles governing biological matter. He postulated a "power of life," an inherent tendency of living systems to become more complex over time, and the "influence of circumstances" over biological form and function. Both these concepts have found a place in evolutionary developmental biology, the first as what scientists, using Kant's term, now refer to as self-organization, and the second as physiological adaptation and phenotypic plasticity.

Darwin also had no understanding of the mechanisms of inheritance. His hypothesis of "pangenesis" (see below), which was his attempt to remedy this, incorporated a conditional plasticity similar to Lamarck's. Darwin's successors however, eliminated any hint of inheritance of acquired characteristics from the standard theory when they were persuaded that the separation of the germ line from the soma (Weismann 1893), despite pertaining mainly to animal species, should be considered paradigmatic (Romanes 1895).

The resulting theory, "neo-Darwinism," required an intergenerational phenotype-specifying medium (see proposition 3, above) that, unlike Darwin's organism-wide pangenesis, was subcellular. When the earlier work of Gregor Mendel (1822–1884) came to the attention of influential biologists at the turn of the twentieth century, this niche was filled by his "elements" or "factors." But Mendel himself had presented no hypothesis for how his elements functioned to influence the choice of one alternative trait or another, and the neo-Darwinists did not attempt to supply one. Therefore, like Darwin's original theory, neo-Darwinism was devoid of a model for generation of phenotypes, particularly morphological phenotypes. What the teleomechanists and rational morphologists, and Lamarck before them, had discerned, however, and what was now slated to be left out of the incipient Modern Synthesis, was the fact that organisms do not perpetuate themselves by passing a set of "particles" on to their offspring, but instead endow them with a repertoire of developmental mechanisms.

Inheritance of Differentiation versus Inheritance of Development

Darwin's theory of pangenesis, introduced in the first edition of *The Variation of Animals and Plants Under Domestication* (Darwin 1868), involved hypothetical particles (gemmules) that determined the character of tissues and organs. In this model, each cell in the body produces gemmules corresponding to its specific type that migrate through the body, accumulate in the gonads, and are thereby transmitted to the next generation. Since specific gemmules were hypothesized to have different degrees of abundance and potency in different individuals, and to change in these capacities in response to the environment, the theory was claimed to account for embryonic development, dominance and blending inheritance, and atavisms, as well as inheritance of acquired characteristics.

Pangenesis was an early attempt to explain determination of species character solely in terms of what modern biologists call cell differentiation. The gemmules were uniquely identified with cell types; in a letter to the botanist Joseph Hooker, Darwin referred to them as "atoms" (Darwin [1868] 1959), that is, analogous to the indivisible units of the chemical elements. The Mendelian theory that would come to replace pangenesis was also approvingly conceived of as "particulate" by the formulators of the Modern Synthesis (Mayr 1982). As we will see, this simplifying impulse continues to exert its effect in certain formulations of modern developmental genetics.

It is illuminating to contrast Darwin's model for the development and transmission of organismal characteristics to the scenario put forward in Lamarck's major biological work, the *Zoological Philosophy* (Lamarck [1809] 1984). In part II of that book, subtitled "An Inquiry into the Physical Causes of Life, the Conditions Required for Its Existence, the Exciting Force of Its Movements, the Faculties Which It Confers on Bodies Possessing It, and the Results of Its Presence in Those Bodies," he asserts that:

> ... the entire work of nature in her spontaneous creation consists in organizing into cellular tissues the little masses of gelatinous or mucilaginous material which she finds at hand under favorable circumstances; in filling these little cellular masses with fluids and in vivifying them by setting these contained fluids in motion by means of the stimulating subtle fluids which are incessantly flowing in from the environment; . . . [t]hat the cellular tissue is the framework in which all organization has been built, and in the midst of which the various organs have successively developed by means of the movement of the contained fluids which gradually modifies the cellular tissue; . . . [t]hat the function of the movement of the fluids in the supple parts of the living bodies which contain them, is to cut out paths and to establish depots and exits, to create canals and afterwards various organs; to cause variation in these canals and organs by means of a diversity in either the movements or in the nature of the fluids which produce and modify them; [f]inally to enlarge, elongate, divide and solidify gradually these canals and organs by substances which are formed and incessantly separated off the essential fluids in movement there; substances of which one part becomes assimilated and united with the organs while the other is thrown out (Lamarck 1809 [1984]:188–189).[3]

Regarding the hereditary transmission of the capacity to generate characters, Lamarck proposed that:

> ... the function of organic movement is not merely the development of organization, and the increase and growth of the parts, but also of the multiplication of organs and the functions they fulfill. (Lamarck 1809 [1984]:189)

Lamarck's view, then, was that forms perpetuate themselves by transmitting specific materials accompanied by their self-organizational capabilities.

Darwin's and Lamarck's notions of development and heredity have counterparts in different strains of evolutionary–developmental modern biology. A widely held view that is akin to Darwin's asserts that cell differentiation is the primary process of development (S.B. Carroll et al. 2004; Davidson 2006). Cell type identity is determined by "gene regulatory networks" (GRNs), circuits composed of a set of transcription factors the targets of which are cis-regulatory sequences of a set of genes that include those specifying the same factors. The cell-type-centered view of development (a successor, in this respect, to Darwin's pangenesis) sees conserved subnetworks of GRNs (kernels), transmitted as essentially autonomous units by virtue of their logical structures, as defining organisms' phenotypes (Davidson 2006). Other transcription-factor-based models for the evolution of developmental systems, though differing in detail, have as their common implicit assumption the centrality of cell differentiation (e.g., Wagner 2007).

A second view, in part a successor to Lamarck's focus on the physics of cell masses and the associated "fluids," sees GRNs as just one part of the story (Newman, Bhat, and Mezentseva 2009). Complementing them are "dynamical patterning modules" (DPMs), combinations of gene products (mainly a set different from the GRN transcription factors) and physical forces, effects and processes they are capable of mobilizing on spatial scales larger than a single cell (Newman and Bhat 2008, 2009). The physics—adhesion and differential adhesion, self-assembly of anisotropic components, diffusion, lateral inhibition of dynamical state, synchronization of oscillatory state, bending instabilities of elastic media—generates a panoply of forms and patterns. Such changes can be continuous deformations of shape, but they can also be abrupt, leading to the appearance of body cavities, tissue layers, segments, and appendages as a result of minor changes in the balance of developmental forces (Newman, Forgacs, and Müller 2006).

As in Lamarck's framework, organismal phenotypes are plastic, that is, subject to the "influence of circumstances." This is likely to have been particularly true early in the evolution of body plans and organ forms, before stabilization and autonomization of phenotypes had set in (Newman, Forgacs, and Müller 2006). Moreover, phenotypes are propagated across generations not primarily by transmittal of particulate or logical determinants of cell type, but by endowing newly forming organisms with developmental mechanisms. These mechanisms comprise the self-organization of multicellular aggregates and their associated materials, via the physical and physicochemical processes they mobilize.

Development and Saltation

Gradualism is a fundamental assumption of Darwinian natural selection (see propositions 7 and 8, above). The mechanism's distinguishing character resides entirely in the premise that macroevolution consists of microevolutionary changes accumulated over time. Lamarck was also a gradualist, stating, for example, that "in all nature's works nothing is done abruptly…but everywhere slowly and by successive stages" (Lamarck 1809 [1984]: 46), but there was nothing intrinsic to his theory that required this doctrine. Indeed, elsewhere he remarked that "often some organ disappears or changes abruptly, and these changes sometimes involve it in peculiar shapes not related with any other by recognizable steps" (Lamarck 1809 [1984]:69). Such saltationist ideas were made more explicit by Lamarck's colleague and intellectual successor, Etienne Geoffroy Saint-Hilaire.

The physical and chemical world is replete with phenomena characterized by abrupt structural reorganization as a result of changes in internal and/or external parameters. Examples include transitions in the phase (solid, liquid, gas) of matter, wave phenomena in excitable media, and the separation of immiscible liquids. Lamarck's focus on the formative effects of the physical properties of cell masses and matrices, and his notion of the influence of circumstances, are compatible with such a "physicalist" framework. The alternative would be to consider living systems as insulated from the physical world and invulnerable to their own physical and physicochemical properties.

The Modern Synthesis concerns itself almost exclusively with quantitative trait loci, or genes of small effect. The theoretical basis for the latter was given by R. A. Fisher (1890–1962), using (among other, quantitative, arguments) the metaphor of focusing a microscope, where small adjustments are more likely to lead to improvements than large ones (Fisher 1930; Dawkins 1986). He contended that sudden phenotypic jumps or "saltations" (which, in the emerging genetic determinist framework of the Modern Synthesis, would inevitably be associated with genes of large effect) would be exceptionally rare and typically lethal occurrences not contributing significantly to the origin of species or higher taxa. It is now recognized, however, that saltational evolution is a fact and that it is indeed often associated with genes of large effect (G. Bell 2009). Examples of saltation include abrupt changes in the presence of pelvic spines in stickleback fish (M. A. Bell 2001) and segment numbers in centipedes (Minelli, Chagas-Junior, and Edgecombe 2009). Saltation can occur when small changes during development lead to disparate phenotypic outcomes, producing morphological novelties such as the bony bridge between the tibia and fibula of the avian hind limb and the related fibular crest of theropod dinosaurs (Müller 1990). The DPM framework described above, which incorporates the physical dynamics of complex materials such as cell aggregates and organ primordia (Newman and Bhat 2008, 2009), is capable of accounting for, and indeed predicts, the sudden emergence of such novelties.

Conclusions: Organisms as Physical Systems

Darwin's main accomplishment, in his own mind and in the view of his modern advocates (Dennett 1995; Dawkins 1996), is having provided a naturalistic theory of biological complexity. From the beginning, this involved incremental changes that occurred via uniform causative processes (so ordinary that they did not warrant being specified) and were preserved by selection for their marginal adaptive advantage. These tenets were not relinquished by the Modern Synthesis, which sought to explain all evolution exclusively by microevolutionary mechanisms.

Although, as we have seen, Darwin himself did not ultimately preclude inheritance of acquired characteristics, Richard Lewontin sees the most "revolutionary" aspect of Darwin's legacy in his "making an absolute separation between the internal processes that generate the organism and the external processes, the environment, in which the organism must operate" (Lewontin 2000:42). He acknowledges the need to understand the relationship between inside and outside, but rejects the possibility that "the forms of heritable variation that arise are ... causally dependent on the world in which organisms find themselves" (Lewontin 2000:47–48). As we have seen, however, the dynamical processes underlying embryonic development incorporate the physics of materials, the laws, and environment dependence which organisms inherit as surely as they inherit the instructions for synthesizing the proteins that mobilize such causal determinants.

But once tissue organization and reorganization are seen to be inherent in living materials in interaction with their environment, there is no particular need for "Darwin's dangerous idea" of complex forms being generated exclusively by numerous cycles of incremental adaptive change (though there is nothing to rule it out in certain cases). The key question then becomes "Will the new phenotype persevere?"

As Lewontin has persuasively argued, organisms do not occupy preexisting niches, but actively create their own ways of life (see, for example, Lewontin 2000). The observable fact of invasive species thriving in environments entirely different from those in which they evolved (S. P. Carroll 2008), shows that novel phenotypes are not necessarily dead on arrival and that they persist. New types spawned by developmental change, particularly if there is an environmental or conditional aspect to the change that transforms multiple progeny at the same time (Gilbert and Epel 2009), would be expected to set forth on their own in a similar fashion (Odling-Smee, Laland, and Feldman 2003).

Phenotypic plasticity and niche construction are two phenomena that can individually be accommodated by the Modern Synthesis. But taken together they obviate the need for gradualist and adaptationist scenarios. Large-scale (i.e., saltational) phenotypic changes can occur in the space of a single lifetime due to the dynamics of development, and the forms that result, not being passive candidates for places in preexisting niches, will invent unique ways of life. A subpopulation of organisms with a novel, possibly environmentally induced, phenotype can come to occupy an entirely new niche or subdivide the original one, and by genetic drift and/or stabilizing selection become reproductively isolated from the originating population. In such cases, all relevant genotypic change, or the most significant portion of it, will follow, rather than precede, the phenotypic change.

Lamarck's own dangerous idea, then, was the (dynamical) materialist one of bringing life into the realm of the physical. In the case of evolving multicellular organisms, the physical incorporates self-organization and self-generating complexity (autopoiesis; Maturana and Varela 1980, corresponding to Lamarck's "power of life"). With increased knowledge of the evolutionary history of developmental systems, therefore, and a better understanding of the physics of complex materials, we can at last appreciate the power of Larmarck's ideas as he speaks to us across the Darwinian divide.

Notes

1. Darwin's uncle, Josiah Wedgwood, is said to have experimented with more than three thousand samples before settling on the blue color of his enormously successful jasperware pottery.

2. In the first edition of *The Origin of Species*, Darwin in fact stated, "[i]f it could be demonstrated that any complex organ existed, which could not possibly have been formed by numerous, successive, slight modifications, my theory would absolutely break down" (Darwin 1859:158).

3. Lamarck distinguished between "containable" fluids, such as gases and water, which can flow through the "containing parts" (e.g., cavities, channels, and tubules) of developing forms, and "uncontainable" or "subtle" fluids (e.g., electricity and heat), which are "naturally capable of passing through the walls of investing membranes, cells, etc." (Lamarck 1809 [1984]:247).

References

Allen G. 1983. T.H. Morgan and the influence of mechanistic materialism on the development of the gene concept. Am Zool. 23: 829–843.

Badyaev AV. 2009. Evolutionary significance of phenotypic accommodation in novel environments: An empirical test of the Baldwin effect. Philos Trans R Soc Lond B Biol Sci. 364: 1125–1141.

Bateson W. *Materials for the Study of Variation.* London: Macmillan; 1894.

Bateson W, Bateson B. *William Bateson, F.R.S., Naturalist; His Essays & Addresses, Together with a Short Account of His Life.* Cambridge: Cambridge University Press; 1928.

Bell G. 2009. The oligogenic view of adaptation. Cold Spring Harb Symp Quant Biol. 74: 139–144.

Bell MA. 2001. Lateral plate evolution in the threespine stickleback: Getting nowhere fast. Genetica 112–113: 445–461.

Carroll SB, Grenier JK, Weatherbee SD. *From DNA to Diversity: Molecular Genetics and the Evolution of Animal Design.* Malden, MA: Blackwell; 2004.

Carroll SP. 2008. Facing change: Forms and foundations of contemporary adaptation to biotic invasions. Mol Ecol. 17,1: 361–372.

Cho H, Jönsson H, Campbell K, Melke, P., Williams JW, Jedynak B, Stevens AM, et al. 2007. Self-organization in high-density bacterial colonies: Efficient crowd control. PLoS Biol. 5,11: e302.

Darwin C. *On the Origin of Species by Means of Natural Selection, or, the Preservation of Favoured Races in the Struggle for Life.* London: John Murray; 1859.

Darwin C. Letter to Joseph Hooker, February 23, 1868. In: F. Darwin, ed. *The Life and Letters of Charles Darwin.* New York: Basic Books; 1959 (originally published in 1887).

Darwin C. *The Variation of Animals and Plants Under Domestication.* 1st ed. 2 vols. London: John Murray; 1868.

Davidson EH. *The Regulatory Genome: Gene Regulatory Networks in Development and Evolution.* London: Academic Press; Amsterdam: Elsevier; 2006.

Dawkins R. *The Blind Watchmaker.* New York: Norton; 1986.

Dawkins R. *Climbing Mount Improbable.* New York: Norton; 1996.

Dennett DC. *Darwin's Dangerous Idea: Evolution and the Meanings of Life.* New York: Simon & Schuster; 1995.

Fisher RA. *The Genetical Theory of Natural Selection.* Oxford: Clarendon Press; 1930.

Forgacs G, Newman SA. *Biological Physics of the Developing Embryo.* Cambridge: Cambridge University Press; 2005.

Gilbert SF, Epel D. *Ecological Developmental Biology: Integrating Epigenetics, Medicine, and Evolution.* Sunderland, MA: Sinauer Associates; 2009.

Gorelick R, Laubichler MD. 2008. Genetic = heritable (genetic ≠ DNA). Biol Theory 3: 79–84.

Griesemer J. 2000. The units of evolutionary transition. Selection 1: 67–80.

Gross MR. 1991. Salmon breeding behavior and life history evolution in changing environments. Ecology 72: 1180–1186.

Hagen EH, Hammerstein P. 2005. Evolutionary biology and the strategic view of ontogeny: Genetic strategies provide robustness and flexibility in the life course. Res Hum Dev. 2,1-2: 87–101.

Jablonka E, Lamb MJ. *Epigenetic Inheritance and Evolution.* Oxford: Oxford University Press; 1995.

Jablonka E, Lamb MJ. *Evolution in Four Dimensions: Genetic, Epigenetic, Behavioral, and Symbolic Variation in the History of Life.* Cambridge, MA: MIT Press; 2005.

Just EE. *The Biology of the Cell Surface.* Philadelphia: P. Blakiston; 1939.

Lamarck J-B. *Zoological Philosophy: An Exposition with Regard to the Natural History of Animals.* [1809]. Elliot H, trans. Chicago: University of Chicago Press; 1984.

Lenin VI. *Materialism and Empirio-Criticism: Critical Comments on a Reactionary Philosophy.* Fineberg A, trans. Moscow: Progress Publishers; [1909] 1972.

Lenoir T. *The Strategy of Life: Teleology and Mechanics in Nineteenth- Century German Biology.* Dordrecht, Netherlands: Reidel; 1982.

Lewontin RC. *The Triple Helix: Gene, Organism, and Environment.* Cambridge, MA: Harvard University Press; 2000.

Marshall Graves JA. 2008. Weird animal genomes and the evolution of vertebrate sex and sex chromosomes. Annu Rev Genet. 42: 565–586.

Maturana H, Varela F. *Autopoiesis and Cognition: The Realization of the Living.* Dordrecht, Netherlands: Reidel; 1980.

Mayr E. *The Growth of Biological Thought: Diversity, Evolution, and Inheritance.* Cambridge, MA: Belknap Press of Harvard University Press; 1982.

Minelli A, Chagas-Junior A, Edgecombe GD. 2009. Saltational evolution of trunk segment number in centipedes. Evol Dev. 11,3: 318–322.

Müller GB. Developmental mechanisms at the origin of morphological novelty: A side-effect hypothesis. In: *Evolutionary Innovations.* Nitecki M, ed. Chicago: University of Chicago Press; 1990:99–130.

Müller GB, Newman SA. 1999. Generation, integration, autonomy: Three steps in the evolution of homology. Novartis Found Symp. 222: 65–79.

Newman SA. 1994. Generic physical mechanisms of tissue morphogenesis: A common basis for development and evolution. J Evol Biol. 7: 467–488.

Newman SA. William Bateson's physicalist ideas. In: *From Embryology to Evo-Devo: A History of Evolutionary Development.* Laubichler M, Maienschein J, eds. Cambridge, MA: MIT Press; 2007:83–107.

Newman SA, Bhat R. 2008. Dynamical patterning modules: Physico-genetic determinants of morphological development and evolution. Phys Biol. 5, 1: 015008. doi:10.1088/1478-3975/5/1/015008. Accessed December 17, 2009.

Newman SA, Bhat R. 2009. Dynamical patterning modules: A "pattern language" for development and evolution of multicellular form. Int J Dev Biol. 53, 5-6: 693–705.

Newman SA, Bhat R, Mezentseva NV. 2009. Cell state switching networks and dynamical patterning modules: Complementary mediators of plasticity in development and evolution. J Biosci. 34,4: 553–572.

Newman SA, Forgacs G, Müller GB. 2006. Before programs: The physical origination of multicellular forms. Int J Dev Biol. 50,2–3: 289–299.

Newman SA, Müller GB. 2000. Epigenetic mechanisms of character origination. J Exp Zool B (Mol Dev Evol). 288,4: 304–317.

Odling-Smee FJ, Laland KN, Feldman MW. *Niche Construction: The Neglected Process in Evolution.* Princeton, NJ: Princeton University Press; 2003.

Paiva JA, Garnier-Géré PH, Rodrigues JC, Alves A, Santos S, Graça G, Le Provost G, et al. 2008. Plasticity of maritime pine (*Pinus pinaster*) wood-forming tissues during a growing season. New Phytol. 179,4: 1080–1094.

Palmer AR. 2004. Symmetry breaking and the evolution of development. Science 306: 828–833.

Romanes GJ. 1895. The Darwinism of Darwin and of the post-Darwinian schools. Monist 6: 1–27.

Salazar-Ciudad I. 2008. Evolution in biological and nonbiological systems under different mechanisms of generation and inheritance. Theory Biosci. 127,4: 343–358.

Torkelson J, Harris RS, Lombardo M-J, Nagendran J, Thulin C, and Rosenberg SM. 1997. Genome-wide hypermutation in a subpopulation of stationary-phase cells underlies recombination-dependent adaptive mutation. EMBO J. 16: 3303–3311.

Wagner GP. 2007. The developmental genetics of homology. Nat Rev Genet. 8: 473–479.

Weismann A. *The Germ-Plasm; A Theory of Heredity.* Parker WN, Rönnfeldt H., trans. New York: Scribner's; 1893.

West-Eberhard MJ. *Developmental Plasticity and Evolution.* Oxford: Oxford University Press; 2003.

17 Behavior, Stress, and Evolution in Light of the Novosibirsk Selection Experiments

Arkady L. Markel and Lyudmila N. Trut

Ever since the Laboratory of Evolutionary Genetics (Siberian Department of the Russian Academy of Sciences, Novosibirsk) was established by Dmitri Belyaev (figure 17.1) in 1958, experimental models of animal domestication, employing artificial selection, have been used to explore evolutionary processes. Although Charles Darwin (1859: 490) coined the term "artificial selection" to highlight the differences between selection by man and natural selection, it should be emphasized that the same genetic processes underlie both types of selection. Artificial selection models can therefore shed light on the patterns and processes of evolution in natural conditions. Nevertheless, there is a difference between natural and artificial selection, because artificial selection may be associated with a reduction in fitness as the animals struggle with the stress of the artificial selection regime. Belyaev (1969, 1979) focused attention on the role of this stress in domestication. Following in Darwin's footsteps, he studied animal domestication, a process undoubtedly resulting from strong selection pressure on an initially wild population.

Belyaev's long-term experiment, which is regarded as ". . . one of the century's most intriguing systematic investigations of the nature of evolutionary processes" (Coppinger and Feinstein 1991:127), was launched more than fifty years ago, and is still ongoing (Trut 1999; Trut, Oskina, and Kharlamova 2009). The subjects of this grand domestication experiment were silver black foxes, *Vulpes vulpes*. Although the foxes with which the experiment started were not taken directly from a wild population, but were fur-farm foxes maintained for years in cages and bred in captivity, these animals retained their wild behavior and wild external appearance. It should be appreciated that Belyaev's goal in pursuing his domestication project went much further than just domesticating a new species of animal. He wanted to explore the process of domestication and obtain data to support his very original theory of evolution (Belyaev 1979). Belyaev believed that reconstructing the early steps of animal domestication could shed light on the evolutionary reorganization that occurs in the stressful conditions accompanying strong selection for behavior. This was because the domestication of wild animals involves, first and foremost,

Figure 17.1
Professor Dmitry K. Belyaev (1917–1985).

profound changes in their behavior in response to humans. In the course of the experiment the foxes were to encounter the same milestones that dogs passed during their domestication.

Weak aggressiveness and decreased fear of humans were essential prerequisites for the fox domestication experiment. Based on simple tests and observational data, individuals at the fox farm showing calm behavior, with curiosity and friendliness overcoming the fearfulness and aggression typically elicited by confrontation with humans, were systematically chosen for breeding (Trut, Plyusnina, and Oksina 2004). After about ten years of selection for tameness, the behavior of the foxes started to become doglike in many ways. Not only did they not flee from humans, they actually followed human tracks, actively sought interactions with humans, emitted positive vocalization, wagged their tails, and responded to their pet names (Trut 1980, 2001).

One of the prominent results of selection was that the behavioral transformations arising in the domestication process were accompanied by a great increase in the variability of many morphological traits and physiological functions that were, so it seemed, of trifling importance and irrelevant to the selected behavior. The number of individuals with unusual external appearance among the domesticated foxes was fairly high. The phenotypic deviants had floppy ears as pups, shorter and curled up (in a circle or semicircle) tails, changes in the shape and size of the face skeleton, shorter legs, and piebaldness (a characteristic white spot on the head, the "star" (Trut 2001) (figure 17.2). These variations arose at rather high frequencies (10^{-2}–10^{-3} per generation)—too high for them to be attributed to spontaneous mutation. Moreover, since some of the newly arisen traits behaved as dominants (the "star"

Figure 17.2
Homologous phenotypes in domestic foxes and dogs. (A) Coat depigmentation—piebaldness; (B) tail carriage—curly tail.

was one good example), the explanation that selection had made preexisting rare recessive alleles homozygous was also ruled out. The question of why such wide-ranging variability emerges during domestication therefore became of great interest.

In fact, the history of domestication abounds with examples of "bursts" of variability (Darwin 1875). The variability was so much greater in domesticated forms than in their wild ancestors that it seemed doubtful that in all these cases the genetic basis for this diversity preexisted in a cryptic form in the wild populations. It was difficult to believe that selection of preexisting genetic variation is the *only* explanation for the remarkably different breeds of domesticated dogs, for example. These doubts gave rise to a reasonable, though a daring, hypothesis: What if the domestication-associated selection not only manipulates the preexisting variability, but also acts as a specific *generator* of genetic variability? One of the remarkable observations was that variations that arose conformed in many ways to a nonrandom pattern. There were remarkable similarities between the variations in different domesticated species. This was particularly true for taxonomically close species such as the dog and the fox. In both species, domestication was associated with the appearance of piebald coat colors and characteristic changes in the shape of ears, tail, and legs, in some behavioral traits, and even in similar vocalization signals.

Belyaev explained this similarity in the changes under domestication by relating them specifically to selection for behavior. Selection for domesticated behavior entailed radical changes in the animals' social environment which brought their relations with humans to the forefront. Individual development was challenged by the novel social surroundings, and this profoundly affected the whole neuroendocrine regulatory network of the organism. There was a reorganization, or "orchestration," of many aspects of physiology, with specific tuning of the regulatory systems at different levels of organization, including the molecular genetic level, along with the triggering of epigenetic regulation of gene activity. This culminated in the "burst" of variability of the particular sort observed in animals under domestication.

Following this line of reasoning, Belyaev arrived at his concept of the important role of neuroendocrine stress in the process of evolution, particularly during domestication (Belyaev 1983). He suggested that stress acts as a factor generating and regulating genetic variability during critical periods in animal evolution, including during evolution through domestication. These ideas were new and pioneering at that time. The concept of genomic stress, and the appreciation of its evolutionary role, were introduced into biology much later.

As Belyaev hypothesized, selection that is associated with stressful conditions affects the major neurohormonal regulatory system (Gulevich, Oksina, Shikhevich, Fedorova and Trut 2004). As a result, the stress resulting from selecting for domesticated behavior triggers cascades of variations. Stress, which destabilizes homeostasis, may be an efficient mechanism for uncovering cryptic variability. Moreover, it became apparent that mutation frequency and the probability of recombination increase in stressful conditions (Belyaev and Borodin 1982; Parsons 1988; Agrawal and Wang 2008). Thus, selection targeting the major systems of developmental homeostasis results in increased heritable variability. Belyaev therefore had good reasons to call this selection "destabilizing" (Belyaev 1979). Its effect, he believed, is to release genetic variability associated with the destabilization of development and the breakdown of established correlations, and to bring about multiple sets of morphological and functional deviations.

The concept of destabilizing selection, which emerged as Belyaev reflected on the significance of his grand fox experiment, provides a novel vantage point for gaining insight into evolutionary change in challenging, unfamiliar, and extreme conditions. Belyaev's theory integrated earlier work of evolutionary biologists and those studying endocrinology and molecular biology. As noted earlier, Darwin drew attention to the fact that plasticity appears in organisms that happen to encounter unusual conditions. He wrote: "There can, however, be no doubt that changed conditions induce an almost indefinite amount of fluctuating variability, by which the whole organization is rendered in some degree plastic" (Darwin 1871:114). On another occasion, Darwin wrote: "As almost every part of the organization becomes highly variable under domestication, and as variations are easily selected both consciously and unconsciously, it is very difficult to distinguish between the effects of the selection of indefinite variations and the direct action of the conditions of life" (Darwin 1875: 414). Ivan Schmalhausen (1949), too, described the recruitment of an indefinite kind of variations when organisms are exposed to unusual conditions. Belayev's ideas were developed within this theoretical framework, and his concept of the destabilizing function of selection under stressful conditions extended and elaborated the ideas of his predecessors. He synthesized their ideas with another suggestion, now generally acknowledged, that under stress, one of the major ways through which hormones exert their effects is the regulation of gene activity. Belyaev wrote about this many years before the idea became generally recognized. As we know today, hormones, particularly the glucocorticoid stress hormones, act as endogenous regulators of DNA methylation (Thomassin, Flavin, Espinás, and Grange 2001). This property of hormones enables them to engage in the generation of a new type of variability during the stress periods that trigger a burst of variation. This variability is epigenetic. The hormonal regulation of epigenetic changes thus plays a role as an effective operational way of altering hereditary traits. The variations in external appearance that arose at high frequencies in the domesticated foxes, and that are homologous to the changes undergone by dogs, might therefore be explained in terms of epigenetic inheritance (Belyaev, Ruvinsky, and Trut 1981).

Further studies in Novosibirsk have focused on the relation between stress and variability (Markel and Borodin 1980). Our experiments, which were conducted with rat and mouse strains and interstrain hybrids, demonstrated that emotional stress can induce phenotypic manifestations of underlying genetic variations. For most of the functional parameters of the major physiological systems that were examined, the genetic variance and its additive component were significantly elevated under stress (Efimov, Kovaleva, and Markel 2005) (figure 17.3). Our studies have demonstrated that stress induces not only variability that may prove to be helpful in dealing with challenges, but also "inexpedient" responses. It is only through subsequent selection that stress-induced changes can be organized into a new specific adaptive response to the challenge, replacing the general adaptive stress response with a specific favorable one. We took advantage of the fact that genetic variability is generated and recruited by stress to develop, by selecting under stressful conditions, different strains of rats: a strain of rats with aberrant behavior (Kolpakov, Barykina, and Alekhina 1996), a strain with disrupted aggressive behavior and neuroendocrine changes (Naumenko, Popova. Nikulina, Dygalo, Shishkina, Borodin, and Markel 1989), and a strain with stress-sensitive arterial hypertension (Markel 1992; Markel, Maslova, Shishkina, Mahanova, and Jacobson 1999).

It has become apparent that stress can both uncover cryptic genetic variability, thereby providing selection with raw material, and be the cause of the de novo formation of genetic variations, that is, enhance mutagenesis (reviewed in Badyaev

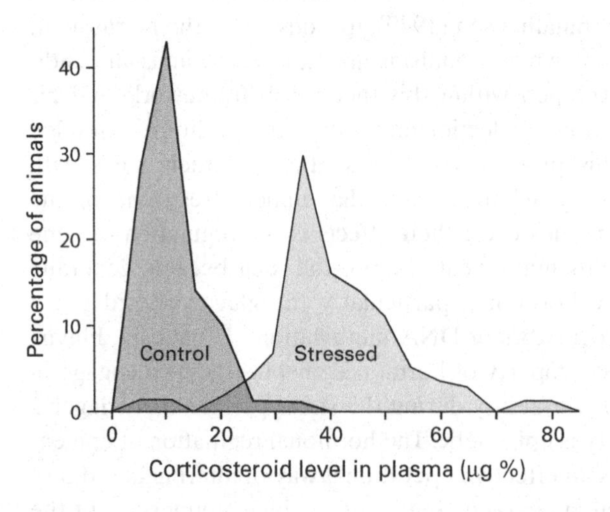

Figure 17.3
Variability in plasma 11-OH-corticosteroid hormone level in gray rats at rest and in stress conditions. In a genetically heterogeneous population, mild emotional stress (daily 0.5 hour restricted movement) significantly increased variation in hormonal blood plasma.

2005). The genetic variations generated by stress may give rise to new organismal features that, when targeted by natural (or artificial) selection, can bring into play more elaborate adaptive responses that replace the less effective ones that were initially used to adapt the organism to the provocative conditions. This seems to be a general type of evolutionary process, which may be important in bacteria as well as in multicellular organisms. Bjedov, Tenaillon, Gérard, Souza, Denamur, Radman, Taddei, et al. (2003) have aptly called bacteria "the champions of evolution" because bacteria have colonized virtually all ecological niches. From their study on 785 *E. coli* strains isolated from natural habitats, they concluded that bacterial success may be due to the ability to increase the mutation rate under the influence of environmental challenges. This observation was consistent with observations on *Chlamydomonas* (a singled-cell eukaryotic alga) demonstrating that various stressors (low temperature, changes in osmotic pressure and pH, starvation, toxic agents) increase the mutation rate (Goho and Bell 2000). It goes without saying that in higher organisms (vertebrates such as mammals) it is far more difficult to analyze stress-induced variability. Generally, the more sophisticated the organism's genetic program, the more plastic and potentially efficient its system of physiological adaptation and homeostasis. This allows higher organisms to adapt to a very wide range of environmental fluctuations by relying on their physiological responses rather than on mutational changes. Nevertheless, the main strategies of generating genetic variability in certain environmental conditions seem to persist in multicellular plants and animals, so stress-induced variability may be a factor in their evolution, too. For example, Badyaev and Foresman (2004) compared ecologically stress-induced morphological variations in the mandibles of six shrew species. Their interesting observation was that there was an increase in the variability of those parameters within which the shrew species differed from each other in natural conditions. The reasonable inferences from these data were that stress-induced changes in mandible size may be involved in evolutionary divergence between species, and that stress-induced variability in traits can serve as the raw material upon which selection and evolutionary reorganization act.

Major insights into stress-induced evolution came following Barbara McClintock's revolutionizing discoveries showing that the genome could no longer be thought of as a static, extremely conserved entity (McClintock 1984). She demonstrated a causal relationship between stress and the modification of the genome, claiming that "Some responses to stress are especially significant for illustrating how a genome may modify itself when confronted with unfamiliar conditions," and suggesting that ". . . stress, and the genome's reactions to it, may underlie many species formations." She arrived at these conclusions on the basis of her studies of the mobile genetic elements (transposons) that are scattered throughout the maize genome, and which, she showed, manifest a high sensitivity to genomic stresses. Vasilyeva, Bubenshchikova, and Ratner (1999) at Novosibirsk subsequently investigated the effect of heat-shock stress on the transposition of mobile elements that are dispersed throughout the *Drosophila* genome.

Stress, which occurs when organisms interact with unusual environments to which they are not adapted, thus seems to be the most common generator of neuroendocrine and genetic changes. It can operate as an important agent of evolution with far-reaching consequences. It seems plausible that any evolutionary innovation of significance can be generated when stress-induced destabilization occurs. When the established form breaks down, new types of organization may be generated. Belyaev (1984) suggested that the appearance of modern humans during hominid evolution occurred through the agency of stress-related neuroendocrine mechanisms. He argued that the changes brought about by the destabilizing effects of a challenged neuroendocrine system opened up new, untrodden evolutionary paths and occasional blind alleys, but eventually led to the rise of *Homo sapiens*. There is reason to believe that increased brain size in hominids was a main cause of their accelerated evolution (Reader and Laland 2002). An enlarged brain and a highly organized nervous system enable a more complex behavioral repertoire, thereby broadening perception and cognition, and allowing expansion to unconquered geographical and ecological areas. However, in doing so, they also increase the probability of encountering new stresses. Hence, metaphorically it may be said that stress is a creative agent of evolution in novel environments, where it plays the dual role of a selection factor and a specific generator of variability and adaptation. The emergence of wide-ranging variability enables selection to create more elaborate organisms with novel adaptive abilities. In this way, stress is an important part of the mechanism of evolution and the development of life.

Acknowledgment

This work was supported by grants from NIH (R03TW008098 and R01MH07781) and from the Russian Academy of Sciences.

References

Agrawal AF, Wang AD. 2008. Increased transmission of mutations by low-condition females: Evidence for condition-dependent DNA repair. PLoS Biol. 6,2:e30.

Badyaev AV, Foresman KR. 2004. Evolution of morphological integration. I. Functional units channel stress-induced variation in shrew mandibles. Am Nat. 163,6: 868–879.

Badyaev AV. 2005. Stress-induced variation in evolution: From behavioural plasticity to genetic assimilation. Proc R Soc Lond B Biol Sci. 272,1566: 877–886.

Belyaev DK. 1969. Domestication of animals. Sci J. 5,1: 47–52.

Belyaev DK. 1979. Destabilizing selection as a factor in domestication. J Hered. 70: 301–308.

Belyaev DK 1983. Stress as a factor of genetic variation and the problem of destabilizing selection. Folia Biol (Prague) 29: 177–187.

Belyaev DK. Genetics, society and personality. In: *Genetics: New Frontiers; Proceedings of the XV International Congress of Genetics.* Chopra VL, Joshi BC, Sharma RP, Bausal HC, eds. Vol. 4. New Delhi: Oxford University Press and IBH; 1984:379–386.

Belyaev DK, Borodin PM. 1982. The influence of stress on variation and its role in evolution. Biol Zentbl. 100: 705–714.

Belyaev DK, Ruvinsky AO, Trut LN. 1981. Inherited activation-inactivation of the star gene in foxes: Its bearing on the problem of domestication. J Hered. 72,4: 267–274.

Bjedov I, Tenaillon O, Gérard B, Souza A, Denamur E, Radman M, Taddei F. et al. 2003. Stress-induced mutagenesis in bacteria. Science 300,5624: 1404–1409.

Coppinger R, Feinstein M. 1991. "Hark! Hark! The dogs do bark . . ." and bark and bark. Smithsonian 21: 119–129.

Darwin CR. *On the Origin of Species by Means of Natural Selection, or the Preservation of Favoured Races in the Struggle for Life.* London: John Murray; 1859.

Darwin CR. *The Descent of Man, and Selection in Relation to Sex.* Vol 1. London: John Murray; 1871.

Darwin CR. *The Variation of Animals and Plants Under Domestication.* Vol 2. 2nd ed. London: John Murray; 1875.

Efimov VM, Kovaleva VY, Markel AL. 2005. A new approach to the study of genetic variability of complex characters. Heredity 94: 101–107.

Goho S, Bell G. 2000. Mild environmental stress elicits mutations affecting fitness in *Chlamydomonas.* Proc R Soc Lond B Biol Sci. 267: 123–129.

Gulevich RG, Oksina IN, Shikhevich SG, Fedorova EV, Trut LN. 2004. Effect of selection for behavior on pituitary–adrenal axis and proopiomelanocortin gene expression in silver foxes (*Vulpes vulpes*). Physiol Behav. 82,2–3: 513–518.

Kolpakov VG, Barykina NN, Alekhina TA. *Some Genetic Animal Models for Comparative Psychology and Biological Psychiatry.* Novosibirsk: Russian Academy of Sciences, Siberian Division; 1996.

Markel AL, Borodin PM. The stress phenomenon: Genetic and evolutionary approach. In: *Problems in General Genetics. Proc XIV Int Congr Genet.* Belyaev D.K., ed. Vol. 2, book 2. Moscow: Mir; 1980: 51–65.

Markel AL. Development of a new strain of rats with inherited stress-induced arterial hypertension. In: *Genetic Hypertension, Colloque INSERM.* Sassard J, ed. vol. 218. London: John Libbey Eurotext; 1992:405–407.

Markel AL, Maslova LN, Shishkina GT, Mahanova NA, Jacobson GS. Developmental influences on blood pressure regulation in ISIAH rats. In: Development of the Hypertensive Phenotype: Basic and Clinical Studies. Handbook of Hypertension. McCarthy R, Blizard D.A., Chevalier Rl, eds. Vol. 19. Amsterdam: Elsevier; 1999:493–526.

McClintock B. 1984. The significance of responses of the genome to challenge. Science 226: 792–801.

Naumenko EV, Popova NK, Nikulina EM, Dygalo NN, Shishkina GT, Borodin PM, Markel AL. 1989. Behavior, adrenocortical activity, and brain monoamines in Norway rats selected for reduced aggressiveness towards man. Pharmacol Biochem Behav. 33: 85–91.

Parsons PA. 1988. Behavior, stress, and variability. Behav Genet. 18,3: 293–308.

Reader SM, Laland KN. 2002. Social intelligence, innovation, and enhanced brain size in primates. Proc Natl Acad Sci USA. 99,7: 4436–4441.

Schmalhausen II. *Factors of Evolution: The Theory of Stabilizing Selection.* Philadelphia: Blakiston; 1949.

Thomassin H, Flavin M, Espinás M-L, Grange T. 2001. Glucocorticoid-induced DNA demethylation and gene memory during development. EMBO J. 20: 1974–1983.

Trut LN. The genetics and phenogenetics of domestic behavior. In: *Problems in General Genetics. Proc XIV Intern Congr Genet.* Belyaev D.K., ed. Vol. 2, book 2. Moscow: Mir; 1980:123–137.

Trut LN. 1999. Early canid domestication: The farm fox experiment. Am Sci. 87(March–April): 160–168.

Trut LN. Experimental studies of early canid domestication. In: *The Genetics of the Dog*, Ruvinsky A, Sampson J, eds. Wallingford, UK: CABI Publishing; 2001:15–41.

Trut L, Oskina I, Kharlamova A. 2009. Animal evolution during domestication: The domesticated fox as a model. BioEssays 31: 349–360.

Trut LN, Plyusnina IZ, Oskina I. 2004. An experiment on fox domestication and debatable issues of evolution of the dog. Russ J Genet. 40,6: 644–655.

Vasilyeva LA, Bubenshchikova EV, Ratner VA. 1999. Heavy heat shock induced retrotransposon transpositions in *Drosophila.* Genet Res. 74,2: 111–119.

18 The Role of Cellular Plasticity in the Evolution of Regulatory Novelty

Erez Braun and Lior David

Meno: "And how will you enquire, Socrates, into that which you do not know? What will you put forth as the subject of enquiry? And if you find what you want, how will you ever know that this is the thing which you did not know?"
—Plato, *Meno* (380 BC; translated by Benjamin Jowett)

As is evident from the biodiversity we see today, organisms are able to adapt to diverse environmental conditions. Complex life-forms evolved from much simpler ones by acquiring novel structures and functions. But how do such novelties emerge, and how are they utilized by organisms during evolution? King and Wilson (1975) suggested that changes in the function of genes, and thus also their evolution, can be driven not only by mutations in coding regions but also by changes in regulatory regions affecting gene expression. Today, the common understanding is that the evolution of regulatory systems is indeed crucially important in generating complexity and novel life-forms. Regulatory modes can be flexible and diverse, and comparative studies have identified cases in which novel phenotypes emerged through genome rewiring (i.e., linking existing genes to foreign regulatory systems), which created a new functional context for an existing gene (Carroll, Grenier, and Weatherbee 2001; Wilkins 2002; Davidson 2006). For example, mutations in cis-regulatory elements (e.g., promoter sequences) have enabled an existing protein to be used in new developmental processes (Wray 2007; Carroll 2005, 2008). However, the impressive set of data on novelty in gene regulation is insufficient to explain how such novelties emerged and became established because very little information exists on the dynamics of processes of adaptation, the role of phenotypic plasticity, and the intermediate states of evolving forms (West-Eberhard 2003). Until recently, relatively little effort was invested in studying directly, through laboratory experiments, the dynamics of evolving phenotypes and the underlying cellular processes that support adaptation.

The neo-Darwinian evolutionary framework attributes adaptation to the selection of advantageous phenotypes that exist in the population due to accumulation of genetic mutations which are rare and random, and occur independently of the selection process (Luria and Delbrück 1943; Drake, Charlesworth, Charlesworth, and Crow 1998; Elena and Lenski 2003). The potential of organisms to evolve depends crucially on the ability of their cells to withstand challenges that they have never before encountered (Gerhart and Kirschner 1997; Kirschner and Gerhart 1998). Mutations usually provide specific solutions to emerging challenges, but the multitude of possible unforeseen challenges raises the question of whether a more general mechanism, such as cellular plasticity, can provide an alternative strategy for evolution. In this chapter we describe and discuss our experimental program, which aims to study population adaptation dynamics simultaneously with the intracellular processes that accompany adaptation, and hence to decipher the mechanisms allowing cells to overcome and adapt to unforeseen challenges.

Experimental Model for the Evolution of Gene Regulation

Our experimental approach to studying the adaptation process in microorganisms allows us to interrelate the dynamic behavior of populations, individuals, and molecules during the short and intermediate timescales of an evolutionary process (Stolovicki, Dror, Brenner, and Braun 2006; Stern, Dror, Stolovicki, Brenner, and Braun 2007). In order to generate an unforeseen challenge, we genetically rewired a genome by linking an essential gene from one biochemical pathway to a regulatory system of a different pathway. In stressful environments, the rewired cells encounter a severe challenge in which the adverse conditions have to be dealt with simultaneously by both the foreign regulatory system and the essential gene that is coupled to it. In many cases these requirements are not easily fulfilled, and might even be contradictory. We used baker's yeast (*Saccharomyces cerevisiae*), a unicellular eukaryote, to study if and how cells can overcome the challenge and adapt to life with such a rewired genome.

Our main working model involves an engineered haploid strain in which the essential gene *HIS3*, which codes for an enzyme from the histidine biosynthesis pathway, was deleted from the chromosome and introduced back into the cell on a plasmid under the promoter of *GAL1*, a gene from the galactose utilization system. Histidine biosynthesis and galactose utilization are two evolutionarily conserved modules that are fundamentally different in their function and regulation. The GAL system, and with it the essential *HIS3*, are strongly repressed in glucose medium; as a result cells are severely challenged to produce histidine and significant adaptation is required for their survival (Stolovicki, Dror, Brenner, and Braun 2006; Stern, Dror, Stolovicki, Brenner, and Braun 2007).

Numerous Cells Can Adapt to Grow on Glucose Plates

We first demonstrated the adaptation capability of such rewired cells by using a colony growth assay on plates of culture medium. In all our experiments, we used histidine-lacking galactose or glucose-containing media (hereafter referred to as galactose or glucose), except for the controls, which were grown on rich medium. Naïve rewired cells, which had not been exposed to glucose previously, were sampled from a single colony and batch cultured in galactose medium. The sample of cells was washed and placed on glucose plates and an equal number of cells were placed on rich-medium plates as a viability control. On rich-medium plates, rewired cells grew like wild-type cells and formed mature colonies within two to three days, since histidine was supplied in the medium. In contrast, on glucose plates (lacking histidine) no colonies were observed until six days post plating. However, with additional incubation, colonies started to form and gradually accumulated until the maximal number of colonies was counted at twenty days post plating. Importantly, cells lacking any copy of *HIS3* could not grow on glucose plates, even after longer incubation times. Thus, cells gradually developed a way to operate *HIS3* from a GAL promoter during their incubation in the repressive glucose environment. Unexpectedly, the fraction of cells that could grow mature colonies on glucose was about 50 percent of the total number of viable cells that were plated, as estimated from colony numbers on rich-medium plates. This large fraction of mature colonies on glucose plates raised the possibility that their growth was merely a transient physiological response to cope with the stressful condition. However, it was not a transient response: cells from these colonies retained their growth capability over hundreds of generations. Furthermore, cells from colonies that grew on glucose and were plated on glucose for the second time formed mature colonies within two to three days, which is similar to their growth on the rich-medium plates, suggesting that the growth phenotype was memorized and did not require another adaptation period. Thus, the growth of 50 percent of the cells on glucose plates was a genuine adaptation process that was inherited across generations, allowing the survival of the genome-rewired strain in this challenging environment.

The large fraction of cells that adapted on the plates is unprecedented. Several evolutionary experiments have tested the rate at which microorganisms develop a new phenotype under stressful conditions. In most cases, genetic mutants arose at extremely low rates of one in 10^5–10^9 cells. However, other "irregular" adaptation phenomena, such as bacterial persistence and phenotypic switching, have been studied in the past, and in some of them relatively high rates at which adaptive phenotypes arose were found (Heidenreich 2007; Levin and Rozen 2006; Hallet 2001). In other classes of phenotypic switching, it was suggested that protein chaperones act as capacitors of phenotypic variation. When the environment changed (for example, following a temperature stress), the normal chaperone activity was modified, leading to the exposure of hidden genetic variation (Rutherford and Lindquist 1998; Queitsch, Sangster, and Lindquist 2002).

Studies of revertants, like in the case of antibiotic resistance, often identified gene mutations that explained the ability to grow in the presence of toxic antibiotics. Since in cases for which causative genetic mutations were identified, rates of phenotypic changes and frequencies of advantageous genetic variants were comparable, these studies established experimentally the role of mutations and genetic variation as the raw material and sole molecular mechanism for evolution. However, the other "irregular" adaptation phenomena suggested that mechanisms other than genetic mutations might be involved. In the present case of genome rewiring, if genetic mutations did indeed provide the mechanism for growth of every other cell on glucose, then either random mutations occurred at the conventional low rate and rare advantageous ones were selected in the population prior to the first exposure to glucose, or else these were adaptive mutations that occurred at a very high rate directly at the advantageous positions. Such adaptive mutations might arise as a response to stressful environment (Foster 2007). However, in spite of major efforts to identify such mutations, the nature of mutations that appear in stressful conditions is still being debated (Cairns, Overbaugh, and Miller 1988; Cairns and Foster 1991; Roth, Kugelberg, Reams, Kofoid, and Andersson 2006). Moreover, none of the reported adaptation phenomena in previous studies came even close to the 50 percent fraction of adapted cells observed in our experiments. Since naïve cells grew colonies only after six to twenty days, while for adapted ones it took only two to three days, the two most probable explanations for the high rate of adaptation are adaptive mutations and mechanisms other than mutations.

The Dynamics of Population Growth during Adaptation

Growth of individual cells on plates does not allow the dynamics of the processes occurring during population adaptation to be studied. We therefore devised a special experimental system allowing us to measure the growth dynamics of cell populations during the adaptation process in real time and at high temporal resolution. We developed a unique reactor, a chemostat, in which large cell populations grow for many generations under controlled and stable environmental conditions (Novick and Szilard 1950). The chemostat consists of a growth vessel, with homogeneous mixing, into which fresh medium is pumped at a constant rate while an efflux of medium and cells flows out at the same rate, thus keeping the culture volume constant. It should be noted that a steady cell density over time signifies a population with a fixed growth rate that equals the chemostat's dilution rate. A change in cell density reflects a change in average cellular metabolism, which, because it affects growth rate, also affects cell density. We combined this classic cell growth method with modern techniques for measuring quantities such as cell density and fluorescence, which can be applied to the culture online (Stolovicki, Dror, Brenner, and Braun 2006). In addition, an automated cell collector facilitated online collection and instantaneous freezing of cell aliquots at precise time intervals, so that the entire population history was available for later analyses at high temporal resolution.

Through a comprehensive series of experiments, we have shown that a population of cells carrying the GAL–*HIS3* rewired genome adapted to grow competitively in glucose medium in the chemostat (Stolovicki, Dror, Brenner, and Braun 2006). As in the plate experiments, the adapted state in the chemostat was memorized and remained stable for hundreds of generations, even under cycles of medium switches between glucose and galactose. Repeated experiments established that the population growth dynamics in glucose has four distinct phases. First, for about eight generations after switching the medium from galactose to glucose, the population increased its growth rate in spite of the challenge posed by the glucose environment (phase I). The observed increase in growth rate is expected for wild-type yeast cells, since glucose is a more efficient carbon source, thus permitting faster proliferation. For rewired cells, however, this is an unexpected behavior because of the observed strong repression of *HIS3* (Stolovicki, Dror, Brenner, and Braun 2006). We still do not fully understand this phase, but it is fair to assume that the prolonged exponential growth relied on existing reservoirs of histidine or His3p from the previous period in galactose, which allowed a limited number of cell divisions. Nevertheless, following the initial growth in phase I, the population collapsed and the cell density dropped exponentially, indicating that histidine production was limited and the cells were facing a severe challenge (phase II). In the next stage, the population reached a turning point in which the cell density started to increase, indicating that most cells were growing faster than the chemostat's dilution rate (phase III). Finally, stable growth was achieved, and the cell density remained constant thereafter (phase IV). The turning point from phase II to phase III indicated the time at which the population contained a majority of adapted cells, and placed an upper limit of ten to thirty generations on the adaptation time observed in repeated experiments. Remarkably, this adaptation time is extremely short compared with those encountered in experiments involving the fixation of spontaneous mutations in microorganisms, which are on the order of hundreds to thousands of generations (Paquin and Adams 1983; Lenski and Travisano 1994; Perfeito, Fernandes, Mota, and Gordo 2007). Thus, the growth dynamics of chemostat populations, like the growth of colonies on plates, suggests that either adaptive mutation or mechanisms other than mutation might be operating.

Plasticity of Gene Regulation

Genome rewiring imposes major challenges on the regulation of gene expression, which should be reflected in changes of mRNA levels. The high temporal resolution cell sampling in our chemostat allowed direct measurements of mRNA levels along the entire time course of the population's adaptation. The regulation of the GAL genes was adaptively modified, as was evident from changes in mRNA expression levels, which were correlated with the population growth dynamics (Stolovicki, Dror, Brenner, and Braun 2006). These measurements proved that the adaptation mechanism had involved utilization of *HIS3* through the emergence of novel regulatory feedbacks. *HIS3*, which is normally expressed in concert with other amino acid biosynthesis genes, was removed from the regulation of its natural functional module, and instead followed the gene expression patterns of the GAL system. In turn, the GAL genes were first repressed in glucose, as expected, but were then de-repressed as adaptation proceeded and allowed expression of *HIS3*. Remarkably, all the GAL genes, which share similar promoters but reside on different chromosomes, responded to the pressure on the histidine pathway and showed similar changes in expression level, suggesting that the adaptation is not a local feature of *HIS3* regulation, but rather a more global regulatory effect. Further evidence for the emergence of such novel regulatory feedbacks was obtained by applying a stronger environmental pressure directly on the His3p enzyme and, thus, on the histidine pathway. We used 3-amino-1,2,4-triazole (3AT), a known competitive inhibitor of His3p enzymatic activity, which has negligible side effects. Applying 3AT enhanced the transcriptional response of the GAL and *HIS3* genes compared with that without 3AT, thus confirming the emergence of a new regulatory feedback enabling tuning of the *HIS3* and the GAL genes according to the requirements of the histidine pathway. These results clearly demonstrate that the emergent ability of cells to grow on glucose was at least in part due to the plasticity of gene regulation and its capacity to rapidly create novel feedback systems.

Adaptation: A Response of Individual Cells to Environmental Change

According to the classical view, evolutionary adaptation is seen as the response of a population to a new and challenging environment. Adaptation relies on selection of individuals with advantageous phenotypes which exist in the population because of the accumulation of random and rare mutations that are independent of the environmental challenge. Our chemostat system allowed us to sample the population at consecutive time points, plate the extracted cells, and follow the growth patterns of individual colonies in order to understand the overall population growth dynamics (David et al. 2010). The plating analysis of samples from phase I revealed that despite the bulk population growth in the chemostat during this phase, none of the cells formed colonies on the plates after two to three days, indicating that no observable fraction of the cells was adapted. Moreover, had adapted cells been present, they would have taken over the culture by selection and prevented the population's collapse in phase II. In contrast, colony counts from the same glucose plates at twenty days post plating indicated that individual cells were able to create adapted colonies, just as in the plate adaptation experiments described earlier. If none of the seed cells were adapted during the first eight generations in the chemostat, but many adapted colonies emerged after further incubation of up to twenty days on the plates, then the adaptation process must have taken place on the plates rather than in phase I in the chemostat. Therefore, the adaptation process must have been a heritable response that was induced in many individual cells by the challenging glucose environment. Since it was only after exposure to glucose that many cells developed an adaptive solution, selection of preexisting, advantageous but rare genetic variants is an improbable explanation of how cells achieved this phenotype. Moreover, the high fraction of cells that developed this response suggests that there may be multiple ways to achieve an adaptive phenotype, and different ways may have been adopted by different cells. Since the adaptive phenotype was induced by the environment and yet was inherited, this genuine adaptation process and the mechanism that enabled it necessitate an extension of the classical neo-Darwinian framework that has far-reaching implications in many areas of biology.

Genetics and Epigenetics Played a Role in Adaptation

Traditionally, heritable phenotypes have been identified with genetic determinants, and since the adaptation we study is heritable, the involvement of DNA mutations in the engineered cells was investigated. First, we sequenced the GAL promoter upstream to *HIS3* in several adapted cells; no mutations were found there. Next, we used crosses and phenotypic analysis of progeny to identify possible mutations. One adapted haploid cell, isolated from the steady-state phase IV of a chemostat experiment, was crossed with its ancestral naive strain to form a diploid that was sporulated to obtain tetrads (i.e., four haploid spores, which are the products of a single meiosis). Phenotypic analysis of the meiotic products revealed a 2:2 segregation ratio between adapted and nonadapted spores, indicating that a single-locus modification could account for the adaptive phenotype. This modification was later identified as a mutation in that particular clone, and further analysis showed similar mutations in other cells from repeated chemostat experiments (unpublished results). However, in several other cells that exhibited a similar adaptive phenotype, such mutations were absent. Moreover, even in cells carrying that mutation, the mutation by itself was insufficient for the adaptive response; propagating adapted cells carrying the mutation on glucose plates revealed that only a fraction of such haploid mutated progeny cells could form new colonies within two to three days. Thus, although the causative genetic change was common to all progeny cells, only a fraction of them exhibited the phenotype. Notably, in proliferating cultures, the adapted state was inherited at a sufficient efficiency to maintain the population's stable growth, as observed in phase IV of the chemostat. Thus, despite the incomplete inheritance of the phenotype, the adaptation we described is a genuine and significant evolutionary phenomenon.

Further crosses between adapted cells not carrying the mutation and a naive ancestral strain have been analyzed. In contrast to the previous tetrad analyses, in which a 2:2 segregation ratio was the rule, in spores of crosses involving cells without the mutation, three major patterns of inheritance were found: (a) 2:2 segregation ratio between adapted and nonadapted spores, (b) none of the meiotic products were adapted (0:4), and (c) all four meiotic products were adapted (4:0). Moreover, in repeated crosses of the same parents, one diploid would show a 2:2 segregation ratio, while another showed a 4:0 ratio, and a third showed a 0:4 ratio (unpublished results). These results indicate that there were multiple adaptive solutions that shared a similar phenotype. No less important, in the absence of the mutation, the phenotype of the adapted parent could either segregate, disappear from, or be gained by all the meiotic segregants of sister diploids, regardless of their identical DNA sequences. This provides strong support for the involvement of an epigenetic mechanism, of as yet unknown nature, that is responsible for the stable transgenerational inheritance of this adaptive state (Lachmann-Tarkhanov and Jablonka 1996; Allis, Jenuwein, and Reinberg 2007). Epigenetic adaptation might generate a heritable state that will bridge the gap in time between the rapid physiological responses to accommodate the stressful condition, and the slow evolutionary adaptation that relies on random and rare genetic variability to create a long-term stable solution under the new genomic and environmental constraints (Jablonka and Lamb 2005; Rando and Verstrepen 2007).

Global Transcriptional Responses in the Process of Adaptation

So far, we have described the regulatory changes directly related to the GAL system, but since regulatory networks are interconnected, we also looked for more global responses that might underlie the adaptation. Measurements of the genome-wide transcriptional response supported the picture of a global regulatory response (Stern, Dror, Brenner, and Braun 2007). The mRNA expression levels of all yeast genes were measured by cDNA microarrays and revealed that one fifth of the genes (>1200) were either induced or repressed upon the first exposure of the rewired cells to glucose. Genes that were induced or repressed retained their state over many generations, during which the population became adapted. Therefore, the global transcriptional response was reflected in changes in the expression levels of many genes, which were maintained for many generations, thus indicating stable transgenerational inheritance that is probably due to a global heritable epigenetic process.

Since many genes coordinately changed their expression, a strong correlation was found between the expression pattern of hundreds of genes representing various metabolic pathways, irrespective of whether they should be up-regulated or down-regulated according to their annotated functionality. Moreover, increasing the environmental pressure on *HIS3* by applying 3AT enhanced these global correlations. Thus, in our experiments, *coexpression* of genes did not imply their *cofunctionality*; we have also shown that *cofunctionality* did not imply *coexpression*, because for most of the metabolic modules both induced and repressed enzymes were found within the same module (Stern, Dror, Stolovicki, Brenner, and Braun 2007). Even more remarkable, comparing repeated chemostat experiments under precisely identical nominal conditions revealed that although the growth dynamics of the population and the global pattern of the transcriptional response were similar, the gene content of that response was different, with the exception of about 15 percent of conserved core genes (Stern, Dror, Stolovicki, Brenner, and Braun 2007). Taken together, it seems that instead of providing specific functional activity to support an adaptive solution, the global transcriptional activity created a response envelope enabling the emergence of adaptation solutions.

Concluding Remarks and Future Outlook

The classical view of evolutionary adaptation lacks the dynamic dimension that, as our experimental studies show, can yield insights that go beyond the current evolutionary framework. In our experiments, numerous cells simultaneously responded to an unforeseen challenge—in this case caused by genome rewiring—with a rapid adaptation process that could then be stably inherited for many generations. The regulatory perturbation evoked strong transcriptional responses to resolve the frustrating requirements both locally, by creating novel regulatory feedbacks, and globally, by providing a response envelope that supported the emergence of adaptive solutions. Our results confirm the plasticity of gene regulation and its capacity to accommodate arbitrary genome rewiring events, and place this plasticity as another type of raw material for evolution besides mutations. In this unique adaptation process, both genetic and epigenetic factors were found to play a role and, in agreement with the concept of cellular plasticity, they created multiple adaptive solutions, all of which shared a similar growth phenotype that was advantageous in the new environment. Following the dynamic processes of evolution in real time has suggested a plausible order of events in evolution. First, physiological accommodation is made possible by the cellular plasticity of gene regulation. Second, the physiological accommodation is stabilized by heritable epigenetic mechanisms that make the phenotype adaptive. Third, the adaptive cell state is ultimately fixed for much longer durations by genetic mutations and modifications. It remains to be seen how general these results are, and what their implications are for multicellular organisms (Jablonka and Raz 2009).

Starting from experiments that aimed to study adaptation to an unforeseen challenge and explore the potential of cells to evolve, we have discovered that the most basic concepts of cell biology require some rethinking. It is necessary to take into consideration the role of epigenetic processes that both maintain cellular homeostasis through changing environments and act as heritable mediators between genotype and phenotype by mitigating incompatibilities and contributing to plastic regulatory responses. The common view that regards the genome as a "program," the environment as an "input signal," and the phenotype as a "logical output" of the cellular "computing device" should therefore be revisited. The alternative, more natural view presented here regards the cell as a flexible dynamical system constrained by both the genome and the environment. This perspective seems to be consistent with the phenotypic plasticity observed in our experiments. Further work along these lines will shed light on the universality of the phenomena we have uncovered and on their implications for our understanding of evolution.

References

Allis CD, Jenuwein T, Reinberg D, eds. *Epigenetics.* Cold Spring Harbor, NY: Cold Spring Harbor Laboratory Press; 2007.

Cairns J, Foster PL. 1991. Adaptive reversion of a frameshift mutation in *Escherichia coli.* Genetics 128: 695–701.

Cairns J, Overbaugh J, Miller S. 1988. The origin of mutants. Nature 335: 142–145.

Carroll SB. 2005. Evolution at two levels: On genes and form. PLoS Biol. 3: 1159–1166.

Carroll SB. 2008. Evo-devo and an expanding evolutionary synthesis: A genetic theory of morphological evolution. Cell 134: 25–36.

Carroll SB, Grenier JK, Weatherbee SD. *From DNA to Diversity: Molecular Genetics and the Evolution of Animal Design.* Malden, MA: Blackwell Science; 2001.

David L, Stolovicki E, Haziz E, Braun E. 2010. Inherited adaptation of genome-rewired cells in response to a challenging environment. HFSP J. 4: 131–141.

Davidson EH. *The Regulatory Genome.* Amsterdam: Elsevier; 2006.

Drake JW, Charlesworth B, Charlesworth D, and Crow JF. 1998. Rates of spontaneous mutation. Genetics 148,4: 1667–1686.

Elena SF, Lenski RE. 2003. Evolution experiments with microorganisms: The dynamics and genetic bases of adaptation. Nat Rev Genet. 4: 457–469.

Foster PL. 2007. Stress-induced mutagenesis in bacteria. Crit Rev Biochem Mol Biol. 42,5: 373–397.

Gerhart J, Kirschner M. *Cells, Embryos, and Evolution.* Malden MA: Blackwell Science, 1997.

Hallet B. 2001. Playing Dr Jekyll and Mr Hyde: Combined mechanisms of phase variation in bacteria. Curr Opin Microbiol. 4,5: 570–581.

Heidenreich E. 2007. Adaptive mutation in *Saccharomyces cerevisiae.* Crit Rev Biochem Mol Biol. 42,4: 285–311.

Jablonka E, Lamb MJ. *Evolution in Four Dimensions.* Cambridge, MA: MIT Press; 2005.

Jablonka E, Raz G. 2009. Transgenerational epigenetic inheritance: Prevalence, mechanisms and implications for the study of heredity and evolution. Q Rev Biol. 84: 131–176.

King MC, Wilson AC. 1975. Evolution at two levels in humans and chimpanzees. Science 188: 107–116.

Kirschner M, Gerhart J. 1998. Evolvability. Proc Natl Acad Sci USA. 95: 8420–8427.

Lachmann-Tarkhanov M, Jablonka E. 1996. The inheritance of phenotypes: An adaptation to fluctuating environments. J Theor Biol. 181: 1–9.

Lenski RE, Travisano M. 1994. Dynamics of adaptation and diversification: A 10,000-generation experiment with bacterial populations. Proc Natl Acad Sci USA. 91,15: 6808–6814.

Levin BR, Rozen DE. 2006. Non-inherited antibiotic resistance. Nat Rev Microbiol. 4: 556–562.

Luria SE, Delbrück M. 1943. Mutations of bacteria from virus sensitivity to virus resistance. Genetics 28: 491–511.

Novick A, Szilard L. 1950. Experiments with the chemostat on spontaneous mutations of bacteria. Proc Natl Acad Sci USA. 36: 708–719.

Paquin C, Adams J. 1983. Frequency of fixation of adaptive mutations is higher in evolving diploid than haploid yeast populations. Nature 302: 495–500.

Perfeito L, Fernandes L, Mota C, Gordo I. 2007. Adaptive mutations in bacteria: High rate and small effects. Science 317,5839: 813–815.

Queitsch C, Sangster TA, Lindquist S. 2002. Hsp90 as a capacitor of phenotypic variation. Nature 417: 618–624.

Rando OJ, Verstrepen KJ. 2007. Timescales of genetic and epigenetic inheritance. Cell 128: 655–668.

Roth JR, Kugelberg E, Reams AB, Andersson DI. 2006. Origin of mutations under selection: The adaptive mutation controversy. Annu Rev Microbiol. 60: 477–501.

Rutherford SL, Lindquist S. 1998. HSP90 as a capacitor for morphological evolution. Nature 396: 336–342.

Stern S, Dror T, Stolovicki E, Brenner N, Braun E. 2007. Genome-wide transcriptional plasticity underlies cellular adaptation to novel challenge. Mol Syst Biol. 3: 106.

Stolovicki E, Dror T, Brenner N, Braun E. 2006. Synthetic gene recruitment reveals adaptive reprogramming of gene regulation in yeast. Genetics 173: 75–85.

West-Eberhard MJ. *Developmental Plasticity and Evolution.* New York: Oxford University Press; 2003.

Wilkins AS, ed. *The Evolution of Developmental Pathways.* Sunderland, MA: Sinauer Associates; 2002.

Wray GA. 2007. The evolutionary significance of cis-regulatory mutations. Nat Rev Genet. 8: 206–216.

19 Evolutionary Implications of Individual Plasticity

Sonia E. Sultan

Phenotypic plasticity is broadly defined as the ability of a single genotype to express different phenotypes in different environments; the term denotes all aspects of an organism's phenotype in which expression varies as a result of variation in environmental conditions. For any given organism, this can include a broad array of developmental, physiological, behavioral, and life-history traits that vary in response to all kinds of abiotic and biotic factors—from temperature, pH, or resource levels to the presence of symbionts, predators, or competing neighbors. Because real-world environments inevitably vary in both space and time, such plasticity is the source of enormous phenotypic diversity among individual organisms in nature, a fact that has recently become a central focus of evolutionary research (Sultan 2000; Pigliucci 2001; DeWitt and Scheiner 2004; Sultan and Stearns 2005).

The Eco-Devo Approach

Studying plasticity requires replacing a simplified, Fisherian view of variation with an ecological developmental or "eco-devo" model (Gilbert 2001; Gilbert and Epel 2009) that explicitly focuses on the range of phenotypes each genotype can express in response to alternative environmental circumstances (figure 19.1). Such developmental flexibility is particularly dramatic in plants and other sessile organisms, but it is true to some extent of all organisms; indeed, biological experiments are carried out in controlled conditions precisely in order to exclude environmental variation that might influence phenotypic expression. To study plasticity, it is necessary to bring environmental variation *into* the experimental design: to determine patterns of plasticity, cloned or inbred replicates of individual genotypes are raised in a range of ecologically relevant environments, and phenotypic traits of interest are measured in each environment. The alternative trait states are plotted against an environmental axis to depict the genotype's "norm of reaction" for that trait—its repertoire of environmentally contingent phenotypic responses (e.g., Gupta and

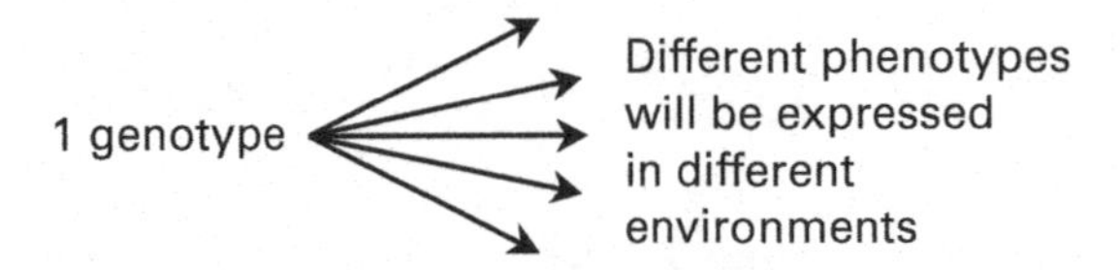

Figure 19.1
An eco-devo view of the genotype–phenotype relation. In contrast to the determinate view of phenotypic expression inherent in the New Synthesis, an ecological development model posits that a given genotype can give rise to various phenotypes, depending on environmental conditions.

Lewontin 1982; Sultan and Bazzaz 1993; Lindroth, Roth, and Nordheim 2001; Kingsolver, Ragland, and Shlichta 2004).

The norm-of-reaction approach makes explicit the key point (proposed a century ago by both the German zoologist Richard Woltereck and the Danish plant geneticist Wilhelm Johannsen) that the phenotype expressed by a given genotype depends on the environmental context, and the precise phenotypic effect of a given environment depends on the genotype in question as a particular developmental response system (discussed by Lewontin 2000; Sarkar 2004). As is now broadly recognized, genotype and environment are not alternative mechanisms of phenotypic determination; rather, the specific interaction of genetic and environmental factors jointly determines phenotypic outcomes. A second foundational point is that these individual response patterns are genomic properties that are inherited and evolve (Via 1987; Gomulkiewicz and Kirkpatrick 1992; Scheiner 1993, 2002; Windig 1994; Nusset, Postma, Gienapp, and Visser 2005). Like other products of evolution, they are necessarily shaped by phylogenetic history and genetic constraints, as well as by natural selection and random drift. Consequently, norms of reaction vary among genotypes, populations, and even closely related species (e.g., Gupta and Lewontin 1982; Sultan 2001; Donohue, Pyle, Messiqua, Heschel, and Schmitt 2001; Heschel, Sultan, Glover, and Sloan 2004).

Inevitable and Adaptive Aspects of Plastic Response

To consider the evolutionary implications of plasticity, it is useful to conceptually distinguish two aspects of environmental response: (1) inevitable, often maladaptive direct effects of resource limitation or other stresses on developmental processes, and (2) functionally adaptive responses—that is, specific adjustments of trait expression that enhance function in the eliciting environments. For instance, although plants of a given genotype are inevitably smaller in total mass if grown in low rather than high light (due to the direct reduction of photosynthetic carbon assimilation under reduced photon flux density), they also allocate proportionately more of their tissue to leaves, and make those leaves larger and thinner. This suite of developmental adjustments steeply increases the ratio of photosynthetic surface area per gram of plant tissue, allowing the plant to maintain fitness in limited light by maximizing its photon harvesting capacity. This functional adjustment aspect of plasticity is particularly interesting, because it allows adaptation to occur at the level of individual organisms. However, both inevitable and adaptive aspects of environmental response influence reproductive fitness, so both contribute to realized fitness differences among individuals (Sultan 2003a). From a norm-of-reaction perspective, the genetic component of fitness variation can be understood as differences in these repertoires of response. Note that inevitable and adaptive aspects of environmental response are not always entirely exclusive. For instance, the greater proliferation of a plant's roots in warmer soil pockets reflects the direct impact of higher temperature on cell division rates, but this response also enhances the functional efficiency of the individual plant's root system since uptake rates will also be higher in these zones of proliferation. Adaptive aspects of individual response likely have been shaped by selection to modulate direct environmental effects, just as selection to minimize direct environmental or genetic perturbation has produced canalized responses to hold certain key developmental traits more constant (Nijhout 2003).

Genetic Diversity for Norms of Reaction

To understand how both inevitable and functionally adaptive plasticity can influence selective evolution, it is necessary to consider genetic diversity among norms of reaction in the context of the environmental variation that elicits different phenotypes and determines their fitness consequences. For simplicity, we can examine norms of reaction in two hypothetical genotypes for a trait that influences fitness (figure 19.2). One possibility is that different genotypes may converge on identical norms (plasticity patterns) for a given trait (figure 19.2a). In such a case, due to strong previous selection or to inherent (developmental or genetic) constraints, there is no genetic variation for the environmental response pattern. Note that although there is no genetic variation for phenotypic expression, phenotypic variation can be elicited *from each genotype*. When genotypes express distinct phenotypes, in the simplest case they will alter phenotypic expression the same way from

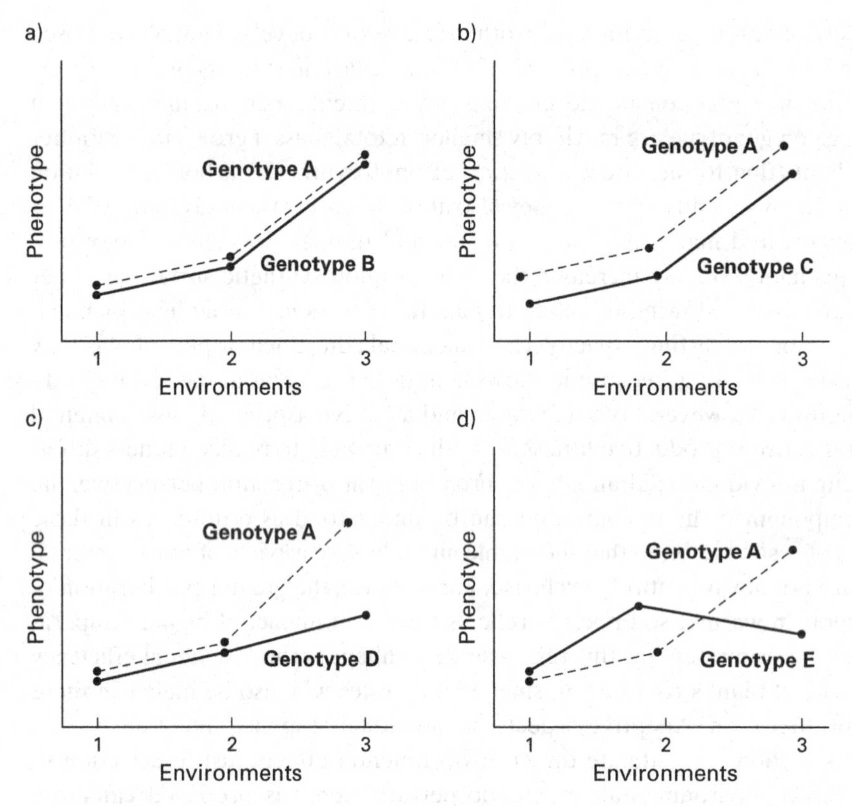

Figure 19.2
Genetic diversity for norms of reaction. Possible patterns of norm-of-reaction diversity are shown for hypothetical genotypes A–D. In each frame, trait values expressed by each genotype in three alternative environments are plotted for some phenotype of interest.

one environment to another, for instance by increasing or decreasing trait value to the same extent. In the case of such common responses, norms of reaction will be parallel across the environmental range (figure 19.2b), and selection will consistently favor the same genotype, and to the same extent, in any environment (in this example, genotype A).

In naturally evolved systems, however, genotypic norms of reaction are rarely parallel—a point made by J.B.S. Haldane in his elegant 1946 paper on "nature and nurture," and confirmed by innumerable quantitative genetics studies in recent decades. Instead, norm-of-reaction data almost always show that genotypes express different patterns of response across a given environmental range, a situation described as *genotype by environment interaction*, or *g x e variation* (Falconer and Mackay 1996). This complex aspect of genetic diversity can be understood as different genomes altering expression differently in response to the same environmental change. It is this type of genetic variation that provides the potential for the evolution of plasticity itself.

Selective Implications of Genetic Variation in Plasticity Patterns

Two distinct aspects of g x e variation affect the outcome of natural selection in key ways. First, the magnitude of phenotypic differences between a given set of genotypes can vary from one environmental state to another, such that genotypes can produce similar or identical phenotypes in certain environments and markedly different ones in others (figure 19.2c). Here, the selective outcome will depend on the relative frequency of the alternative environmental states that the population encounters, which is determined by both the spatial/temporal distribution of environments and gene flow across populations. For instance, if this two-genotype population encountered only environment (3), then genotype A would be selectively fixed. Conversely, if the environment varied from state (1) to state (2) to state (3), the elimination of genotype D would be slowed (the precise rate of selective elimination will depend on the frequency of the environment where the fitness difference is expressed). One important real-world consequence of this property is that adding a novel environment to a population's experience—increased concentration of atmospheric carbon dioxide, for instance, or higher seawater temperatures—can either *trigger* a selective event by revealing g x e variation, or *buffer* selection, depending on whether norms of reaction are similar or different in that environment (Sultan 2007). This aspect of g x e interaction is one reason why the evolutionary outcomes of altered environments are so unpredictable—or, more precisely, are predictable only in taxa for which norm-of-reaction diversity is known in the new conditions.

A second aspect of g x e variation that influences selective outcomes is that the rank order of genotypes for a given trait may change or "cross over" from one environmental state to another (figure 19.2d). In such a case, a genotype may have the highest fitness in certain environments, for instance, genotype E in environments 1 and 2 (figure 19.2d), but the lowest fitness elsewhere, for example, in environment 3 (figure 19.2d). As in the previous situation, the selective outcome in such a case will depend on which environments are encountered by the population. In this example (figure 19.2d), if environments 2 and 3 represent isolated sites, genotype E would be predicted to be fixed in environment 2 and genotype A in environment 3—that is, selection should result in the evolution of genetically distinct local specialists. If instead the environment varies within a site, or there is gene flow between sites, both genotypes would be maintained; in these instances, selection will favor the pattern of plasticity that results in the highest net fitness across the entire set of environments (given their relative frequencies), leading to broadly adaptive norms of reaction rather than to specialists for each particular environment (De Jong 1995; Sultan and Spencer 2002).

Norm-of-reaction data reveal a complex picture of genetic diversity, since it is very rare for one genotype to be superior or inferior in all environments (Haldane 1946). Although certain environments can be relatively favorable or unfavorable for all genotypes, the magnitude of those environmental effects is likely to vary among genotypes, and for a given set of genetic individuals both their degree of difference for a trait and their rank order for that trait can vary from one set of conditions to another. The ubiquity of such g x e interaction, together with the fact that real environments vary, may partly explain why such high levels of genetic variation are found in naturally evolved populations, even in fitness-related traits that should be under strong selection (Gillespie and Turelli 1989).

Plasticity Differences at the Population and Species Levels

When different populations of a species occupy contrasting habitats, both genetic constraints (due to founder effects or drift) and local selective pressures may lead to the evolution of distinct patterns of plasticity, with particular adaptive consequences. Schmitt and colleagues have carefully studied how plant developmental responses to shade cast by neighbors (perceived as a change in the ratio of red to far-red wavelengths when light passes through or reflects off nearby plant tissues) influence selective change in natural populations of the herbaceous annual *Impatiens*. Their studies show that plasticity to elongate shoots in the presence of neighbors (presumably to compete more effectively for light) is under strong selection in populations inhabiting open sites, but not in woodland sites, where such a response to shade cues from a tree canopy would be both futile and maladaptive (e.g., Donohue, Pyle, Messiqua, Heschel, and Schmitt 2001). Ecotypes of a species thus may be characterized not by distinct trait states but by habitat-specific plasticity patterns.

In other cases, populations consisting of highly plastic "generalist" genotypes may occupy diverse habitats based on similar patterns of adaptive plasticity. For instance, populations of the weedy colonizing plant *Polygonum persicaria* from variably dry versus consistently moist sites showed broadly similar patterns of individual plasticity in response to drought stress: when grown in dry soil, genotypes from both types of population increased tissue allocation to roots (for water uptake) as well as physiological water use efficiency (the ratio of photosynthetic carbon fixation to water loss), and were equally effective at maintaining reproductive output despite limited water (Heschel, Sultan, Glover, and Sloan 2004). Specific patterns of plasticity did differ in these populations, however. Genotypes from the consistently moist site expressed a greater physiological adjustment but a less dramatic change in root tissue allocation compared with genotypes from variably dry sites. Possibly a genetic constraint on the physiological response to drought in certain populations resulted in stronger selection for sharply increased root allocation, or vice versa. The equivalent drought tolerance of these populations despite these plasticity differences illustrates how genotypes can achieve similar levels of fitness across environments by means of different combinations of underlying trait adjustments.

At the species level, too, phylogenetic and genetic constraints along with previous selection have resulted in distinct patterns of plastic response that form an important dimension of ecological diversity. To return to the example of *Polygonum* plants, naturally evolved genotypes sampled from four closely related species in this monophyletic group express contrasting patterns of allocational, morphological, physiological, and reproductive plasticity in response to light, moisture, and nutrient availability (Sultan 2003b and references therein). The species differ in their capacities for functionally adaptive response to these resource limits, such as proportionally increasing leaf area in low-light conditions and boosting water uptake capacity in full sun. These species differences in plasticity may be an important factor in shaping their contrasting environmental distributions.

Species Differences in Transgenerational Plasticity

The range of environments a species inhabits may also be influenced by a particularly intriguing aspect of adaptive plasticity that has been recognized only recently: ways that parent (generally maternal) plants and animals alter traits of their *offspring* in response to specific environmental stresses so as to enhance offspring success under those stresses (Mousseau and Fox 1998; Rotem, Agrawal, and Kott 2003; Mondor, Rosenheim, and Addicott 2005; Galloway and Etterson 2007). In such cases, if the offspring encounter an environment similar to that of their parent, this transgenerational aspect of plasticity constitutes a remarkable, nongenetic mode of inherited adaptation—a well-substantiated biological phenomenon that many might consider "Lamarckian."

For instance, in a 1999 *Nature* article, Agrawal and colleagues reported a transgenerational experiment with the well-studied planktonic crustacean *Daphnia* (Agrawal, Laforsch, and Tollrian 1999). These water fleas develop a defensive "helmet" morphology in the presence of invertebrate predators that release a chemical that acts as a cue for this adaptive plastic response (helmeted individuals suffer less mortality, since predators cannot as successfully capture them). In this dramatic study, *Daphnia* mothers exposed to the chemical cue produced offspring that expressed the helmet morphology as neonates, even if the predator cue was absent from the offsprings' own environment.

Similarly, drought-stressed parent plants of *Polygonum persicaria* produced seedling offspring with longer, more rapidly extending root systems, resulting in significantly greater seedling growth in dry soil, compared with the offspring of genetically identical parents that had been given ample water (Sultan, Barton, and Wilczek 2009). In this experiment, a difference in parental environment alone led to significant changes in offspring development and total mass, a strong correlate of reproductive output. In contrast, the closely related species *P. hydropiper* expressed a direct, maladaptive effect of parental drought stress on offspring; in this species, drought-stressed parents simply produced smaller seedlings with correspondingly slower-extending root systems that grew less in both moist and dry soil environments, compared with the offspring of well-watered parents (Sultan, Barton and Wilczek 2009). Since the survival of seedlings or neonates is a critical factor in the ability of species to establish populations, such species differences in transgenerational plasticity could influence relative success in establishing populations in resource-poor or variable sites, and hence could shape the ecological breadth of taxa in nature (Fox and Savalli 2000; Sultan 2000). Indeed, this appears to be the case with *Polygonum*. Although *P. persicaria* occurs in a broad range of moisture conditions in the field, *P. hydropiper* does not establish populations in either dry or variably dry sites. Note that selection on transgenerational norms of reaction shows complex dynamics because of the lag period and potential adaptive mismatch between the parental environment and the offspring phenotype (Donohue 2005).

Conclusions: Plasticity and Future Evolution

Individual patterns of environmental response vary among genotypes, populations, and species in ways that influence both selective diversification and ecological distribution. These response patterns can thus be understood as a central aspect of adaptive diversity. Differences among taxa in individual norms of reaction are likely to bear important consequences for biodiversity, as humans continue to alter and disrupt both abiotic and biotic conditions in natural systems. Organisms with broad repertoires of adaptive plasticity may be better able to withstand these rapid environmental changes, which generally occur too quickly to allow for an evolutionary response in macroorganisms. Consequently, differences among species in existing, previously evolved norms of reaction may influence which taxa are able to persist in the face of human-mediated environmental changes. Plasticity patterns may also affect evolutionary diversification, since taxa with functionally plastic individuals may be less likely to form locally specialized populations that can give rise to allopatric species. Just as the capacity for individual plasticity can enhance a species' ecological breadth, so broadly adaptive norms of reaction can promote a species' invasiveness in a novel geographic range; indeed, many invasive plant and animal species are such ecological generalists.

Norm-of-reaction insights also inform predictions about potential evolutionary response to the new environments that humans are so rapidly creating. Recall that any evolutionary change in plasticity depends on the presence of genotype x environment variation—that is, genetic variation for plasticity patterns. For instance, if certain genotypes are less phenotypically disrupted than others in new environments, selection will favor these genotypes, leading to the evolution of more canalized or homeostatic norms of reaction in response to those new stresses. However, if all genotypes are equally disrupted under a novel stress and express equally low fitness, no selective response can take place.

More generally, environmental changes will be predicted to result in the selection of plasticity patterns that best accommodate the new conditions. Necessarily, though, the amount of norm-of-reaction diversity will vary among populations and species, which will consequently hold different potentials for selective accommodation of new environments. Additional constraints may arise from the presence of shared regulatory mechanisms that may preclude the evolution of certain plastic responses, and from the disruption of environmental cues for plasticity as particular abiotic and biotic components of natural habitats are increasingly altered. It is only by studying the cues, developmental pathways, and phenotypic outcomes of individual plastic response that the interaction of genetic diversity with future environments can be understood and predicted.

Acknowledgments

I am grateful to Eva Jablonka and Snait Gissis for the invitation to contribute to this volume, and to the Israeli and British funding agencies that supported the international workshop, Transformations of Lamarckism, from which it emerged.

References

Agrawal AA, Laforsch C, Tollrian R. 1999. Transgenerational induction of defences in animals and plants. Nature 401: 60–63.

De Jong G. 1995. Phenotypic plasticity as a product of selection in a variable environment. Am Nat. 145: 493–512.

DeWitt T, Scheiner SM, eds. *Phenotypic Plasticity: Functional and Conceptual Approaches.* New York: Oxford University Press; 2004.

Donohue K. 2005. Niche construction through phenological plasticity: Life history dynamics and ecological consequences. New Phytol. 166: 83–92.

Donohue K, Pyle EH, Messiqua D, Heschel MS, Schmitt J. 2001. Adaptive divergence in plasticity in natural populations of *Impatiens capensis* and its consequences for performance in novel habitats. Evolution 55: 692–702.

Falconer DS, Mackay TFC, eds. *Introduction to Quantitative Genetics.* 4th ed. London: Longman Group Limited ; 1996.

Fox CW, Savalli UM. 2000. Maternal effects mediate host expansion in a seed-feeding beetle. Ecology 81,1: 3–7.

Galloway LF, Etterson JR. 2007. Transgenerational plasticity is adaptive in the wild. Science 318,5853: 1134–1136.

Gilbert SF. 2001. Ecological developmental biology: Developmental biology meets the real world. Dev Biol. 233: 1–12.

Gilbert SF, Epel D. *Ecological Developmental Biology: Integrating Epigenetics, Medicine, and Evolution.* Sunderland, MA: Sinauer Associates; 2009.

Gillespie JH, Turelli M. 1989. Genotype-environment interactions and the maintenance of polygenic variation. Genetics 121,1: 129–138.

Gomulkiewicz R, Kirkpatrick M. 1992. Quantitative genetics and the evolution of reaction norms. Evolution 46: 390–411.

Gupta AP, Lewontin RC. 1982. A study of reaction norms in natural populations of *Drosophila pseudoobscura.* Evolution 36: 934–948.

Haldane JBS. 1946. The interaction of nature and nurture. Ann Eugen. 13: 197–205.

Heschel MS, Sultan SE, Glover S, Sloan D. 2004. Population differentiation and plastic responses to drought stress in the generalist annual *Polygonum persicaria.* Int J Plant Sci. 165: 817–824.

Kingsolver JG, Ragland GJ, Shlichta JG. 2004. Quantitative genetics of continuous reaction norms: Thermal sensitivity of caterpillar growth rates. Evolution 58,7: 1521–1529.

Lewontin RC. *The Triple Helix: Gene, Organism and Environment.* Cambridge, MA: Harvard University Press; 2000.

Lindroth RL, Roth S, Nordheim EV. 2001. Genotypic variation in response of quaking aspen to atmospheric CO2 enrichment. Oecologia 126: 371–379.

Mondor EB, Rosenheim JA, Addicott JF. 2005. Predator-induced transgenerational phenotypic plasticity in the cotton aphid. Oecologia 142: 104–108.

Mousseau TA, Fox C, eds. *Maternal Effects as Adaptations.* New York: Oxford University Press; 1998.

Nijhout HF. 2003. Development and evolution of adaptive polyphenisms. Evol Dev. 5,1: 9–18.

Nusset DH, Postma E, Gienapp P, Visser, ME. 2005. Selection on heritable phenotypic plasticity in a wild bird population. Science 310,5746: 304–306.

Pigliucci M. *Phenotypic Plasticity: Beyond Nature and Nurture.* Baltimore: Johns Hopkins University Press; 2001.

Rotem K, Agrawal AA, Kott L. 2003. Parental effects in *Pieris rapae* in response to variation in food quality: Adaptive plasticity across generations? Ecol Entomol. 28,2: 211–218.

Sarkar S. From the *Reaktionsnorm* to the evolution of adaptive plasticity: A historical sketch. In: *Phenotypic Plasticity: Functional and Conceptual Approaches.* DeWitt TJ, Scheiner SM, eds. New York: Oxford University Press; 2004:10–30.

Scheiner SM. 1993. Genetics and evolution of phenotypic plasticity. Ann. Rev Ecol Systemat. 24: 35–68.

Scheiner SM. 2002. Selection experiments and the study of phenotypic plasticity. J Evol Biol. 15,6: 889–898.

Sultan SE. 2000. Phenotypic plasticity for plant development, function and life history. Trends Plant Sci. 5,12: 537–542.

Sultan SE. 2001. Phenotypic plasticity for fitness components in *Polygonum* species of contrasting ecological breadth. Ecology 82,2: 328–343.

Sultan SE. 2003a. The promise of ecological developmental biology. J Exp Zoolog B Mol Dev Evol. 296B: 1–7.

Sultan SE. 2003b. Phenotypic plasticity in plants: A case study in ecological development. Evol Dev. 5: 25–33.

Sultan SE. 2007. Development in context: The timely emergence of eco-devo. Trends Ecol Evol. 22,11: 575–582.

Sultan SE, Barton K, Wilczek A. 2009. Contrasting patterns of trans-generational plasticity in ecologically distinct congeners. Ecology 90: 1831–1839.

Sultan SE, Bazzaz FA. 1993. Phenotypic plasticity in *Polygonum persicaria*. I. Diversity and uniformity in genotypic norms of reaction to light. II. Norms of reaction to soil moisture and the maintenance of genetic diversity. III. The evolution of ecological breadth for nutrient environment. Evolution 47: 1009–1031, 1032–1049, 1050–1071.

Sultan SE, Spencer HG. 2002. Metapopulation structure favors plasticity over local adaptation. Am Nat. 160,2: 271–283.

Sultan SE, Stearns S. Environmentally contingent variation: Phenotypic plasticity and norms of reaction. In: *Variation: A Central Concept in Biology.* Hallgrimsson B, Hall B, eds. New York: Academic Press; 2005:303–332.

Via S. Genetic constraints on the evolution of phenotypic plasticity. In: *Genetic Constraints on Adaptive Evolution.* Loeschke V, ed. Berlin: Springer; 1987:47–71.

Windig JJ. 1994. Genetic correlations and reaction norms in wing pattern of the tropical butterfly *Bicyclus anynana.* Heredity 73: 459–470.

20 Epigenetic Variability in a Predator-Prey System

Sivan Pearl, Amos Oppenheim, and Nathalie Q. Balaban

Epigenetic variability, here defined as the variability observed between organisms despite identical genetic and environmental conditions, can have far-reaching consequences. In particular, it has been shown that the stochastic differentiation of a population of genetically identical cells into two distinct phenotypes can provide a survival advantage in unpredictable and fluctuating environments (Lachmann and Jablonka 1996; Kussell et al. 2005; Acar et al. 2008). The phenomenon of bacterial persistence, which plays a major role in the failure of various antibiotics to eliminate pathogens, is a striking example of the advantage of variability.

The phenomenon of bacterial persistence was observed as early as 1944, when *Staphyloccocal* infections treated with penicillin recurred, even after high doses and extended treatments (Bigger 1944). In contrast to resistance, which is genetically acquired and passed on to the next generations, persistence is a transient phenotype. It is typically observed when a population of cells is exposed to antibiotics and the number of survivors monitored (Moyed and Bertrand 1983). Such a killing curve is shown in figure 20.1a. The initial steep decrease in survival, characterized by a fast killing rate, is followed by a much slower decrease, revealing the existence of persister cells. These cells have not acquired genetic resistance; when they regrow a population, it is still sensitive to the antibiotic treatment. Exposing the population regrown from these survivors to the same antibiotic results in an identical killing curve (Moyed and Broderick 1986; Keren et al. 2004). For many years, the phenomenon has been overlooked, mainly because in most bacterial infections, the few remaining persistent bacteria are eliminated by the immune system. However, it has become increasingly obvious that persistence is the main problem in diseases such as tuberculosis, in which a single bacterium can start an infection (Stewart, Robertson and Young 2003). Apart from its wide clinical relevance, the study of bacterial persistence reveals the inherent epigenetic heterogeneity of bacterial populations, which might be a general adaptation to variable environments (Lachmann and Jablonka 1996). Such strategies seem indeed to exist in nature (Bull 2000; Meyers and Bull 2002). For instance, *Salmonella* can change the serotype of its flagella by

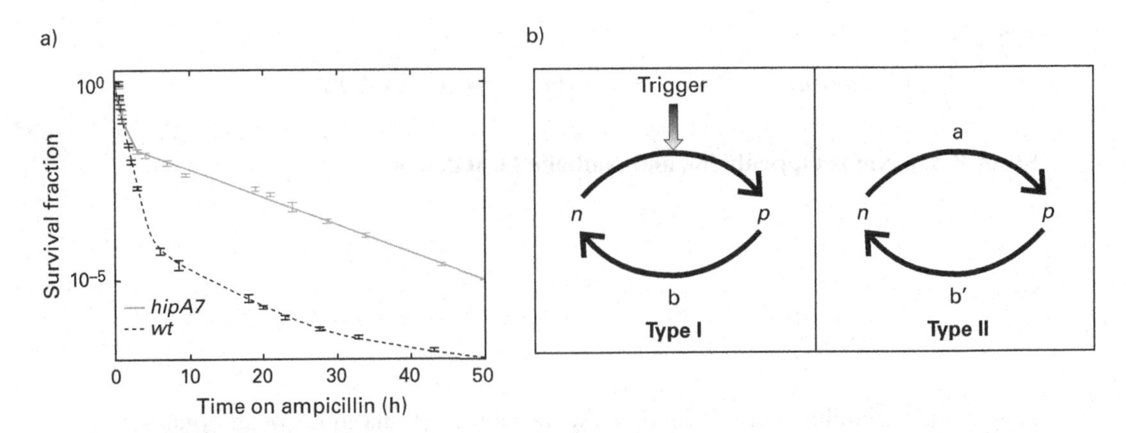

Figure 20.1
Persistence is due to a preexisting heterogeneity of growth rates. (a) Killing-curves under antibiotics for wild-type (dashed) and high persistence mutant cells (solid). (b) Persistence type have different dynamics of two sub-populations, here denoted as "normal" (*n*) and "persister" (*p*) cells. Adapted from Balaban et al. 2004.

"phase variation," resulting in a heterogeneous population. An immune response of the host against one phenotype will not be effective against another (Lederberg 1956).

Persistent bacteria are difficult to study because they do not possess a genetic mutation that can be propagated and characterized. Most studies rely on macroscopic phenotypes such as colony shape variants (Drenkard and Ausubel 2002). Another challenge in the study of persistence comes from the typically low frequency of persistence in the population. In order to increase the number of persisters and identify putative genes involved in persistence, Moyed and Bertrand (1983) developed a screen to identify mutants of *Escherichia coli* K–12 that exhibit increased persistence rate to ampicillin treatment. The screen identified a mutation in a gene termed *hipA* for *high persistence*. Interestingly, the *hipA7* mutation conferring high persistence was later shown to be linked to persistence to other stresses, such as thymine starvation, norfloxacin treatment, and cold sensitivity (Scherrer and Moyed 1988; Wolfson et al. 1990), showing that persistence is a broader phenomenon than the response to a specific antibiotic treatment. It has also been observed in many different bacteria including *S. aureus* (Massey, Buckling and Peacock 2001), *P. aeruginosa*, and *M. tuberculosis* (Wallis et al. 1999). Despite its first observation almost sixty years ago, bacterial persistence remains a puzzle. Several hypotheses have been put forward to explain the heterogeneous response to antibiotics in an apparently uniform population (Lewis 2000). It has been proposed, for instance, that persistent bacteria are in some "protected" part of the cell cycle at the time of exposure to antibiotics. Alternatively, persistent bacteria have been viewed as adapting rapidly to the antibiotic stress, or being in a "dormant" state prior to the exposure to antibiotics. It has also been suggested that persistence is linked to a defective programmed cell-death module (Lewis 2000; Sat et al. 2001). Recent studies show that certain antibiotics might trigger a stress response and enhance persistence (Miller et al. 2004). We have recently shown that high persistence mutants of *E. coli* spontaneously generate persister cells *prior* to the antibiotic treatment. Using quantitative measurements and single cell observations in microfluidic devices, we precisely measured single cell behavior and showed that persistence in *E. coli* comes from heterogeneity of growth rates which existed before the antibiotic treatment (Balaban et al. 2004). Two types of persistence have been described: Types I and II (figure 20.1b). Type I persisters are formed under specific conditions, such as stationary phase, and do not grow. Type II persisters, are continually formed in a growing population at a rate a, and have decreased growth rate (Gefen and Balaban 2009). In an exponential culture, both types switch back to the normal state, with constant rates b, b'.

However, while it is appealing to speculate that persistence to antibiotics might have evolved for facing transient antibiotic stress periods, it seems that concentrations of antibiotics encountered by bacteria in nature might be too low to provide a strong enough selection for the evolution of persistence. Therefore, searching for a stress that is both strongly lethal and often encountered by bacteria, we chose to investigate the interaction between bacteria and phages. Specifically, our interest is in the role played by the epigenetic variability of growth rates, which is observed in persistence, in this host–phage system. The abundance of phages in various ecological niches (Breitbart and Rohwer 2005) suggests that they represent one of the common stresses that bacteria have encountered during evolution. Furthermore, the ability of phages to replicate in their hosts makes them potent killers, even at very low doses. Recent studies have indeed revealed the important role played by phages in the evolution and ecology of bacteria (Pal et al. 2007). For example, phages are believed to play an important role in regulating bacterial density in oceans (Angly et al. 2006). However, phages need a metabolically active host to replicate their genetic material. This requirement, not fulfilled by Type I persisters, might have driven phages to develop mechanisms to cope with persisters.

Our goal was to study the effect of the dormant sub-population of Type I persisters on the bacterial-phage interaction, and to determine whether bacteria persistent to antibiotic treatment are also protected from phage attacks (Pearl et al. 2008). For this purpose, we chose to study the interaction of *E. coli* and its bacteriophage lambda (λ). λ is a temperate phage that infects *E. coli*. Following injection of the phage DNA into the bacterial cell, the phage is able to proceed to a lytic or to a lysogenic life cycle (figure 20.2). In a lysogenic cycle, the bacteriophage's genome integrates into the chromosome of the host bacterium. When the bacterium repli-

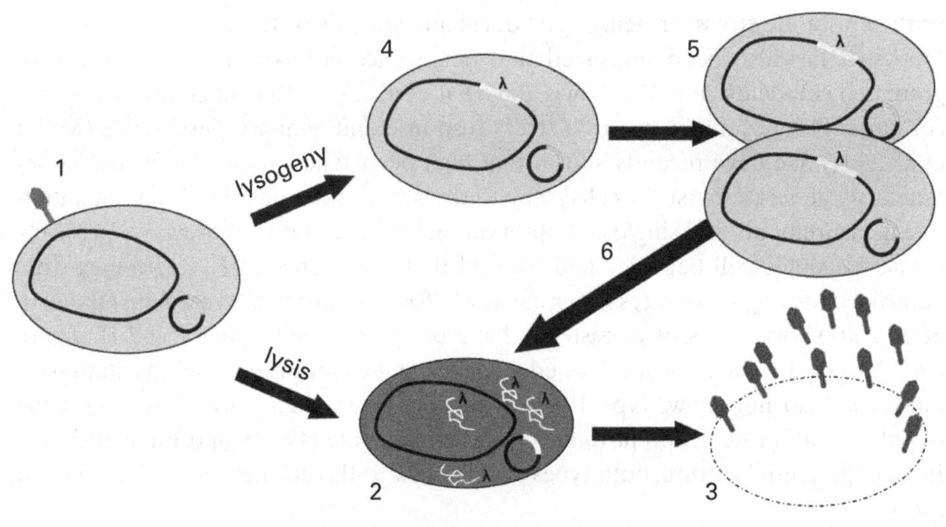

Figure 20.2
The λ phage life cycles used for our experiments. (1) A phage particle attaches to the cell surface and injects the phage DNA into the cell. (2-3) The lytic cycle: phage DNA is transcribed, translated and replicated to form new phages. (3)The cell lyses to liberate a burst of phage progeny. (4-5) The lysogenic cycle: phage development is repressed and phage DNA integrates into the bacterial chromosome. (5) The resulting lysogenic cell and its prophage can replicate indefinitely. (6) Prophage activation: the prophage can be induced to switch to the lytic cycle with the excision of phage DNA from the chromosome.

cates, the integrated phage DNA is also replicated, but no active viruses are produced. In this state the phage, called a *prophage*, is stably replicated as part of its host genome and transmitted to its descendants. In contrast, in a lytic cycle, new phages are formed and released by lysing the host cell. The lysogenic cycle can end in the lytic cycle if the prophage is triggered to exit the host genome, for example when exposed to stressful conditions such as UV. This process is called prophage induction (figure 20.2).

Experimental Characterization of the Interaction between Persister Bacteria and the Lambda Phage

Using long-term single-cell observations, we compared both prophage induction and lytic infection stresses on bacterial populations with high and low persistent rates. We used the wild-type (*wt*) *E. coli* and its *hipA7* derivative, which was shown to confer a high level of Type I persistence (Moyed and Broderick 1986).

We found that high persistence to antibiotics of the *hipA7* population also protects bacteria from prophage induction (table 20.1). Quantitative analysis of single

Table 20.1
Typical survival following phage assaults

hipA7	*wt*	Survival (%)
50	0.1	Prophage induction
0.05	0.05	Lytic infection

Note: Survival percentage of *wt* and *hipA7* cells shown following 80 min of prophage induction, or following incubation with lytic phages. An increase in survival rates of the high persistence strain is found for prophage induction but not for lytic infection.

cell gene expression revealed that the expression of lytic genes is suppressed in these persistent bacteria. After prophage-inducing-conditions are removed, lysogenic persisters can switch to normal growth, as if no stress was encountered. Thus, Type I persistence enables lysogenic bacteria to survive transient prophage-inducing-conditions (Pearl et al. 2008).

In contrast to the high survival of persisters to prophage induction, persisters infected by lytic phages eventually lyse. No increase in survival was observed following lytic infections for the *hipA7* strain when compared to the *wt* (table 20.1). This result was at first surprising: why would phages that need to attach to the cells and penetrate them in order to replicate and lyse them, be more effective than prophages that are already inside the cell? The answer to that puzzle came from the direct observation of the dynamics of those lytic phages' replication: we found that lytic phages wait for persister bacteria to switch back to normal growth in order to kill them. Thus, while persisters are not immediately killed upon lytic infection, they do lyse once they switch back to the normal state.

Fitness Advantage of Persistence under Prophage-Inducing Conditions: Mathematical Analysis and Competition Experiments

It was previously shown that the optimal rate of switching between normal and persister cells, defining the frequency of persisters in the population, is found to depend strongly on the frequency of environmental changes (Lachmann and Jablonka 1996; Kussell et al. 2005). From these works it appears that persistence constitutes an adaptation that is tuned to the distribution of environmental changes rather than to a specific single environment. Since persister cells are found to be protected from prophage induction it is expected that in an environment in which prophage-inducing conditions are often encountered, a population with high persistence level would have an advantage over a population with very low persistence. In order to test this prediction, we performed a competition experiment between a high persistence strain (*hipA7*) and a low persistence strain (*wt*), both carrying a prophage. The two populations were mixed and exposed to cycles of prophage-inducing conditions. As predicted, the high persistence strain rapidly took over the low persistence strain, showing the advantage conferred by persistence under cyclic prophage-inducing conditions (Pearl et al. 2008). These observations, suggest that Type I persistence, namely the ability of part of the population to enter a state of transient dormancy, might have been selected under prophage induction pressure to prevent the eradication of lysogenic bacterial population. Interestingly, the bet-hedging strategy that persister bacteria adopt probably also benefits the prophage population. Without persister bacteria, all prophages would have been induced and released to the environment, where new hosts may be scarce. Prophages that remained dormant together with their persister host have a better chance of being passed on to the next generation of hosts.

The Effect of Persistence on Predator-Prey Population Dynamics under Lytic Conditions: Predictions from Mathematical Analysis

At first glance, it seems that our experimental results show that persistence provides no advantage under lytic infections, in contrast to the strong advantage it provides under prophage-inducing conditions (table 20.1). However, persistence introduces a new time-scale in the interaction between bacteria and phages: Only when persister bacteria switch to normal growth do phages replicate and spread out to find new hosts. This time-scale of the release of phages by persister bacteria can be as long as a few days, significantly longer than the release of phages by normal bacteria, which occurs within minutes. In order to understand the implications of this extremely long delay introduced by persister bacteria in the interaction between phages and bacteria, we developed a mathematical model of population dynamics of phages and normal and persister bacteria (Pearl et al. 2008).

In our model the switching between persister and normal cells can be mapped on an effective "migration" model between two "patches" representing not spatial differences but different phenotypic states. Patch 1, where growth of the prey is fast (normal bacteria-*n*) and patch 2, where growth of the prey is slow (persisters-*p*). In the wild, prey can switch back and forth between the "patches" with rates that correspond to the switching rates between persistent and normal states. In our experimental system, Type I persisters are generated at stationary phase and, when diluted in fresh medium, switch to normal growth on a timescale of 10 hours. Until the population reaches a stationary state again (a case we do not consider in this simplified model), no additional persisters are formed. The equations below, describing this system, take into account the unidirectional migration from persister to normal states. We start with a certain percentage of persisters following stationary phase and consider only the transitions of persisters to the normal state (patch 1-to-2 transitions). Predators (phages) "diffuse" fast between the patches, similarly infecting normal and persister bacteria. However, normal and persister bacteria react differently to the phage infection. While normal bacteria burst immediately after phage attachment, releasing new phages to the system, infected persisters (p_i) burst and release new phages on a significantly longer timescale (b). Note that b is the same timescale for the switching from uninfected persisters to normal cells. The use of the same rate to describe both processes reflects the fact that persisters are killed by phages only when they switch back to normal growth, as measured in our experiments (Pearl et al. 2008). We show below simplified dynamics, illustrating the potential effects of Type I persisters on the host-phage interaction:

$$\left.\begin{aligned} \frac{dn}{dt} &= \mu n - \alpha\lambda n + bp \\ \frac{dp}{dt} &= -bp - \alpha\lambda p \\ \frac{dp_i}{dt} &= -bp_i + \alpha\lambda p \\ \frac{d\lambda}{dt} &= burst\,(\alpha\lambda n + bp_i) - d\lambda \end{aligned}\right\}$$

where n = normal bacteria, p = persister bacteria, p_i = infected persister bacteria, μ = growth rate of normal bacteria, b = switching rate from p to n, α = rate of phages attachment, λ = phages, *burst* = burst size, namely the number of released phages per bacterium, and d = dilution/death rate of phages.

A typical phage infection starts with an initial number of phages of the order of the burst size. Free phages are rapidly diluted out. Simulations of infected populations without and with persisters are shown in figure 20.3a and b, respectively. For similar initial conditions, the bacterial and phage population undergoes very large oscillations in the absence of persisters. When Type I persisters are introduced to the system they continuously release phages on a slow time scale. The release of new phages at a very slow rate alters the dynamics of the system and has a stabilizing effect, by preventing large oscillations. For wild-type *E. coli*, a Type I persister fraction of nearly 1 percent has been measured (Keren et al. 2004). In our model, this fraction was sufficient to reduce the amplitude of the oscillations by more than 2 orders of magnitude. This indicates that population dynamics of wild-type populations can be affected by the host heterogeneity and that this heterogeneity should be taken into account in complete models of the predator-prey interactions.

The oscillations in the population size of the predator and prey predicted from Lotka-Volterra equations are not always found experimentally, with many systems ending up either in extinction or stability. Our model suggests one explanation for the lack of strong oscillations in the wild: the inherent heterogeneity of the prey that mimics the class of models considering the stabilizing effect of spatial hetero-

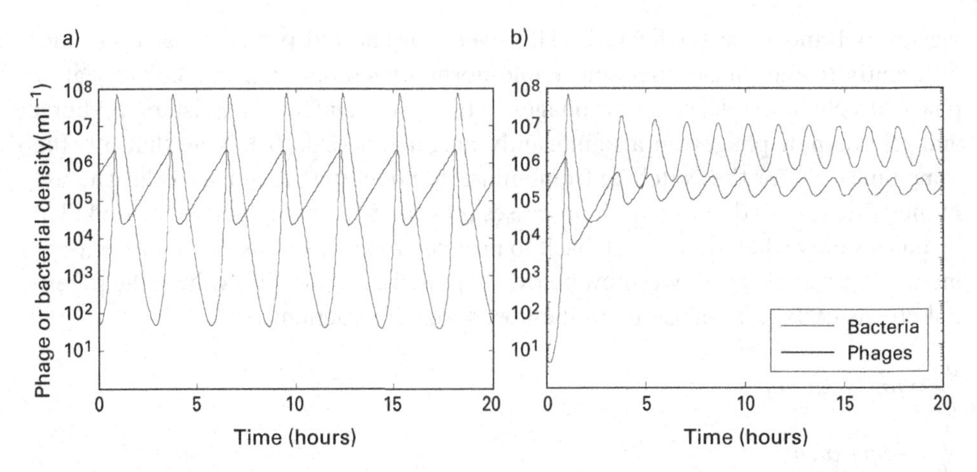

Figure 20.3
Effect of persistence on the dynamics in a Lotka-Volterra predator–prey model. (a) On the left, population dynamics without persisters; (b) On the right, population dynamics with 1 percent persisters, as there is wild-type *E. coli*. The population oscillations are significantly reduced. Adapted from Pearl et al. 2008.

geneity on the population dynamics (Briggs and Hoopes 2004). Here migration occurs in the phenotypes space but the stabilizing effect has the same origin. Kuang and Takeuchi (1994) have analyzed the equilibrium points of the general equations describing similar processes for symmetrical diffusion between patches as well as their stability. It would be interesting to expand their analytical work to the case of non-symmetrical diffusion, which is more appropriate for the description of migration in phenotype space.

Conclusions

Persistence has been studied mostly in the context of antibiotics' failure to eradicate a bacterial population. While this has a clear clinical value, we showed that the nongenetic heterogeneity imposed by persistence has far-reaching effects on the population dynamics of bacteria and phages. This suggests that persistence might have evolved under the ubiquitous stresses imposed by phages in the wild, with persistence acting as a bet-hedging strategy for facing unpredictable stress.

While being persistent or not *is not* genetically determined, the abundance of persisters in the population *is* a genetic trait. In a fluctuating environment, the frequency of stressful conditions will have a key role in tuning the switching rates between normal and persistent states. These rates ultimately ascribe the frequency of persisters in a given population, a trait that might be under selective pressure.

Recently, experimental evidence suggests that such bet-hedging strategies can be evolved in the lab (Beaumont et al. 2009).

Moreover, we showed that persistence can affect not only the survival of a bacterial population, but change the outcome of a predator–prey system, such as lytic phages and their bacterial host. The epigenetic variability of the prey was shown to stabilize both predator and prey populations, thus reducing the probability of extinction, a phenomenon that should have clear implications for the ecology and evolution of bacteria and phage populations in the wild.

Our experimental and theoretical analysis has focused on Type I persistence, which protects bacteria through dormancy. It would be interesting to apply the same approach to evaluate the effect of Type II persistence on the predator–prey interaction of phages and bacteria, as Type II persistence constitutes a clearer example of epigenetic inheritance (Jablonka and Raz 2009).

The study of persistence in particular, and epigenetic variability in microorganisms in general, should uncover the many roles played by epigenetic variability in the ecology and evolution of populations.

References

Acar MJ, Mettetal T, van Oudenaarden A. 2008. Stochastic switching as a survival strategy in fluctuating environments. Nat Genet 40: 471–475.

Angly FE, Felts B, Breibart M, Salamon P, Edwards RA, Carlson C, Chan AM, et al. 2006. The marine viromes of four oceanic regions. PLoS Biol 4: e368.

Balaban NQ, Merrin J, Chait R, Kowalik L, Leibler S. 2004. Bacterial persistence as a phenotypic switch. Science 305: 1622–1625.

Beaumont HJ, Gallie J, Kost C, Ferguson GC, Rainey PB. 2009. Experimental evolution of bet hedging. Nature 462: 90–93.

Bigger JW. 1944. Treatment of staphylococcal infections with penicillin by intermittent sterilisation. Lancet 2: 497–500.

Breitbart M, Rohwer F. 2005. Here a virus, there a virus, everywhere the same virus? Trends Microbiol 13: 278–284.

Briggs CJ, Hoopes MF. 2004. Stabilizing effects in spatial parasitoid-host and predator-prey models: a review. Theor Popul Biol 65: 299–315.

Bull JJ. 2000. Evolutionary biology—Déja vu. Nature 408: 416–417.

Drenkard E, Ausubel FM. 2002. Pseudomonas biofilm formation and antibiotic resistance are linked to phenotypic variation. Nature 416: 740–743.

Gefen O, Balaban NQ. 2009. The importance of being persistent: heterogeneity of bacterial populations under antibiotic stress. FEMS Microbiol Rev 33: 704–717.

Jablonka E, Raz G. 2009. Transgenerational epigenetic inheritance: prevalence, mechanisms, and implications for the study of heredity and evolution. Q Rev Biol 84: 131–176.

Keren I, Kaldalu N, Spoering A, Wang Y, Lewis K. 2004. Persister cells and tolerance to antimicrobials. FEMS Microbiol Lett 230: 13–18.

Kuang Y, Takeuchi Y. 1994. Predator-prey dynamics in models of prey dispersal in 2-patch environments. Math Biosci 120: 77–98.

Kussell E, Kishony R, Balaban NQ, Leibler S. 2005. Bacterial persistence: a model of survival in changing environments. Genetics 169: 1807–1814.

Lachmann M, Jablonka E. 1996. The inheritance of phenotypes: An adaptation to fluctuating environments. J Theor Biol 181: 1–9.

Lederberg J. 1956. Phase variations in *Salmonella*. Genetics 41: 743–756.

Lewis K. 2000. Programmed death in bacteria. Microbiol Mol Biol Rev 64: 503–514.

Massey RC, Buckling A, Peacock, SJ. 2001. Phenotypic switching of antibiotic resistance circumvents permanent costs in *Staphylococcus aureus.* Curr Biol 11: 1810–1814.

Meyers LA, Bull JJ. 2002. Fighting change with change: Adaptive variation in an uncertain world. Trends Ecol Evol 17: 551–557.

Miller C, Thomsen LE, Gaggero C, Mosseri R, Ingmer H, Cohen SN. 2004. SOS response induction by beta-lactams and bacterial defense against antibiotic lethality. Science 305: 1629–1631 Epub 2004 Aug 12.

Moyed HS, Bertrand KP. 1983. *hipA*, a newly recognized gene of *Escherichia coli* K-12 that affects frequency of persistence after inhibition of murein synthesis. J Bacteriol 155: 768–775.

Moyed HS, Broderick SH. 1986. Molecular cloning and expression of *hipA*, a gene of *Escherichia coli* K-12 that affects frequency of persistence after inhibition of murein synthesis. J Bacteriol 166: 399–403.

Pal C, Macia MD, Antonio Oliver A, Schachar I, Buckling A. 2007. Coevolution with viruses drives the evolution of bacterial mutation rates. Nature 450: 1079–1081.

Pearl S, Gabay C, Kishony R, Oppenheim AB, Balaban NQ. 2008. Nongenetic individuality in the host-phage interaction. PLoS Biol 6: e120.

Sat B, Hazan R, Fisher T, Khaner H, Glaser G, Engelberg-Kulka H. 2001. Programmed cell death in *Escherichia coli*: some antibiotics can trigger mazEF lethality. J Bacteriol 183: 2041–2045.

Scherrer R, Moyed HS. 1988. Conditional impairment of cell-division and altered lethality in Hipa mutants of *Escherichia coli* K-12. J Bacteriol 170: 3321–3326.

Stewart GR, Robertson BD, Young, DB. 2003. Tuberculosis: A problem with persistence. Nat Rev Microbiol 1: 97–105.

Wallis RS, Patil S, Cheon S-H, Edmonds K, Phillips M, Perkins MD, Joloba M, et al. 1999. Drug tolerance in *Mycobacterium tuberculosis.* Antimicrob Agents Chemother 43: 2600–2606.

Wolfson JS, Hooper DC, McHugh GL, Bozza MA, Swartz MN. 1990. Mutants of *Escherichia coli* K-12 exhibiting reduced killing by both Quinolone and Beta-Lactam antimicrobial agents. Antimicrob Agents Chemother 34: 1938–1943.

21 Cellular Epigenetic Inheritance in the Twenty-First Century

Eva Jablonka

Since the turn of the twenty-first century, epigenetic inheritance—the inheritance of cellular phenotypic variations that are not dependent on differences in DNA base sequence—has become an important aspect of pure and applied biological research. In this chapter I present a brief overview of the history of cellular epigenetic inheritance, outline recent evidence showing that it is ubiquitous, and suggest how it legitimizes the notion that soft inheritance is part of heredity and evolution.

A (Very) Short Historical Overview: From Cell Heredity to Epigenetic Inheritance

In the mid-twentieth century, biologists became increasingly interested in several aspects of cell heredity (Sager and Ryan 1961). Studies of cell differentiation in animals showed that cells could "remember" their embryonically established developmental fates. For example, Hadorn's research on *Drosophila* imaginal disks (groups of undifferentiated larval insect cells that eventually give rise to adult structures) showed that these disk cells usually retain their determined state for many cell generations (Hadorn 1967). Similarly, Lyon (1961) found that during the early embryonic development of female mammals, one of the two X chromosomes becomes inactivated; which X is inactivated in a particular cell is a stochastic process, but once the decision is made, the functional states of the X chromosomes are faithfully inherited by subsequent cell generations.

At about the same time, microbiologists and botanists were describing many cases of non–Mendelian heredity (reviewed in Sapp 1987). As was shown by Sonneborn and his students for ciliates (e.g., Sonneborn 1950), and by Ephrussi (1958) for yeast, some of the variations exhibiting non–Mendelian inheritance are inducible. Botanists, too, found that heritable variations could be induced: Durrant (1962), for example, showed that exposing flax to certain fertilizers could produce heritable changes, and similar phenomena were described for other species (see Lamb, chapter 11 in this volume). Work with maize revealed some very different types of non–

Mendelian inheritance: Brink discovered a new type of heritable change, paramutation, in which one allele modifies the other allele of the same gene (reviewed in Brink 1973), and McClintock demonstrated that some genetic elements move around the genome, a phenomenon that she believed was an adaptive response to stress (discussed in McClintock 1984).

An insightful, but generally ignored, discussion of many of these early studies was provided by David Nanney (1958). He called the mechanisms that led to persistent and hereditary cellular changes, yet did not seem to depend on DNA changes, "epigenetic control systems." However, the microbiologists' and botanists' studies of unconventional patterns of inheritance and the zoologists' work on differentiation remained dissociated, and the molecular basis of the phenomena they investigated was unknown. The phenomena were regarded either as developmental processes having nothing to do with heredity (e.g., X-inactivation) or, when "soft" hereditary transmission was clearly manifested (e.g., in the experiments with flax and ciliates), as theoretically unimportant and unrelated oddities. *The Cold War in Biology* (Lindegren 1966), which highlighted these neglected aspects of genetical research and discussed the social and political background to this neglect, was, like Nanney's ideas, ignored by mainstream biologists.

Modern studies of epigenetic inheritance can be said to have begun in the mid-1970s (see Jablonka and Lamb 2002, 2010; Haig 2004; and Holliday 2006 for historical detail). In two important theoretical papers, Holliday and Pugh (1975) and Riggs (1975) independently suggested a molecular mechanism for cellular differentiation and cell heredity. Both papers put forward the idea that DNA modification, specifically cytosine methylation or demethylation (addition or removal of CH_3 groups), affects the activity of genes, and that patterns of methylation can be enzymatically copied following DNA replication. DNA methylation was thus seen both as part of the regulatory system of the cell and as a cellular memory system. The authors of the papers proposed that DNA methylation was the molecular basis of cell determination, X-inactivation, cancer, and developmental clocks. Soon afterward, in the 1980s, investigations of DNA methylation became central to studies of normal and pathological cellular inheritance. Research on cell lines, for example, showed that the heritable variations appearing in cells of genetically homogeneous lines are not always variations in DNA base sequence (mutations), but rather are often variations in methylation patterns (Harris 1984). Study of the cellular inheritance of altered patterns of DNA methylation was soon incorporated into research on aging, and the early 1980s saw the first experimental work showing that changes in DNA methylation are involved in carcinogenesis (reviewed by Feinberg and Tycko 2004).

In an important paper published in 1987, Holliday used the term "epigenetic inheritance" for the transmission of patterns of DNA methylation (Holliday 1987). In this paper he also suggested that not only are inherited epigenetic variations important for normal and pathological development, but also that epigenetic defects (epimutations) in germ line cells may sometimes be transmitted to the next generation of organisms.

The general realization that cell heredity and the "strange" patterns of inheritance found in plants and microorganisms may share common mechanisms began to dawn in the late 1980s, largely as a result of the development of molecular technologies that allowed foreign genes (transgenes) to be introduced into fungi, plants, and animals (reviewed in Jablonka and Lamb 1995, 2010). The insertion of transgenes often yielded surprising patterns of inheritance: in many cases, the activity of the introduced transgene depended on the sex of the parent from which it was inherited (a phenomenon known as genomic imprinting), and this parent-of-origin-dependent activity was often associated with methylation differences. Moreover, some studies of transgenes revealed phenomena that went beyond classical genomic imprinting: the transgenes became inactivated through methylation, and this inactive state was transmitted to descendants, irrespective of parental sex. In other words, sometimes the epigenetic change in the transgene was stably transmitted between generations.

Epigenetic Inheritance Systems

As the studies of cellular epigenetic inheritance based on the transmission of DNA methylation patterns emerged in the late 1980s, Marion Lamb and I (Jablonka and Lamb 1989, 1995; Jablonka, Lachmann, and Lamb 1992) suggested that the various mechanisms that lead to the transmission of cellular epigenetic variants—those discussed by Nanney as well as those discussed by Holliday—should be united within an evolutionary framework that incorporates the developmental reconstruction of all types of hereditary variations and encompasses all taxa. We referred to the factors and processes that underlie different types of epigenetic transmission as "epigenetic inheritance systems" (abbreviated to EISs by Maynard Smith 1990), and called for the recognition of the Lamarckian aspects of heredity and evolution that they bring about (Jablonka and Lamb 1989, 1995; Jablonka, Lachmann, and Lamb 1992). We initially characterized three types of EISs: the EIS based on self-sustaining loops, the EIS based on structural inheritance, and the chromatin-marking EIS (Jablonka, Lachmann, and Lamb 1992). In the late 1990s, we added a fourth type of EIS: one that was mediated through RNA (Jablonka and Lamb 2005). These systems are described in more detail in appendix B, so here I describe them only briefly.

Self-Sustaining Regulatory Loops

This type of epigenetic inheritance occurs when, following the induction of gene activity, the gene product acts as a positive regulator that maintains the gene's own activity and, when transmitted during cell division, causes the same state of gene activity to be reconstructed in daughter cells (see Hirschbein, chapter 37 in this volume).

Structural Templating

Preexisting three-dimensional cellular structures act as templates for the production of similar structures, which then become components of daughter cells. Structural templating includes a wide spectrum of mechanisms, the best understood being that which leads to the propagation and cellular inheritance of prions.

Chromatin-Marking Systems

Chromatin marks are small chemical groups (such as the methyl group, CH_3) that are covalently bound to DNA, and the modifiable, histone and nonhistone proteins that are noncovalently bound to DNA. These marks are involved in the control of gene activity, and some segregate semiconservatively or conservatively during DNA replication, nucleating the reconstruction of similar marks on the daughter DNA molecules that are inherited at cell division.

RNA-Mediated Inheritance

Silent transcriptional states are initiated and actively maintained through repressive interactions between small RNA molecules and the mRNA or DNA to which they are complementary (Bernstein and Allis 2005). Transcriptional silence can be transmitted by cells and organisms (1) through an RNA replication system; (2) when small RNAs interact with chromatin in ways that cause heritable modifications of chromatin marks; and (3) when small RNAs lead to targeted gene deletions and amplifications, which are then inherited (see chapter 22 in this volume).

The study of epigenetic inheritance received a great boost in the late 1990s when medical research in epigenetics, which included the epigenetics of cancer as well as the study of the epigenetic basis of metabolic and environmental diseases, became a very hot topic (reviewed in Gilbert and Epel 2009). There was also a very fruitful interaction between psychobiologists, who had an independent behavioral-epigenetic approach, and molecular developmental biologists, which began to uncover fascinating links between behavior, gene expression in the brain, and the transmission of phenotypes to the next generation(s) (reviewed by Crews 2008). This transmission can occur either through the germ line (e.g., see Skinner, Anway, Savenkova, Gore, and Crews 2008) or through somatic, behaviorally mediated interactions between mother and offspring, which are constructed anew in every generation (Weaver, Cervoni, Champagne, D'Alessio, Sharma, Sekl, Dymov et al. 2004; Weaver, Champagne, Brown, Dymov, Sharma, Meaney, and Szyf 2005).

Recently, ecologists have also become concerned about some of the implications of epigenetic inheritance. They recognize that disturbed environments could have wide-ranging epigenetic effects on human development as well as the development of plants and other animals. For example, animal studies have already shown that chemical pollutants that interact with or mimic reproductive hormones, thereby disrupting endocrine functions, can have long-term transgenerational effects (Gilbert and Epel 2009). Furthermore, the realization that heritable epigenetic changes are likely components of the responses to global changes in climate and may accompany the introduction of invasive species of plants has led ecologists to take a more active interest in epigenetic changes in plant populations which exhibit a large amount of epigenetic variation (reviewed by Bossdorf, Richards, and Pigliucci 2008).

The Ubiquity of Transgenerational Epigenetic Inheritance

The expanding field of epigenetics has altered the framework within which soft inheritance is examined, and is leading to its legitimization. However, it is clear that the importance of epigenetic inheritance in both heredity and evolution depends critically on its frequency, its stability, and its inducibility. In a literature survey published in 2009, Gal Raz and I identified more than one hundred well-documented cases of epigenetic inheritance in forty-two species (Jablonka and Raz 2009). In summary, we found the following:

1. Twelve cases of epigenetic inheritance in bacteria and one in phage λ of *Escherichia coli*. Some of these, such as the epigenetically based heritable antibiotic resistance found in *E. coli*, represent a type of inheritance that is thought to occur in many other pathogenic bacteria and with other antibiotics. Self-sustaining loops were a common EIS in bacteria, but examples of chromatin marking and structural inheritance were also found (detailed examples are described by Balaban in chapter 20 and by Hirschbein in chapter 37 of this volume). Some of the observed epigenetic variations were the results of developmental noise, and others were the outcome of experimentally induced challenges or of naturally occurring environmental stress. Their stability was variable, seeming to depend on the inducing conditions, the characteristics of the locus or element, and the EIS involved.

2. Eight cases in protists. Most cases involved ciliates, among which structural inheritance (the transmission of cortical morphologies) is common, and all loci can be modified through the RNA-mediated EIS that is fundamental to their normal life cycle. The other two EISs were also found in protists. Some novel variations were generated by developmental noise, and others were induced by environmental agents (e.g., chemicals, temperatures) or experimental manipulations (for example,

surgical alterations to the cortex). A range of transmission stabilities has been observed; stability depends on environmental conditions, on the characteristics of the locus or element, and on the EIS involved.

3. Seventeen cases in fungi, involving many phenotypes and loci. Examples of all four types of EISs were found. As with protists, stressful environmental conditions as well as developmental noise have been found to induce epigenetic variations. The stability of variants was variable, depending on the conditions of induction, the locus or region involved, and the type of EIS employed. Often stability was very high: for example, switches between prion and nonprion forms occurred at a frequency of 10^{-5} in *Saccharomyces cerevisiae* in normal (nonstressful) conditions. Braun and Lior (chapter 18 in this volume) have shown how, in yeast cells, new adaptive solutions to environmental challenges which seem to have epigenetic components are transmitted for hundreds of generations.

4. Thirty-six cases in plants, involving many loci and many traits. Among the thirty-six cases, four (a sample only) were in plant hybrids, and in all of these, many loci were heritably modified. Genomic stresses such as hybridization and polyploidization, especially allopolyploidization, seem to induce genome-wide epigenetic changes, some of which are transmitted between generations through the chromatin-marking and the RNA-mediated EISs (see Feldman and Levy, chapter 25 in this volume). No evidence was found for gametic, between-generation inheritance based on self-sustaining loops and structural templating. The conditions that induced heritable epigenetic changes included chemical treatments, various types of physical (heat, pressure, radiation) stresses, and genetic shocks. As with other taxa, the stability of transmission depended on the type of inducing conditions, the locus, and the EIS involved.

5. Twenty-eight cases in animals, some of which involved many loci. Epigenetic variations were induced in a variety of ways, including transient chemical treatment, novel physical conditions, and the temporary introduction of EIS-destabilizing mutations. Stress seems to induce multiple epigenetic changes in the animal epigenome, just as it does in plants. For example, the administration of vinclozolin, a widely used fungicide and an androgen receptor antagonist, to a pregnant female rat (generation F_0) resulted in various abnormalities in her male and female offspring in the next three offspring (F_1–F_4) generations (Anway, Leathers, and Skinner 2006; Nilsson, Anway, Stanfield, and Skinner 2008). Furthermore, the patterns of gene expression in the brains of vinclozolin-treated males and females were drastically altered, and were also inherited for the next three generations. In both males and females, hundreds of genes in the hippocampus and in the amygdala were affected. The changed gene expression patterns were different in males and females, and were associated with different behaviors: F_3 generation males showed a decrease

in anxiety-like behavior, whereas in females this type of behavior increased (Skinner, Anway, Savenkova, Gore, and Crews 2008).

All of the epigenetic variations that have been found in animals were transmitted transgenerationally through the chromatin-marking and the RNA-mediated EISs; so far, there is no evidence for inheritance through gametes that is based on self-sustaining loops or structural templating (Jablonka and Raz 2009). The stability of epigenetic variations is variable; systematic studies like those made in plants (see later) have not yet been performed in animals.

Our analysis of the data available in early 2009 shows that heritable epigenetic variations can be generated stochastically (i.e., they may result from developmental noise), or they can be induced. In theory, environmental factors could induce heritable epigenetic variations either (1) directly in cells of the germ line (direct induction), or(2) indirectly through somatic effects on germ line cells (somatic induction), or (3) in both the soma and the germ line (parallel induction). All three types of induction have been found (Jablonka and Raz 2009). Induced epigenetic variations were often associated with new phenotypes, although the relationship between the epigenetic variation and the overt phenotype was not always direct. Some of the new phenotypes were clearly adaptive (e.g., antibiotic resistance in bacteria, resistance to bacterial infection in plants) or potentially adaptive (alterations of flower shape and color, and flowering time).

Since Raz and I wrote our review, additional and important studies of transgenerational epigenetic inheritance have been reported for both plants and animals. Paszkowski's group in Switzerland and Colot's group in France have used the plant *Arabidopsis* in systematic studies to evaluate how many methylation patterns can be independently inherited, and for how long (Johannes, Porcher, Teixeira, Saliba-Colombani, Simon, Agier, Bulski et al. 2009; Reinders, Wulff, Mirouze, Marí-Orodóñez, Dapp, Rozhon, Bucher et al. 2009; Teixeira, Heredia, Sarazin, Roudier, Boccara, Ciaudo, Cruaud et al. 2009). Both groups employed similar methodology: they crossed wild-type plants from a line lacking genetic variability with genetically similar plants that differed only in that they had a mutation disabling the gene for a key enzyme involved in DNA methylation. The mutant plants had very low levels of DNA methylation, and hybrid F_1 plants had methylation levels intermediate between those of the two parental strains. In the F_2 generation, the individual plants had chromosomes with recombinant methylation patterns. From these F_2 plants, the researchers chose those that did not carry the destabilizing mutation and continued to breed from them by self-fertilization, thus establishing many separate, nonmutant, methylation-variable lines, which they followed for eight sexual generations. They analyzed the stability of the methylation patterns in these genetically almost identical but epigenetically different epi-RILs (epigenetic recombinant inbred lines) and studied some of their heritable, visible phenotypes. The results of both studies show that a large fraction of the *Arabidopsis* genome can be differentially methylated, and that the methylation patterns can be stably and independently inherited for at least eight generations (Johannes, Porcher, Teixeira, Saliba-Colombani, Simon, Agier, Bulski et al. 2009; Reinders, Wulff, Mirouze, Marí-Orodóñez, Dapp, Rozhan, Bucher et al. 2009). Both studies also showed that different epi-RILs have new heritable phenotypes, such as resistance to pathogenic bacteria, altered flower height, and delayed flowering time. Johannes and his colleagues found that plant height and flowering time have about 30 percent heritability, which is similar to values obtained for genetically determined traits.

Both sets of experiments uncovered similar patterns of stability and instability of epigenetic variations. Some loci that were demethylated in the F_1 generation become remethylated within five generations: these loci were usually those associated with transposable elements (TEs) and centromeric regions, where the demethylated states induced by the mutation do not persist (Johannes, Porcher, Teixeira, Saliba-Colombani, Simon, Agier, Bulski et al. 2009; Reinders, Wulff, Mirouze, Marí-Orodóñez, Dapp, Rozhon, Bucher et al. 2009; Teixeira, Heredia, Sarazin, Roudier, Boccara, Ciaudo, Cruaud et al. 2009). With other loci that were demethylated in the F_1 generation, the demethylated state was stably retained. Almost all of the loci that retained the stably demethylated state were in the euchromatic regions of the genome (i.e., regions with potentially active coding genes). In both studies it was also found that genome-wide demethylation had activated transposable elements in the epi-RILs. However, analysis of the data by Johannes and his colleagues (2009) showed no significant effect of the activated TEs on heritable phenotypic variations in plant height and flowering time.

Although there were some differences between the two sets of studies, possibly caused by the different destabilizing mutations that were used, both showed that a large number of differentially methylated sites can be stably inherited. Reinders and colleagues (2009) found 6532 such sites in one of their lines. Moreover, these sites seemed to be independently inherited, since in both studies there were different heritable patterns of epigenetic variations following sexual segregation. Crucially, the two studies showed that the epi-RILs had different heritable phenotypes which could be adaptive in some ecological conditions. The importance of these results for understanding the dynamic changes in natural and experimental populations did not escape the scientists' notice: ". . . the demonstration that numerous epialleles across the genome can be stable over many generations in the absence of selection or extensive DNA sequence variation highlights the need to integrate epigenetic information into population genetics studies" (Johannes, Porcher, Teixeira, Saliba-Colombani, Simon, Agier, Bulski et al. 2009:1).

A Molecular Interpretation of Soft Inheritance

The suggestion that epigenetic inheritance is a form of soft inheritance has a definite Lamarckian ring to it, making it sound anachronistic and out of tune with twenty-first-century evolutionary thinking. Modern molecular epigenetics and Lamarck's subtle fluids are certainly very remote from one another. So does the molecular framework of epigenetics really make soft inheritance more plausible and acceptable?

Ernst Mayr (1980) used the term "soft inheritance" for genetic changes brought about "either by use or disuse, or by some internal progressive tendencies, or through the direct effect of the environment," and consistently maintained that it did not take place (see part II of this volume). However, these three aspects of soft inheritance—direct environmental effects, modification through use and disuse, and internal biasing processes—can, in some cases, be reinterpreted in modern epigenetic terms.

Direct Effects of the Environment

Environmental factors can directly induce epigenetic variations in the germ line, which are then inherited through the chromatin-marking and the RNA-mediated EISs. This would be direct induction or, if similar changes occurred in the soma as well, parallel induction. The example given earlier of the hormonally mediated effects of the androgen disruptor vinclozolin is an instance of direct induction of an epigenetic change in the germ line of the embryos (Anway, Leathers, and Skinner 2006; Skinner, Anway, Savenkova, Gore, and Crews 2008).

Use and Disuse

Environmental factors may alter an animal's behavior and physiology, including its neural and endocrine activity. Such changes in parents can influence, through epigenetic mechanisms, the offspring's development, leading to similarity between generations. For example, exposure of *Daphnia cucullata* to predators leads to dramatic changes in its morphology, which are transmitted to offspring who have never encountered the predator (Agrawal, Laforsch, and Tollrian 1999). Such cases correspond to what I described earlier as somatic induction.

Internal Biasing Factors

The structure of the genome can bias and direct the generation of genetic and epigenetic variations during the genomic shocks of polyploidization and hybridization, and during periods of ecological stress (Lamm and Jablonka 2008). Epigenetic mechanisms such as RNA interference are involved when wide-ranging genomic changes occur, and the changes they bring about are not random: they are targeted to particular loci and are reproducible (see Buiatti, chapter 24 in this volume; Feldman and Levy, chapter 25 in this volume; Lamm, chapter 32 in this volume). Moreover, as Newman and Bhat argue in chapter 16 of this volume, the physical forces involved in development constrain and channel the evolution of morphology. Natural selection is therefore not the only direction-giving process in evolution.

A Forgotten Form of Soft Inheritance: Heredity as Memory

One specific notion of soft inheritance, that developed by Richard Semon during the early twentieth century (Semon 1921), is now being revived. Semon saw fundamental commonalities between heredity and memory, and interpreted embryonic induction, as well as the evolution of development, in terms of a common cellular/neural/genetic memory principle, which he called the "mneme" (Semon 1921). At present, research in epigenetic inheritance focuses on induced phenotypes that are reconstructed in descendants without the need for repeated induction, but Simona Ginsburg and I have suggested that there may be more subtle (and probably more common) types of heritable epigenetic effects. An inducing agent may elicit a phenotypic response that leaves a persistent epigenetic trace (e.g., a pattern of methylation marks), which, in the absence of the inducer, partially decays; when a second stimulus of the same type is applied later, then, because of the traces that remain, either a smaller stimulus is required to elicit the response or the response is faster. We called such an effect "epigenetic learning," and argued that this type of cellular learning, and other more complex cases involving more genes and more interactions, may occur in single cells (including neurons) and in nonneural organisms such as protists and plants (Ginsburg and Jablonka 2009). This means that, like Semon, we see memory and recall as aspects of soft inheritance.

The soft inheritance that occurs with epigenetic variants is both an integral part of development and a factor in evolution. Adaptive evolution within populations can occur on the epigenetic axis and, in conditions of stress, epigenetic mechanisms are involved in the generation of macrovariations that can lead to saltational evolution. Dobzhansky's (1973) famous dictum "Nothing in biology makes sense except in the light of evolution" is correct, but it needs to be extended and qualified, for nothing in evolution makes sense except in the light of development.

References

Agrawal AA, Laforsch C, Tollrian R. 1999. Transgenerational induction of defences in animals and plants. Nature 401,6748: 60–63.

Anway MD, Leathers C, Skinner MK. 2006. Endocrine disruptor vinclozolin induced epigenetic transgenerational adult-onset disease. Endocrinology 147: 5515–5523.

Bernstein E, Allis CD. 2005. RNA meets chromatin. Genes Dev. 19: 1635–1655.

Bossdorf O, Richards CL, Pigliucci M. 2008. Epigenetics for ecologists. Ecol Lett. 11,2: 106–115.

Brink RA. 1973. Paramutation. Annu Rev Genet. 7: 129–152.

Crews D. 2008. Epigenetics and its implications for behavioral neuroendocrinology. Front Neuroendocrinol. 29,3: 344–357.

Dobzhansky T. 1973. Nothing in biology makes sense except in the light of evolution. Am Biol Teach. 35: 125–129.

Durrant A. 1962. The environmental induction of heritable change in *Linum*. Heredity 17: 27–61.

Ephrussi B. 1958. The cytoplasm and somatic cell variation. J Cell Comp Physiol. 52(suppl. 1): 35–53.

Feinberg AP, Tycko B. 2004. The history of cancer epigenetics. Nat Rev Cancer 4: 143–153.

Gilbert SF, Epel D. *Ecological Developmental Biology: Integrating Epigenetics, Medicine, and Evolution.* Sunderland, MA: Sinauer Associates; 2009.

Ginsburg S, Jablonka E. 2009. Epigenetic learning in non-neural organisms. J Biosci. 34: 633–646.

Hadorn E. Dynamics of determination. In: *Major Problems in Developmental Biology.* Locke M, ed. New York: Academic Press; 1967:85–104.

Haig D. 2004. The (dual) origin of epigenetics. Cold Spring Harb Symp Quant Biol. 69: 67–70.

Harris M. 1984. High-frequency induction by 5-azacytidine of proline independence in CHO-K1 cells. Somat Cell Mol Genet. 10: 615–624.

Holliday R. 1987. The inheritance of epigenetic defects. Science 238,4824: 163–170.

Holliday R. 2006. Epigenetics: A historical overview. Epigenetics 1,2: 76–80.

Holliday R, Pugh JE. 1975. DNA modification mechanisms and gene activity during development. Science 187,4173: 226–232.

Jablonka E, Lachmann M, Lamb MJ. 1992. Evidence, mechanisms and models for the inheritance of acquired characters. J Theor Biol. 158,2: 245–268.

Jablonka E, Lamb MJ. 1989. The inheritance of acquired epigenetic variations. J Theor Biol. 139: 69–83.

Jablonka E, Lamb MJ. *Epigenetic Inheritance and Evolution: The Lamarckian Dimension.* Oxford: Oxford University Press; 1995.

Jablonka E, Lamb MJ. 2002. The changing concept of epigenetics. Ann NY Acad Sci. 981: 82–96.

Jablonka E, Lamb MJ. *Evolution in Four Dimensions: Genetic, Epigenetic, Behavioral, and Symbolic Variation in the History of Life.* Cambridge, MA: MIT Press; 2005.

Jablonka E, Lamb MJ. Transgenerational epigenetic inheritance. In: *Evolution: The Extended Synthesis.* Pigliucci M, Müller GB, eds. Cambridge, MA: MIT Press; 2010: 137–174.

Jablonka E, Raz G. 2009. Transgenerational epigenetic inheritance: Prevalence, mechanisms, and implications for the study of heredity and evolution. Q Rev Biol. 84: 131–176.

Johannes F, Porcher E, Texeira FK, Saliba-Colombani V, Simon M, Agier N, Bulski A et al. 2009. Assessing the impact of transgenerational epigenetic variation on complex traits. PLoS Genet. 5,6: e1000530. doi:10.1371/journal.pgen.

Lamm E, Jablonka E. 2008. The nurture of nature: Hereditary plasticity in evolution. Philos Psychol. 21,13: 305–319.

Lindegren CC. *The Cold War in Biology.* Ann Arbor, MI: Planarian Press; 1966.

Lyon MF. 1961. Gene action in the X-chromosome of the mouse (*Mus musculus* L.). Nature 190: 372–373.

Maynard Smith J. 1990. Models of a dual inheritance system. J Theor Biol. 143: 41–53.

Mayr E. Prologue: Some thoughts on the history of the evolutionary synthesis. In: *The Evolutionary Synthesis: Perspectives on the Unification of Biology.* Mayr E, Provine W, eds. Cambridge, MA: Harvard University Press; 1980:1–48.

McClintock B. 1984. The significance of responses of the genome to challenge. Science 226,4676: 792–801.

Nanney D. 1958. Epigenetic control systems. Proc Natl Acad Sci USA. 44: 712–717.

Nilsson EE, Anway MD, Stanfield J, Skinner MK. 2008. Transgenerational epigenetic effects of the endocrine disruptor vinclozolin on pregnancies and female adult onset disease. Reproduction 135: 713–721.

Reinders J, Wulff BB, Mirouze M, Marí-Orodóñez A, Dapp M, Rozhon W, Bucher E et al. 2009. Compromised stability of DNA methylation and transposon immobilization in mosaic *Arabidopsis* epigenomes. Genes Dev. 23: 939–950.

Riggs AD. 1975. X inactivation, differentiation, and DNA methylation. Cytogenet Cell Genet. 14,1: 9–25.

Sager R, Ryan FC. *Cell Heredity.* New York: John Wiley and Sons; 1961.

Sapp J. *Beyond the Gene: Cytoplasmic Inheritance and the Struggle for Authority in Genetics.* Oxford: Oxford University Press; 1987.

Semon R. *The Mneme.* London: Allen and Unwin; 1921.

Skinner MK, Anway MD, Savenkova MI, Gore AC, Crews D. 2008. Transgenerational epigenetic programming of the brain transcriptome and anxiety behavior. PLoS ONE 3,11: e3745. doi:10.1371/journal.pone.0003745.

Sonneborn T. 1950. The cytoplasm in heredity. Heredity 4,1: 11–36.

Teixeira FK, Heredia F, Sarazin A, Roudier F, Boccara M, Ciaudio C, Cruaud C et al. 2009. A role for RNAi in the selective correction of DNA methylation defects. Science 323,5921: 1600–1604.

Weaver ICG, Cervoni N, Champagne FA, D'Alessio AC, Sharma S, Sekl JR, Dymov S et al. 2004. Epigenetic programming by maternal behavior. Nat Neurosci. 7: 847–854.

Weaver ICG, Champagne FA, Brown SE, Dymov S, Sharma S, Meaney MJ, Szyf M. 2005. Reversal of maternal programming of stress responses in adult offspring through methyl supplementation: Altering epigenetic marking later in life. J Neurosci. 25,47: 11045–11054.

22 An Evolutionary Role for RNA-Mediated Epigenetic Variation?

Minoo Rassoulzadegan

The powerful scenario of genetic variation combined with natural selection provides us with a comfortable and satisfying picture of evolution. According to this view, mutations, which are the raw material of evolution, are continuously arising and either accumulate or are eliminated. The following quotations from Darwin indicate how he saw the process of selection and the formation of new species: "This preservation of favorable individual differences and variations, and the destruction of those which are injurious, I have called Natural Selection, or the Survival of the Fittest"(1872:63). ". . . according to my view, varieties are species in process of formation, or are, as I have called them, incipient species. How, then, does the lesser difference between varieties become augmented into the greater difference between species?" (1872:86). "That natural selection will always act with extreme slowness, I fully admit" (1872:84).

Darwin was right in thinking that the selection of random mutations is a slow process, although today we know that a new species could arise with only a single mutation in a master gene. Nevertheless, in general, random mutation and selection operating in complex organisms require long periods of gradual change. There have been, and still are, many debates around these issues. One possibility that has been considered is that selection for fast mutators under critical conditions could accelerate the rate of evolution (Wylie, Ghim, Kessler, and Levine 2009). However, this does not account for all aspects of evolution. The "fast mutator" phenotype can itself be silenced or induced, depending on the conditions of selection at the time. Mutation introduced by such a system can create novelty and introduce variations that could provide adaptation to new conditions.

Mutation is not the only source of variation. As François Jacob (1977) put it, "to create is to recombine." From bacteria to higher organisms, in addition to the genetic modifications (mutation and recombination) within the individual's own germ line, recombination with genetic material from other entities may also allow adaptations to new conditions. Recombination is the foundation of change, which, by itself, provides tremendous possibilities for switching between types. However, mutation and recombination may not be enough. We still do not know all the possible mechanisms of generating variation. Lamarck asked whether it is possible "that nature has created single-handed that astonishing diversity of powers, artifice, cunning, foresight, patience and skill, of which we find so many examples among animals," and concluded "that she has multiplied and diversified within unknown limits the organs and faculties of the organized bodies whose existence she subserves or propagates" (Lamarck [1809] 1963: 40, 41). We are finding such multiplicity and diversity in every realm of life, including in the mechanisms for generating variations.

It now appears that it is possible to generate variations in a different way which is faster and less costly for the species. We know that there are molecular processes in addition to mutation and the processes that mobilize DNA transposable elements that could be important to ensure that variations are produced. Among them are those related to epigenetics—the activities of the gene expression networks operating during development and differentiation. The contribution of epigenetic changes should be taken more seriously, because they might allow changes to be maintained not only from one developmental stage to another, but also from one generation to the next.

Epigenetic Selection?

Differentiation in the male germ line produces abundant sperm during the animal's life span. A large variety of epigenetic modifications may therefore occur in the male. In contrast, the female germ line consists of a very limited number of germ cells, each maintaining the same genome for a long period of time, but they, too, can be subject to epigenetic modification during their development. After fertilization, for a limited time embryonic life also provides "a workshop" that can test and allow local changes. Epigenetic remodeling plays an important role at all these stages of development. Moreover, there is also increasing evidence that epigenetic modifications can be induced by events occurring during embryogenesis, just as they can occur in the adult (Zernicka-Goetz, Morris, and Bruce 2009).

Epigenetic programming and reprogramming are powerful and routine processes through which a mouse is made following the fusion of sperm and egg. Twin formation is another important manifestation of the way in which epigenetic mechanisms can produce developmental changes. Recently, "animal cloning" with nuclei that are derived from various cell types, but always have the same genome, has opened up a wider range of experimental possibilities for making new animals. The reports of these experiments show that there are many technical difficulties, and there have been very low rates of success in most of the species tested so far. This highlights the importance of epigenetic modification in the ordering of development (Smith and Murphy 2004). If we can find the epigenetic rules (if there are any) for the orchestration of development, they should help us to know more about the range of variations that can be induced during organogenesis.

We can now ask what may sound like a science-fiction question: Is it possible to produce a different, *novel* type of animal with the same genome? If it is, this could mean that epigenetic modification is strong enough to reverse or to push the mouse genome, or any other genome, from its present state and turn it into something different and new. Multiple epigenetic changes through DNA methylation, chromatin modifications, and RNA-mediated modifications can induce and influence phenotypic changes. These deep and persistent epigenetic changes are possible through the use of several families of cellular factors that have become important tools for modifying development in mammals. Subtle changes in phenotype are correlated with changes in gene expression brought about by modification in DNA methylation and/or chromatin. Epigenetic variations generated by widely distributed epigenetic control mechanisms may underlie such variations in traits. Then, if compatible with life, the variant animal would develop further or, if incompatible with life, the changes would be reversed. Fast reversibility, in two or three generations of a complex organism, could be a considerable advantage.

The Mouse as a Model System

Epigenetic Variation Directed by Long, Short, Double-, and Single-Stranded RNAs

A number of developmental processes and their pathological deviations—for instance, cancer—have been found to depend on epigenetic modifications, that is, on mitotically stable changes that do not involve alterations of the genetic text, but rather are modulations of gene expression due to modified chromatin structures. A recent breakthrough has been the recognition of a role for long, noncoding RNAs in the establishment and maintenance of these chromatin structures (for example, in X-chromosome inactivation). In addition to the long, noncoding RNAs, increasingly complex populations of small RNAs have been identified in mammalian cells. MicroRNAs (miRNAs) and small, interfering RNAs (siRNAs) are generated from double-stranded precursors through the action of RNaseIII family members, including the enzymes Dicer and Drosha (Jinek and Doudna 2009). Other distinct families of small RNA species are not, however, generated from double-stranded precursors, and the list of these noncoding transcripts is far from complete. Moreover, deciphering the specific functions of these small RNAs is likely to be one of the keys to understanding complex phenotypes. Small RNAs interfere with several pathways, and in so doing they regulate gene expression, mediate host defenses (against viruses and transposons), shape chromatin, and transfer information in somatic and germ cells transgenerationally (see below). In germ cells, piRNAs (piwi-interacting RNAs) are another specific class of small noncoding RNAs; they are produced from clusters of genomic loci that give rise to long, single-stranded precursor transcripts. They are present in both nuclei and cytoplasm. In nuclei, chromatin is enriched with RNAs, and in the cytoplasm they are detected in both ribonucleoproteins and polysomes. The piRNAs seem to have conserved roles in the suppression of mobile genetic elements in *Caenorhabditis elegans*, *Drosophila*, zebra fish, and *Xenopus* (Hartig, Tomari, and Förstemann 2007).

The complexity and importance of these noncoding RNA systems is illustrated by the discovery of complex families of proteins that interact with them. One example is the mouse Piwi family (MILI, MIWI, and MIWI2); until recently, this was known simply to have some role in male germ cell development. A distinct role for two of its members, MILI and MIWI2, has now been described: they specify epigenetic marks by imposing de novo DNA methylation on transposon sequences in embryonic germ cells, leading to the transcriptional repression of these transposon sequences (Aravin and Hannon 2008).

These systems have been maintained during evolution, having the same function in a variety of species although there is always the possibility of new functions arising during evolution. Variety is seen in the maintenance and amplification strategies for producing small RNAs. Whereas yeast and plants use RNA-dependent RNA polymerases to produce and amplify antisense repeat sequences, *Drosophila* and mammals encode them from piRNA loci. Amplification of piRNA in these cells begins with a transposon-rich piRNA cluster, which produces a variety of piRNAs (O'Donnell and Boeke 2007).

There is certainly more to discover on the generation of RNAs and their amplification and maintenance in mammalian cells. For instance, a new mode of generating small RNAs has recently been described, and this has suggested that RNA polymerization may occur in a complex with telomerase (Maida, Yasukawa, Furuuchi, Lassmann, Possemato, Okamoto, Kasim et al. 2009).

Small RNAs are at the center of regulatory networks during development. Although much remains to be learned about their possible functions, one cannot help wondering whether they might provide a powerful means for generating diversity and new species.

Epigenetic Inheritance

Gene silencing by small RNAs is being extensively studied today, but the story is definitely not complete. We have also begun recently to consider the role of small RNAs in the induction of gene expression during development and in transgenerational effects.

The observation that epigenetic variations are in some instances meiotically stable—in other words, they are inherited following the formation of sex cells—was first reported in plants and, more than fifty years ago, was designated *paramutation*. Paramutation is a special type of heritable epigenetic variation which is induced when two different alleles of a single locus (in a heterozygote) interact, and one allele alters the expression of the other allele (rendering the cell functionally homozygote; see Brinks 1973). Recently, a role for RNA in the control of epigenetic states, including those involving paramtutations, has been suggested to explain several observations suggestive of similar phenomena in mice and human pathology. My laboratory reported the first clear instance of RNA-mediated, epigenetic, non–Mendelian heredity in the mouse, which involved the *Kit* receptor gene and a characteristic color phenotype. These initial studies used a mutant of the gene encoding the *Kit* receptor. Unexpectedly, we observed that the white-tail phenotype characteristic of the mutant was maintained by wild-type homozygotes in the progeny of heterozygotes and in the subsequent generations. The Kit^+ (wild-type) alleles thus underwent an epigenetic modification (indicated *Kit**) resulting in the expression of a modified phenotype that was similar to that of a mutant with reduced *Kit* mRNA expression. Once established, the epigenetic modification was transmitted with high efficiency, both paternally and maternally, in a clearly non–Mendelian mode. Although stable in successive intercrosses of modified animals, the *Kit** condition eventually reverted to the normal phenotype within three to four generations of backcrosses with normal parents. Furthermore, we showed that *Kit** can be efficiently induced by injecting into fertilized normal eggs either *Kit* RNA fragments or the *Kit*-specific microRNAs. Taken together with the presence of *Kit* RNA in the spermatozoa of the male transmitting the *Kit** phenotype, these observations led us to conclude that this was a novel form of "RNA-mediated heredity" (Rassoulzadegan, Grandjean, Gounon, Vincent, Gillot, and Cuzin 2006).

We further extended this concept to a pathological situation in the mouse that mimics hypertrophic cardiomyopathy, a dangerous human disease known for its frequent familial occurrence. We injected fertilized mouse eggs with miR-1 microRNAs that target the Cdk9 protein, a major regulator of cardiac growth. This led to high levels of expression of the homologous RNA, resulting in an epigenetic defect, cardiac hypertrophy. This defect was inherited, and its transmission between generations was correlated with the presence of miR-1 in the sperm nucleus. Interestingly, in this case, the paramutation increased rather than decreased the expression of the Cdk9 protein (Wagner, Wagner, Ghanbarian, Grandjean, Gounon, Cuzin, and Rassoulzadegan 2008). More recently, we have reported yet another distinct case of heritable epigenetic variation, which modulates the growth of the embryo and the size of adult bodies (figure 22.1). Following injection of miR-124 microRNA

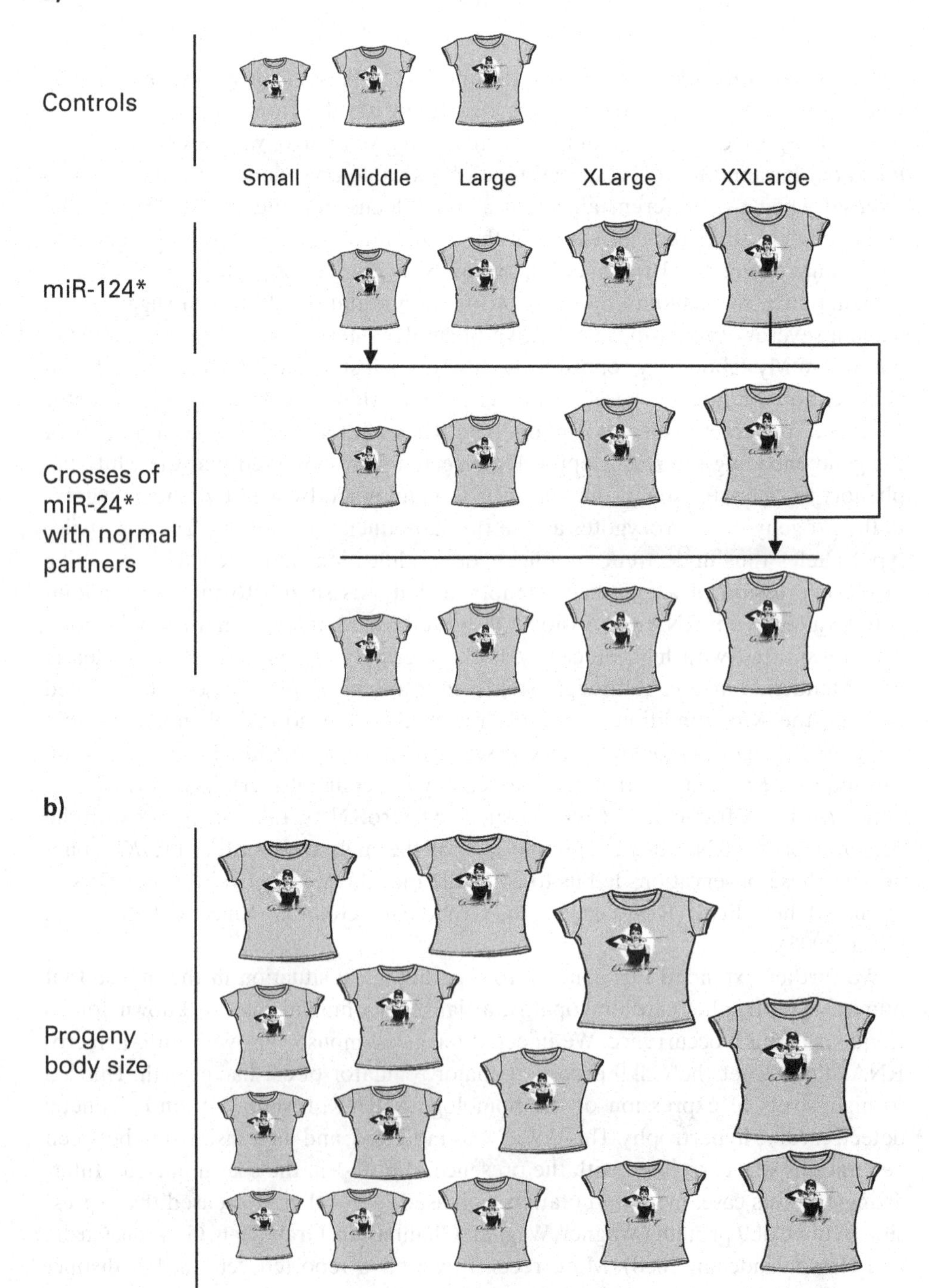
a)
Controls
Small
Middle
Large
XLarge
XXLarge
miR-124*
Crosses of
miR-24*
with normal
partners
b)
Progeny
body size
Parental phenotype

into fertilized eggs, mouse pups that were born showed a 30 percent increase in size (Grandjean, Gounon, Wagner, Martin, Wagner, Bernex, Cuzin et al. 2009). Each of these distinct, RNA-induced phenotypic changes probably reveals only little of the vast number of possible RNA-mediated changes, which, we believe, must almost certainly extend far beyond our present level of experimentation.

In a laboratory mouse with a stable genome, phenotypic variations can be easily induced by microinjecting synthetic short RNA molecules into the pronuclei of fertilized eggs (Rassoulzadegan, Grandjean, Gounon, Vincent, Gillot, and Cuzin 2006). Although the mechanisms of their action are yet to be elucidated, there are important questions regarding the primary induction by RNA and the subsequent maintenance of the new phenotypes for generations without DNA modification. One question in particular is whether such a process could happen in nature. In other words, do small RNA molecules naturally find their way into eggs? So far, we have considered only RNA accumulation in the head of sperm and the experimental introduction of RNA molecules into the developing embryo.

Several new concepts of interest are already emerging from these data. One of them is that of a "rheostat" mode of inheritance (a rheostat is a *continuously variable* electrical resistor used to regulate a current, such as a dimmer), as opposed to an "on-off switch" model (Beaudet and Jiang 2002). The rheostat model suggests that some genes (for example, imprinted genes and paramutable alleles) have variable levels of expressivity in different tissues, and that this allows functional tissue-specific adjustments during development. Moreover, as such quantitative variants arise, functional haploidy could allow for selection among individuals with different levels of tissue-specific variability. It has been observed that large quantitative variations are seen between individual mice of the same family, even in the same litters, and this diversity is created anew in each generation. These large variations may be related to a selective advantage for the underlying paramutagenic events. We suggested that one possible reason for their occurrence is an imperfect partitioning of the RNA molecules in the germ cells at the haploid stages, leading each haploid cell to differ from its sister cells with respect to their RNAs content. From an evolutionary point of view, it is clear that such a mode of inheritance is favorable in the many situations in natural conditions that show

Figure 22.1
RNA-mediated paramutation of body weight variation in mice. (a) Body weight variation is represented by T-shirts: S, M, and L correspond to classes of body weight in a control laboratory mouse colony. After microinjection of miR-124, larger mice are found, as indicated by XL and XXL, while S size mice are missing from the colony. Large mice are also seen when miR-124 M and XXL are backcrossed with a normal partner. Note that in lines with miR-124 paramutated founder animals, the M, L, XL, and XXL phenotypes are transmitted and there are no S size mice. (b) A hypothetical case: when the paramutagenic agent is present (in lines with XL and XXL individuals), the average size of the progeny will be larger.

a large amount of spatial and temporal variability. Along with the generation of a sufficiently large number of progeny, it may ensure that, in each generation, a fraction of them will survive. This brings us to a second striking feature of paramutation, the accumulation of RNAs in the spermatozoon head, which is considered to be responsible for hereditary transmission of the epigenetic state. This is of special interest because a large quantity of RNA has been observed in human sperm. The amount of RNA in the sperm head of the "normal" laboratory mouse is minute compared with that in human sperm and, interestingly, also in our paramutant mice. On the basis of the occurrence of paramutation associated with a high RNA content in the sperm of progeny of the *Kit* heterozygotes, we currently favor the hypothesis that, in our fully outbred species, the presence of the RNAs is the result of the large degree of heterozygosis. Of particular importance, we believe, is heterozygosity which leads to mispairing between homologous regions, for example, when one homologue has a significant deletion or insertion. Such mispairing triggers the formation of small RNAs from the unpaired region, and this leads to the modulation of gene expression and possibly to transmission between generations (Rassoulzadegan, Grandjean, Gounon, and Cuzin 2007). If so, this suggests the possibility of an extended spectrum of RNA-directed epigenetic regulation in outbred populations. Deep nucleic-acid-sequencing technologies may make it possible in the near future to study the inventory of sperm RNA pools in various species, and this will be an important step toward ascertaining the causes and extent of epigenetic heredity.

Concluding Remarks

It is now apparent that it is important to take into consideration the power of transgenerationally transmitted epigenetic modifications to modulate gene expression and thereby modify the phenotype of organisms. Small regulatory RNAs, which induce either gene expression or gene silencing, are active determinants of such differentiation. By generating a pool of variations, through the "rheostat effect" these subtle changes can endow organisms with a novel means of responding to natural selection.

We suggest that the role of small RNAs in evolution is to "knock genomes into shape," correcting the effects of newly arisen deleterious mutations and orienting changes in new directions. By generating variable phenotypes, epigenetic, heritable, yet potentially reversible variations increase the chances of population survival. It seems reasonable to conjecture that by modifying cell functions during germinal and embryonic development, small RNAs may contribute to the evolution of species.

Acknowledgments

I thank François Cuzin and Kenneth Marcu for critical reading and improving of the manuscript. This work was funded by grants from Agence Nationale de la Recherche (ANR-06-BLAN-0226 PARAMIR and ANR-08-GENO-011-01, EPIPATH-PARAPATH), France.

References

Aravin AA, Hannon GJ. 2008. Small RNA silencing pathways in germ and stem cells. Cold Spring Harb Symp Quant Biol. 73: 283–290.

Beaudet AL, Jiang YH. 2002. A rheostat model for a rapid and reversible form of imprinting-dependent evolution. Am J Hum Genet. 70,6: 1389–1397.

Brink RA. 1973. Paramutation. Annu Rev Genet. 7: 129–152.

Darwin C. *On The Origin of Species.* 6th ed. London: John Murray; 1872.

Grandjean V, Gounon P, Wagner N, Martin L, Wagner KD, Bernex F, Cuzin F, Rassoulzadegan M. 2009. The miR-124-Sox9 paramutation: RNA-mediated epigenetic control of embryonic and adult growth. Development 136,21: 3647–3655.

Hartig JV, Tomari Y, Förstemann K. 2007. piRNAs—the ancient hunters of genome invaders. Genes Dev. 21,14: 1707–1713.

Jacob F. 1977. Evolution and tinkering. Science n.s.196,4295: 1163.

Jinek M, Doudna JA. 2009. A three-dimensional view of the molecular machinery of RNA interference. *Nature* 457,7228: 405–412.

Lamarck J-B. *Zoological Philosophy*. [1809]. Elliot H, trans. New York and London: Hafner; 1963.

Maida Y, Yasukawa M, Furuuchi M, Lassman T, Possemato R, Okamoto N, Kasim V et al. 2009. An RNA-dependent RNA polymerase formed by TERT and the RMRP RNA. Nature 461,7261:230–235.

O'Donnell KA, Boeke JD. 2007. Mighty Piwi defend the germline against genome intruders. Cell 129,1:37–44.

Rassoulzadegan M, Grandjean V, Gounon P, Cuzin F. 2007. Sperm RNA, an "epigenetic rheostat" of gene expression? Arch Androl J Reprod Med. 53: 235–238.

Rassoulzadegan M, Grandjean V, Gounon P, Vincent S, Gillot I, Cuzin F. 2006. RNA-mediated non-Mendelian inheritance of an epigenetic change in the mouse. Nature 441,7092: 469–474.

Smith LC, Murphy BD 2004. Genetic and epigenetic aspects of cloning and potential effects on offspring of cloned mammals. Cloning Stem Cells; 6,2: 126–132.

Wagner KD, Wagner N, Ghanbarian H, Grandjean V, Gounon P, Cuzin F, Rassoulzadegan M. 2008. RNA induction and inheritance of epigenetic cardiac hypertrophy in the mouse. Dev Cell 14,6: 962–969.

Wylie CS, Ghim CM, Kessler D, Levine H. 2009. The fixation probability of rare mutators in finite asexual populations. Genetics 181: 1595–1612.

Zernicka-Goetz M, Morris SA, Bruce AW. 2009. Making a firm decision: Multifaceted regulation of cell fate in the early mouse embryo. Nat Rev Genet. 10: 467–477.

23 Maternal and Transgenerational Influences on Human Health

Peter D. Gluckman, Mark A. Hanson, and Tatjana Buklijas

Models of disease are tools that medicine uses to order the complex natural world of pathologies. Around 1800, the humoral model that defined disease as an imbalance of bodily fluids was replaced with the pathological-anatomical one, according to which the morbid state arose from a well-defined lesion (Nutton 1993; Bynum 1993). This easily accommodated infectious diseases, but it could not explain noncommunicable pathological conditions. In the "lifestyle transition" of the twentieth century, those conditions have come to lead to morbidity and mortality statistics, first in developed countries and then worldwide. In medical literature, "risk factors"—from physiological parameters such as blood pressure to "lifestyle" behaviors such as salt intake, smoking, and lack of exercise—replaced etiology. But their relative contributions to the development of disease remained contested and their mechanistic explanation was often inadequate (LaBerge 2008, Timmermann 2006), so the quest for a better explanatory model continued. Genetics, which shifted the focus from the externally inducing factor to the heritable change in DNA, promised an answer, but genome-wide association studies (GWAS) have succeeded in explaining only a small portion of risk of noncommunicable disease (Goldstein 2009). A new model is emerging: one in which heredity, development, and environment are brought together within an evolutionary framework. The model states that noncommunicable disease is an outcome of the mismatch between the developing organism's prediction of the future environment and the actual conditions in which the adult organism lives (Gluckman and Hanson 2004a). In mammals, the prediction is made from cues received from the maternal environment and is effected by epigenetic alterations that change the expression of genes which then influence how the individual responds to later challenges (Gluckman, Hanson, Cooper, and Thornburg 2008). This model explained better than a purely genetic one the differences in individual disease risks in spite of similar lifestyles and the same environment. The epigenetic effects, new research indicates, may be transmitted across generations (Jablonka and Raz 2009; chapter 21 in this volume).

In his *Philosophie zoologique* (1809), the French biologist Jean-Baptiste Lamarck espoused the concept of organic evolution that proceeded by somatic modifications resulting from the development of particular habits (Burkhardt 1995). At the same time, Lamarck's advocate and colleague at the Muséum d'Histoire Naturelle, Étienne Geoffroy Saint-Hilaire, conceived teratology, the science of pathological development, as an experimental tool for elucidating how environmental change affects evolutionary pathways (Richards 1994). While for much of the nineteenth century the idea that environmentally induced modifications can be transmitted across generations was widely accepted, from the 1880s onward, development, evolution, and environment went separate ways, and it is only in recent decades that they have been brought together (Gilbert 2001). Following a brief historical overview, this chapter reviews recent research to show how the maternal environment shapes the disease risk of the next and later generations, and how—in the reverse of Geoffroy Saint-Hilaire's logic—evolutionary concepts may be profitably applied to explain the emergence of disease.

Maternal Environment and Offspring Disease

Around 1800, new disciplines for studying normal and pathological development, namely embryology and teratology, emerged. They were part of a broader movement that replaced the hitherto prevailing static view of the natural world with a dynamic one. The founder of teratology, Étienne Geoffroy Saint-Hilaire, is best known for his theory of evolution according to which all animals possessed the same basic ground plan and could therefore be ordered in a single evolutionary chain of increasing complexity (Appel 1987). The evolutionary change, in his view, was a result of direct environmental influence. A doctrine of parallelism, arguing that embryonic development recapitulated evolution, was formulated in this period by Geoffroy Saint- Hilaire's disciple Étienne Serres from an earlier proposition by the German anatomist Johann Friedrich Meckel. In 1818 Serres argued, based on his studies of embryonic and adult vertebrate brains, that however different adult brains might appear, they were essentially the same: by following them down both the animal scale and the embryonic development, one could see the brain "decomposing" (Appel 1987). Today this is known as the law of recapitulation and is associated with its later advocate, Ernst Haeckel, yet embryology and evolution had been connected since their early days, around 1800, and Haeckel's merit is in setting recapitulation into Darwinian terms and making it durably famous. Accordingly, Geoffroy Saint-Hilaire viewed embryos as proxies for our evolutionary ancestors. For him, the experimentally produced "monsters"—produced, for instance, by varnishing or heating chicken eggs—were "developmental arrests" that could teach us how new species could arise in nature (Richards 1994). Experimental study of pathological development flourished, especially in France in the nineteenth century (Fischer 1990), but with the separation of developmental and evolutionary studies in the late nineteenth century (Bowler 1989), it fell exclusively under the aegis of medicine.

Although teratology was a marginal pathological-anatomical field, around 1900 its findings stimulated some obstetricians such as the Scottish pioneer of "antenatal hygiene," John William Ballantyne, worried about perceived biological and moral degeneration of humankind, to replace their exclusive focus on childbirth and reduction in maternal mortality with one on the health of the offspring (Al-Gailani 2009). Building on the compelling evidence from conditions such as congenital syphilis, they refused to draw a sharp line between transmitted diseases and the environment more generally, on the one hand, and heredity, on the other. In their view, study of the antenatal period not only complemented the eugenics' focus on heredity but indeed surpassed it in value. Eugenics, according to them, looked "very far back for its causes and very far forward for its results, while ante-natal and gestational hygiene treats the present and immediate future" (Ballantyne 1914, xi). Maternal body provided the immediate point of medical and research interest as "although he (infant) is hidden from sight in the womb of his mother, he is not beyond the influences of her environment, nay, her body is his immediate environment" (Ballantyne 1914, xii).

This conceptual change was part of a larger shift in medicine. Moving away from the late nineteenth-century reductionism, exemplified by the bacteriological model of disease causation, medicine now focused on the organism as a whole—including its socioeconomic and biological environment (Lawrence and Weisz 1998). In the 1930s and 1940s, teratology expanded through seminal research into the disruptive effects on development of viruses, vitamin deficiency, radiation, and chemical toxins. The major environmental factor that received attention in this period was nutrition. Yet, while interwar nutritional surveys focused on maternal health outcomes (Williams 1997), studies that followed up the survivors of the Second World War famines looked closely at the offspring. Along with the number of abortions, stillbirths, and malformations, the recorded effects included subtle changes in birth weight and length (Smith 1947).

These insights were important, but they all focused on the early postnatal period and did not venture into explaining longer-term consequences. It was only in the early 1970s that the East German endocrinologist Günter Dörner proposed that pre- and early postnatal conditions were related to later risks of arteriosclerosis, obesity, and diabetes mellitus (Gluckman, Hanson, and Buklijas 2010). Simultaneously, in the United States Norbert Freinkel developed the hypothesis of fuel-mediated teratogenesis, arguing that maternal metabolic state influenced offspring and so led to possible intergenerational transmission of the risk of diabetes (Freinkel 1980). The greatest impetus came from epidemiological studies, most influentially from those by David Barker. He showed a strong geographical relation between incidence of ischemic heart disease (IHD) in 1968–1978 and infant mortality between 1921 and 1925; he went on to reveal, on a large sample of men born in Hertfordshire between 1911 and 1930, that those born small had a higher risk of IHD and chronic obstructive pulmonary disease in adulthood (Barker and Osmond 1986; Barker, Winter, Osmond, Margetts, and Simmonds 1989).

Theoretically formulated as "fetal origins of adult disease," the proposition that poor environment in early life, manifested as low birth weight, causes chronic disease was tested and refined in animal experiments as well as prospective clinical studies. Subsequently, the view of the poor developmental environment was expanded to include pathological conditions in which fetal nutrition was compromised, such as placental and maternal disease. There was also discussion of physiological causal mechanisms, such as stress and maternal constraint. The latter encompasses a set of poorly defined maternal mechanisms through which, in order to secure maternal survival, the fetal growth is limited. Forms of maternal constraint include maternal size, parity, young age, diet, and nutrition (Gluckman and Hanson 2004b).

It was soon clear that refinements of this hypothesis were necessary. Barker's own and others' (Curhan, Chertow, Willett, Spiegelman, Colditz, Manson, Speizer et al. 1996) data showed that the relation between size at birth and disease risk is not a matter of extremes but that it extends in a graded manner across the normal birth weight range (Godfrey 2006). Indeed, the realization that the curve representing the relationship between birth weight and the risk of NCDs is U-shaped, rather than simply an inverse relation, helped accommodate the findings that high birth weight (>9.5 pounds) may also indicate future risk. A well-known example concerns macrosomic children of mothers with gestational diabetes, in whom fetal overstimulation of pancreatic β cells and increased insulin secretion is followed by reduced insulin secretion and diabetes in later life (Van Assche, Holemans, and Aerts 2001). Furthermore, obese mothers may give birth to large children who later become obese. The processes underpinning this outcome are unclear; both toxic and adaptive explanations have been advanced (Kuzawa, Gluckman, and Hanson 2007).

While birth weight served as a useful proxy for those studies where no other measure of developmental course was available, "normal" birth weight could mask pathological effects of suboptimal maternal nutrition. One study found that carotid artery intima-media thickness (a marker of early atherosclerosis) at the age of nine was greater in those children whose mothers had lower energy intake during pregnancy in relation to their basal metabolic rate (Gale, Jiang, Robinson, Godfrey, Law, and Martyn 2006). That postnatal environment may further amplify the effects of poor developmental environment was shown by a study in which pregnant rat dams were fed either ad libitum (AD group) or at 30 percent of AD (undernutrition or UN group) (Vickers, Breier, Cutfield, Hofman, and Gluckman 2000). At weaning, offspring from both groups were assigned to either a control or a hypercaloric diet. Compared with the UN group fed a control diet, the UN offspring on a hypercaloric diet showed significantly higher systolic blood pressure, fasting blood insulin, and leptin levels in spite of concurrent hyperphagia, thus suggesting a state of leptin and insulin resistance. A modification in postnatal environment thus revealed hidden metabolic alterations.

The mechanistic basis of such modifications is now becoming clearer, in particular the epigenetic processes which mediate over- or underexpression of relevant genes. A protein-restricted diet during pregnancy reduces methylation of the hepatic glucocorticoid receptor (GR) 1_{10} promoter and peroxisomal proliferator-activated receptor (PPAR)α promoter, and consequently causes higher expression of these receptors in the offspring. These changes affect fatty acid oxidation and gluconeogenesis by the liver. These effects were not recorded when folic acid, a methyl-group donor cofactor, was added to the maternal diet (Lillycrop, Phillips, Jackson, Hanson, and Burdge 2005). Changes in the expression of GR 1_{10} have been explained by lower expression of DNA methyltransferase 1 (Dnmt1), the enzyme responsible for de novo methylation of CpG islands (Lillycrop, Slater-Jefferies, Hanson, Godfrey, Jackson, and Burdge 2007). As GR 1_{10} has importance in the regulation of blood pressure, and PPARα is central to lipid and carbohydrate homeostasis, their overexpression may have a role in the development of cardiovascular and metabolic diseases.

Developmental Plasticity and Origins of Disease

The finding that the offspring of malnourished mothers may be well when fed sparsely, yet show signs of metabolic disease in a nutritionally rich environment, led to a suggestion that the developmentally induced adult chronic disease may be a consequence of immediate effects of a low nutritional plane on the fetus (Hales and Barker 1992). This model, termed the "thrifty phenotype," viewed the effect on the fetus as pathological. However, it was later shown that it should be interpreted as an evolutionarily appropriate reaction to a difficult environment, setting the fetal phenotype for later adaptive advantage (Bateson 2001; Bateson, Barker, Clutton-Brock, Deb, D'Adine, Foley, and Gluckman 2004). Developmental responses were then formally divided into potentially adaptive and nonadaptive (teratogenic) responses. Depending on the timing of the adaptive advantage—whether it was concurrent with the cue or delayed—the potentially adaptive responses were then formally divided into immediate and delayed (predictive) adaptive responses (Gluckman and Hanson 2004a).

Environmental cues that can disrupt the developmental genetic program require an immediate survival response, such as premature delivery or another of the possible pathways that is best suited to the expected future; but should the prediction turn out to be incorrect, the adaptive response would make the phenotype of the individual less well suited to the environment and could lead to greater risk of disease (Gluckman and Hanson 2006a). Such predictive adaptive responses are found in all taxa; they constitute an expression of developmental plasticity, a set of developmental mechanisms that enhance the reproductive success and survival rate by better matching the organism to the environment.

The predictive adaptive model lends itself to experimental testing. If the value of the maternal cue lies in informing the offspring of its future environment, then the maladaptive outcome of mismatch should be remedied by a change in the environment during the developmentally plastic period. That was found for at least some of the aspects of the offspring phenotype in rodents and large mammals (Cleal, Poore, Boullin, Khan, Chau, Hambidge, Correns et al. 2007; Khan, Dekou, Hanson, Poston, and Taylor 2004). Also, it should be possible to rescue the phenotype by using a pharmacological or endocrine intervention during the plastic period. Vickers, Gluckman, Coveny, Hofman, Cutfield, Gertler, Breier et al. (2005) showed that leptin administered between neonatal days 3 and 13 to the offspring of dams undernourished during pregnancy prevented the detrimental effects of the predictive adaptive response when the animals were subsequently fed an obesogenic diet. The observed parameters included caloric intake, locomotor activity, body weight, fat mass, and fasting plasma glucose, insulin, and leptin concentrations. Similarly, a statin given in late pregnancy to dams fed a high-fat diet prevented the detrimental cardiovascular and metabolic effects of feeding a similarly high-fat diet to their offspring after weaning (Elahi, Cagampang, Anthony, Curzen, Ohri, and Hanson 2006). While most research has been on animal models, initial studies testing the predictive adaptive hypothesis in humans, comparing the relationship between the ponderal index at birth and the adult reproductive hormone levels, have now been conducted (Jasienska, Thune, and Ellison 2006).

The predictive adaptive hypothesis is further supported by demographic observations that the age of puberty in developed countries decreased from 17 to 12.5 years in the twentieth century (Gluckman and Hanson 2006b). Following clinical studies that showed that girls who are born small but become large at eight years reach menarche earlier (Sloboda, Hart, Doherty, Pennell, and Hickey 2007), animal studies confirmed that caloric restriction before birth or during lactation accelerates the age of sexual maturation (Sloboda, Howie, Pleasants, Gluckman, and Vickers 2009). The earliest onset of menarche was found in the group undernourished in pregnancy and lactation but fed a high-fat diet afterward. Interestingly, the group which received a high-fat diet through all stages also entered puberty early in comparison with controls. From an evolutionary viewpoint, this effect may be for a different reason. While the offspring of undernourished dams matured early to hasten reproduction in expectation of a short life span in the predicted poor nutritional environment, in those of dams fed the high-fat diet, the goal might be to extend the length of the reproductive period. The former can be viewed as an attempt to maintain fitness in a difficult environment, and the latter as an opportunistic increase in fitness in response to a predicted abundance. This hypothesis was supported by finding divergent progesterone levels in the two groups: low in the former and high in the latter. Perhaps the most intriguing proposition of this study is that noncommunicable diseases are a secondary outcome of the strategies of early maturation.

The practical importance of this developmental model of chronic disease is high as the world undergoes nutritional transition characterized by increased access to high-fat, high-calorie food (Drewnowski and Popkin 1997). The current obesity epidemic in the West is a consequence of mismatch to the change in diet post Second World War (Gluckman and Hanson 2006a), but we should not assume that normal or overweight (as assessed by the body mass index, BMI) means optimal nutrition. A study on 6125 nonpregnant women aged twenty to thirty-four years in the United Kingdom showed that educational achievement was closely linked with the quality of nutrition (Robinson, Crozier, Borland, Hammond, Barker, and Inskip 2004). Even women who had access to plentiful nutrition consumed an unbalanced diet. Interviews with women who would go on to become pregnant within the next three months showed that only 2.9 percent complied with recommendations on alcohol consumption and adequate folic acid intake (Inskip, Crozier, Godfrey, Borland, Cooper, and Robinson 2009). The situation is even more serious in developing countries, in which the nutritional transition has been abrupt and recent. Once again, the provision of more abundant nutrition unfortunately does not mean that an unbalanced diet is not consumed. With somewhat improved financial circumstances and under the influence of advertising, many families in developing countries will consume more of the increasingly available sugar, fat, and salt, red meat, and highly processed foods rather than increasing amounts of fruits, vegetables, and fish. The tragedy is that the increase in chronic noncommunicable disease, caused by a mismatched developmental predictive adaptive response, occurs in developing societies at a much lower level of affluence than in the developed world. The consequences of improved socioeconomic conditions following reductions in infectious disease may therefore lead to increased incidence of noncommunicable disease.

Transgenerational Influences

A major question for developmental and epigenetic science today is whether developmentally induced effects may be transmitted to subsequent generations (Gluckman, Hanson, and Beedle 2007). Epidemiological research offers intriguing if inconclusive evidence. A study that combined records of early twentieth-century harvest and food prices with the clinical follow-up of children born in the early 1990s, found sex-specific male-line transgenerational effects (Pembrey, Bygren, Kaati, Edvinssin, Northstone, Sjöstrom, Golding et al. 2006). Good food supply during the paternal grandfather's growth period preceding the prepubertal peak was associated with the grandson's increased relative mortality risk, while good food supply in early childhood of the paternal grandmother doubled the granddaughter's relative mortality risk. Studies of the offspring (F_1) born during a famine under German siege in the Netherlands in the winter of 1944/1945 ("Dutch Hunger Winter") followed effects to the grandchildren (F_2) generation. It has been shown that grandmaternal exposure to famine is associated with neonatal adiposity and poorer health over the course of life in F_2 women (Painter, Osmond, Gluckman, Hanson, Phillips, and Roseboom 2008). Rodent studies have shown that nutritional and endocrinological interventions in pregnant F_0 animals resulted in phenotypic and/or epigenetic changes that persisted to at least F_2 (Drake, Walker, and Seckl 2005; Burdge, Slater-Jefferies, Torrens, Phillips, Hanson, and Lillycrop 2007; Jimenez-Chillaron, Isganatis, Charalambous, Gesta, Pentiat-Pelegrin, Faucette, Otis et al. 2009).

One of the proposed mechanisms of nongenomic inheritance concerns direct influence on the F_2 through fetal oocyte exposure in F_0, as the ovary starts to differentiate and primary oogenesis occurs in the early fetal life of the woman. Another mechanism involves indirect transmission by re-creating the environment which induces the phenotype (akin to niche construction) in the early postnatal period. The latter process has been demonstrated in the studies of maternal behavioral induction of epigenetic changes in the brains of infant mice: while the epigenetic mark may not survive gametogenesis, in each generation the inducing behavior is repeated (Weaver, Cervoni, Champagne, D'Alessio, Sharma, Seckl, Dymov et al. 2004; Champagne and Curley 2008). Similarly, development of girls may depend on the size of the mother's uterus; if it is small, they may grow to have a smaller uterus and, in turn, smaller offspring (Ibáñez, Potau, Enriquez, Marcos, and De Zegher 2003). Both generations might have similar epigenetic changes present without germ line transmission having occurred. Thus gametic transgenerational epigenetic inheritance generally requires evidence of transmission to at least F_3 in females (unless the induction in F_0 is prior to ovulation) but only F_2 in males, and so far has been studied in animal—mostly rat and mouse—models only.

The challenge of the current research is, then, in proving the multigenerational transmission of prenatally (and early postnatally) induced effects and in elucidating their epigenetic mechanisms; this will be mainly via transfer of methylation and histone marks, and microRNAs. For example, in rats, endocrine disruptors (substances that influence gonadal development and sex determination), administered to gravid dams during the sensitive developmental period, increased spermatogenic cell apoptosis and reduced sperm numbers and motility in the offspring. Clinical and histological effects persisted to the F_4 generation, but it remains uncertain if alterations in the methylation pattern are sufficiently strong to explain the extensive phenotypic effect (Anway, Cupp, Uzumcu, and Skinner 2005). Another study tested the proposition that maternal obesity promotes obesity in children by using a genetic model of obesity in agouti viable yellow (A^{vy}) mice (Waterland, Travisano, Tahiliani, Rached, and Mirza 2008). The A^{vy} allele produces agouti molecule that induce yellow color of the coat and antagonize satiety signaling in the hypothalamus. In the study, A^{vy}/aa (slightly mottled yellow and prone to overeating) F_0 females were weaned onto either a standard or a methyl-supplemented diet and were then mated with *a/a* males; the same procedure was repeated in F_1 and F_2. While supplementation did not significantly affect adult body weight in F_1 and F_2, by F_3 the cumulative transgenerational effect was evident, with the unsupplemented group significantly more obese than its ancestors as well as its methyl-supplemented counterparts. The precise mechanism, however, remains unknown: the lack of association between body weight and coat color indicates that an increase of methylation that prevented obesity over generations did not happen at the A^{vy} locus but somewhere else in the genome.

The best explained epigenetic hereditary mechanism is sperm- or oocyte-mediated microRNA transfer, which has been demonstrated in two mouse models. In the *Kit* gene paramutation (meiotically stable epigenetic modification) model, the engineering of a mutant phenotype followed by the crossing of F_1 heterozygotes with wild types as well as heterozygotic intercrossing produced a mutant phenotype in some of the F_2 offspring—although they did not have the corresponding (homo- or heterozygotic) genome (Rassoulzadegan, Grandjean, Gounon, Vincent, Gillot, and Cuzin 2006; chapter 22 in this volume). The observation of a decrease in mature mRNA and an increase of abnormal RNA suggested that RNA might be involved. Moreover, the accumulation of RNA in sperm cells encouraged the idea of testing the transmission of paramutations by injecting RNAs from *Kit* homo- or heterozygotes into wild-type embryos. Indeed, offspring with a mutant phenotype were born which then transmitted the paramutation to the next generation. In the second model, fertilized mouse eggs were injected with RNAs that targeted the heart growth regulator Cdk9 (Wagner, Wagner, Ghanbarian, Grandjean, Gounon, Cuzin, and Rassoulzadegan 2008). The intervention resulted in cardiac hypertrophy, transmitted epigenetically by microRNA in the sperm nucleus.

While the early results are encouraging, the mechanisms by which epigenetic modifications emerge and, for the most part, the nature of these changes remain poorly understood. Yet, these first confirmations of non-Mendelian inheritance in mammals are of potential significance to human biology and medicine, as they might elucidate some hitherto unexplained forms of familial inheritance of disease.

Conclusions

The recent reemergence of interest in the ways in which the maternal environment shapes the disease risk of the offspring of the next and subsequent generations is part of a larger intellectual shift that, two centuries after Lamarck, has brought together environment, development, and evolution. Its echoes are felt across biomedicine, from the new fields of evo-devo and eco-devo, to physiology and medicine, as shown here. These shared influences encourage productive exchanges. The predictive adaptive response hypothesis has demonstrated how medicine has profited from evolutionary biology, which can explain the clinical and experimental evidence about the consequences—from the age of puberty to predisposition to disease—of prenatal and early postnatal environments in terms of subtle adjustments in the adult phenotype. In turn, medical problems, in particular forms of familial inheritance of disease that cannot be explained in terms of fixed genetic variation, offer models with which to tackle the more general problem of non-Mendelian transgenerational inheritance.

References

Al-Gailani S. 2009. Teratology and the clinic: John William Ballantyne and the making of antenatal life. Wellcome History 42: 2–4.

Anway MD, Cupp AS, Uzumcu M, Skinner MK. 2005. Epigenetic transgenerational actions of endocrine disruptors and male fertility. Science 308,5727: 1466–1469.

Appel TA. *The Cuvier-Geoffroy Debate: French Biology in the Decades Before Darwin.* New York: Oxford University Press; 1987.

Ballantyne JW *Expectant Motherhood: Its Supervision and Hygiene.* London: Cassell; 1914.

Barker DJ, Osmond C. 1986. Infant mortality, childhood nutrition, and ischaemic heart disease in England and Wales. The Lancet; 1,8489: 1077–1081.

Barker DJP, Winter PD, Osmond C, Margetts B, Simmonds SJ. 1989. Weight in infancy and death from ischaemic heart disease. The Lancet 2: 577–580.

Bateson P. 2001. Fetal experience and good adult design. Int J Epidemiol. 30: 928–934.

Bateson P, Barker D, Clutton-Brock T, Deb D, D'Adine B, Foley RA, Gluckman P et al. 2004. Developmental plasticity and human health. Nature 430,6998: 419–421.

Bowler PJ. *The Mendelian Revolution: The Emergence of Hereditarian Concepts in Modern Society.* London: Athlone Press; 1989.

Burdge GC, Slater-Jefferies JL, Torrens C, Phillips ES, Hanson MA, Lillycrop KA. 2007. Dietary protein restriction of pregnant rats in the F_0 generation induces altered methylation of hepatic gene promoters in the adult male offspring in the F_1 and F_2 generations. Br J Nutr. 97,3: 435–439.

Burkhardt RW. *The Spirit of System: Lamarck and Evolutionary Biology*. 2nd ed. with new introduction. Cambridge, MA.: Harvard University Press; 1995.

Bynum WF. Nosology. In: *Companion Encyclopedia of the History of Medicine.* Bynum WF, Porter R, eds. Vol. 1. London: Routledge; 1993:335–356.

Champagne FA, Curley JP. 2008. Maternal regulation of estrogen receptor α methylation. Curr Opin Pharmacol. 8: 1–5.

Cleal JK, Poore KR, Boullin JP, Khan O, Chau R, Hambidge O, Correas C et al. 2007. Mismatched pre- and postnatal nutrition leads to cardiovascular dysfunction and altered renal function in adulthood. Proc Natl Acad Sci USA. 104,22: 9529–9533.

Curhan GC, Chertow GM, Willett WC, Spiegelman D, Colditz GA, Manson JE, Speizer FE et al. 1996. Birth weight and adult hypertension and obesity in women. Circulation; 94,6: 1310–1315.

Drake AJ, Walker BR, Seckl JR. 2005. Intergenerational consequences of fetal programming by in utero exposure to glucocorticoids in rats. Am J Physiol Regul Integr Comp Physiol. 288: R34–R38.

Drewnowski A, Popkin BM. 1997. The nutrition transition: New trends in the global diet. Nutr Rev. 55,2: 31–43.

Elahi M, Cagampang FR, Anthony FW, Curzen N, Ohri S, Hanson MA. 2005. Statins during pregnancy and lactation in mice on hypercholesterolaemic diet prevents obesity, hypertension, and sedentary behaviour in adult offspring. Early Hum Dev. 82,8: 559.

Fischer J-L. 1990. Experimental embryology in France (1887–1936). Int J Dev Biol. 34: 11–23.

Freinkel N. 1980. Banting Lecture 1980. Of pregnancy and progeny. Diabetes 29,12: 1023–1035.

Gale CR, Jiang B, Robinson SM, Godfrey KM, Law CM, Martyn CN. 2006. Maternal diet during pregnancy and carotid intima-media thickness in children. Arterioscler Thromb Vasc Biol. 26,8: 1877–1882.

Gilbert SF. 2001. Ecological developmental biology: Developmental biology meets the real world. Dev Biol. 233: 1–12.

Gluckman PD, Hanson MA. 2004a. Living with the past: Evolution, development, and patterns of disease. Science 305,5691: 1733–1736.

Gluckman PD, Hanson MA. 2004b. Maternal constraint of fetal growth and its consequences. Semin Fetal Neonatal Med. 9,5: 419–425.

Gluckman PD, Hanson MA. *Mismatch: Why Our World No Longer Fits Our Bodies.* Oxford: Oxford University Press; 2006a.

Gluckman PD, Hanson MA. 2006b. Evolution, development and timing of puberty. Trends Endocrinol Metab. 17,1: 7–12.

Gluckman PD, Hanson MA, Beedle AS. 2007. Non-genomic transgenerational inheritance of disease risk. BioEssays 29,2: 149–154.

Gluckman PD, Hanson MA, Buklijas T. 2010. A conceptual framework for the developmental origins of health and disease. J DOHaD 1,1: 6–18.

Gluckman PD, Hanson MA, Cooper C, Thornburg KL. 2008. Effect of in utero and early-life conditions on adult health and disease. N Engl J Med. 359,1: 61–73.

Godfrey K. The "developmental origins" hypothesis: Epidemiology. In: *Developmental Origins of Health and Disease.* Gluckman PD, Hanson MA, eds. Cambridge: Cambridge University Press; 2006:6–32.

Goldstein DB. 2009. Common genetic variation and human traits. N Engl J Med. 360,17: 1696–1698.

Hales CN, Barker DJ. 1992. Type 2 (non-insulin-dependent) diabetes mellitus: The thrifty phenotype hypothesis. Diabetologia 35: 595–601.

Ibáñez L, Potau N, Enriquez G, Marcos MV, De Zegher F. 2003. Hypergonadotrophinaemia with reduced uterine and ovarian size in women born small-for-gestational-age. Hum Reprod. 18,8: 1565–1569.

Inskip HM, Crozier SR, Godfrey KM, Borland SE, Cooper C, Robinson SM. 2009. Women's compliance with nutrition and lifestyle recommendations before pregnancy: General population cohort study. BMJ. 338: b481.

Jablonka E, Raz G. 2009. Transgenerational epigenetic inheritance: Prevalence, mechanisms, and implications for the study of heredity and evolution. Q Rev Biol. 84,2: 131–176.

Jasienska G, Thune I, Ellison PT. 2006. Fatness at birth predicts adult susceptibility to ovarian suppression: An empirical test of the Predictive Adaptive Response hypothesis. Proc Natl Acad Sci USA. 103,34: 12759–12762.

Jimenez-Chillaron JC, Isganaitis E, Charalambous M, Gesta S, Pentiat-Pelegrin T, Faucette RR, Otis JP et al. 2009. Intergenerational transmission of glucose intolerance and obesity by in utero undernutrition in mice. Diabetes 58,2: 460–468.

Khan I, Dekou V, Hanson M, Poston L, Taylor P. 2004. Predictive adaptive responses to maternal high-fat diet prevent endothelial dysfunction but not hypertension in adult rat offspring. Circulation 110: 1097–1102.

Kuzawa CW, Gluckman PD, Hanson MA. Developmental perspectives on the origin of obesity. In: *Adipose Tissue and Adipokines in Health and Disease*. Fantuzzi G, Mazzone T, eds. Totowa, NJ: Humana Press; 2007:207–219.

LaBerge AF. 2008. How the ideology of low fat conquered America. J Hist Med Allied Sci. 63,2: 139–177.

Lawrence C, Weisz G, eds. *Greater Than the Parts: Holism in Biomedicine, 1920–1950*. Oxford: Oxford University Press; 1998.

Lillycrop KA, Phillips ES, Jackson AA, Hanson MA, Burdge GC. 2005. Dietary protein restriction of pregnant rats induces and folic acid supplementation prevents epigenetic modification of hepatic gene expression in the offspring. J Nutr. 135: 1382–1386.

Lillycrop KA, Slater-Jefferies JL, Hanson MA, Godfrey KM, Jackson AA, Burdge GC. 2007. Induction of altered epigenetic regulation of the hepatic glucocorticoid receptor in the offspring of rats fed a protein-restricted diet during pregnancy suggests that reduced DNA methyltransferase-1 expression is involved in impaired DNA methylation and changes in histone modifications. Br J Nutr. 97,6: 1064–1073.

Nutton V. Humoralism. In: *Companion Encyclopedia of the History of Medicine*. Bynum WF, Porter R, eds. Vol. 1. London: Routledge; 1993:281–291.

Painter RC, Osmond C, Gluckman P, Hanson M, Phillips DIW, Roseboom TJ. 2008. Transgenerational effects of prenatal exposure to the Dutch famine on neonatal adiposity and health in later life. BJOG. 115,10: 1243–1249.

Pembrey ME, Bygren LO, Kaati G, Edvinsson S, Northstone K, Sjöstrom M, Golding J et al. 2006. Sex-specific, male-line transgenerational responses in humans. Eur J Hum Genet. 14: 159–166.

Rassoulzadegan M, Grandjean V, Gounon P, Vincent S, Gillot I, Cuzin F. 2006. RNA-mediated non-Mendelian inheritance of an epigenetic change in the mouse. Nature 441,7092: 469–474.

Richards E. 1994. A political anatomy of monsters, hopeful and otherwise: Teratogeny, transcendentalism, and evolutionary theorizing. Isis 85,3: 377–411.

Robinson SM, Crozier SR, Borland SE, Hammond J, Barker DJ, Inskip HM. 2004. Impact of educational attainment on the quality of young women's diets. Eur J Clin Nutr. 58: 1174–1180.

Sloboda DM, Hart R, Doherty DA, Pennell CE, Hickey M. 2007. Age at menarche: Influences of prenatal and postnatal growth. J Clin Endocrinol Metab. 92,1: 46–50.

Sloboda DM, Howie GJ, Pleasants A, Gluckman PD, Vickers MH. 2009. Pre- and postnatal nutritional histories influence reproductive maturation and ovarian function in the rat. PLoS ONE 4,8: e6744.

Smith CA. 1947. Effects of maternal undernutrition upon the newborn infant in Holland (1944–45). J Pediatr. 30: 229–243.

Timmermann C. To treat or not to treat: Drug research and the changing nature of essential hypertension. In: *The Risks of Medical Innovation: Risk Perception and Assessment in Historical Context*. Schlich T, Tröhler U, eds. London: Routledge; 2006:133–147.

Van Assche FA, Holemans K, Aerts L. 2001. Long-term consequences for offspring of diabetes during pregnancy. Br Med Bull. 60: 173–182.

Vickers MH, Breier BH, Cutfield WS, Hofman PL, Gluckman PD. 2000. Fetal origins of hyperphagia, obesity, and hypertension and postnatal amplification by hypercaloric nutrition. Am J Physiol Endocrinol Metab. 279,1: E83–E87.

Vickers MH, Gluckman PD, Coveny AH, Hofman PL, Cutfield WS, Gertler A, Breier BH et al. 2005. Neonatal leptin treatment reverses developmental programming. Endocrinology 146,10: 4211–4216.

Wagner KD, Wagner N, Ghanbarian H, Grandjean V, Gounon P, Cuzin F, Rassoulzadegan M. 2008. RNA induction and inheritance of epigenetic cardiac hypertrophy in the mouse. Dev Cell 14,6: 962–969.

Waterland RA, Travisano M, Tahiliani KG, Rached MT, Mirza S. 2008. Methyl donor supplementation prevents transgenerational amplification of obesity. Int J Obes. 32: 1373–1379.

Weaver ICG, Cervoni N, Champagne FA, D'Alessio C, Sharma S, Sekl JR, Dymov S et al. 2004. Epigenetic programming by maternal behavior. Nat Neurosci. 7: 847–854.

Williams AS. Relief and research: The nutrition work of the National Birthday Trust Fund, 1935–9. In: *Nutrition in Britain: Science, Scientists and Politics in the Twentieth Century*. Smith DF, ed. London: Routledge; 1997:99–122.

24 Plants: Individuals or Epigenetic Cell Populations?

Marcello Buiatti

According to classical neo-Darwinian theory, individuals and genes are the sole units of selection, with all or most cells of single organisms bearing the same genotype. Moreover, as Weismann suggested (Weissman 1892), the germ line is supposed to be completely independent from the soma. Within this framework, the adaptation strategy in all organisms is assumed to be based on mutation-driven variability and the selection of the fittest genotypes. In this chapter, which focuses on plants, I suggest several challenges to this dominant view. First, the fuzzy sequestration of the germ line in plants, where germ cells are continuously generated and somatic cells may contribute to the germ line, allows the transmission of genetic and epigenetic variations to the next generation, a process that contradicts Weismann's assertion about the impossibility of transmitting somatically generated variation. Second, plants have great phenotypic plasticity, and their epigenetic control mechanisms can generate heritable genetic and epigenetic variations in response to altered conditions. Third, selection for coordinating processes at different levels of organization is important in plant evolution. Understanding the structure and dynamics of heritable, somatic, genetic mutations and epimutations in plants may therefore open up new ways of looking at selection and evolution.

Plants May Be Thought of as Cell Populations

Many features differentiate plant development from that of animals. Animals are "closed systems" in the sense that they reach a "maturity" stage after which somatic stem cells are essentially used only for recovery from damage. Plants, on the other hand, are "open systems" due to the permanent presence of meristems (undifferentiated tissues from which new cells are formed) and a high regeneration capacity. Moreover, variability can be accumulated through vegetative propagation, which allows adaptation to the many environmental changes with which plants are confronted during their life cycles. It is therefore not surprising that plant somatic cells have been shown to be more genetically and epigenetically heterogeneous than animal ones. Lamarck anticipated these findings, writing in his *Philosophie zoologique*: "But every bud in a plant is itself an individual plant, which shares in the common life of all the rest, develops its flower or inflorescence once a year, then produces fruit and may finally give rise to a branch already supplied with other buds, that is, other individual plants. Each of these individual plants either produces fruits, in which case it does so only once, or produces a branch which gives rise to other similar plants" (Lamarck 1830). According to Lamarck, single ramets (individual members of a clone) may differ from each other, and their variations become heritable through vegetative propagation; this situation occurs not only in clonal plants that reproduce vegetatively but also in seed-producing species when a mutated somatic cell enters the germ line.

Histological studies in plants confirm the imperfect sequestration of the germ line. Shoot meristems (see Medford 1992; Klekowski 2003) are organized into two layers: the L1 layer that produces epidermal cells and the L2 that produces endodermal cells and gametes. Below those two layers (the tunica), L3 cells, particularly those from the central part of the meristem (the corpus), may enter the L2 layer, thus contributing to gamete formation. Somatic heterogeneity, particularly in vegetatively propagated plants, may be organized into layers or show mosaicism. The components of this patchiness, however small they are, are liable to contribute to shoots and to inflorescences. Data from studies of in vitro cultures, and artificial propagation and selection through meristem multiplication, show that in many instances there is indeed both genetic and epigenetic intraclonal, somatic, heritable variability. Notably, a large part of this "somaclonal variation" in plants that have been regenerated from tissue culture is derived from the proliferation of genetic and epigenetic variants that were present in the original tissue (see, for example, Bennici, Buiatti, and D'Amato 1968). This genetic and epigenetic somatic variability responds to natural and artificial selection (Singh and Saroop 2004; Bellini, Giordani, and Rosati 2008; Schellenbaum, Mohler, Wenzel, and Walter 2008). Monro and Poore (2009), for example, successfully increased variation in plant proliferation and altered the plasticity of the response to light in the red seaweed *Asparagopsis armata* by selecting from meristems, which in seaweeds are all derived from a single cell.

Genetic and Epigenetic Plasticity in Plants

Unlike animals, plants do not have a real nervous system, and are unable to actively move from one environment to another. Since behavioral plasticity based on movement is largely precluded, adaptive plasticity is based on cellular plasticity. However, such cellular plasticity, which occurs on both the genetic and the epigenetic levels, could destabilize the organism's development. Multiple structural and functional processes in plants allow their reproduction and maintenance in spite of the extraordinarily high level of both quantitative and qualitative variability that results from chromosomal, genetic, and epigenetic changes. These changes can occur at the somatic level during development, and may be transmitted to the next generation because of the incomplete germ line sequestration, but probably also because of inefficient epigenetic resetting during gametogenesis (Finnegan and Whitelaw 2008; Gehring and Henikoff 2007).

In most cases, the genetic and epigenetic variations that occur during development have a functional and adaptive value. Gene activation or silencing is the outcome of epigenetic mechanisms (such as DNA methylation and RNA interference, RNAi) whose activity is induced by internal and external signals from living and nonliving sources. These epigenetic mechanisms also seem to be part of DNA-modifying systems, and developmental plasticity in plants therefore typically involves both genetic and epigenetic variations. Somatic plant cells may be diploid, aneuploid, or polyploid, and this usually allows the retention of mutations in the heterozygote state, as is shown by the high genetic variability seen when vegetatively propagated plants are self-fertilized (Gimelli, Venturo, Buiatti, et al. 1984). Changes in the chromosome complement, particularly increases in the number of single chromosomes or entire sets of chromosomes, can occur at the somatic level through processes such as endoreduplication, polyteny, and allopolyploidy. These chromosomal variations are often localized in specific areas of the plant or in specific groups of cells, where they have functional effects. An increase in chromosome numbers generally leads to an increased cell size, and seems to occur when a higher level of production of particular RNAs or proteins is required.

External stimuli such as infection (Callow 1975), as well as specific developmental signals, lead to genomic changes in plants. Reviews by Nagl (1978) and Barow (2006) show that endopolyploidy can be induced by developmental signals such as hormones, by the positions of the cells in the plant, and by environmental signals such as light. Polytene chromosomes (which are also present in animals, particularly in dipteran insects) are concentrated in plants in areas where a very high synthesis of RNA and protein is required, such as the endosperm, the embryo suspensor, and the anthers. Although the mechanisms that induce variation in chromosome numbers are still unclear, the data show that the variations are developmentally regulated in the sense that specific cells are differentiated not only with respect to gene expression but also with respect to chromosome numbers. In addition to the somatic effects, polyploidy also has a direct impact on the evolutionary process in plants, as it may lead to the formation of new fertile, allopolyploid species when it follows hybridization.

As Feldman and Levy (chapter 25 in this volume) have shown, speciation through allopolyploidization is not easy because, if they are to survive, the putative new species must coordinate preexisting genomes which until then had evolved separately. In other words, allopolyploid genomes face a "genetic shock," and typically undergo massive genome rearrangements and persistent epigenetic changes in gene expression. A much weaker version of this epigenomic stress response occurs in new polyploids. According to Lavania et al. (2006), plants respond to the shock of polyploidization by an increase in heritable variability subject to internal selection processes, resulting in the establishment of a new, coherent, genetic network, the result of which is subject to selection by the external environment. A similar two-step process—induction of heritable variations followed by selection—also occurs in bacteria and animals when they are challenged by stress. Bacteria activate the gene *RpoS*, and this in turn induces the expression of "mutator" genes, thereby increasing the mutation rates (reviewed in Buiatti and Buiatti 2008). Animals and plants, on the other hand, respond to stress by activating mobile genetic elements, transposons. Transposons, which were discovered by Barbara McClintock in the middle of the twentieth century, are a heterogeneous class of nucleic acid sequences. They are able to move from one region of DNA to another and to induce genome shuffling. They are very common, making up around 40 percent of the human genome, 50 percent of grasses' genomes, and 90 percent of *Liliaceae* genetic complexes. In plants, their insertion into other sequences may alter the protein complement produced, as shown, for instance, in coffee plants by Lopes, Carazzolle, Pereira, Colombo, and Carareto (2008). Their insertion may also modify, and even silence, the expression of neighboring genes. Transposons and particularly a class of them, the "helitrons," which are able to transfer large DNA sequences from one site of the genome to another, may lead to massive genome changes not only in polyploids but also in diploids (Lal, Oetjens, and Hannah 2009). Morgante, Brunner, Pea, Fengler, Zuccolo, and Rafalski (2005) compared the structure of a single locus in two maize cultivars and found astounding structural differences in the order and localization of DNA fragments, which were the result of transposition. This type of event has been confirmed in other cases, and has led to the proposal that the term "pangenome" be used for the ensemble of genomes of the same species (Morgante 2006).

Transposons often remain quiescent for long periods. They may be activated not only in response to a genomic shock but also by other types of stress, such as cold or heat shock, infections, mechanical damage, and exposure to cell wall hydrolisates (Mansour 2007). In several cases, transposons have been shown to contain in their promoters sequences "boxes" that are capable of recognizing the transcription factors induced by stress. For example, in *Lycopersicum chilense*, a Tnt1 transposon that is normally methylated and inactive is activated by fungal extracts (Salazar,

González, Casaretto, Casacuberta, and Ruiz-Lara 2007). These data suggest that transposons are by no means "junk DNA" or "selfish DNA," as was thought for many years, but are part of an adaptive system that increases variability during periods of stress, a system that probably all organisms need in order to cope with stress conditions (discussed in Buiatti and Buiatti 2008).

Genome shuffling is one of the best examples of epigenetically regulated genetic modifications because it occurs in somatic cells, is developmentally controlled, and influences both genome structure and gene expression. Another type of epigenetically regulated modification, DNA amplification, which changes the number of copies of specific DNA sequences, was discovered in *Drosophila* by Ritossa (1968). He described the "magnification" process, which involves increases and decreases in the numbers of ribosomal DNA sequences. Amplified copies may be released into the cytoplasm and eventually get lost, but they are also liable to become integrated into the chromosomes as tandem repeats, thus changing genomic DNA sequences.

The first molecular demonstration of functional sequence amplification in plants was obtained some years later by Parenti, Guillé, Grisvard, Durante, Giorgi, and Buiatti (1973), who found that the amplification of repeated DNA sequences was needed to initiate cell proliferation in in vitro cultures of *Nicotiana glauca*. Subsequently it was shown that in vitro cell proliferation occurs without amplification in a tumorous hybrid, *Nicotiana glauca-langsdorffii*, but not in a nontumorous "mutant" of it. This turned out to be the result of an "epimutation": the silencing of an ancient, introgressed *Agrobacterium rhizogenes* gene (*rolC*). This gene could restore tumorous behavior when genetically engineered into nontumorous plants (Buiatti and Bogani, unpublished results). It should be noted that the genotypes analyzed were all derived from hybrids between two species of *Nicotiana*, one (*N. glauca*) that conserved the ancestrally introgressed *Agrobacterium rhizogenes* complement of genes, which are known to heavily influence the plant hormonal system, and the other (*N. langsdorffii*) being devoid of the bacterial gene complex. The introgression of *A. rhizogenes* genes is known to have influenced the taxonomic organization of the whole genus, which has been divided by Näf (1958) into two well-defined groups of species, plus and minus, according to the presence or absence of introgressed genes. Only crosses between species belonging to different groups give rise to genetically tumorous hybrids. This implies that the ancestral spontaneous insertion of genes has profoundly influenced the evolution of the group carrying them, because the harmful effects of tumorogenesis introduced a barrier to sexual crossing (Bogani et al. 1997; Intrieri and Buiatti 2001). The original horizontal transfer of the bacterial gene probably occurred in somatic cells, since *A. rhizogenes* is known to infect the roots of plants. The transformed cell lineage must then have reached the inflorescence and resulted in "genetically modified" gametes. All this deeply changed the physiology of the members of the *Nicotiana* genus, since the tumorous hybrid cells acquired the capacity to divide autonomously, as is shown by the lack of the need for repetitive DNA amplification for in vitro cell proliferation.

Another complex epigenetic/genetic system was found in flax by Durrant (1962) and studied at the molecular level by Cullis (Cullis 1990; Schneeberger and Cullis 1991; Oh and Cullis 2003). In the original experiments, flax plants of several cultivars were grown in fields with different fertilizer treatments. One "plastic" cultivar (Stormont Cirrus, PL) was found to be particularly responsive to the different treatments, developing small and large variants that differed not only in height but also in other characters, such as the presence of hairs on seed capsule septa, the number of branches, and isoenzyme mobility in relation to hormone levels. The small (S) and large (L) variant strains were found to be stable for at least the next six generations, even when they were grown in the standard environment (*not* exposed to differing fertilizer levels). Cullis and coworkers carried out a large number of analyses which showed a wide range of differences among the three strains (PL, L, and S) at the biochemical and molecular levels. A particularly relevant change was in the amount of ribosomal RNA, the result of r-DNA amplification (Cullis 1990; Oh and Cullis 2003; Schneeberger and Cullis 1991), but sequence modifications and rearrangements that were not due to transposon mobility were also observed. However, in spite of the impressive amount of data, so far it has been impossible to find which, if any, single initial process is the trigger for such a wide range of apparently independent variations in the originally plastic, stress-responding PL genotype (the L and S variants having completely lost the initial plasticity of PL). It is possible that genetic or epigenetic element(s) responsible for the genetic and epigenetic plasticity of PL are lost (or more probably blocked) in the derived L and S strains. Such "plasticity elements" could be acting like inducible transposons which have become silenced in the two derived variants.

In all eukaryotes, most epigenetic processes involve, at one point or another, DNA and/or histone methylation or demethylation leading to the activation or inactivation of genes. One of these processes, paramutation, involves an interaction between two alleles of a single locus in which one allele induces a heritable change in the other allele. It was discovered in plants by Brink (1958), and has only recently been observed in animals (see chapter 22 in this volume). Paramutations are heritable and have effects similar to transgene silencing, since in both cases the final result is the regulation of the number of active copies of a single gene, a process that highlights the importance of a specific gene dosage. It has been thoroughly studied in two maize genes affecting pigmentation, *R* and *B* (reviewed by Chandler and Alleman 2008). The r1 alleles that undergo silencing contain multiple copies of the coding region, are expressed in the aleurone layer (the outermost cell layer of the seed coat) of seeds, and their silenced state is unstable. The *B* gene, on the other hand, has a single coding region, is expressed in epidermal layers, and its silenced state is stable. Moreover, with *R*, the silencing power of the paramutagenic allele (the allele that transforms the other one to its own silent state) is proportional to the number of multiple tandem repeats in the associated regions, while in *B* the number of repeats (853) is stable. Two more genes are involved in this complex process, one of which codes for an RNA-dependent RNA polymerase, and the second of which is involved in chromatin remodeling through an RNAi system. It is interesting that transgene silencing, which has a gene dosage regulatory role similar to the effects of paramutation, is also mediated by the RNAi system. RNAi was discovered in plants transgenic for a chimeric chalcone synthase by Napoli, Lemieux, and Jorgensen (1990), and is generally observed when an inserted sequence has a DNA fragment that is homologous to an existing DNA region in the transformed plant.

Toward a Broader View of Selection?

In the traditional view, selection is a process whereby fitter genotypes are selected by the environment, the conditions in which organisms live and reproduce. The data from plants suggest that selection operates at several levels of the hierarchical organization of life, and that fitness is determined by the effects of both epigenetic and genetic factors. I suggest that, since at every level (cell, organism, population, species, ecosystem) there are internal and external factors that affect fitness, there is a need for organisms to coordinate the networks of interactions that occur at the different levels. This view of selection for coordination challenges the classical neo-Darwinian view, which suggested that the components at different levels of organization evolved separately, and requires a co-evolutionary view based on cooperation between epigenetic and genetic processes at different levels of biological organization. Evidence of the need for all changes in the system to be "harmoniously integrated" within the framework of the genetic/physiological networks of a cell comes from the system of *A. rhizogenes* genes' integration into *Nicotiana* species that has been discussed in this chapter. As Intrieri and Buiatti (2001) showed, different parts of the inserted "transgenic" complex have been integrated into the genomes of different species, where they have differentially influenced the evolution of their physiological networks (Bogani, Liò, Intrieri, and Buiatti 1997). An analysis comparing the phylogeny of each of the bacterial genes in the *Nicotiana* species with that of the same genes in the bacterial *A. rhizogenes* strains showed that integrated bacterial genes have coevolved with *Nicotiana*, while the same genes in bacteria formed a distinct cluster in the overall phylogenetic tree (Intrieri and Buiatti 2001).

This coevolutionary process between bacterial genes and plant genomes is in line with Darwin's idea of "correlated variation" and the importance of a harmonious organization of the components of living systems. The stress on coordination and co-evolution is at variance with the traditional mechanistic neo-Darwinian conception, which sees organisms as machines evolving through the addition of independent parts.

Selection within the organism, at the somatic cell level, involves competition between genetically and/or epigenetically heterogeneous cells in the same individual. In this case, somatic selection pressure comes both from the internal physiological network and from the effects of the external environment on it. Somatic selection and its response to environmental factors were widely tested in the 1960s and 1970s, when mutagenesis was used to increase the genetic variability of crops that were to be subject to artificial selection. In cereals, mutation frequency was shown to depend on the position of seeds on the spike, decreasing from bottom to top. This behavior was interpreted as the effect of random drift and somatic selection of the stem cells forming the gametes (Gaul 1965; Klekowski 2003). Data on physiological and environmental factors affecting somatic selection came mainly from studies on vegetatively propagated plants and in vitro cell cultures. An analysis of mutation frequencies and of the width of the mutated sectors in flowers at different positions on the spikes developing from gamma-ray-irradiated gladiolus corms grown in the greenhouse or in open air showed that in both environments mutation frequencies decreased from bottom to top, whereas the opposite was true for sector width (Buiatti, Tesi, and Molino 1969). The environment had an influence on the process: in the open air sector width was greater at all levels of the spike than that in the greenhouse, with maximal values observed in the first flowers from the bottom. On the other hand, mutation frequencies at the bottom of the spike were lower in the open air than in the greenhouse, but decreased slowly moving toward the top, where more mutated sectors were observed. This differential behavior can be explained by postulating that fewer initials (stem cells giving rise to a tissue or an organ) were developing in the open air than in the greenhouse, and therefore the chance of having a mutation in one flower was lower but the relative sector widths were greater. In this way the environment influenced both parameters by modulating the number of initials of petals. The influence of the physiological environment has also been tested in long-term in vitro cultures of *Haplopappus gracilis* and tomatoes grown on media with different hormone (auxin/cytokinin) balances. These experiments showed that different genotypes survived, depending on the culture media. In particular, with *Haplopappus* cells, clones grown in different physiological conditions were found to have different chromosome complements (Bennici, Buiatti, D'Amato, and Pagliai 1971). In the tomato, DNA copy numbers and methylation levels were found to differ according to the hormone balance in subclones derived from the same initials (Bogani, Simoni, Bettini, Mugnai, Pellegrini, and Buiatti 1995; Bogani, Simoni, Liò, Scialpi, and Buiatti 1996).

All of these experiments show that internal physiological conditions affect genotype evolution in a way that seems similar to the effect of the physiological changes induced by *A. rhizogenes* on the evolution of the genus *Nicotiana*. In plants, the selection of epigenetic, genetic, and chromosomal variants plays a role in "harmonizing" cellular networks, which are influenced by physiological conditions and are at least partially regulated by the external conditions. The resulting phenotype is not stable; it is the ever-changing result of its dynamic interactions with the environment(s). Ultimately, the internal networks are confronted with the external ones, that is, with ecosystemic interactions between species and individuals, and selection for coordination among the networks occurs. This view, which is strongly supported by the data from plants, but which could and should be extended to all living systems, suggests that the neo-Darwinian view of a "fragmented" living nature should be revisited so that a richer Darwinian/Lamarckian synthesis can be constructed.

References

Barow M. 2005. Endopolyploidy in seed plants. BioEssays 28: 271–281.

Bellini E, Giordani E, Rosati A. 2008. Genetic improvement of olive from clonal selection to cross-breeding programs. Advan Hortic Sci. 22,2: 73–86.

Bennici A, Buiatti M, D'Amato F. 1968. Nuclear conditions in haploid *Pelargonium in vivo* and *in vitro*. Chromosoma 24: 194–201.

Bennici A., Buiatti M, D'Amato F, Pagliai M. Nuclear behaviour in Haplopappus gracilis callus grown in vitro on different culture media. In: *Colloques Internationaux CNRS*. 193: *Les cultures de tissus de plantes*. 1971; 245–253.

Bogani P, Simoni A, Bettini P, Mugnai M, Pellegrini MG, Buiatti M. 1995. Genomic flux in tomato auto- and auxo-trophic cell clones cultured in different auxin/cytokinin equilibria. I: DNA multiplicity and methylation levels. Genome 38,5: 902–912.

Bogani P, Simoni A, Liò P, Scialpi A, Buiatti M. 1996. Genome flux in tomato cell clones cultured in vitro in different physiological equilibria. II: A RAPD analysis of variability. Genome 39,5: 846–853.

Bogani P, Liò P, Intrieri C, Buiatti M. 1997. A physiological and molecular analysis of the genus *Nicotiana*. Mol Phylogenet Evol. 7: 62–70.

Brink RA. 1958. Paramutation at the R locus in maize. Cold Spring Harb Symp Quant Biol. 54: 25–40.

Buiatti M. DNA amplification and tissue culture. In: *Applied and Fundamental Aspects of Plant Cell Tissue and Organ Culture*. Reinert J, Bajaj YPS, eds. Berlin: Springer; 1977:358–374.

Buiatti M, Buiatti M. 2008. The chance vs. necessity antinomy and third millennium biology. Biol Forum. 101: 29–66.

Buiatti M, Tesi R, Molino M. 1969. A developmental study of induced somatic mutations in *Gladiolus*. Radiat Bot. 9: 39–48.

Callow J. 1975/2006. Endopolyploidy in maize: Smut neoplasm induced by the maize smut fungus *Ustilago maydis*. New Phytol. 75,2: 253–257.

Chandler V, Alleman M. 2008. Paramutation: Epigenetic instructions passed across generations. Genetics 178: 1839–1844.

Cullis CA. 1990. DNA rearrangements in response to environmental stress. Adv Genet. 28: 73–97.

Durrant A. 1962. The environmental induction of heritable changes in *Linum*. Heredity 17: 27–61.

Finnegan EJ, Whitelaw E. 2008. Leaving the past behind. PLoS Genet. 4,10: e1000248.

Gaul H. Selection in M2 generation after mutagenic treatment of barley seeds. In: *Induction of Mutations and the Mutation Process*. Veleminsky J, Gichner T, eds. Prague: Czech Academy of Sciences; 1965:62–73.

Gehring M, Henikoff S. 2007. Methylation dynamics in plant genomes. Biochim Biophys Acta 1769: 276–286.

Gimelli F, Ginatta G, Venturo R, Positano S, Buiatti M. 1984. Levels of heterozygosity in a vegetatively propagated plant: The Mediterranean carnation (*Dianthus caryophillus* L.). Riv ortofloro frut ital. 1984; 68:367–374.

Intrieri MC, Buiatti M. 2001. The horizontal transfer of *Agrobacterium rhizogenes* genes and the evolution of the genus *Nicotiana*. Mol Phylogenet Evol. 20,1: 100–110.

Klekowski EJ. 2003. Plant clonality, mutation, diplontic selection and mutational meltdown. Biol J Linn Soc Lond. 79,1: 61–67.

Lal S, Oetjens M, Hannah LC. 2009. Helitrons: Enigmatic abductors and mobilizers of host genome sequences. Plant Sci. 176: 181–186.

Lamarck J-B. Zoological Philosophy [1809]. Paris: Baillière Librairie; 1830.

Lavania UC, Misra NK, Lavania S, Basu S, Srivastava S. 2006. Mining *de novo* diversity in palaeopolyploids. Curr Sci. 90,7: 938–941.

Lopes F, Carazzolla M, Pereira G, Colombo C, Carareto C. 2008. Transposable elements in *Coffea* (*Gentianales, Rubiaceae*) transcripts and their role in the origin of protein diversity in flowering plants. Mol Genet Genomics 279,4: 385–401.

Mansour A. 2007. Epigenetic activation of genomic retrotransposons. J Cell Mol Biol. 6,2: 99–107.

Medford JI. 1992. Vegetative apical meristems. Plant Cell 4: 1029–1039.

Monro K, Poore AGB. 2009. The potential for evolutionary responses to cell-lineage selection on growth form and its plasticity in a red seaweed. Am Nat. 173: 151–163.

Morgante M. 2006. Plant genome organisation and diversity: The year of junk! Curr Opin Biotechnol. 17: 168–173.

Morgante M, Brunner S, Pea G, Fengler K, Zuccolo K, Rafalski A. 2005. Gene duplication and exon shuffling by helitron-like transposons generate intra-species diversity in maize. Nat Genet. 37: 997–1002.

Näf U. 1958. Studies on tumor formation in *Nicotiana* hybrids: I. The classification of the parents into two etiologically significant groups. Growth 22: 167–180.

Nagl W. *Endopolyploidy and Polyteny in Differentiation and Evolution*. Amsterdam: Elsevier North Holland; 1978.

Napoli C, Lemieux C, Jorgensen R. 1990. Introduction of a chimeric chalcone synthase gene into petunia results in reversible co-suppression of homologous genes in trans. Plant Cell 2,4: 279–289.

Oh TJ, Cullis CA. 2003. Labile DNA sequences in flax identified by combined sample representational difference analysis (csRDNA). Plant Mol Biol. 52: 527–536.

Parenti R, Guillé E, Grisvard J, Durante M, Giorgi L, Buiatti M. 1973. Transient DNA satellite in dedifferentiating pith tissue. Nat New Biol. 246,155: 237–239.

Ritossa F. 1868. Unstable redundancy of genes for ribosomal DNA. Proc Natl Acad Sci USA. 60: 509–516.

Salazar M, González E, Casaretto JA, Casacuberta JM, Ruiz-Lara S. 2007.The promoter of the *TLC1.1* retrotransposon from *Solanum chilense* is activated by multiple stress-related signaling molecules. Plant Cell Rep. 26,10: 1861–1868.

Schellenbaum P, Mohler V, Wenzel G, Walter B. 2008. Variation in DNA methylation patterns of grapevine somaclones (*Vitis vinifera L.*). BMC Plant Biol. 8: 78–88.

Schneeberger RG, Cullis CA. 1991. Specific DNA alterations associated with the environmental induction of heritable changes in flax. Genetics 128: 619–630.

Singh M, Saroop J. 2004.Detection of intra-clonal genetic variability in vegetatively propagated tea using RAPD markers. Biol Plant 4: 113–115.

Weismann A. *Das Keimplasma. Eine Theorie der Vererbung*. Jena: Gustav Fischer; 1892.

25 Instantaneous Genetic and Epigenetic Alterations in the Wheat Genome Caused by Allopolyploidization

Moshe Feldman and Avraham A. Levy

Allopolyploidy is a state whereby two or more different genomes are brought together into the same nucleus by inter-specific or inter-generic hybridization, followed by genome duplication (figure 25.1). It has played a major evolutionary role in the formation of many plant species (Manton 1950; Stebbins 1950, 1971; Grant 1971; Soltis and Soltis 1993, 1995). Recent molecular studies, mainly comparative genomics and whole genome sequencing, along with the use of novel molecular and computational tools has shown that allopolyploidy is found among many groups of plants and is even more widespread than previously believed (Leitch and Bennett 1997; Wendel 2000; Soltis, Soltis, and Tate 2004). These studies have also clarified how allopolyploidy effects speciation, genome structure, and gene expression (Leitch and Bennett 1997; Feldman, Liu, Segal, Abbo, Levy, and Vega 1997; Comai 2000; Soltis and Soltis 2000; Pikaard 2001; Levy and Feldman 2002, 2004; Feldman and Levy 2005, 2009). Thus, allopolyploidy is one of the major processes that has driven and shaped the evolution of higher plants.

Allopolyploidization, being a radical, rapid mode of speciation, produces a new species in single step: a novel plant is formed that is isolated genetically from its parental species. Two processes are involved in the formation of an allopolyploid species: hybridization and genome doubling. This results in the formation of a hybrid species with two or more different genomes enveloped within the same nucleus.

As a result of this drastic genetic construction, the newly formed allopolyploid species faces several challenges: it must secure an exclusively intragenomic pairing at meiosis, which will lead to increased fertility, orchestrate intergenomic gene expression and DNA replication, and reduce the energy and material costs of supporting large genomes (Levy and Feldman 2002, 2004; Feldman and Levy 2005). To meet these challenges, thereby ensuring its increased fitness and successful establishment in nature, the newly formed allopolyploid genome must undergo immediate changes. Divergence of the homeologous chromosomes, that is, related chromosomes of the different genomes, must be rapidly achieved so that they cannot pair and recombine at meiosis. There is also a need to quickly alter gene expression to

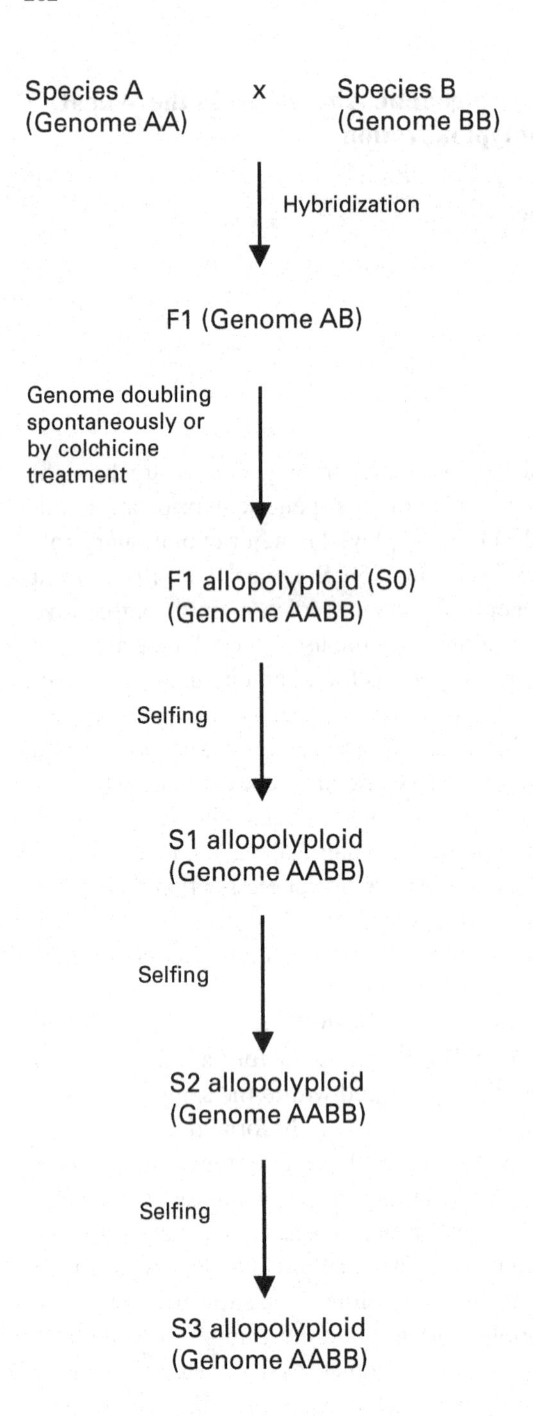
Species A
(Genome AA)
x
Species B
(Genome BB)
Hybridization
F1 (Genome AB)
Genome doubling
spontaneously or
by colchicine
treatment
F1 allopolyploid (S0)
(Genome AABB)
Selfing
S1 allopolyploid
(Genome AABB)
Selfing
S2 allopolyploid
(Genome AABB)
Selfing
S3 allopolyploid
(Genome AABB)

enable harmonic intergenomic coexistence along with reduction in genome size. The above changes can be achieved through alterations in the DNA or in chromatin structure.

Allopolyploidization, which radically changes nuclear structure, exerts a major genetic stress which presumably triggers a variety of cardinal genetic and epigenetic changes. These are presumably critical for the successful establishment of the newly formed species in nature as a competitive entity. Thus, the successful allopolyploidizations—which involve a small number of individuals over a major span of time—are those that have managed to trigger, during or soon after the completion of this process, an array of revolutionary genomic changes that are incongruent with Mendelian principles.

To study the nature of these changes, we used bread wheat (*Triticum aestivum*) as a model. This wheat is a young allohexaploid (2n=6x=42) that was formed about ten thousand years ago from the hybridization of allotetraploid wheat and a diploid *Aegilops* species (Feldman, Lupton, and Miller 1995; Feldman 2001; figure 25.2). The genome of bread wheat is very large, with the amount of DNA in the haploid genome being 1.7×10^{10} bp (1C = 17.8 pg, 1.78×10^{-12} of a gram; Eilam, Anikster, Millet, Manisterski, Sagi-Assif, and Feldman 2008), which is about 5.6 times larger than the human genome (3.0×10^{9} bp; 1C = 3.19 pg). It contains three distinct genomes, A, B, and D, which derived from three different diploid species of the genera *Triticum* (genome A) and *Aegilops* (genomes B and D). Each genome consists of seven pairs of homologous chromosomes. On the basis of the genetic similarity of different pairs of chromosomes from the parental genomes, the twenty-one pairs of homologous chromosomes were classified into seven homeologous (= orthologous) groups of three pairs each (Sears 1954). The homeologous chromosomes in each group contain similar genetic loci.

On the basis of specificity, the DNA sequences of bread wheat were classified into four groups: nonspecific, those occurring in all or most chromosomes (these are mainly retrotransposons and other repetitive sequences); group-specific, those present in chromosomes of one homeologous group (these are mainly coding sequences); genome-specific, those occurring in several chromosome pairs of one genome; and chromosome-specific those present in only one pair of homologous chromosomes (Feldman, Liu, Segal, Abbo, Levy, and Vega 1997). It was assumed

Figure 25.1
Evolution of an allopolyploid species. Two diploid species, one with genome AA and the other with genome BB, hybridize to form a sterile F_1 hybrid. Spontaneous chromosome doubling (or colchicine treatment) gives rise to a fertile allopolyploid having the genomes AABB. The two diverged genomes have to undergo immediate changes (cytological and genetic diploidization) in order to achieve harmonious functioning—a necessary condition for the successful establishment in nature of the newly formed allopolyploid. The three first generations of the allopolyploid are produced by self-pollination.

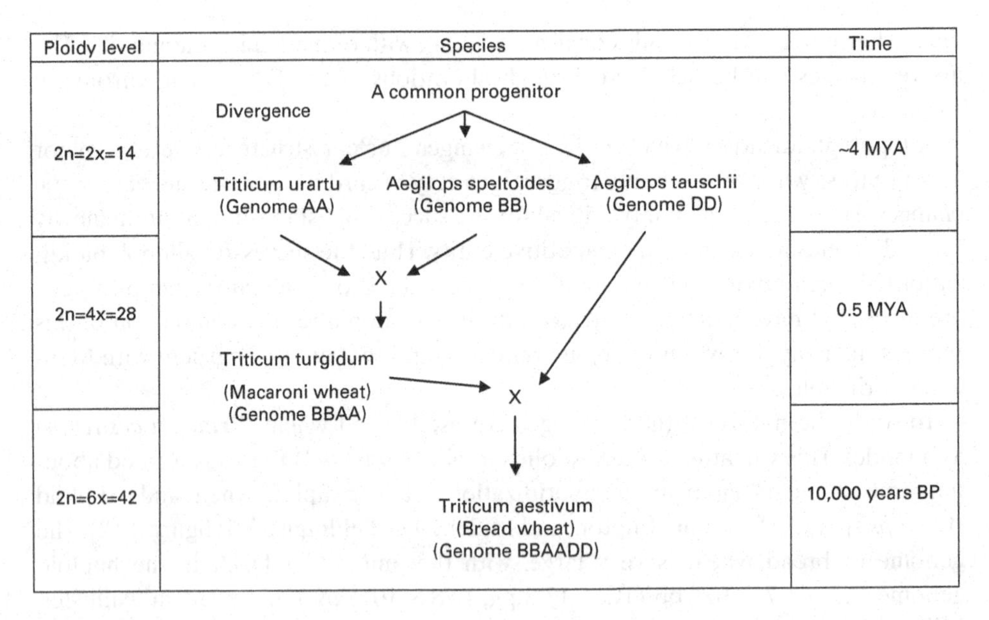

Figure 25.2
Evolution of the diploid and allopolyploid species of wheat. The diploid ($2n$=$2x$=14) species from the genera *Aegilops* and *Triticum* diverged about 4 million years ago (Mya) from a common diploid progenitor. Intergeneric hybridization between the diploid *T. urartu* (genome AA) as male and the donor of the B genome (an unknown species close to *Ae. speltoides*) as female, followed by spontaneous chromosome doubling, gave rise (about 0.5 Mya) to the allotetraploid *Triticum turgidum* (macaroni wheat) with $2n = 4x = 28$ (genome BBAA). A second round of intergeneric hybridization between allotetraploid wheat and the donor of the D genome, *Ae. tauschii* ($2n = 2x = 14$, genome DD), followed by spontaneous chromosome doubling, gave rise to the allohexaploid *T. aestivum* (bread wheat) with $2n = 6x = 42$ chromosomes (genome BBAADD).

that chromosome-specific sequences (CSSs) are those that determine homology. They are therefore assumed to be involved in recognition and initiation of chromosome pairing at meiosis.

Further Divergence of Homeologous Chromosomes (Cytological Diploidization)

Eighteen CSSs were isolated from bread wheat through amplification of DNA sequences from a chromosome that was isolated by micromanipulation using the degenerate oligonucleotide-primed polymerase chain reaction (PCR) (Vega, Abbo, Feldman, and Levy 1994, 1997; Liu, Segal, Vega, Feldman, and Abbo 1997). We then investigated the presence of these isolated sequences in the various polyploid and the diploid species of the wheat group (the genera *Aegilops* and *Triticum*) (Feldman, Liu, Segal, Abbo, Levy, and Vega 1997; Liu, Vega, Segal, Abbo, Redova, and Feldman 1998; Liu, Vega, and Feldman 1998). It was seen that while some of the sequences occur in only a few of the diploids related to the genome contained in the bread wheat sequence, others occur in all the diploid species of the group, but in only one pair of homologous chromosomes of the allopolyploids. Evidently, these sequences were eliminated from one of the two genomes in tetraploid wheat and from two of the three genomes in hexaploid wheat.

To study the extent, pattern and duration of this elimination, a number of F_1 hybrids and synthetic allopolyploids were produced and analyzed (Ozkan, Levy, and Feldman 2001, 2002). All plants were cytologically analyzed at mitosis and meiosis and their integral chromosome constitutions were verified. In F_1 and in each allopolyploid generation, DNA, extracted from young leaves, was examined for the presence or absence of the isolated CSSs by the Restriction Fragment Length Polymorphism (RFLP) technique (Ozkan, Levy, and Feldman 2001). It was found that elimination of several sequences already occurred in the F_1 generation, whereas most sequences were eliminated in the first two allopolyploid generations. Thus, it was concluded that the CSSs, present in all the diploid species of the group, undergo elimination from one of the constituent genomes in the allopolyploids. This elimination is rapid and occurs during or soon after the formation of the allopolyploids. This elimination is reproducible, as the same sequences were eliminated in newly formed and in natural allopolyploids having the same genomic combinations. No further elimination of sequences occurs during the successive generations of the allopolyploids.

The above study was carried out on the CSSs, which are a selected set of DNA sequences. To learn more about the types of changes that occur, how rapidly they occur, and the types of sequences involved, a random and large set of loci was analyzed in the parental diploid lines of *Aegilops* and *Triticum*, in a few F_1 hybrids and in newly formed allopolyploids derived from crosses between these diploid lines. For these studies, we took advantage of the Amplified Fragment Length Polymorphism (AFLP) and DNA gel blot analysis (Shaked, Kashkush, Ozkan, Feldman, and Levy 2001). It was found that sequence elimination is one of the major and immediate responses of the wheat genome to inter-specific or inter-generic hybridization and/or allopolyploidization, that the elimination affects a large fraction of the genome, and that it is reproducible. In one allotetraploid, *Ae. sharonensis-Ae. umbellulata*, 14 percent of the loci from *Ae. sharonensis* were eliminated compared with only 0.5 percent from *Ae .umbellulata,* with most changes occurring in the F_1 hybrid (Shaked, Kashkush, Ozkan, Feldman, and Levy 2001). In contrast, the allotetraploid *Ae. longissima-T. urartu* showed greater sequence elimination following genome duplication than in the F_1.

Further assessment of the amount of DNA that is eliminated due to allopolyploidization was obtained by comparing the amount of nuclear DNA in natural and in newly formed allopolyploids. Nuclear DNA amount was determined by flow cytometry in twenty-seven natural allopolyploid species, as well as in fourteen newly synthesized allopolyploids and in their parental plants (Eilam, Anikster, Millet, Manisterski, Sagi-Assif, and Feldman 2008). In most of the allopolyploids, the quantity of nuclear DNA was significantly less than the sum of the DNAs of the parental species (between 2 percent and 6 percent less). Newly synthesized allopolyploids exhibited a similar decrease in nuclear DNA in the first generation, indicating that genome downsizing occurs during and (or) immediately after the formation of the allopolyploids and that there are no further changes in genome size in the following generations of the allopolyploids. This is supported by the observations that a similar amount of DNA is found in natural allopolyploids and newly formed allopolyploids (after downsizing) (Eilam, Ankister, Millet, Manisterski, Sagi-Assif, and Feldman 2008). Genome downsizing was also found in triticales, the man-made allohexaploid and allooctoploid between wheat and rye (Boyko, Badaev, Maximov, and Zelenin 1984, 1988). Leitch and Bennett (2004) determined the quantity of nuclear DNA in a large number of diverse polyploids, and found that genome downsizing characterizes most polyploids.

From the above studies it was concluded that instantaneous elimination of DNA sequences in F_1 and/or in the first two generations of the newly formed allopolyploids of *Triticum* and *Aegilops*, is a general phenomenon. The similarity between natural and synthetic allopolyploids with the same constituent genomes indicates that elimination of DNA sequences is a reproducible process, and that there is no further change in the amount of DNA over the evolutionary life of the allopolyploids. Mainly low-copy number and, to some extent, high-copy number, non-coding sequences are involved.

Elimination of DNA sequences from one genome further augments the divergence of the homeologous chromosomes and leads to cytological diploidization. This suppresses intergenomic pairing (also known as multivalent-type pairing), and facilitates exclusive intragenomic pairing at meiosis (bivalent pairing). Thus, cytological diploidization increases fertility (up to the maximum), maintains positive intergenomic gene combinations, increases heterosis (hybrid vigor) across the homeoallelic population, and sustains genomic asymmetry (operational specialization) among the constituent genomes.

Orchestration of Gene Action (Genetic Diploidization)

A central question that must be answered regarding newly formed allopolyploids is how the two divergent genomes present in a single nucleus, are driven to operate in a harmonious manner. For instance, in some cases an extra genetic dose, mainly in loci coding for functional proteins, can have a positive effect. In one example, the occurrence of isoenzymes may increase the biochemical potential of the allopolyploids (Hart 1987). In addition, expression of all the duplicated genes may produce desirable inter-genomic interactions and heterotic effects among homeoalleles. On the other hand, activity of all the homeoalleles that code for storage protein genes, 45S rRNA genes, 5S RNA genes, tRNA genes and various structural protein genes, might be redundant, thereby resulting in overproduction and inefficiency. There are several genetic and epigenetic ways to reduce duplicated gene activity. Epigenetic silencing can be engendered by cytosine methylation of DNA sequences, chromatin modifications or remodeling, or by altering the activity of small RNA molecules. Genetic suppression or reduction of gene activity can also be caused by DNA elimination (Kashkush, Feldman, and Levy 2002) or via novel regulatory interactions such as inter-genomic suppression (Galili and Feldman 1984). On the other hand, demethylation of silenced genes can lead to their activation. Genome-wide rewiring of gene expression, through novel cis-trans interactions, can cause both over and under expression of genes (Tirosh, Reikhav, Levy, and Barkai 2009).

We studied genome wide changes caused by cytosine methylation using Methylation-Sensitive Amplified Polymorphism (MSAP) (Shaked, Kashkush, Ozkan, Feldman, and Levy 2001). In this study, we took advantage of the isoschizomers HpaII and MspI, which differ in their ability to sever cytosine-methylated DNA sequences. The enzyme HpaII is sensitive to methylation of either of the cytosine residues at the 5′-CCGG recognition site, whereas its isoschizomer MspI is sensitive to methylation of the external cytosine alone. Therefore, methylation of the internal cytosine would lead to a different MSAP cleavage pattern for the two enzymes as seen in the sequencing gel. MSAP was applied to the parents, the F_1 and the allotetraploid of the cross *Ae. sharonensis* x *T. monococcum* ssp. *aegilopoides* (Shaked, Kashkush, Ozkan, Feldman, and Levy 2001). 501 bands were resolved in the two parental plants. Of the 501, 159 were methylated in one or both parents (about 30 percent methylation). Altogether, 13.1 percent altered methylation pattern were observed, of which 1.8 percent were found only in F_1, 6.9 percent both in the F_1 and the allopolyploid, and 4.4 percent occurred in the allopolyploid alone. Methylation alteration in the F_1 involved mainly repetitive DNA (Shaked, Kashkush, Ozkan, Feldman, and Levy 2001) while that in the allopolyploid involved changes in protein-coding sequences (Liu, Vega, Segal, Abbo, Rodova, and Feldman 1998). It was concluded that one of the main types of change in the synthetic allopolyploids is epigenetic, altering the pattern of cytosine methylation in F_1 plants or in the first generation of the newly formed allopolyploids. Alteration in cytosine methylation affects about 13 percent of loci and involves low-copy number and repetitive sequences. Demethylation, mainly of retrotransposons was also observed in the F_1 and the allopolyploid.

Our next step was to assess the extent of alterations in gene expression by determining the percentage of transcript loss and transcript activation due to allopolyploidization. This analysis was carried out in a newly formed allotetraploid *Ae. sharonensis – T. monococcum* ssp. *aegilopoides* by cDNA-AFLP analysis (Kashkush, Feldman, and Levy 2002). Alterations in gene expression amounted to about 2 percent, of which 1.6 percent resulted from the inactivation of active genes and 0.4 percent from activation of genes that were silent in the diploid parents. The gene expression alterations were then validated by reverse Northern and RT-PCR analysis of transcripts that disappeared in cDNA-AFLP gels (Kashkush, Feldman, and Levy 2002). It was found that transcript losses in the allopolyploid were the result of gene silencing or gene elimination in one or both parental genomes.

Allopolyploidization may also activate retrotransposons that are silent in the parental plants. Retrotransposons constitute most of the DNA of allopolyploid wheat but they are normally transcriptionally silent and can be activated under environmental or genetic stress. It was found that allopolyploidization activated the Wis 2-1A retrotransposons, which subsequently silenced or activated adjacent genes (Kashkush, Feldman, and Levy 2003).

Concluding Remarks

Allopolyploidization accelerates genome evolution by triggering instantaneous radical genomic upheavals through the generation of the following genetic and epigenetic changes: (1) nonrandom elimination of a large number of low- and high-copy coding and non-coding DNA sequences; (2) epigenetic changes such as DNA methylation of coding and non-coding DNA sequences that produce, among other things, modified chromatin conformations and gene silencing; (3) activation of genes, some of which are presumably stress tolerant, due to demethylation; and (4) activation of retroelements which, in turn, alters the expression of adjacent genes. These nonrandom, reproducible changes, occurring in the F_1 hybrids or in the first generation(s) of the nascent allopolyploids, might be a natural defense mechanism exhibited by the novel allopolyploid genome to the genetic shock of hybridization and genome doubling. Similar genomic changes were observed in wild and newly synthesized allopolyploids, indicating that genome alterations occur during and (or) immediately after the formation of the allopolyploids and that there are no further major genomic changes during the life of the allopolyploids. These non–Mendelian changes are heritable and have an important role in the adaptation of the nascent allopolyploids to their environment.

Elimination of DNA sequences from a single polyploid genome increases the differentiation of homeologous chromosomes, thereby providing a physical basis for cytological diploidization that prevents inter-genomic recombinations, and hence leading to full fertility and sustaining plausible inter-genomic interactions. In addition, changes in gene structure and expression bring about rapid genetic diploidization, which ameliorates the harmonious coexistence of the two or more genomes that reside in the same allopolyploid nucleus. Such regulation of gene expression may also lead to improved intergenomic interactions due to gene inactivation, whereas activation of genes through demethylation or through transcriptional activation of retroelements may lead to novel expression patterns. These phenomena, which indicate the plasticity of the allopolyploid genome with regards to structure and function, presumably improved the adaptability of the newly formed natural allopolyploids and facilitated their rapid and successful establishment in nature as new competitive species.

References

Boyko EV, Badaev NS, Maximov NG, Zelenin AV. 1984. Does DNA content change in the course of *Triticale* breeding? Cereal Res Commun. 12: 99–100.

Boyko EV, Badaev NS, Maximov NG, Zelenin AV. 1988. Regularities of genome formation and organization in cereals. I. DNA quantitative changes in the process of allopolyploidization. Genetika 24: 89–97.

Comai L. 2000. Genetic and epigenetic interactions in allopolyploid plants. Plant Mol Biol. 43: 387–399.

Eilam T, Anikster Y, Millet E, Manisterski J, Sagi-Assif O, Feldman M. 2008. Nuclear DNA amount and genome downsizing in natural and synthetic allopolyploids of the genera *Aegilops* and *Triticum*. Genome 51: 616–627.

Feldman M. 2001. The origin of cultivated wheat. In: *The World Wheat Book*. Bonjean A, Angus W, eds Paris: Lavoisier Tech and Doc; 2001:3–56.

Feldman M, Levy AA. 2005. Allopolyploidy—a shaping force in the evolution of wheat genomes. Cytogenet Genome Res. 109: 250–258.

Feldman M, Levy AA. 2009. Genome evolution in allopolyploid wheat—a revolutionary reprogramming followed by gradual changes. J Genet Genomics 36: 511–518.

Feldman M, Liu B, Segal G, Abbo S, Levy AA, Vega JM. 1997. Rapid elimination of low copy DNA sequences in polyploid wheat: A possible mechanism for differentiation of homoeologous chromosomes. Genetics 147: 1381–1387.

Feldman M, Lupton FGH, Miller TE. 1995. Wheats. In: *Evolution of Crop Plants*. Smartt J, Simmonds NW, eds. 2nd ed. London: Longman Scientific; 1995:184–192.

Galili G, Feldman M. 1984. Intergenomic suppression of endosperm protein genes in common wheat. Can J Genet Cytol 26: 651–656.

Grant V. *Plant Speciation.* New York: Columbia University Press; 1971.

Hart GH. Genetic and biochemical studies of enzymes. In: *Wheat and Wheat Improvement*. Heyne EG, ed. 2nd ed. Madison, WI: American Society of Agronomy; 1987.

Kashkush K, Feldman M, Levy AA. 2002. Gene loss, silencing and activation in a newly synthesized wheat allotetraploid. Genetics 160: 1651–1659.

Kashkush K, Feldman M, Levy AA. 2003. Transcriptional activation of retrotransposons alters the expression of adjacent genes in wheat. Nat Genet. 33: 102–106.

Leitch IJ, Bennett MD. 1997. Polyploidy in angiosperms. Trends Plant Sci 2: 470–476.

Leitch IJ, Bennett MD. 2004. Genome downsizing in polyploidy plants. Biol J Linn Soc Lond. 82: 651–663.

Levy AA, Feldman M. 2002. The impact of polyploidy on grass genome evolution. Plant Physiol. 130: 1587–1593.

Levy AA, Feldman M. 2004. Genetic and epigenetic reprogramming of the wheat genome upon allopolyploidization. Biol J Linn Soc Lond. 82: 607–613.

Liu B, Segal G, Vega JM, Feldman M, Abbo S. 1997. Isolation and characterization of chromosome-specific DNA sequences from a chromosome-arm genomic library of common wheat. Plant J. 11: 959–965.

Liu B, Vega JM, Feldman M. 1998. Rapid genomic changes in newly synthesized amphiploids of *Triticum* and *Aegilops*. II. Changes in low-copy coding DNA sequences. Genome 41: 535–542.

Liu B, Vega JM, Segal G, Abbo S, Rodova M, Feldman M. 1998. Rapid genomic changes in newly synthesized amphiploids of *Triticum* and *Aegilops*. I. Changes in low-copy non-coding DNA sequences. Genome 41: 272–277.

Manton I. *Problems of Cytology and Evolution in the Pteridophyta.* Cambridge: Cambridge University Press; 1950.

Ozkan H, Levy AA, Feldman M. 2001. Allopolyploidy-induced rapid genome evolution in the wheat (*Aegilops-Triticum*) group. Plant Cell 13: 1735–1747.

Ozkan H, Levy AA, Feldman M. 2002. Rapid differentiation of homoeologous chromosomes in newly formed allopolyploid wheat. Isr J Plant Sci. 50: S-65–S-76.

Pikaard CS. 2001. Genomic change and gene silencing in polyploids. Trends Genet. 17: 675–677.

Sears ER. 1954. The aneuploids of common wheat. Res Bull Univ Missouri Agric Exp Sta. 572: 1–59.

Shaked H, Kashkush K, Ozkan H, Feldman M, Levy AA. 2001. Sequence elimination and cytosine methylation are rapid and reproducible responses of the genome to wide hybridization and allopolyploidy in wheat. Plant Cell 13: 1749–1759.

Soltis DE, Soltis PS. 1993. Molecular data and the dynamic nature of polyploidy. Crit Rev Plant Sci. 12: 243–273.

Soltis DE, Soltis PS. 1995. The dynamic nature of polyploidy genomes. Proc Natl Acad Sci USA 92: 8089–8091.

Soltis DE, Soltis PS, Tate JA. 2004. Advances in the study of polyploidy since plant speciation. New Phytol. 161: 173–191.

Soltis PS, Soltis DE. 2000. The role of genetic and genomic attributes in the success of polyploids. Proc Natl Acad Sci USA 97: 7051–7057.

Stebbins GL. *Variation and Evolution in Plants.* New York: Columbia University Press;1950.

Stebbins GL. *Chromosomal Evolution in Higher Plants.* London: Edward Arnold; 1971.

Tirosh I, Reikhav S, Levy AA, Barkai N. 2009. Gene expression divergence between related yeast species and its rewiring in their hybrid. Science 324: 659–662.

Vega JM, Abbo S, Feldman M, Levy AA. 1994. Chromosome painting in plants: *In situ* hybridization with a DNA probe from a specific microdissected chromosome arm of common wheat. Proc Natl Acad Sci USA 91: 12041–12045.

Vega JM, Abbo S, Feldman M, Levy AA. 1997. Chromosome painting in wheat. Chromosomes Today 12: 319–332.

Wendel JF. 2000. Genome evolution in polyploids. Plant Mol Biol. 42: 225–249.

26 Lamarckian Leaps in the Microbial World

Jan Sapp

The neo-Darwinian synthesis of the twentieth century was concerned with the visible world of plants and animals over the last 550 million years of evolution. It said nothing of the microbial world, where the greatest biochemical and genetic diversity and largest biomass are found. The modes of evolution in the microbiosphere, exposed by molecular phylogenetic methods, have led to a revolution in our understanding of evolutionary processes with the recognition that (1)there are not two, but three, primary lineages or domains of life: the Archaea, the Bacteria, and the Eukarya; (2) horizontal gene transfer (between taxa) is pervasive in the microbial world; and (3) symbiosis is a fundamental mode of evolutionary change, ubiquitous in the microbial world, among plants, and in the most species-rich phyla of the animal world. In the microbial realm, central tenets of Darwinism do not apply, and the mode and tempo of evolution resemble "Lamarckian processes" coupled with saltational change.

Chaos Infusoria

In *Philosophie zoologique*, Lamarck ([1809] 1984:56) posited that the aim of a general classification is not simply to "possess a convenient list for consulting" but also to have an order of things that "represents as nearly as possible the actual order followed by nature in the production of animals." The affinities among the great classes of animals could be established by comparisons of "the essential system of organs," those highly preserved ancient features far removed from the everyday life of the individual organism.

Darwin made a similar argument with regard to classification, though the principal mechanisms of evolution he proposed were different. The similarities between species would reveal their common descent, whereas differences would reveal species' modifications.

Like Lamarck, Darwin argued that genealogical trees of great size and scope required comparisons of those organizational characteristics that had become ever-integrated within the organism as evolution proceeded, and thus were far removed from those "parts of structure which determined the habits of life, and the general place of each being in the economy of nature" (Darwin [1859] 1969:414). The course of evolution could thus be represented as a bifurcating tree, and organisms could be classified hierarchically, group within group, just as naturalists had been doing since the time of Linnaeus.

Microbes posed the greatest difficulty for such a hierarchical arrangement. Darwin said virtually nothing about them in *The Origin of Species*. Linnaeus had packed all of them into one species that he humorously named *Chaos infusoria* in his *Systema Naturae*. In the nineteenth century, comparative embryology, comparative anatomy, and the fossil record were the means of inferring relationships among species of plants and animals. These were lacking for microbes.

Biology's Scandal

Microbes were again not included in the Darwinian revival of the middle of the twentieth century. That "Modern Synthesis" was based on gene mutations and recombination between individuals of a species as fuels for evolution by natural selection. Because of bacteria's apparently active response to environmental change, they were typically thought to belong to "a Lamarckian rather than a Darwinian world, or at least a world in which both Lamarckian and Darwinian evolutionary mechanics are operative" (Manwarring 1934:470). It was far from certain whether bacteria even possessed genes. As Julian Huxley (1942:131–132) commented:

> They have no genes in the sense of accurately quantized portions of hereditary substances. . . . The entire organism appears to function as soma and germplasm, and evolution must be a matter of alteration in the reaction-system as a whole. . . . We must, in fact, expect that the processes of variation and evolution in bacteria are quite different from the corresponding processes in multicellular organisms. But their secret has not yet been unraveled.

It was somewhat surprising, then, that bacteria (or at least a few representatives) became the model organisms for examining the molecular structure and function of genes after the Second World War. Still, molecular biologists were generally no more interested in the natural history of bacteria than *Drosophila* geneticists were interested in entomology.

Moreover, "the scandal of bacteriology," as it was called, was that there was no satisfactory definition of a bacterium (Stanier and Van Niel 1962). Bacteriology was primarily an applied field. But when not viewed through the lenses of pathology and industry, bacteria were generally understood to be a class of plants: the fission fungi, *Schizomycetes*. There were, however, proposals that bacteria, along with the blue-green algae, be assigned to their own kingdom, Monera, on the basis that they were devoid of membrane-bound nuclei, plastids, and sex. But critics of this kingdom pointed to its almost entirely negative definition (like invertebrates), and they doubted that it was a coherent monophyletic grouping and that it should include the blue-green algae. It was also not certain how smaller bacteria such as *Rickettsia* and *Chlamydia* could be distinguished from viruses (see Sapp 2009).

At the Pasteur Institute, André Lwoff (1957) made the distinction between the virus and a cell unequivocal on the basis of electron microscopy and biochemistry. Viruses contained DNA or RNA but never both; they lacked ribosomes; and they were virtually devoid of enzymes. They also did not reproduce by division like a cell; their replication occurred only within a susceptible cell. Stanier and van Niel (1962) subsequently made a hard-and-fast distinction between bacterial cells and those of other cells on the same basis when summoning the terms "prokaryote" and "eukaryote," coined decades earlier (Sapp 2005). Eukaryotic cells possessed a membrane-bound nucleus and a cytoskeleton, divided by mitosis, and possessed mitochondria (and chloroplasts in the case of algae and plant cells). Prokaryotic cells were smaller and lacked these characteristics. Soon there were proposals for two new kingdoms or superkingdoms, *Prokaryotae* and *Eukaryotae.*

Stemming from the Progenotes

There were two untested assumptions underlying the prokaryote–eukaryote dichotomy: that prokaryotes preceded and gave rise to eukaryotes, and that prokaryotes themselves derived from other prokaryotic cells. Both were challenged with the rise of molecular phylogenetics. The field of "molecular evolution" revolutionized much of classification, but especially bacterial taxonomy. Its methods in the 1960s were based on the percentage of GC content, DNA–RNA hybridization, and amino acid sequences; they were extended to include sequencing of segments of ribosomal RNAs in the 1970s, whole genes in the 1980s, and whole genome analysis in the 1990s.

The revolution in microbial phylogenetics began in the 1970s when Carl Woese at the University of Illinois developed a method of analyzing the sequences of short segments of ribosomal RNAs. He focused on the small-subunit ribosomal RNA known as SSU rRNA or 16S rRNA, claiming it was the ultimate molecular chronometer at the organismal core. In 1977 a "third form of life," the archaebacteria, was announced (Woese and Fox 1977b). Based on 16S rRNA comparisons, they were as different from bacteria as from eukaryotes. They comprised methanogens, living in the guts of rumens and in swamps; salt-loving halophiles, known for rotting salted fish; and extreme thermoacidophiles, found in smoldering coal-mine refuse piles.

These "extremophilic" organisms had been considered to be completely unrelated before they were reconceived as an ur-kingdom apart and as ancient organisms at the dawn of life. Their walls and the lipids in their membranes were unlike those of all other bacteria; so, too, were their transfer RNAs, transcription enzymes, and viruses. When the outline of the tree of life was rooted by using ancient gene duplications, the stem split archaebacteria and eukaryotes on one side, and eubacteria on the other, a divide that was supported by other molecular differences among the three taxa. Three domains—the Eukarya, the Bacteria, and the Archaea—were formally proposal so as to reflect that phylogenetic ordering (Woese, Kandler, and Wheelis 1990). While molecular phylogeneticists who explored the roots of life typically searched for a universal single prokaryotic-like cell, Woese conceptualized three primary lineages emerging from precellular entities, called "progenotes," in the throes of developing the relationship between nucleic acids and proteins (Woese and Fox 1977b; Woese 1982, 1998).

Innovation Sharing among Bacteria

Bacteria's Lamarckian ways seemed more evident after the Second World War. Since the 1950s, it has been known that bacteria can acquire genes by various modes: (1) by transformations, picking up foreign genetic material from the environment; (2) by conjugation; and (3) by transduction. Conjugational transfer of DNA requires physical contact and is unidirectional from donor to recipient. Bacterial fertility is dependent on the presence of a specific fertility factor plasmid in the donor cell. Conjugation triggers the transfer of the F factor plasmid, along with any other DNA which might become integrated with that plasmid. In transduction, bacteriophages (bacterial viruses) transmit bacterial genes. Bits of host DNA may be incorporated into the bacteriophage: when the particle infects a new host cell; it injects not only its own DNA, which may become integrated into the host's genome, but also segments of the DNA of its former host.

Horizontal gene transfer was often referred to as a form of hybridization. And hybridization accomplished in the laboratory said nothing of its prevalence in nature. Some leading microbiologists of the 1950s and 1960s maintained that it would indeed be a rare event, with no more implications for classification than the artificial hybridization of two higher plants or animals (Stocker 1955; Luria and Burrous 1957). Horizontal gene transfer acquired great medical significance with the recognition of its importance in outbreaks of antibiotic resistance. Multiple drug resistance did not arise in a series of discrete steps, each corresponding to a single drug, but usually appeared fully developed as groups of resistances to some or all of the drugs (Watanabe 1963). It took longer for horizontal gene transfer to be considered as an important aspect of evolution. That it was a powerful force in bacterial evolution was voiced loudly by a few bacteriologists of the 1970s (Sapp 2009). The biological species concept would not apply to the bacterial world if gene transfer was rife, and because genomes are shared within and between bacterial populations, bacteria would possess superorganismal properties (Sonea and Panisset 1983).

The rise of genomics brought with it substantial evidence that bacterial evolution did not occur gradually, in the neo-Darwinian manner, but rapidly, by leaps resulting from horizontal gene transfer across the taxonomic spectrum. The data were of two kinds. One was based on incongruities of trees generated from different genes (Doolittle 1999, 2000): phylogenies of genes for various proteins often conflicted not only with the 16S rRNA tree but also with themselves. The other was based on anomalous G+C content and codon usages. The genetic code is highly redundant: all but two amino acids are encoded by more than one codon (e.g., UAU and UAC for tyrosine). There were often taxon preferences for one of the several codons that encode the same amino acid. On this basis, about 18 percent of the genes of *E. coli* were relatively recent acquisitions, and all the genes that distinguished *Escherichia coli* from *Salmonella enterica* had been horizontally transferred (Lawrence and Ochman 1998:9416). Horizontal gene transfer not only could confer antibiotic resistance but also could transform bacteria from benign to pathogenic in a single saltational step.

The great majority of genes in bacteria are subject to horizontal transfer as bacteria actively alter their genomes in response to environmental stresses or opportunities. Gene swapping among bacteria represents a "Lamarckian world" (Goldenfeld and Woese 2007; Buchanan 2009), albeit one that is also saltational. Only a small number of *essential* genes can trace cell lineages. These are informational genes involved in transcription and translation of DNA and related ancient, essential, and complex processes that involved the coevolution and interaction of hundreds of gene products. The hierarchical conception of groupings within groupings, based on common descent, does not hold for the bacterial world. The "tree" of bacterial life is highly reticulated, a worldwide web.

Lamarckian–Saltational Symbiosis

Lamarckism and saltationism have also been attributed to the inheritance of acquired bacteria among eukaryotic organisms. The idea is a venerable one, though it still remains on the margins of biology. When Anton de Bary introduced the term "symbiosis" to a group of naturalists and physicians at Cassel in 1878, he understood it to be a mode of sudden evolutionary change in juxtaposition to Darwinian gradualism: "Whatever importance one wants to attach to natural selection for the gradual transformation of species," he said, "it is desirable to see yet another field opening itself up for experimentation" (De Bary 1879:29–30). De Bary was referring to the recent experimental evidence that lichens resulted from the symbiosis of fungi and algae; of blue-green algae living inside special cavities of the leaves of the aquatic fern *Azolla*; and of blue-green algae which produced nodules inside the cells of the roots of the palmlike cycads in which they lived.

Within a decade, various other classical instances of microbes causing physiological and morphological changes advantageous to their hosts accumulated: nitrogen-fixing bacteria in the root nodules of legumes, mycorrhizae of forest trees and orchids, algae living in protozoa, and algae inside the tissue of translucent animals—sea anemones, hydra, and coral. Symbiotic conceptions were also extended to the evolution of the organelles of eukaryotic cells (Sapp 1994). That chloroplasts might be symbionts was suggested the moment the term was coined (Schimper 1883), and that the nucleus and cytoplasm represent a division of labor between two different kinds of organisms soon followed (Watasē 1893). The concept of "symbiogenesis" for "the origin of organisms by the combination or by the association of two or several beings which enter into symbiosis" was developed further by Constanin Merezhkowsky (Sapp, Carrapiço, and Zolotonosov 2002).

The idea that mitochondria originated as symbiotic bacteria was championed by Paul Portier (1918) at the Institut Océanographique de Monaco, and by Ivan Wallin at the University of Colorado. In his book *Symbionticism and the Origin of Species*, Wallin (1927) postulated three principles of evolution: (1) symbionticism, mainly responsible for the origin of species; (2) natural selection, responsible for the retention and destruction of species; and (3) an unknown principle, responsible for evolution's direction toward an ever more complex end. He also proposed that repeated acquisition of mitochondria would account for the origin of new genes. That infectious microbes might be the source of new genes had been previously postulated by one of the founding Mendelian geneticists, William Bateson, a saltationist known for his opposition to Darwinian gradualism. In his *Problems of Genetics* (Bateson 1913: 88), he suggested that while recessive traits result from the loss of a genetic constituent, new dominant genes might originate from infection.

Hereditary symbiosis had been discussed by leading neo-Lamarckian French-speaking biologists (Sapp 1994). Félix d'Hérelle associated the many cases of microbial symbiosis in plants when he wrote of bacteria that harbor viruses ("lysogenic bacteria"); he called them "microlichens." "Symbiosis," he said, "is in large measure responsible for evolution" (D'Hérelle 1926:320). As it was for De Bary and others, so it was for D'Hérelle: symbiosis entailed sudden change. "Botany offers many examples of such sudden mutations brought about by symbiosis, and even in man such modifications occur, as exemplified by the changes in the skeleton observed in congenital syphilis" (D'Hérelle 1931: 56). Still, there were few systematic research programs on hereditary symbiosis between the two world wars. Paul Buchner, who also denied natural selection and gene mutations as the primary basis of evolution, studied the morphological and physiological effects of microbes transmitted through the cytoplasm of the eggs of many species of insects (Sapp 2002).

Neo-Darwinians dismissed all of it. Examples of such symbiosis were rare curiosities, exceptions, anomalies to the Hobbesian struggle of all against all. That microbial infections could be the source of evolutionary change conflicted with the overwhelming reports of microbial disease-causing infections. As Wallin (1927: 8) commented: "It is a rather startling proposal that bacteria, the organisms which are popularly associated with disease, may represent the fundamental causative factor in the origin of species."

The concept of an organism as a symbiotic complex opposed the one-germplasm–one-organism model of the individual, and symbionts inherited through the egg opposed the all-exclusive role of nuclear genes in inheritance upheld by classical geneticists (Sapp 1994). As the leader of the *Drosophila* school and the chromosome theory of heredity, T. H. Morgan (1926:491) asserted: "In a word, the cytoplasm may be ignored genetically." By definition, hereditary infections were contaminations. As the Harvard corn geneticist E.M. East commented, they were a source of experimental error; biologists needed to be wary, lest they be passed off as evidence of the inheritance of acquired characteristics:

> . . . there are several types of phenomena where there is direct transfer, from cell to cell, of alien matter capable of producing morphological changes. It is not to be supposed that modern biologists will cite such instances when recognized, as examples of heredity. But since an earlier generation of students used them, before their cause was discovered, to support arguments on the inheritance of acquired characteristics, it is well to be cautious in citing similar, though less obvious, cases as being illustrations of non-Mendelian heredity. (East 1934:431)

Debates over the scope and significance of cytoplasmic inheritance became entangled in Cold War rhetoric. On the one hand, Lysenkoists in the Soviet Union used the evidence for non-Mendelian cytoplasmic inheritance to debunk the whole of Western genetics. On the other hand, leading geneticists in the West dismissed the evidence for cytoplasmic inheritance as being due to infections, and therefore of little significance for genetics (Sapp 1987). The concept of the inheritance of acquired characteristics became politicized and discredited as associated with frauds and ideology, not science.

Joshua Lederberg (1952:403) was an exception among geneticists when he sought to broaden genetic concepts to include "infective heredity." In so doing, he allied cytoplasmic inheritance with the maligned hereditary symbiosis in insects and protists, with viral symbiosis (lysogeny), transduction, and genetic particles in the cytoplasm of bacteria. He thus proposed the term "plasmid" "for any extrachromosomal hereditary determinant" and extended the boundaries of what was to count as an individual: "The cell or the organism is not readily delimited in the presence of plasmids whose coordination grade from plasmagenes to frank parasites." Rather than as an absolute ontological category, he suggested that what should count as an individual should be defined in terms of "the practices of specialties" (Lederberg 1952: 425–426).

The confrontation between symbiosis and neo-Darwinism started up again when DNA was found in chloroplasts and mitochondria. Did these organelles arise by "a gradual and continuous evolutionary process" within cells, or "rather abruptly, as a result of endosymbiosis?" (Uzzell and Spolsky 1974:343). In their framing of the question, Uzzell and Spolsky grouped symbiotic origins with Special Creation in opposition to Darwinian gradualism:

> The endosymbiosis hypothesis is retrogressive in the sense that it avoids the difficult thought necessary to understand how mitochondria and chloroplasts have evolved as a result of small evolutionary steps. Darwin's *On the Origin of Species* first provided a convincing evolutionary viewpoint to contrast with the special-creation position. The general principle that organs of great perfection, such as the eye, can evolve provided that each small intermediate step benefits the organism in which it occurs seems appropriate for the origin of cell organelles as well. (Uzzell and Spolsky 1974:343)

Then again, neither side, Darwinian or symbiotic, could offer definitive proof for its claims which, in effect, were then outside the purview of empirical science. Within a decade, the study of organellar origins would be turned into an empirical science and emerge as one of the great pillars of microbial evolutionary biology (Sapp 2007, 2009). Molecular phylogenetics based on 16S rRNA confirmed that mitochondria evolved as symbionts from alpha-proteobacteria, and chloroplasts from cyanobacteria (blue-green algae) (Gray and Doolittle 1982). That the nucleus and the outlying cytoplasm of eukaryotic cells are also relics of a symbiosis is a concept that is well considered today (Sapp 2009).

At the decline of the Cold War, hereditary symbiosis was reinterpreted as a Lamarckian process, as was horizontal gene transfer among bacteria. As F.J.R. Taylor (1987:13) commented: "The mechanism of organellar acquisition, 'inheritable symbiosis' . . . seems to be a fine example of the 'evolution of acquired characteristics,' even if not of the type Lamarck had in mind." Paul Nardon and Ann-Marie Grenier (1991:162) allied their studies of hereditary symbiosis in insects and all other cases of hereditary symbiosis with a redefinition of "neo-Lamarckism" as "a theory of evolution that takes into account the possibility of acquiring exogenous characteristics or genes. Symbiosis is a 'neo-Lamarckian' mechanism of evolution." This, in their view, was why it was underrated by neo-Darwinian biologists. Lynn

Margulis and Dorion Sagan (2002:41) went further, asserting that the inheritance of acquired genes is the only meaning that can be given to Lamarckism today:

> Lamarck was correct: Acquired traits can be inherited not as traits but as genomes. . . . The only "character" or trait that can be passed down (vertical inheritance, or acquired (horizontal inheritance) and then be propagated from generation to generation, is a character encoded in genes.

Gene-encoded characters are certainly not the only traits passed on vertically. A theory of the organism and of heredity cannot be reduced to genes and gene products, horizontal gene transfer, and symbiosis. It must include epigenetic inheritance, discussed in other chapters in this volume, as well as non–DNA-based inheritance reflecting the supramolecular structure of cells, as demonstrated in "cortical inheritance" in ciliates. The best-known form of structural inheritance is the organization of the longitudinal ciliary rows that cover the surfaces of most ciliates. In a now classical experiment, Janine Beisson and Tracy Sonneborn (1965) demonstrated that an inversion (180° rotation) of one or more of the ciliary rows was nongenically inherited. This demonstration was repeated in *Tetrahymena* and for other ciliates (see Nanney 1980; Frankel 1989). Sonneborn (1963: 202) coined the term "cytotaxis" for "this ordering and arranging of new structures under the influence of preexisting cell structure. . . . Without cytotaxis an isolated nucleus could not make a cell even if it had all the precursors, tools, and machinery for making DNA and RNA and the cytoplasmic machinery for making peptides." When discussing the inheritance of acquired characteristics and cortical inheritance, John Maynard Smith (1983:39) commented: "Neo-Darwinians should not be allowed to forget these cases for they constitute the only significant threat to our views." This remark today is a gross understatement in light of more recent evidence for epigenetic inheritance described in this volume, and hereditary symbiosis and horizontal gene transfer among bacteria.

Hereditary symbiosis is also ubiquitous among protists. Still, classical evolutionists who focus on animals consider it to be rare, as "the quirky and incidental side" of evolution (Gould 1986:310). Some neo-Darwinian theorists, basing their views on selfish gene theory, kinship, or the nature of ecological relationships, have conjectured that hereditary symbiosis must be rare in animals, which have a sequestered germ line; and that all symbionts transmitted hereditarily should be considered as slaves (Maynard Smith and Szathmáry 1999). Yet, molecular techniques for screening have shown that this is not the case among the great majority of insects, nematodes, and the many species of mites and spiders in which bacteria of the genus *Wolbachia* are inherited through the egg (Werren 2005). Far from being "slaves," *Wolbachia* cause profound morphological and behavioral modifications in their hosts, including cytoplasmic incompatibility between strains and parthenogenesis induction; they can also convert genetic males into reproductive females (and produce intersexes). Widespread and sometimes wholesale transfer of *Wolbachia* genes to their host genome is also known (Hotopp, Clark, Oliveira, Foster, Fischer, Torres, Giebel et al. 2007; 2008). Gladyshev, Meselson, and Arkhipova When considering symbiosis in animals, we would also do well to remember how much of the human genome is of viral origin (Ryan 2007).

Horizontal gene transfer among bacteria, and hereditary symbiosis in eukaryotes, are still not taught as central principles of evolution alongside Darwinian theory. On the contrary, in the biopolitical context of Darwinism versus Special Creation, breaking from ranks is discouraged, and non–Darwinian processes remain effectively under house arrest as "exceptions." To fully explore the panoply of evolutionary processes, and for the continued advance of science, evolution must be uncoupled from its anachronistic adjective "Darwinian."

References

Bateson W. *Problems of Genetics.* New Haven, CT: Yale University Press; 1913.

Beisson J, Sonneborn TM. 1965. Cytoplasmic inheritance of the organization of the cell cortex in *Paramecium aurelia.* Proc Natl Acad Sci USA. 53: 275–282.

Buchanan M. 2009. Collectivist revolution in evolution. Nat Phys. 5,8: 531.

D'Hérelle F. *The Bacteriophage and Its Behavior.* Smith GH, trans. Baltimore: Williams & Wilkins; 1926.

D'Hérelle F. 1931. Bacterial mutations. YJBM. 4:55–57.

Darwin C. *On the Origin of Species* [1859]. Cambridge, MA: Harvard University Press; 1964.

De Bary A. Die Erscheinung der Symbiose. *Vortrag auf der Versammlung der Deutscher Naturforsher und Ärtze zu Cassel.* Strassburg: Verlag von Karl J. Trubner; 1879:1–30.

Doolittle WF. 1999. Phylogenetic classification and the universal tree. Science 284,5423: 2124–2128.

Doolittle WF. 2000. Uprooting the tree of life. Sci Am. 282: 90–95.

East EM. 1934. The nucleus-plasma problem. Am Nat. 68:289–303, 402–439.

Frankel J. *Pattern Formation: Ciliate Studies and Models.* New York: Oxford University Press; 1989.

Gladyshev EA, Meselson M, Arkhipova IR. 2008. Massive horizontal gene transfer in bdelloid rotifers. Science 320,5880: 1210–1213.

Goldenfeld N, Woese C. 2007. Biology's next revolution. Nature 445: 369.

Gould SJ. *Wonderful Life. The Burgess Shale and the Nature of History.* London: Hutchison Radius; 1986.

Gray MW, Doolittle WF. 1982. Has the endosymbiont hypothesis been proven? Microbiol Rev. 46,1: 1–42.

Hotopp JCD, Clark ME, Oliveira DC, Foster JM, Fischer P, Torres MC, Giebel JD et al. 2007. Widespread horizontal gene transfer from intracellular bacteria to multicellular eukaryotes. Science 317,5845: 1753–1756.

Huxley J. *Evolution: The Modern Synthesis.* London: Allen & Unwin;1942.

Lamarck J-B. *Zoological Philosophy* [1809]. Elliot H, trans. Chicago: University of Chicago Press; 1984.

Lawrence JG, Ochman H. 1998. Molecular archaeology of the *Escherichia coli* genome. Proc Natl Acad Sci USA. 95: 9413–9417.

Lederberg J. 1952. Cell genetics and hereditary symbiosis. Physiol Rev. 32: 403–430.

Luria SE, Burrous J. 1957. Hybridization between *Escherichia coli* and *Shigella*. J Bacteriol. 74,4: 461–476.

Lwoff A. 1957. The concept of virus. J Gen Microbiol. 17: 239–253.

Manwarring WH. 1934. Environmental transformation of bacteria. Science 79: 466–470.

Margulis L, Sagan D. *Acquiring Genomes: A Theory of the Origin of Species.* New York: Basic Books; 2002.

Maynard Smith J. Evolution and development. In: *Development and Evolution.* Goodwin BC, Holder N, and Wylie CC, eds. Cambridge: Cambridge University Press; 1983:33–45.

Maynard Smith J, Szathmáry E. *The Origins of Life. From the Birth of Life to the Origin of Language.* New York: Oxford University Press; 1999.

Morgan TH. 1926. Genetics and the physiology of development. Am Nat. 60: 489–515.

Nanney DL. *Experimental Ciliatology.* New York: Wiley; 1980.

Nardon P, Grenier A-M. Serial endosymbiosis theory and weevil evolution: The role of symbiosis. In: *Symbiosis as a Source of Evolutionary Innovation.* Margulis M, Fester R, eds. Cambridge, MA: MIT Press; 1991:153–169.

Portier P. *Les symbiotes.* Paris: Masson; 1918.

Ryan F. 2007. Viruses as symbionts. Symbiosis 44,1–3: 11–21.

Sapp J. *Beyond the Gene: Cytoplasmic Inheritance and the Struggle for Authority in Genetics.* New York: Oxford University Press; 1987.

Sapp J. *Evolution by Association: A History of Symbiosis.* New York: Oxford University Press; 1994.

Sapp J. 2002. Paul Buchner and hereditary symbiosis in insects. Int Microbiol. 5: 145–160.

Sapp J. 2005. The prokaryote-eukaryote dichotomy: Meanings and mythology. Microbiol Mol Biol Rev. 69,2: 292–305.

Sapp J. Mitochondria and their host: Morphology to molecular phylogeny. In: *Mitochondria and Hydrogenosomes.* Martin W, Müller M, eds. Heidelberg: Springer; 2007:57–84.

Sapp J. *The New Foundations of Evolution: On the Tree of Life.* New York: Oxford University Press; 2009.

Sapp J, Carrapiço F, Zolotonosov M. 2002. Symbiogenesis: The hidden face of Constantin Merezhkowsky. Hist Philos Life Sci. 24,3–4: 413–440.

Schimper, A F W. 1883. Ueber die Entwicklung der Schlorophyllkörner und Farbkörper. Botan. Zeit. 41: 105–114, 121–131, 137–146, 153–160.

Sonea S, Panisset P. *A New Bacteriology.* Boston: Jones and Bartlett; 1983.

Sonneborn TM. Does preformed cell structure play an essential role in cell heredity? In: *The Nature of Biological Diversity.* Allen JM, ed. New York: McGraw-Hill; 1963:165–221.

Stanier RY, Van Niel CB. 1962. The concept of a bacterium. Arch Mikrobiol. 42: 17–35.

Stocker BAD. 1955. Bacteriophages and bacterial classification. J Gen Microbiol. 12: 375–381.

Taylor FJR. 1987. 1. An overview of the status of evolutionary cell symbiosis theories. Ann N Y Acad Sci. 503: 1–16.

Uzzell T, Spolsky C. 1974. Mitochondria and plastids as endosymbionts: A revival of special creation? Am Sci. 62: 334–343.

Wallin IE. *Symbionticism and the Origin of Species.* Baltimore: Williams & Wilkins; 1927.

Watanabe T. 1963. Infective heredity of multiple drug resistance in bacteria. Bacteriol. Rev. 1963; 27,1: 87–115.

Watasē S. 1893. On the nature of cell organization. Woods Hole Biol Lect. 83–103.

Werren JH. Heritable microorganisms and reproductive parasitism. In: *Microbial Evolution and Phylogeny: Concepts and Controversies.* Sapp J, ed.. New York: Oxford University Press; 2005:290–316.

Woese CR. Archaebacteria and cellular origins: An overview. In: *Archaebacteria.* Kandler O, ed. Stuttgart: Gustav Fischer; 1982:1–17.

Woese CR. 1998. The universal ancestor. Proc Natl Acad Sci USA. 95: 6854–6859.

Woese CR, Fox GE. 1977a. The concept of cellular evolution. J Mol Evol. 10,1: 1–6.

Woese CR, Fox GE. 1977b. Phylogenetic structure of the prokaryotic domain: The primary kingdoms. Proc Natl Acad Sci USA. 74,11: 5088–5090.

Woese CR, Kandler O, Wheelis M. 1990. Towards a natural system of organisms: Proposal for the domains Archaea, Bacteria, and Eucarya. Proc Natl Acad Sci USA. 87,12: 4576–4579.

27 Symbionts as an Epigenetic Source of Heritable Variation

Scott F. Gilbert

Evolution arises from heritable changes in development. Most evolutionary developmental biology has focused on changes in the regulatory components of the genome. However, development also includes interactions between organisms and their environments. One area of interest concerns the importance of symbionts for the production of the normal range of phenotypes. Many, if not most, organisms have "outsourced" some of their developmental signals to a set of symbionts that are expected to be acquired during development. Such intimate interactions between species are referred to as codevelopment, the production of a new individual through the coordinated interactions of several genotypically different species. Several research programs have demonstrated that such codevelopmental partnerships can be selected. Here I focus on symbioses in coral reef cnidarians and pea aphids, wherein the symbiotic system provides thermotolerance for the composite organism, and on mice, whose gut symbionts provide critical signals for host gut development, immune function, and fat storage.

The theory of evolution by natural selection is predicated on the existence of widespread variation within species. But from whence does this variation arise? Darwin (1859) realized that selection could not act upon characters that had not yet appeared, noting that "characters may have originated from quite secondary sources, independently from natural selection." He continued this line of reasoning in his book *The Variation of Animals and Plants Under Domestication* (1868), in which he concludes (p. 351, in his discussion of the origin of nectarines from several different varieties of peach, each in a different environment), "the external conditions of life are quite insignificant, in relationship to any particular variation, in comparison with the organization and constitution of the being which varies. We are thus driven to conclude that in most cases the conditions of life play a subordinate part in causing any particular modification…" The sources of variation remained obscure.

Population genetics provided the first set of answers to Darwin's quandary. Alleles of the protein-encoding regions of the genome were shown to be major sources of variation. Later, allelic differences in the cis-regulatory regions provided developmental genetic mechanisms of variation (Arthur 2004; Gilbert and Epel 2009): heterochrony (change in the timing of gene expression), heterotopy (change in the cells in which genes are expressed), and heterometry (change in the amount of gene expression). Examples have been found for each of these mechanisms.

There are also environmentally induced components of developmental variation. These include developmental plasticity, developmental symbioses, and epialleles caused by environmentally induced chromatin modification. It is therefore important to determine if these environmental mechanisms produce selectable variation. The selectability of epialleles and developmental plasticity has been discussed by Jablonka and colleagues (Jablonka and Lamb 1995; Jablonka and Raz 2009) and by West-Eberhard (2003, 2005). Thus, in addition to genetic variation, there is also selectable epigenetic variation. This chapter will focus on one of those epigenetic inheritance systems, developmental symbiosis.

Developmental Partnerships

Darwin's idea of the "struggle for existence," in which competition exists between "one individual with another of the same species, or with the individuals of distinct species" (1859) sets up a framework in which each individual is essentially singular, competing only for itself and the survival and propagation of its lineage. But this situation changes if the "individual" is actually a "team" or a "consortium" of cells with different genotypes. Gilbert (2002) referred to this chimeric mode of development as "interspecies epigenesis," emphasizing the developmental roles played by symbionts and the notion that the fertilized egg is not an autopoietic, self-creating, entity. Rosenberg, Koren, Reshef, Efrony, and Zilber-Rosenberg (2007) referred to this phenomenon of variation through symbiosis as "the hologenome theory of evolution." They called the host and its full symbiont population the *holobiont*, and they named the combination of the host genome and the genomes of all its symbiotic organisms the *hologenome*. However, this original hologenome concept did not include development as an aspect, and I would like to expand it to include not only symbiosis but also *symbiopoiesis*—the codevelopment of the holobiont.

More and more, symbiosis appears to be the "rule," not the exception (McFall-Ngai 2002; Saffo 2006; Gilbert and Epel 2009). One well-studied example of developmental symbiosis is the colonies of the bacterium *Vibrio fischeri* residing within the mantle of the Hawaiian bobtail squid, *Euprymna scolopes. Euprymna* prey on shrimp in shallow water, but they run the risk of alerting predatory fish to their presence if the moon casts the squid's shadow onto the seafloor. The bobtail squid deals with this potential threat by emitting light from its underside, thereby hiding its shadow from potential predators. The squid does not accomplish this feat alone;

the presence of *Vibrio fischeri* in the squid's light organ is required to generate the squid's characteristic glow. Both the squid and the bacterium benefit from this mutualistic relationship. The squid gains protection from predators and the bacterium is able to live safely within the host's light organ, an environment free of predators and adverse environmental conditions. More significantly, *Vibrio fischeri* actually constructs the light organ in which it will reside. The newly hatched squid collects bacteria from the seawater. Only members of the species *V. fischeri* are allowed to adhere to the underside of the squid and to induce apoptosis in the tissue that will become the light organ. And only when they have reached a certain density do they begin to emit light (Nyholm and McFall-Ngai 2004; Visick and Ruby 2006).

The development of numerous insects involves obligate symbiosis with bacterial partners. Normal female development in the wasp *Asobara tabida* is dependent on *Wolbachia* infection. If *A. tabida* females are treated with antibiotics that kill their symbiotic bacteria, the ovaries of the female wasps undergo apoptosis, and eggs are not produced (Dedeine, Vavre, Fleury, Loppin, Hochberg, and Boulétreau 2001; Pannebakker, Loppin, Elemans, Humblot, and Vavre 2007). Unlike the squids, which receive their symbionts "horizontally" from the seawater, *A. tabida* infects its juveniles "vertically" through the egg cytoplasm. Thus, *Wolbachia* bacteria become an epigenetically transmitted source of critical developmental signals. Such essential developmental relationships are common throughout the animal kingdom, and they are well known throughout the plant kingdom (McFall-Ngai 2002; Gilbert and Epel 2009). An orchid may produce thousands of seeds, but these seeds have no carbon reserves. Only those seeds that find a fungal partner can get the carbon they need to germinate (Waterman and Bidartondo 2008). Symbiosis is a major player in the evolutionary game. It is not just for lichens.

The host and symbiont species interact in ways that are vital to the proper functioning of both organisms. Disruption of that interaction can lead to illness and death. For example, Mazmanian, Liu, Tzianabos, and Kasper (2005) showed that mice raised without gut microbiota have deficient proliferation of helper T cells, but the introduction of *Bacteroides fragilis* into the gut was enough to stimulate T cell expansion. They were also able to demonstrate that *B. fragilis* protects mice from experimental bowel colitis normally induced by a second symbiotic bacterium, *Helicobacter hepaticus* (Mazmanian, Round, and Kasper 2008). In exchange for these benefits, the mice provide the bacterium with a relatively safe and nutrient-rich environment. This has important medical implications for the health of humans, considering that the human digestive tract harbors over five hundred species of bacteria (Gilbert and Epel 2009).

These symbionts and hosts do not lead independent existences. Rather, each is the cause of the other's development. The *Bacteroides* in the mammalian gut induce the expression of genes in the intestinal epithelium, resulting in the proper development of the mammalian gut, gut vasculature, and host immune system (Hooper, Wong, Thelin, Hansson, Falk, and Gordon 2001; Rhee, Sethupathi, Driks, Lanning, and Knight 2004). *Bacteroides*, for instance, induce gene expression in the intestinal Paneth cells to produce Angiogenin-4 and RegIII. These proteins provide benefits to both the mammalian body and the *Bacteroides*. Angiogenin-4 helps induce blood vessel development in the villi, and both Angiogenin-4 and RegIII are selective bacteriocidal proteins that kill competitors of *Bacteroides*, such as *Listeria*. (Hooper, Stappenbeck, Hong, and Gordon 2003; Cash, Whitman, Benedict, and Hooper 2006). In inducing gene expression in its host's intestinal epithelium, *Bacteroides* does well by doing good. It helps construct its own niche by creating mutually favorable conditions in the gut (see Laland, Odling-Smee, and Gilbert 2008). In return, human intestinal cells instruct the bacteria to produce biofilms, allowing the bacteria to continue residence therein. Thus, as expected in development, there are reciprocal inductions. Only here, they are between different species residing in the same body. Kauffman (1995) famously said that "All evolution is coevolution." The situation may actually be more intimate. Almost all development may be codevelopment. By "codevelopment" I refer to the ability of the cells of one species to assist the normal construction of the body of another species.

Codevelopment

It has been proposed that symbiotic relationships are unstable over evolutionary time, and thus are both rare and evolutionarily transient, because organisms with genotypes that confer advantages to non-kin are at a disadvantage in comparison with organisms with "selfish" genotypes that do not provide other species with such benefits (see Douglas 2007). However, the persistence of symbioses such as the coral–algae symbiosis that evolved about 240 million years ago and continues to this day, indicates otherwise. Most symbiotic relationships involve microorganisms that have fast growth rates and thus can change more rapidly under environmental stresses than invertebrates or vertebrates. Rosenberg, Koren, Reshef, Efrony, and Zilber-Rosenberg (2007) describe four mechanisms by which microorganisms may confer greater adaptive potential to the hologenome than the host genome can alone. First, the relative abundance of microorganisms associated with the host can be changed due to environmental pressure. Second, adaptive variation can result from the introduction of a new symbiont into the community. Third, changes to the microbial genome can occur through recombination or random mutation, and these changes can occur in a microbial symbiont more rapidly than in the host. Fourth, there is the possibility of horizontal gene transfer between members of the holobiont. This possibility was shown to be realized in the symbiosis of humans with different species of gut bacteria (Hehemann et al. 2010).

In a symbiotic relationship, the interactions among partners can affect the evolutionary fitness of both the symbiont and the host. While the genomes of the individual symbionts affect the development of each organism, development of symbiotic species is also regulated by interactions of the symbiont genomes within the holobiont (Gilbert and Epel 2009). This in turn can alter the fitness of the organisms involved in the symbiosis, which would make the symbiotic relationship an integrated evolutionary unit. In this sense, the individual is actually a community of organisms behaving as an ecosystem. In group selection theory, the group is usually treated as an individual. Here, the individual is treated as a group. Nature may be selecting "relationships" rather than individuals or genomes. What we usually consider to be an "individual" is often a multispecies group that is under selection.

If the relationship between symbiotic species is so important, then perhaps the environment selects not only on each species in the relationship but also among variants of the holobiont. The fitness traits would therefore be not merely those of the host but also the traits of the group per se. Therefore, it may prove useful to look at the evolution of a species in the context of the hologenome. This view of evolution would link these hosts and symbionts together as a single coevolving unit, because the fitness of each species would rely on its interactions with the other species in the symbiosis. The two cases presented in this chapter focus on the thermotolerance of hologenomes due to changes in one of the symbiont genomes in the symbiosis. These cases include corals with zooxanthellae and pea aphids with the bacteria *Buchnera aphidicola*.

Selectable Thermotolerance in the Coral and *Symbiodinium* Partnership

Symbiodinium is a genus of photosynthetic endosymbiotic dinoflagellates. The genus comprises multiple species of zooxanthella algae which have been found to inhabit the tissues of scleractinia (stony) coral. These coral are largely dependent on their endosymbionts for survival and in return provide the zooxanthella with protection, nutrients, and a supply of carbon dioxide for photosynthetic products. Under stressful environmental conditions, corals undergo a bleaching event in which they expel or digest their endosymbiont populations, leaving behind a white skeletal structure. Such events have increased in recent decades and are expected to occur more frequently in the near future due to global warming (Hoegh-Guldberg, Mumby, Hooten, Steneck, Greenfield, Gomez, Harvell et al. 2007).

Within the *Symbiodinium* genus there exists a great deal of genetic diversity, and six clades form symbiotic relationships with corals (Baker 2003; Pochon, Montoya-Burgos, Stadelman, and Pawlowski 2006). *Symbiodinium* clades can differ in traits such as thermal tolerance and the photosynthetic response to light (Robinson and Warner 2006). Clade D zooxanthellae, for instance, are less heat sensitive than Clade

C zooxanthellae and can tolerate higher temperatures (Fabricius, Mieog, Colin, Idip, and Van Oppen 2004). Genetically distinct coral colonies can have unique zooxanthellae DNA fingerprints (Goulet and Coffroth 2003), and real-time PCR methods that can detect background symbionts at levels as low as 0.001 percent have shown that most coral colonies harbor multiple strains of *Symbiodinium*. These techniques have shown that coral colonies from four scleractinian species (*Acropora millepora, Acropora tenuis, Stylophora pistillata,* and *Turbinaria reniformis*) previously thought to harbor only a single *Symbiodinium* clade actually harbor multiple strains (Berkelmans and Van Oppen 2006; Mieog, Van Oppen, Cantin, Stam, and Olsen 2007).

The ability of coral hosts to support several different clades of *Symbiodinium* has led to theories of "symbiont shuffling" (Baker 2003; Goulet and Coffroth 2003; Goulet 2006). Here, the resident *Symbiodinium* algae can compete with each other and create a new combination of the coral and zooxanthellae hologenome from strains that are already within the coral. Low-level background symbionts have the ability to outcompete the dominant clade, given the right environment (Baker 2003). With symbiont shuffling, no new symbionts are introduced from the environment. Rather, the environment places selective pressure on the different types of *Symbiodinium* cells already within the coral tissue. Berkelmans and Van Oppen (2006) have shown that such symbiont shuffling can occur in transplanted populations of *A. millepora* in the Great Barrier Reef. The corals originally have a large population of Type C *Symbiodinium* and minor populations of Type D. Once the faster-reproducing Type C symbionts are expelled from the corals during heat stress, the thermally tolerant Type D zooxanthellae are able to dominate in that particular colony of *A. millepora* transplants.

Moreover, when the surviving *A. millepora* population changes the symbiont from Type C to Type D, their thermal tolerance and photosynthetic yields increase appreciably. It is possible that the thermal tolerance of zooxanthellae is due to the stability of the thylakoid or other lipid membranes of their chloroplasts (Berkelmans and Van Oppen 2006). It is hypothesized that the thermally tolerant D strain of zooxanthellae possesses more stable thylakoid membranes that enable it to cope better with rapid rates of global warming (Tchernov, Gorbunov, De Vargas, Narayan Yadav, Milligan, Häggblom, and Falkowski 2004).

Alternatively, other investigators have proposed that "symbiont switching" could be the major way of changing the dominant population of endosymbionts. Symbiont switching is achieved through the elimination and replacement of the dominant clade of *Symbiodinium* by a new strain of endosymbionts from the surrounding environment. The environment selects which cells survive within the body. While the above-mentioned experiments supported the symbiont shuffling hypothesis, a subsequent study by the same researchers provided evidence for symbiont switching among corals that did not appear to contain a minor, more heat- tolerant, population of *Symbiodinium* (Jones, Berkelmans, Van Oppen, Mieog, and Sinclair 2008). This may have important ecological consequences, since symbiont shuffling to more heat-resistant types may not be efficient enough to keep up with global climate change, and it may not be possible for many species. Over the next one hundred years, it is predicted that average tropical sea temperatures will increase by 1–3°C (Berkelmans and Van Oppen 2006). Therefore, in order to adapt to the changing environment, coral colonies would greatly benefit from evolving a method of symbiont shuffling or switching.

Pea Aphids and *Buchnera aphidicola:* Taking the Heat

The pea aphid *Acrythosiphon pisum* and its bacterial symbiont *Buchnera aphidicola* have become a widely accepted model for a mutually obligate symbiosis. That is, neither the aphids nor the bacteria will flourish without their partner. *Buchnera* provides essential amino acids that are absent from the phloem sap diet of the pea aphids (Baumann 2005), and the pea aphids supply nutrients and intracellular niches that permit the *Buchnera* to grow and reproduce (Sabater Muñoz, Van Ham, Martínez Torres, Silva Moreno, Latorre Castillo, and Moya Simarro 2001). Because of this interdependence, aphids are highly constrained to the ecological tolerances of *Buchnera* (Dunbar, Wilson, Ferguson, and Moran 2007). In the field, temperatures ranging from 25° to 30°C result in pea aphids with lower densities of *Buchnera* (Montllor, Maxmen, and Purcell 2002).

A recent study (Dunbar, Wilson, Ferguson, and Moran 2007) showed that heat tolerance of pea aphids and *Buchnera* holobiont could be destroyed with a single nucleotide deletion in the promoter of the *Buchneria ibpA* gene. This microbial gene encodes a small heat shock protein, and the small deletion eliminates the transcriptional response of *ibpA* to heat. *Buchnera* are at least partly able to survive at high temperatures because of constitutive expression of genes that are normally up-regulated in response to heat (Wilcox, Dunbar, Wolfinger, and Moran 2003). It is important to note that secondary symbionts, such as *Serratia symbiotica,* have also been implicated in *A. pisum* response to heat shock (Russell and Moran 2006), suggesting a complex interplay of multiple genomes under thermal stress.

Clones (or "lines") with this deletion can be maintained in the laboratory, and the deletion is present in field populations, suggesting a selective advantage under certain environmental conditions (Dunbar, Wilson, Ferguson, and Moran 2007). Although pea aphids harboring *Buchnera* with the short *ibpA* promoter allele suffer from decreased thermotolerance, they experience increased reproductive rates under cooler temperatures (15°–20°C). Aphid lines containing the short-promoter *Buchnera* produce more nymphs per day during the first six days of reproduction compared with aphid lines containing long-allele *Buchnera*. This trade-off between thermotolerance and fecundity allows the pea aphids and *Buchnera* to diversify. Moreover, the holobiont can survive due to changes in the symbiont's genome. As Rosenberg, Koren, Reshef, Efrony, and Zilber-Rosenberg (2007) have pointed out, advantageous mutations will spread more quickly in bacterial genomes than in host genomes because of the rapid reproductive rates of bacteria. In an environment where heat stress is less common, a mutation that increases the reproductive rates of the host (at the cost of heat tolerance) will provide advantages to both organisms. The pea aphids and *Buchnera* both produce more progeny. Depending on the conditions, the survival of the holobiont depends on the type of *Buchnera* inherited. In this manner, variant *Buchnera* genomes can be thought of as alleles for the larger hologenome. Just as certain alleles in a species population may be more advantageous, so certain genomes may be more advantages for the holobiont. Variation in the symbiont genome may be especially important when the host has limited variability, as in the clonal, parthenogenetic populations of aphids.

Conclusions

The examples in this chapter provide evidence that symbiosis and evolution are not separate phenomena. Evolution shapes and selects for symbiosis, while organisms in symbiotic relationships evolve to accommodate one another. Although there is tension between the needs of the individual organisms and the relationships among the symbionts, symbioses continue to exist, implying that symbiosis increases the overall fitness of the individual species involved. The evidence presented here shows that different symbiont subgroups (either clades or mutations) can be selected and affect the fitness of certain populations of holobionts (i.e., what we have traditionally considered as the large individual). I have tried to document several evolutionary ramifications of widespread symbiotic associations.

First, developmental symbiosis appears to be a widespread phenomenon, found throughout arthropods and vertebrates. It is not relegated to remarkable exceptions, such as lichens or squids. Codevelopment may prove to be the rule, not the exception.

Second, symbionts can provide their hosts with signals for development (as when *Wolbachia* provides anti-apoptotic signals for the wasp ovary or *Bacteroides* induces gene expression in the mammalian gut) and for homeostasis (as in the heat tolerance provided by various symbionts).

Third, such symbioses can provide selectable variation. The symbioses of corals with their dinoflagellates and of aphids with their bacteria indicate that genotypic variants of the symbiont can be selected by the environment.

While we have documented that symbionts can provide selectable epigenetic variation for *homeostatic* functions (i.e., thermotolerance), and we have documented cases of developmental symbioses, we have not documented cases wherein allelic or clade differences in the symbiont population effects the *development* of the host in different ways. However, experiments on mice and wasps are pointing in this direction. When mice with mutations in their leptin genes become obese, their guts contain a 50 percent higher proportion of *Firmicutes* bacteria and a 50 percent reduction in *Bacteroides* bacteria than wild-type mouse guts. Moreover, when the gut symbionts from the leptin-deficient mice were transplanted into genetically wild-type germ-free mice, these mice gained 20 percent more weight than those germ-free mice receiving gut microbes from wild-type mice (Ley et al. 2005; Turnbaugh, Ley, Mahowald, Magrini, Mardis, and Gordon 2006). Thus, there appear to be interactions between the genotype of the host and the types of microbial symbionts that are selected by that host environment. Together, a particular symbiont population and a particular host genotype generate a particular phenotype, in this case, obesity. Sinilarly, Dedeine, Boulétreau, and Vavre (2005) reported that different genotypes of *Asobara* interact differently with *Wolbachia*.

Terrestrial webs of life are predicated on symbioses between plants and their rhizobacterial, endophytic, and mycorrhizal symbionts. As developmental biologists begin appreciating how important symbionts are for animal development, *symbiopoiesis*, rather than *autopoiesis*, appears to predominate. Moreover, the host–symbiont partnership can be a codeveloping, coevolving entity that can be selected by the environment. The symbionts provide an epigenetic source of heritable variation parallel to that of the host genome. They are an acquired inheritance that can produce selectable variation.

Acknowledgments

Many of the ideas in this chapter originated with the members of the 2009 Swarthmore College evolutionary developmental biology seminar. SFG is funded by a grant from the NSF.

References

Arthur W. *Biased Embryos and Evolution.* Cambridge: Cambridge University Press; 2004.

Baker AC. 2003. Flexibility and specificity in coral-algal symbiosis: Diversity, ecology, and biogeography of *Symbiodinium*. Annu Rev Ecol Evol Syst. 34:661–689.

Baumann P. 2005. Biology of bacteriocyte-associated endosymbionts of plant sap-sucking insects. Annu Rev Microbiol. 59: 155–189.

Berkelmans R, Van Oppen MJH. 2006. The role of zooxanthellae in the thermal tolerance of corals: A "nugget of hope" for coral reefs in an era of climate change. Proc R Soc Lond. B273,1599: 2305–2312.

Cash HL, Whitman CV, Benedict CL, Hooper LV. 2006. Symbiotic bacteria direct expression of an intestinal bactericidal lectin. Science 313,5790: 1126–1130.

Darwin C. *On the Origin of Species.* London: John Murray; 1859.

Darwin C. *The Variation of Animals and Plants Under Domestication.* 2 vols. London: John Murray; 1868.

Dedeine F, Boulétreau M, Vavre F. 2005. *Wolbachia* requirement for oogenesis: Occurrence within the genus *Asobara* (Hymenoptera, Braconidae) and evidence for intraspecific variation in *A. tabida.* Heredity 95: 394–400.

Dedeine F, Vavre F, Fleury F, Loppin B, Hochberg ME, Boulétreau, M. 2001. Removing symbiotic *Wolbachia* bacteria specifically inhibits oogenesis in a parasitic wasp. Proc Natl Acad Sci USA. 98: 6247–6252.

Douglas AE. 2007. Conflict, cheats and the persistence of symbioses. New Phytol. 177,4: 849–858.

Dunbar HE, Wilson AC, Ferguson NR, Moran NA. 2007. Aphid thermal tolerance is governed by a point mutation in bacterial symbionts. PLoS Biol. 5,5: e96.

Fabricius KE, Mieog JC, Colin PL, Idip D, Van Oppen MJH. 2004. Identity and diversity of coral endosymbionts (zooxanthellae) from three Palauan reefs with contrasting bleaching, temperature and shading histories. Mol Ecol. 13,8: 2445–2458.

Gilbert SF. 2002. The genome in its ecological context: Philosophical perspectives on interspecies epigenesis. Ann NY Acad Sci. 981: 202–218.

Gilbert SF, Epel D. 2009. *Ecological Developmental Biology*. Sunderland MA: Sinauer Associates; 2009.

Goulet T. 2006. Most corals may not change their symbionts. Mar Ecol Prog Ser. 321: 1–7.

Goulet T, Coffroth MA. 2003. Stability of an octocoral-algal symbiosis over time and space. Mar Ecol Prog Ser. 250: 117–124.

Hehemann J-H, Correc G, Barbeyron T, Helbert W, Czjzek M, Michel G. 2010. Transfer of carbohydrate-active enzymes from marine bacteria to Japanese gut microbiota. Nature 464: 908–912.

Hoegh-Guldberg O, Mumby PJ, Hooten AJ, Steneck RS, Greenfield P, Gomez E, Harvell CD et al. 2007. Coral reefs under rapid climate change and ocean acidification. Science 318,5857: 1737–1742.

Hooper LV, Stappenbeck TS, Hong CV, Gordon JI. 2003. Angiogenins: A new class of microbiocidal proteins involved in innate immunity. Nat Immunol. 4: 269–273.

Hooper LV, Wong MH, Thelin A, Hansson L, Falk PG, Gordon JI. 2001. Molecular analysis of commensal host-microbial relationships in the intestine. Science 291,5505: 881–884.

Jablonka E, Lamb MJ. *Epigenetic Inheritance and Evolution: The Lamarckian Dimension.* Oxford: Oxford University Press; 1995.

Jablonka E, Raz G. 2009. Transgenerational epigenetic inheritance: Prevalence, mechanisms, and implications for the study of heredity and evolution. Q Rev Biol. 84,2: 131–176.

Jones AM, Berkelmans R, Van Oppen MJH, Mieog JC, Sinclair W. 2008. A community change in the algal endosymbionts of a scleractinian coral following a natural bleaching event: Field evidence of acclimatization. Proc R Soc Lond. B275,1641: 1359–1365.

Kauffman SA. *At Home in the Universe: The Search for the Laws of Self-Organization and Complexity.* New York: Oxford University Press; 1995.

Laland KN, Odling-Smee J, Gilbert SF. 2008. EvoDevo and niche construction: Building bridges. J Exp Zoolog B, Mol Dev Evol. 310: 549–566.

Ley RE, Bäckhed F, Turnbaugh P, Lozupone CA, Knight RD, Gordon JI. 2005. Obesity alters gut microbial ecology. Proc Natl Acad Sci USA. 102,31: 11070–11075.

Mazmanian SK, Liu CH, Tzianabos AO, Kasper DL. 2005. An immunomodulatory molecule of symbiotic bacteria directs maturation of the host immune system. Cell 122,1: 107–118.

Mazmanian SK, Round JL, Kasper DL. 2008. A microbial symbiosis factor prevents intestinal inflammatory disease. Nature 453: 620–625.

McFall-Ngai MJ. 2002. Unseen forces: The influence of bacteria on animal development. Dev Biol. 242: 1–14.

Mieog J, Van Oppen M, Cantin N, Stam W, Olsen J. 2007. Real-time PCR reveals a high incidence of *Symbiodinium* clade D at low levels in four scleractinian corals across the Great Barrier Reef: Implications for symbiont shuffling. Coral Reefs 26,3: 449–457.

Montllor CB, Maxmen A, Purcell AH. 2002. Facultative bacterial endosymbionts benefit pea aphids *Acyrthosiphon pisum* under heat stress. Ecol Entomol. 27,2: 189–195.

Nyholm SV, McFall-Ngai MJ. 2004. The winnowing: Establishing the squid-vibrio symbiosis. Nat Rev Microbiol. 2: 632–642.

Pannebakker BA, Loppin B, Elemans CPH, Humblot L, Vavre F. 2007. Parasitic inhibition of cell death facilitates symbiosis. Proc Natl Acad Sci USA. 104,1: 213–215.

Pochon X, Montoya-Burgos JI, Stadelman B, Pawlowsli J. 2006. Molecular phylogeny, evolutionary rates, and divergence timing of the symbiotic dinoflagellate genus *Symbiodinium.* Mol Phylogenet Evol. 38,1: 20–30.

Rhee K-J, Sethupathi P, Driks A, Lanning DK, Knight KL. 2004. Role of commensal bacteria in development of gut-associated lymphoid tissue and preimmune antibody repertoire. J Immunol. 172: 1118–1124.

Robison JD, Warner ME. 2006. Differential impacts of photoacclimation and thermal stress on the photobiology of four different phylotypes of *Symbiodinium* (*Pyrrhophyta*). J Phycol. 42,3: 568–579.

Rosenberg E, Koren O, Reshef L, Efrony R, Zilber-Rosenberg I. 2007. The role of microorganisms in coral health, disease, and evolution. Nat Rev Microbiol. 5: 355–362.

Russell JA, Moran NA. 2006. Costs and benefits of symbiont infection in aphids: Variation among symbionts and across temperatures. Proc R Soc Lond. B273,1586: 603–610.

Sabater Muñoz B, Van Ham RCHJ, Martínez Torres D, Silva Moreno F, Latorre Castillo A, Moya Simarro A. 2001. Molecular evolution of aphids and their primary (*Buchnera sp.*) and secondary endosymbionts: Implications for the role of symbiosis in insect evolution. Interciencia 26,10: 508–512.

Saffo MB. Symbiosis: The way of all life. In: *Life As We Know It.* Seckbach J, ed. New York: Springer; 2006:325–339.

Turnbaugh PJ, Ley RE, Mahowald MA, Magrini V, Mardis ER, Gordon JI. 2006. An obesity-associated gut microbiome with increased capacity for energy harvest. Nature 444: 1027–1031.

Visick KL, Ruby EG. 2006. Vibrio fischeri and its host: It takes two to tango. Curr Opin Microbiol. 9: 1–7.

Waterman RJ, Bidartondo MI. 2008. Deception above, deception below: Linking pollination and mycorrhizal biology of orchids. J Exp Bot. 59,5: 1085–1096.

West-Eberhard MJ. *Developmental Plasticity and Evolution.* Oxford: Oxford University Press; 2003.

West-Eberhard MJ. 2005. Phenotypic accommodation: Adaptive innovation due to developmental plasticity. J Exp Zool (Mol Dev Evol). 304B:610–618.

Wilcox JL, Dunbar HE, Wolfinger RD, Moran NA. 2003. Consequences of reductive evolution for gene expression in an obligate endosymbiont. Mol Microbiol. 48,6: 1491–1500.

IV PHILOSOPHY

ANAT 09

28 Introduction: Lamarckian Problematics in the Philosophy of Biology

Snait B. Gissis and Eva Jablonka

The new developmental orientation in studies of heredity and evolutionary biology were assimilated and sometimes instigated by philosophers of biology. Although with a few notable exceptions philosophers have mentioned Lamarckism rather rarely, the developmental, phenotype-first approach to evolution has been central to much of their work, and they have advanced bold ideas about the relationships among heredity, development, and evolution. The contributors to this part of the volume represent this broad agenda, and several of them pioneered it. Although they approach it in different ways and from different perspectives, they are all informed by the latest developments in molecular, developmental, and evolutionary biology. In order to understand how this philosophical style has developed, it is worth looking at the recent history of the philosophy of biology. Obviously, we cannot supply a comprehensive review of this burgeoning field and its subdivisions, but we would like to highlight a few points of relevance to the discussion in the present volume.

Philosophy of biology is, relatively speaking, a newcomer to modern philosophy of science, and the beginning of its flourishing as an autonomous subfield can be dated to the 1970s. The first-generation practitioners were mainly philosophically oriented biologists, such as Theodosius Dobzhansky, Ernst Mayr, and George Gaylord Simpson, who struggled to establish biology as a science independent from physics and chemistry. They therefore highlighted the unique philosophical problems that biology presents, stressing the historical (evolutionary) nature of biology, the centrality of the notion of function, and the irreducibility of biology to physics and chemistry. Their philosophy of biology was based on the Modern Synthesis and the concepts of fitness, function, and teleology; the possibility of theoretical reductionism dominated their writing. The second generation included a number of influential biologists–philosophers, among them Francisco Ayala, Richard Lewontin, and Steven J. Gould, whose work greatly influenced the next generation of philosopher–biologists. Concepts such as adaptation, units of selection, and hierarchy of levels of selection were introduced, deepened, and criticized through their work. Lewontin's view of the organism as an active agent, interacting with and constructing its environment, is clearly relevant to the discussions in this volume. So was the work of Gould, who, with Niles Eldredge, tackled fundamental evolutionary questions concerning gradual versus saltational (punctuated) evolution. This generation also included philosophers of science with a strong biological orientation, such as Michael Ruse, David Hull, and William Wimsatt. Some of the philosophers worked with innovative biologists: for example, Eliott Sober collaborated with Richard Lewontin and David Sloan Wilson, and Daniel Dennett aligned himself with Richard Dawkins; these interactions led to the philosophical elaboration of biological themes such as levels of selection by Sober and gene-oriented views by Dennett. Several topics became central to this philosophical discussion: fitness; the roles of models, prediction, probability, and selection in evolutionary theories; the role of contingency; the status of the species concept; and the evolution of social behavior. Biological ethics, particularly the scientific, political, and moral standing of sociobiology, which was launched by the biological–theoretical work of Edward O. Wilson, received considerable attention. These topics were often associated with arguments about the possibility or impossibility of genetic reductionism.

Since the 1990s, several issues which are especially relevant to the discussions in this volume have been assuming increasing importance. On the one hand, John Beatty and others have advanced the thesis of evolutionary contingency, and consequently the apparent lack of laws in biology. On the other hand, developmental problems, and questions focused on evolutionary origins and on the nature and evolution of heredity, have begun to assume a central role in the philosophy of biology. While the origin-oriented questions have drawn their inspiration from debates about the origin of life and the theoretical study of evolutionary transitions, the development-oriented questions have been influenced not only by the dramatic rise of developmental molecular biology and computational biology, but also by developmental psychology and the philosophy of mind. The latter disciplines are, not surprisingly, the original home disciplines of several influential third-generation philosophers of biology (e.g., Susan Oyama, Paul Griffiths, Kim Sterelny, and Peter Godfrey-Smith). These third-generation philosophers brought into the philosophy of biology new approaches and new questions, some of which were, at the time they took them up, peripheral to biological research. Many of them challenged the nature–nurture dichotomy by undermining the boundaries between heredity and development, and between development and evolution. They mingled with biologists, and directly affected theoretical work in evolutionary biology.

The winds of pluralism that started to blow in biology in the 1990s also influenced the plurality of methodologies that philosophers of biology began to use. *Experimental philosophy of biology*, which is based on obtaining empirical data in order to find out how key biological–philosophical concepts are understood by particular communities of biologists and non-biologists, and how such concepts are reflected in actual practices, became increasingly popular. The present, fourth generation of philosophers of biology blur disciplinary boundaries by using observational and experimental methodologies (represented in chapters 29 and 30), by employing formal and descriptive models (for example, Samir Okasha's 2006 work on levels of selection and Omri Tal's 2009 studies on heritability), and by conceptual analysis based on intimate knowledge of the biological processes (chapters 31, 32, and 33). Most take for granted the Lamarckian problematics for whose very legitimacy the third generation had to argue, and have begun to scrutinize its philosophical–biological potential.

The way of doing philosophy of biology that characterizes this fourth generation leads to a disciplinary structure that is philosophical–theoretical–biological. As Pigluicci noted, it is a way of doing science through and with philosophy that is "creating a borderline field of interaction between science and philosophy" (2009:559). This is reflected in this volume, in which the philosophy chapters could belong in the biology section, and several of those that are in the biology part could be in with the philosophy.

Studying Developmental Systems: Niche Construction and the Construction of Human Nature

A developmental perspective has profound implications for the philosophy of biology—it is a particular way of thinking about biology, and it impinges upon actual biological practices. Developmental Systems Theory (DST) is a framework for thinking about biological processes that was developed by Susan Oyama, Paul Griffith, and Russel Gray. It goes beyond the dichotomies genes = nature and experience/environment = nurture. DST stresses the many determinants that influence biological processes in general, and behavioral ones in particular (e.g., a bird's parenting behavior). The causally relevant determinants are seen as *developmental resources*. Some of the resources are internal (e.g., particular genes and physiological starting conditions) and others, such as the specific ecological conditions and environmental factors, are external. The DST theorists stress the extensive plasticity of organisms; the fact that the relevant (selective) environment is partially constructed by the organism itself and by other interacting individuals; the importance of ecological inheritance (legacies such as bird nests and beaver dams); and the multiple ways in which developmental resources can be transmitted between generations. Genes have no explanatory priority from the DST point of view: genetic differences can be very important resources in some cases, and irrelevant in others. Evolution is not defined as a change in gene frequencies, but as a change in developmental systems, so genes are one, nonexclusive, source of evolutionary change.

Niche construction—the modification of ecological and geographic features of the environment by living organisms, which then affect the selection regimes that their offspring and other organisms will experience—is clearly an important aspect of DST. In chapter 29 Ayelet Shavit and James Griesemer report on their analysis of the way in which biodiversity researchers conceptualize space when they survey changes in species abundance and distribution. For several years, Shavit and Griesemer closely followed and monitored scientists' survey practices in the field, and showed that they used two different practices representing two different concepts of species locality. The first practice does not consider the niche construction activities of organisms, and uses an exogenous regular grid of latitude and longitude with randomly sampled points on that grid for the location of species. The second practice is one that does use information about niche construction, and the sampling points are based on signs of animal activity, including visible niche-constructing activities. The two practices represent different methodologies with different research commitments, and have important effects on how information is recorded and how it can be used in the future. Shavit and Griesemer show that both methodologies are in fact used by field workers, and that the explicit incorporation of the niche-construction perspective into empirical ecological research, although seemingly more complex than the "objective" method, can in fact clarify, simplify, and make sense of complex patterns of spatial and temporal ecological changes. They argue that if this perspective is not explicitly acknowledged, the possibility of follow-up ecological surveys will be hampered. Their experimental approach to their philosophical subject matter is one good example of the ways in which experimental philosophy is done and the insights that it can lead to.

Paul Griffiths analyzes the notion of "human nature," and more generally the concept of "inner nature" from a DST perspective. His analysis, another fine but very different example of experimental philosophy of biology, is based on questionnaires that expose the ways in which lay people classify and describe different types of behavior. The data reveal how the concept of "inner nature" is constructed, and unravel the problems and ambiguities inherent in it. Griffiths shows that three features—fixity (lack of plasticity), species typicality (all individuals in a species exhibit the trait), and teleology (the assumption that the behavior has a goal, usually that it is adaptive)—are associated with traits that are supposed to manifest the "inner nature" of an animal. People usually believe that if one of these feature is shown to be present (for example, typicality), the other two are also present, although, as he shows, this is very often an invalid assumption. Since, as he claims, the animal is a dynamic developmental system and the environment can modify the organism in ways which can be inherited, the assumption that it is a static entity with clearly delineated "innate" and "acquired" characters reflects deep-seated stereotypes which have to be actively overthrown. Moreover, in humans, socialization, which is a diversity-producing process, is absolutely essential for making humans what they are: an amazingly plastic species. What is important is to understand *the shape of plasticity* and to realize that the resources on which this plasticity depends, especially in the case of humans, are distributed far beyond the individual and its genes. Griffiths suggests that the focus on adaptation, which is typically associated with strong commitment to the view that there are "genetic programs" and with folk notions of inner self, should be replaced by a view that highlights both adaptation and diversity. He argues that homology (similarity based on common ancestry) can provide a conceptual anchor for such a perspective because it enables patterns of similarity and difference in traits—including different patterns of plasticity—between species, between populations, and among individuals to be uncovered.

Models, Mechanisms, and the Significance of Evolutionary Epigenetics

The chapter by James Griesemer looks at another aspect of plasticity: epigenetic inheritance, something that is potentially troubling for population geneticists and evolutionary theorists. The debate about the significance of epigenetic inheritance in biology has moved a long way since the late 1980s. Skepticism about its very existence, and then about its ubiquity, have quietly disappeared, and its importance in medicine and ecology is beginning to be widely recognized. However, even if epigenetic inheritance is acknowledged to be ubiquitous and to have great importance for understanding ontogeny, its direct significance for evolutionary biology is debated and usually resisted by population geneticists. Some of this resistance can be countered by empirical data about the stability of developmentally induced phenotypes and the relation between epigenetic and genetic inheritance, issues that are addressed in the Final Discussion. However, before such data are analyzed, there have to be accepted theoretical frameworks for both their analysis and the evaluation of their significance. Can epigenetic inheritance be fitted into models of population genetics and quantitative genetics, and, if it can, what will the consequences be? Will it have conservative or transformative effects on evolutionary theory?

James Griesemer suggests how such questions can be approached and answered. As a philosopher who is doing philosophy *for* biology rather than the more traditional philosophy *of* biology, he regards his task as to provide new descriptions of phenomena and processes that can organize theoretical work in biology, as well as to expose the assumptions that are used by different biological disciplines and to suggest research heuristics. He points out that the significance of the discoveries about epigenetic inheritance is very different for mechanistically oriented molecular biologists (the MM community) and for quantitative evolutionary biologists (the

QE community). For the MMs, the present-day discoveries about epigenetic inheritance mechanisms are an interesting but not earthshaking extension of their framework. The introduction of epigenetic control does not alter, eliminate, or substitute for other mechanisms the MMs are familiar with; on the contrary, it enriches the framework and uses the same biochemical language. No reframing of old discoveries in terms of the new ones is, at present, necessary. In contrast, for the QEs, recognizing that soft epigenetic inheritance can support evolutionary change requires a revision of their models and has potentially transformative effects. Here the incompatibility is between theoretical perspectives: one that endorses only hard heredity and one that permits soft heredity. This seems to be the reason why the QEs downplay the role of epigenetic inheritance in evolution: they find a host of reasons to argue against the need to incorporate epigenetic inheritance into their models. (Some of their objections, which are all contested, are discussed in the Final Discussion.) QE models that incorporate epigenetic inheritance are therefore exceedingly rare, and only recently has there been some recognition that model construction by QE theorists has to take epigenetic inheritance on board.

Griesemer's analysis clarifies the empirical and theoretical implications that the work on epigenetic inheritance has for these two different research communities. He does this by elucidating MMs' and QEs' theoretical commitments, clarifying what theory reduction means for each community, and suggesting how they may go about incorporating new mechanisms and parameters into their models. As he points out, the incorporation of epigenetic inheritance into QE models can begin in a conservative way. Additions to existing models need first to be explored, and if this additive strategy works, the QE framework is extended but not fundamentally altered (although it may be necessary to reformulate the old models as a special case of the new ones). If, however, the additive strategy fails, different types of models will be needed to accommodate and test the evolutionary effects of the new mechanisms, and such models may require deep structural changes. At present, the conservative or transformative effects on evolutionary theory of the MMs' discoveries about epigenetic control mechanisms are an open question. They depend, as Griesemer argues convincingly, not just on the data found or on the development of new conceptual perspectives, which in turn rely on funding priorities, but also on the evolutionary dynamics that their modeling reveals.

Conceptual Extensions: The Genome as a Developmental System and the Broadening of the Notion of Function

The last two chapters of this section explore some of the conceptual innovations that a focus on Lamarckian problematics illuminates. The chapter by Ehud Lamm extends the notion of a developmental system to the genome, and reverses the usual way that biologists think about the relations between genes and genomes, by arguing that the genes supervene on (depend on, are a manifestation of) genomes rather than the other way around. The last chapter, by Evelyn Fox Keller, extends the concept of function, a concept that is central in the philosophy of biology. Keller discusses the reasons for and repercussions of this extension for another major concept that is traditionally associated with function: the concept of natural selection. Lamm and Keller overturn entrenched intuitions about familiar biological and philosophical concepts, and this has far-reaching ramifications both for theoretical biology and for philosophy.

The usual view of the genome is as an ensemble of genes, with genes being prior to genomes both historically (genes preceded genomes in evolution) and epistemologically (we should explain genomes in terms of genes). Lamm suggests that we reverse this relation when considering ways of explaining biological phenomena. Genomes, he proposes, are prior to genes, and genes can be thought of as manifestations of the physiology of genomes. First, properties of genes such as recessivity and dominance are not intrinsic genic properties, but rather are properties of the developmental system; second, the behavior of chromosomes during asexual and sexual cycles determines how genes are inherited and delineated; and third—and crucially from the point of view developed in the chapter—the genome is a responsive system rather than a passive repository of genes. As Buiatti (chapter 24) pointed out for plants, and as is also known for animals (e.g., cells of the immune system), changes in the organization of DNA are part of ontogenetic development. The mechanisms underlying these developmental genomic changes are part of what Lamm calls GEMs (genomic epigenetic mechanisms), and these include epigenetic control mechanisms that underlie epigenetic inheritance (reviewed in appendix B). GEMs are activated under conditions of ecological or genomic stress, and the genomic changes that they induce are inherited across generations. *Genomic plasticity* is therefore a property not only of individuals but also of lineages, and it is an aspect of the lineage's evolvability. Lamm suggests that the view of the genome as a metastable developmental system that can move between several attractors (these attractors are genomic configuration and activation patterns of GEMs), and whose developmental stability must be actively maintained and controlled, can shed light on problematic biological concepts such as stress and genomic information. The notion of stress becomes focused on organismal responses rather than on stress conditions, and the distinction between genetic and epigenetic information is replaced by a single and more coherent notion of unified genetic–epigenetic (genomic) information. This view of the relation between genes and the genome is the reason for the fundamental problems of defining the gene that Keller and others have pointed to. If, as Lamm suggests, genes are best considered as the physiological manifestations of genomes and are dependent on the genome's large repertoire of plastic behaviors, it is inevitable that the gene is an ill-defined entity.

Another set of conceptual extensions is suggested by Keller, who looks at one of the foundational concepts of the philosophy of biology past and present—the concept of function. The most influential modern notion of function is linked to natural selection: function is seen as the property that originated through, or is maintained by, natural selection. However, argues Keller, this notion of function overlooks the fact that for primitive life to exist, components of the living entity must already have had functions. She proposes that self-organization, growth by accretion, and reciprocal transformation are the platform on which selection for persistence operates and from which function emerges.

Self-organization, as Newman and Bhat's chapter (chapter 16) illustrates, is a fundamental Lamarckian notion. It is crucial for understanding the origins of animal form, and it is also the basis for understanding the origins of adaptive novelties such as those that Braun and David emphasize (chapter 18). Crucially, Keller argues, it is the basis for understanding the origins of life and its growth in complexity, a Lamarckian viewpoint that needs to be couched in twenty-first-century scientific terms. However, it is not sufficient to posit complex nonlinear dynamics that can lead to homeostasis, as Stuart Kaufmann and many others suggest. For understanding how systems with components that can be said to have functions appeared, additional principles have to be added. Keller draws on the ideas of Herbert Simon, who proposed that the compositionality—growth by addition of new simple and homeostatically organized components—is crucial for the formation of complex organized systems. This is a kind of growth in complexity through molecular symbiosis. Such growth by accretion is made possible by the inherent properties of large organic molecules that have stickiness—binding sites that can make molecules stick together to form large complexes. However, such growth is necessary but not sufficient for organized complexity to emerge, and therefore Keller introduces the third component, selective stabilization: selection for stability, for persistence through time. This stabilization is mediated by binding interactions among the molecules that during the binding process change each other (prions are a striking example of such transforming interactions). Such selective stabilization thus draws both on the chemical properties of the molecular complexes and on the self-sustaining networks of interactions that they form. Those molecular complexes that persist longer than others eventually dominate the population of molecular complexes. The structural properties of the molecular components that contribute to persistence of the complex as a whole can be said to have functions: they not only exhibit self-sustaining feedbacks, but their feedbacks also contribute to temporal stability of the containing molecular system. Keller therefore suggests that function is any property of components that contributes to its own maintenance as well as to the persistence of the dynamic containing system. Function is no longer the special domain of living entities (or the domain constructed by living entities). Moreover, natural selection is no longer the organizing concept behind function. Natural selection of already replicating and reproducing entities is a special case of selective stabilization.

Keller applies this analysis to the pre-life era, when chemical evolution reigned; when complexity grew; when matter became "smart" through processes of compositionality, reciprocal transformation, and feedback; and when selection for persistence contributed to the cumulative growth of complexity. It also applies, as she points out, to the evolution of very primitive life before modern cells existed, when molecular interactions among porous protocells that exchanged and shared large molecules, including nucleic acids, were widespread and uncontrolled.

The chapters in this section give a taste of the fruitfulness of a viewpoint that takes Lamarckian problematics seriously. There are many different recent areas in the philosophy of developmental evolutionary biology that are not even mentioned here. This is frustrating, but it also testifies to the productiveness of this way of looking at evolution.

Further Reading

Keller EF. The Century of the Gene. Cambridge, MA: Harvard University Press; 2000.

Keller EF. Making Sense of Life. Cambridge, MA: Harvard University Press; 2002.

Keller EF. 2008. Organisms, machines and thunderstorms: A history of self-organization, part one. Hist Stud Nat Sci. 38,1:45–75.

Keller EF. 2009. Organisms, machines and thunderstorms: A history of self-organization, part two. Complexity, emergence and stable attractors. Hist Stud Nat Sci. 39,1:1–31.

Odling-Smee FJ, Laland KN, Feldman MW. Niche Construction: The Neglected Process in Evolution. Princeton, NJ: Princeton University Press; 2003.

Okasha S. Evolution and the Levels of Selection. New York: Oxford University Press; 2006.

Oyama S, Griffiths PE, Gray RD, eds. Cycles of Contingency: Developmental Systems and Evolution. Cambridge, MA: MIT Press; 2001.

Pigliucci M, Okasha S. 2009. Evolution and the levels of selection. Biol Philos. 24:551–560.

Tal O. 2009. From heritability to probability. Biol Philos. 24:81–105.

29 Mind the Gaps: Why Are Niche Construction Models So Rarely Used?

Ayelet Shavit and James Griesemer

What Are "Niche Construction" and "Ecological Inheritance"?

An eco-evo process of niche construction has been convincingly argued to be an important theoretical possibility that is typically ignored (Odling-Smee, Laland, and Feldman 2003; Laland and Sterelny 2006). We argue here that models of niche construction are ignored, in part, because their acceptance seems to require additional and more complex fieldwork, yet we also demonstrate—by following a survey of species distribution—an unexpected flaw in this heuristic. It turns out that recording and storing data on niche construction is both necessary and practical for studying species distribution.

The term *niche construction* refers to a process by which organisms, through their metabolism, physiology, behavior, and dispersal, causally affect abiotic and biotic features of their environments. By doing so, organisms generate feedback from their environments that may operate on both ecological and evolutionary timescales (Odling-Smee, Laland, and Feldman 2003). Whereas neo-Darwinism typically depicts each organism as a key that has to fit into the environment's lock, niche construction theory sees reciprocal causation that undermines dichotomous thinking, which tears them apart (Lewontin 1983; Laland, Odling-Smee, and Gilbert 2008).

Populations of organisms that construct ecological legacies, each generation anew, out of persisting materials in their environment—such as turtles' nests or gophers' burrows—can modify selection pressures on subsequent generations. Sometimes, modified selection pressures cause the constructing trait to spread in successive populations along with its environmental material. When a niche construction process has evolutionary consequences such that ecological legacies are propagated to offspring generations in parallel with organism legacies (including genes), then *ecological inheritance* has taken place.

If *inheritance* refers to a causal process in which the heritable capacity is realized, and if *development* refers to the acquisition of the capacity to reproduce (Griesemer

2000), then *ecological inheritance* refers to a recurring developmental process that caused an evolutionary consequence. Ecological inheritance is not a mere persistent environmental legacy (such as the sun), but a process of organism–environment ecological interaction with developmental consequences that recurs across different population generations.

Ecological inheritance is not an epiphenomenon of genetic inheritance, but rather a different mechanism constrained by additional factors—such as species and niche distribution in addition to genetic and other factors—and it often transmits selection pressures to other species as well.

For example, pocket gophers (Grinnell 1923) and earthworms (Darwin 1881) aerate the soil. The former construct burrows—an environmental legacy—that propagate across generations of gophers and their predators, modifying not only the landscape but also species distributions and community compositions. The gophers' tendency to construct burrows is not merely a result of being born into a preexisting environment that selects for this trait, but also and more significantly a cause of selection on this trait—and many other traits linked to it—as adaptive in this environment. In such cases, old distinctions between development and inheritance, ecology and evolution, need to be reconfigured.

Problems for Accepting the Phenomenon of Niche Construction

But reconfiguration is hard. For classical neo-Darwinists to admit niche construction and incorporate it into their practices, a detailed mathematical model of this process is required which represents organisms as causes of environmental states and dependence of selection pressures feeding back to organisms on those same environmental parameters. Such a model will need to harness resources from both population genetics and population ecology, yet the model parameters that one discipline traditionally held constant, the other discipline traditionally manipulated. While classical population genetics assumed a constant environment in its simplest models (via assumption of constant fitnesses) and tracked within-population variance across generations, classical population ecology tracked N genetically identical individuals within a population across variable environments (see Roughgarden 1978:311). The problem is, of course, that for modeling niche construction one must track both the variance in an environmental resource across a gradient or pattern and the variance in allele frequency within and among generations, and test for causal linkage of the two.

As to the good news, more than one mathematical model (Odling-Smee, Laland, and Feldman 2003; Hastings, Byers, Crooks, Cuddington, Jones, Lambrinos, Talley et al. 2007; Kylafis and Loreau 2008) is currently available for exactly that task, and the literature has posed no severe objection to the accuracy, testability, or novelty of these models' predictions. It seems the attentive reception that biologists gave the *concept* of niche construction is mostly due to the analytical achievement of formulating such sophisticated models rather than to acceptance of the empirical claim that reciprocal causation between biotic and abiotic elements generates results that the models predict.

Indeed, if one accepts the limited theoretical claim that niche construction is a possible important cause of evolution, one should record it whenever it may empirically occur (compare Griesemer 1998). But recording even the most basic field data of species locality seems practically complicated when that locality is constructed by organisms' activity. Clearly, an *exogenous* grid of lines—latitudes and longitudes conventionally located with respect to the Earth's poles, equator, and Greenwich, England, as prime meridian—plus elevations above or below sea level (at some arbitrary date)—cannot suffice to specify locality. Another, *interactionist*, concept of space which also depends on what the organisms in question do, must also be used in the field. Organism locations studied by scientists are intersections of human and organism actions. These interactions can be located in space exogenously, on a grid, but the localities of the organisms, constructed by organism–environment interaction, are inadequately represented this way. In this sense, space is neither fully imposed by human investigators nor fully independent of them (Shavit and Griesemer, 2010 in press). For recording species localities in the field that are pertinent to niche construction theory one needs, in addition to lat/long coordinates, species microhabitats and data relevant to selection pressures, such as distribution and abundance of ecologically inherited legacies. In the case we discuss (a biodiversity resurvey), we show that to record species localities that are relevant to niche construction theory, evidence of organism–environment interaction also requires a recorded history of research: of concepts, tools, methods, and collecting efforts as these change across time and usage, because they concern the history of human–organism interactions.

If one is satisfied with asking only the same questions that classical evolutionists and ecologists did, it seems rational and practically justified not to invest in more elaborate and detailed fieldwork of this kind. We argue against this practical reasoning, not because it is mistaken in principle but because it does not apply to the case of niche construction. Based on a biodiversity resurvey we followed at Lassen National Park, conducted by the Museum of Vertebrate Zoology at Berkeley, California (Shavit and Griesemer 2009), we will show here that (a) recording phenomena pertinent to niche construction is practically necessary for collecting rigorous data on species locality for biodiversity purposes, and (b) that backgrounding the niche-construction context through which the data were obtained actually adds complications and time-consuming drawbacks for the next rigorous biodiversity survey. In other words, *not* taking account of the phenomena of niche construction in the field is scientifically impractical. If one wishes to continue disregarding niche construction, one will need to explicate flaws in the theoretical models. Appeals to impracticality do not work.

A Case Study

The Museum of Vertebrate Zoology (MVZ) was established in 1908 by the patron and entrepreneur Annie Alexander and the scientific director Joseph Grinnell (Griesemer and Gerson 1993; Stein 2001). Between 1911 and 1929, as part of an intensive and extensive series of surveys across California (Kohler 2006), MVZ researchers collected and recorded small mammals in the Lassen region. This survey took a total of 673 person-days in the field and resulted in 3592 specimens, thousands of field notes and photographs, and a published monograph, *Vertebrate Natural History of a Section of Northern California Through the Lassen Peak Region* (Grinnell, Dixon, and Linsdale 1930). The reason for recording species distribution was a faunal change Grinnell had noticed and attributed to human activity. He envisioned others tracking this change after his lifetime, and hence he intended his museum to produce distribution facts that would be widely useful to researchers of the future, whatever their questions might be:

> In the Lassen section there are under way profound modifications of earlier conditions, modifications due directly to the increasing activity of man—grazing and over-grazing by domestic stock, lumbering, road-building. . . . Facts as to the status of species in the region at a given stage in the general faunal changes engendered by these processes will become significant in their bearing with regard to the conditions reached in the future. (Grinnell, Dixon, and Linsdale 1930:iv)

In 2001, the museum director decided the time had come for these distribution facts to be brought to bear on our understanding of faunal conditions via a resurvey of Grinnell's original sites. Grinnell's vision materialized, but in a technological context he could not have foreseen. The resurvey aimed at data that would be Internet-accessible and *interoperable* with data from other databases,[1] to provide new data on old localities and new questions about old data. The old questions about the Lassen region were explicitly stated:

> What kinds (species and subspecies) of land vertebrates are present within the section arbitrarily outlined; the frequency of observed occurrence and the relative abundance of these kinds; the local or habitat distribution of each kind; the factors which determine the presence and habitat distribution of each kind. (Grinnell, Dixon, and Linsdale 1930:2)

In the Lassen resurvey Grinnell's causal question about factors of distribution was narrowed to a specific causal question on the role of climate in producing faunal change. Grinnell was well aware that it would not be easy to replicate visits to the same localities; hence an elaborate protocol for recording a species locality was developed by the MVZ (Grinnell 1938). The gap between two different ways to regard the organism's environment—as exogenous or interactionist—was reflected in two different modes of recording a locality: in a specimen tag versus a field journal.

The *tag*, a small piece of paper attached to a specimen with its information exactly copied to the collector's catalog and the MVZ's catalog (computer database since the 1970s), named the species, the collector, and the date of collection, and provided one or two sentences giving a *descriptive locality* of the research site where the animal was collected. This text typically encompasses the air distances from the locality to two orthogonal reference points within a hierarchy of geopolitical units (e.g., two miles north of the intersection of highways 37 and 89, Lassen County, California). In the MVZ database, this descriptive locality is accompanied with longitude and latitude coordinates extracted from a handheld GPS (global positioning system) used and calibrated in the field.

A second source of locality information was the *field journal* each researcher kept. These field notes, used today as well, contain detailed descriptions of the path: its landscape, microhabitat and macrohabitat, snow level, signs of animal activity, human land use and collecting effort, trap type, trap number, and the time each trapline was open. The MVZ's tradition of *comprehensive* record taking and trap positioning aimed to reflect all microhabitats in the relevant environments, previous species distributions, dispersal mechanisms, and signs of species-specific niche construction (e.g., runways, clippings, fresh burrows).[2] Grinnell (1923) not only noticed construction activity of pocket gophers (*Thomomys monticola*) but also quantified their effect:

> It was found that the average amount of earth put up in the form of winter cores was, on a selected area, 1.64 pounds per square yard. Assuming that, on the average, gopher workings covered only one-tenth percent of the land surface, there would be 3.6 tons of earth put up per square mile, or 4,132 tons over the whole Park. This is for a single winter! (Grinnell 1923:137)

The resurvey attempted to revisit the same localities via the same field note techniques and, whenever possible, use the same trapping method. Most trapping devices had changed during the past century, but for gophers, Grinnell and his academic descendants used the exact same Macabee gopher trap made in central California for over a century by the same company. A pocket gopher rarely ventures outside its ecologically inherited niche, so the way to collect it is to place a trap underground, which draws the gopher to remove the foreign object from its burrow. Thus, noticing burrows is a necessary part of collecting data on gopher's presence. Here an episode from our short story begins.

On May 17, 1924, Grinnell arrives at 9:30 at a "glacial lake, ice and snow covered, 8500 ft. set—perhaps Lake Helen" (Grinnell's unpublished Field Journal 1924:2344).[3] He points out that here "[t]he U.S.G.S. map is much at fault" (Grinnell's unpublished field journal [1924]: 2344) and further records that on the northern slope "there is winter work of pocket gophers under the snow, as shown by the presence of earth-cores and plugged burrows. I saw no evidence that the leaves are eaten; only the smallest terminal twigs, 2 to 3 inch in diameter are gone."[4] This observation of niche construction was considered a reliable enough indication of species presence at Lassen to be fit for publication (Grinnell, Dixon, and Linsdale 1930:499).

Eighty-three years later, on September 4, 2007, a young—though experienced—field researcher began to work his trapline on the same northern slope from Lake Helen to Lake Emerald. He "saw *no* mammals in the 2 hours it took to set the line + return to my truck. There is much sign this year though: many holes showing signs of recent activity . . . after we set the line (along lake edge, up talus slope, and up into saddle), and we fetch Carla's truck, we set 14 Macabees at fresh signs" (Perrine's unpublished field journal 2007). Indeed, the construction signs were reliable, and two specimens were obtained (catalog numbers 220463, 220464). Their specific locality (Lake Emerald) was stored in the MVZ database, although most other types of data remained in the researcher's journal. Finally, it seems, a rigorous comparison of species distribution between 1924 and 2007 is within reach.

Yet it is exactly the rigorous mood of both survey and resurvey that prevents the latter from using all this information on gopher presence! The resurvey uses a sophisticated statistical tool to analyze data on species occupancy (for details see Moritz, Patton, Conroy, Parra, White, and Beissinger 2008; Shavit and Griesemer 2009). The resurvey's modeling goal, to compare animal presence and absence at the same localities across almost a century of climate change, is validated via statistical control for variance in trapping effort across different eras. In other words, replicating one's trapping effort is part of replicating one's record of species locality. To that end, the present-day researcher uses a standardized trapping protocol, to be applied in the same way for all species in the resurvey. Macabee traps are species-specific, depending upon a specific organism–environment interaction, and hence, alas, not part of the resurvey's standardized trapline. Presence and absence data on pocket gophers and their constructed legacies cannot be used for comparison.

The young MVZ's field researcher endorses rigorous sampling that would be *representative* of a larger area; hence he does not put Macabees on the trapline even as he notices fresh gopher burrows on his path (Shavit 2007 personal observation). The trapline has a standardized structure—forty Sherman traps for smaller animals and ten live Tomahawk traps for bigger ones—so adding Macabees would violate the field protocol he himself has put together.[5] Yet that same researcher also endorses rigorous comprehensive detection of all species that he suspects to be present in this habitat. Hence, after the trapline is set according to protocol, he puts 14 Macabee traps on a nearby saddle, far enough from the trapline not to affect its trapping results yet within the specific locality of Emerald Lake. The researcher did not explicitly intend to acknowledge the phenomenon of niche construction; he simply could not do his job without assuming its importance for the species distributions modeled in the resurvey. It is important to note that recording information of organism–environment interaction does not delay or complicate his fieldwork; it has been standard MVZ procedure for the past century (Sunderland, submitted).[6]

In the museum, the director, the field researcher, and the theoretical population ecologist all agreed on how to conduct the research when writing the grant proposal. At that stage, the comprehensive and representative standards did not seem to conflict, yet when he arrived in the field, the field researcher was suddenly forced to choose between exogenous and interactionist records of species locality. Common ground was lost. For the theoretical ecologist, placing a trap according to your expected result is a sign of circular reasoning. Statistical sampling requires an unvarying sampling procedure with error estimation: one looks in the same way in each spot with the same effort. If an unvarying procedure using a fixed grid has a low probability of detecting certain species—such as the pocket gophers—one should use that same procedure, and either explain the low results or exclude these species from the survey as outliers. However, for the naturalist researcher "grid trapping" these species is biologically meaningless, yet they must be detected and recorded if they are there. Since ecologically inherited burrows attest to gopher presence, these gophers should be detected by other, possibly irregular, effort and method.

The field researcher used an interactionist concept of space to record gopher locality on the scales of a single trap and the whole lake area, while he employed an exogenous, grid-type concept of space for the scale of a trapline. There was no single *locality* and no general solution to merge both practices. The researcher could not avoid scale-dependent, species-dependent, and human-dependent compromise, yet this did not stop the work. We have argued (Shavit and Griesemer 2009) that while there is no genuine conceptual integration to be found, *alternation* between exogenous and interactionist concepts of space was found to be a workable resolution.

The real problems lie elsewhere. We argued previously that the museum's exceptionally rigorous research was based on the heuristic that a single exogenous concept of space suffices to represent a species locality well enough (Shavit and Griesemer 2009). So while Grinnell's observations from 1924, which included evidence pertinent to niche construction, were foregrounded for his peers (Grinnell, Dixon, and Linsdale 1930), the same type of interactionist information was marginalized in the modern museum's database (Shavit and Griesemer 2010 in press). To be sure, the

MVZ eventually made an interactionist view of the data available by workable alternation between its main database and its researchers' field journals, yet given the domination of online databases as the main memory practice in biodiversity (Bowker 2006) and its ongoing pressing problem of interoperability (NRC 1995), it is not certain that current users, let alone the next resurvey, will have access to such information, practically speaking.

We argued (Shavit and Griesemer 2009) that backgrounding an interactionist concept of space comes with a heavy practical price: (a) a local database, such as the Lassen or Yosemite database, is not interoperable, nor is it workable by alternation, with the museum's main database; and since local databases occupy a common and substantial part of current fieldwork, accurately revisiting localities outdoors for the next resurvey will probably require even more time and effort than today; (b) the resurvey's data model—the protocol specifying how and what data are to be recorded in the field—was designed to fit into the main database model,[7] but also designed to be supplied to specific statistical models;[8] hence, using these data in the future for answering different questions and testing different statistical models will likely be more difficult, although collecting data that is hypothesis-neutral was Grinnell's and his descendants' intention; and finally, (c) ecological models of species response to climate change will be more difficult to test, since comprehensive information on the ways in which humans, animals, and environments interacted in the production of data is necessary for combining various locality data sets in any meaningful way. Such information is often backgrounded.

Since the results of MVZ's resurvey are widely recognized (Moritz, Patton, Conroy, Parra, White, and Beissinger 2008), since the MVZ's georeferencing guidelines (Wieczorek, Guo, and Hijmans 2004) are followed by most natural history research museums around the world, and since ignoring the process of niche construction and backgrounding its relevant data has generated practical predicaments in this case, it could generate problems in analogous cases. Therefore the practical rationale that ignoring the phenomenon of niche construction will ease the fieldwork and not affect the model appears to be flawed in general.

Conclusion

In his 1923 paper "The Burrowing Rodents of California as Agents in Soil Formation," Grinnell's very first lines are:

> The interactions between vertebrate animals and their environments are exceedingly variable and far-reaching. To base any conclusion upon a contrary assumption has proven dangerous, for in specific cases such procedure has led people to expend effort and substance not only needlessly but definitely against their own best interests. (1923:137)

In a sense, we have argued that assuming a purely exogenous, noninteractive concept of locality amounts to, or at least entails, the assumption that the ecological and developmental processes contributing to niche construction have no far-reaching effects. This assumption works against the practical interests of those researchers who describe and predict species distributions, for instance, as part of biodiversity research into climate change, as in our case study. In the field, biodiversity researchers cannot establish rigorous protocols to revisit species localities if they assume a locality is not partly constructed by its occupying species (Shavit and Griesemer 2009). The interactive locality assumption, and the kinds of information one needs to collect because of it, have been standard MVZ procedure for over a century, so no extra fieldwork is needed for accepting niche construction. In attempting to predict species localities based on theoretical and model considerations (e.g., due to climate change), many practical complications accompany those who aim to rigorously record and represent species locality in computer databases while backgrounding its interactionist context (Shavit and Griesemer 2009). In short, those who argue against niche construction on practical or heuristic grounds should look elsewhere. In the case we consider, it is actually the *failure* to consider niche construction that undermines research on practical grounds. We conclude that it is essential that biodiversity research *mind* the gaps between practical and theoretical conceptions of organism interactions and species interactions rather than *ignore* or *deny* them.

We have illustrated the gap between practical and theoretical understandings of what a species locality is, and shown that what *locality* refers to in biodiversity research is not specific to those who examine niche construction models but part of a wider, and more basic, ambiguity shared by all biological disciplines that use surveys. Not minding the gaps between two different concepts of space can partly explain why niche construction processes are still so rarely examined in biodiversity research.

Acknowledgment

We would like to thank the MVZ's staff for their scientific integrity, ability and rigorous note taking and digitizing, Eva Jablonka and Snait Gissis for inviting us to their inspiring workshop, and A.S. would like to thank the Israel Science Foundation for its support.

Notes

1. "Interoperability is the ability of two or more systems or components to exchange information and to use the information that has been exchanged" (IEEE 1990: 42).

2. "Write full notes, even at risk of entering much information of apparently little value"(Grinnell 1938: 5; underscore in original). Present-day senior naturalists have expressed similar views (August 25, 2007; May 5, 2008).

3. For Grinnell's field notes, see http://bscit.berkeley.edu/cgi-bin/mvz_volume_query?special=page&scan_directory=v1326_s2§ion_order=2&page=51&orig_query=814375.

4. For Grinnell's field notes, see http://bscit.berkeley.edu/cgi-bin/mvz_volume_query?special=page&scan_directory=v1326_s2§ion_order=2&page=57&orig_query=814375.

5. John Perrine, "Data Fields to Capture for Grinnell Resurvey Project." Unpublished MVZ document, May 5, 2007.

6. We found the same kind of tension—between different practices related to different concepts of space—in another biodiversity survey that tested a type of niche construction model (Shavit and Griesemer, in preparation).

7. Resurvey staff meeting on March 17, 2008 (Shavit, personal observation).

8. In our case study, program *MARK* analyzed the species probability of occupancy in a locality already sampled, and *Maxent* was often used for predicting species distribution.

References

Bowker GC. *Memory Practices in the Sciences.* Cambridge, MA: MIT Press; 2006.

Commission on Physical Sciences, Mathematics, and Applications. *Finding the Forest in the Trees: The Challenge of Combining Diverse Environmental Data.* Washington, DC: National Academies Press (NRC); 1995.

Darwin C. *The Formation of Vegetable Mould through the Action of Worms, with Observations on their Habits*. London: John Murray; 1881.

Griesemer JR. 1998. Commentary: The case for epigenetic inheritance in evolution. J Evol Biol. 11: 193–200.

Griesemer JR. 2000. Development, culture, and the units of inheritance. Philos Sci. 67: S348–S368.

Griesemer JR, Gerson EM. 1993. Collaboration in the Museum of Vertebrate Zoology. J Hist Biol. 26,2: 185–204.

Grinnell J. 1923. The burrowing rodents of California as agents in soil formation. J Mammal. 4:137–149.

Grinnell J. Suggestions as to collecting. Unpublished document, MVZ Archive, located at the MVZ main gallery, top left cabinet, file name Museum Methods—Historical; 1938.

Grinnell J, Dixon J, Linsdale JM. *Vertebrate Natural History of a Section of Northern California Through the Lassen Peak Region.* University of California Publications in Zoology 35. Berkeley: University of California Press; 1930.

Hastings A, Byers JE, Crooks JA, Cuddington K, Jones CG, Lambrinos JG, Talley TS et al. 2007. Ecosystem engineering in space and time. Ecol Lett. 10: 153–164.

IEEE. IEEE Standard Computer Dictionary. New York: Institute of Electrical and Electronics Engineers; 1990.

Kohler RA. *All Creatures: Naturalists, Collectors and Biodiversity 1850–1950.* Princeton, NJ: Princeton University Press; 2006.

Kylafis G, Loreau M. 2008. Ecological and evolutionary consequences of niche construction for its agent. Ecol Lett. 11,10: 1072–1081.

Laland, KN, Sterelny K. 2006. Seven reasons (not) to neglect niche construction. Evolution 60(9): 1751–1762.

Laland KN, Odling-Smee J, Gilbert SF. 2008. EvoDevo and niche construction: Building bridges. J Exp Zool. B310,7:549–566.

Lewontin RC. 1983. The organism as the subject and object of evolution. Scientia 188: 65–82.

Moritz C, Patton JL, Conroy CJ, Parra JL, White GC, Beissinger SR. 2008. Impact of a century of climate change on small-mammal communities in Yosemite National Park, USA. Science 322,5899: 261–264.

Odling-Smee JF, Laland KN, Feldman MW. *Niche Construction: The Neglected Process in Evolution.* Princeton, NJ: Princeton University Press; 2003.

Roughgarden J. *Theory of Population Genetics and Evolutionary Ecology: An Introduction.* New York: Prentice Hall; 1978.

Shavit A, Griesemer JR 2009.There and back again, or, the problem of locality in biodiversity research. Philos Sci. 76,3:273–294.

Shavit A, Griesemer JR. Transforming objects into data: How minute technicalities of recording species location entrench a basic theoretical challenge for biodiversity. In: *Science in the Context of Application.* Nordmann A, Carrier M, Schwartz A, eds. Bielefeld, Germany: Zentrum für Interdisziplinäre Forschung; 2010 in press.

Stein BR. *On Her Own Terms: Annie Montague Alexander and the Rise of Science in the American West.* Berkeley: University of California Press; 2001.

Sunderland M. Teaching Natural History at the Museum of Vertebrate Zoology. Br J Hist Sci, submitted.

Wieczorek J, Guo Q, Hijmans RJ. 2004. The point-radius method for georeferencing locality descriptions and calculating associated uncertainty. Int J Geogr Inf Sci. 2004; 18,8: 745–767.

30 Our Plastic Nature

Paul Griffiths

In one sense the idea of human nature should be uncontroversial. Human nature is simply what human beings are like. If there were no human nature, then the human sciences could study only individuals and particular social groups. But many disciplines study human beings generally, and human society generally, and this is surely reasonable. However, human nature is commonly understood in a second, causal sense. Human nature is something that causes us to have these human characteristics. It is this sense of human nature that has frequently been controversial, and which we need to rethink substantially in the light of current biology.

The first idea that needs to be revised is that human nature is inside us. I will argue that human nature results from the whole organism–environment system that supports human development. There is no special part, such as our blood or our genes, where human nature resides. That is to say, I will take a developmental systems perspective on human nature (Griffiths and Gray 1994; Oyama 1985; Oyama, Griffiths, and Gray 2001).

Another connotation of "human nature" is that it is something we all share: human universals. But a major theme in recent evolutionary biology has been phenotypic plasticity, both as an adaptive phenomenon and as a potential source of evolutionary novelty (Schlichting and Pigliucci 1998; Sultan 1992, 2000; Uller 2008; West-Eberhard 2003). Hence human nature in the causal sense includes the causes of difference as well as of uniformity. Human nature in the descriptive sense includes patterns of human difference, not merely human universals.

Before exploring these two themes I want to outline recent work by a group of collaborators and myself suggesting that people's idea of human nature arose from an intuitive "folkbiology" whose conceptual structure is fundamentally inhospitable to a developmental perspective on biology.

The Folkbiology of Human Nature

The study of "folkbiology" is part of the broader field of cognitive anthropology (Atran 1990; Berlin 1992; Medin and Atran 1999). People everywhere identify at least three general levels of biological classification: a "generic species" category (e.g., cats and oaks), a superordinate category of biological domains (e.g., animals and plants), and a subordinate category of species varieties (e.g., Siamese cats or holm oaks). The generic species level is of particular importance. Membership in a generic species is associated with "psychological essentialism" (Medin and Atran 2004). This is the belief that members of a species share an essence or inner nature which causes them to share the typical properties of that kind.

Psychological essentialism explains two findings. First, adults believe that membership in a species is inherited by descent and is not changed if something changes the observable properties of an individual, even if this change makes the individual more similar to members of another generic species than to members of its original generic species. Second, adults believe that the development of species-typical traits does not depend on environmental influences. If asked to imagine a cow that has been raised by a family of pigs, adults assume that the cow will display the normal bovine trait of mooing instead of oinking like the pigs (Atran, Medin, Lynch, Vapnarsky, Edilberto, and Sousa 2001). Scott Atran has proposed that folkbiology has another core feature: the tendency to explain traits teleologically (Atran 1995). That is, people tend to explain the traits possessed by animals and plants by asserting that these traits have a purpose, but this additional proposal remains controversial.

I have argued elsewhere that terms like "innate," "instinctive," and "human nature" are expressions of this kind of psychological essentialism (Griffiths 2002). They express aspects of a folkbiological (implicit) theory of "animal natures" (Griffiths, Machery, and Linquist 2009). Animal natures are transmitted from parent to offspring. When an individual develops, some of its traits are expressions of this inner nature, while others are imposed by the environment. Human nature as commonly understood is the application of this implicit theory to human beings.

Folkbiological classifications of plants and animals have an important practical role. They provide a framework for inductive inference. Generalizations from one organism to another are made in proportion to the distance between those organisms in the folk taxonomy. The folkbiological distinction between traits that are expressions of an animal's nature and those that are imposed by the environment has a related role. It specifies the range of traits for which inductive inference within a species or larger folk classification is supposed to be reliable. If a trait is an expression of an animal's nature, then it make sense to expect other members of the species to share it. It also provides a source of expectations about heredity. Traits that reflect an animal's nature will be inherited but those imposed on it by the environment will not, or at least not in the short term. It may be that folkbiology allows that an organism's nature can, ultimately, be altered by the environment in which it finds itself. Belief in the inheritance of acquired characteristics was widespread in the history of biology, as other chapters in this volume make clear.

I have suggested that three features are particularly associated with traits that are expressions of the inner nature that organisms inherit from their parents (Griffiths 2002; Griffiths, Machery, and Linquist 2009). These features are the following:

Fixity The trait is hard to change; its development is insensitive to environmental inputs in development; its development appears goal-directed or resistant to perturbation.

Typicality The trait is part of what it is to be an organism *of that kind;* every individual has it, or every individual that is not malformed, or every individual of a certain age, sex, or other natural subcategory.

Teleology This is how the organism is *meant* to develop; to lack the innate trait is to be malformed; environments that disrupt the development of this trait are themselves abnormal.

The three features are described in very general terms because they are supposed to be broad themes that will be expressed differently in different cultures.

Whether a trait has these three features will influence whether it is thought to stem from an animal's inner nature. Conversely, if a trait is thought to stem from an animal's inner nature, it will be expected to have these three features. At a practical level, this means that evidence that a trait has one of the three features will lead to the expectation that it has the others.

Griffiths, Machery, and Linquist (2009) tested this model of the folkbiology of animal natures by examining the effect of the three features on naive subjects' application of the distinction between innate and acquired characteristics. We argued that subjects will label a trait as innate to the extent that they take it to stem from the animal's inner nature rather than being imposed on the animal by its environment. In that case, the three-feature model predicts that judgments of innateness will be influenced by all three features, and that they will influence those judgments independently, since each of the three features is evidence for a further, underlying property, not for one of the other features.

To test these predictions, the three features were systematically covaried in eight vignettes describing the development of birdsong. Birdsong was chosen because it was possible to provide realistic examples of birdsong development manifesting every possible combination of the three features (for details of the experimental designs and data analysis, see Griffiths, Machery, and Linquist 2009). The results were broadly supportive of the three-factor theory. The greatest portion of variance was explained by Fixity, a slightly smaller amount by Typicality, and a very small amount by Teleology. Most significantly, there was no interaction among the three features. Each had an independent effect on whether the song was judged to be an innate trait. A more recent series of studies, which we believe address some deficiencies in the original version, confirmed our previous results, with the exception that the effect of Teleology was not significant (Linquist, Machery, Griffiths, and Stotz in press). They also showed that replacing the term "innate" with other terms commonly used to discuss the issue of innate versus acquired alters the relative significance of the three features. When the term "innate" is replaced with "in its DNA" all three features explain a substantial proportion of the variance, with Teleology the second largest factor.

The folkbiology of animal natures is undoubtedly complex, and it may vary significantly between individuals and between groups. Nevertheless, the three-feature model does seem to capture something of how biologically naive subjects understand animal natures and, by extension, human nature.

How Folkbiology Obstructs Thinking about Development

Scientists focused on behavioral development have long been critical of the distinction between innate and acquired characteristics. At the heart of this critique is the idea that "innateness" confounds several important, but essentially independent, biological properties. There have been many proposed definitions of "innate" which try to pick out just one of these properties (Mameli and Bateson 2006). But these efforts have not been successful because the other connotations of the word are entrenched in everyday usage and inevitably creep back in. The research just described shows that Typicality, Fixity, and Teleology each have an *independent influence* on folk judgments about innateness. This means that people need not know whether a trait is fixed, or has a function, to decide whether it is innate on the basis of evidence about typicality (and vice versa). Thus, if they are told that a trait is typical of a species, people may well infer that it is innate. But all three features are connotations of the term "innate." Having judged that a trait is innate, people are likely to infer that the trait has all three features. The problem is that traits that are Typical are not necessarily Fixed; traits that are Teleological are not necessarily Typical; and so forth.

In order to make this last claim, I need to show that Fixity, Typicality, and Teleology have meaningful interpretations in terms of real biology, so that there is a fact of the matter as to whether they are strongly associated with one another. A meaningful biological interpretation of a feature is some property which manifests the basic conceptual themes of that feature and also makes sense in terms of current biology.

The obvious biological interpretation of Teleology is Darwinian adaptation (Pittendrigh 1958). An organism is malformed if it fails to develop a phenotype which it was designed to develop by natural selection.

The obvious biological interpretation of Fixity is "canalization" (Waddington 1942). A trait is *environmentally* canalized if its development is relatively insensitive to the manipulation of environmental parameters. A trait is *genetically* canalized if its development is relatively insensitive to the manipulation of genetic parameters. Fixity in the current sense corresponds to a high degree of environmental canalization.

The obvious biological interpretation of Typicality is simply being a species-typical characteristic: one for which a scientific description of the species has no reason to mention variation.

Given these interpretations of the three features, I can argue the following:

Adaptation (Teleology) does not imply developmental canalization (Fixity).
Typicality does not imply environmental canalization (Fixity).
Adaptation (Teleology) does not imply Typicality.

Adaptation does not Imply Developmental Canalization

It is easy to suppose that traits which have evolved by natural selection must develop independently of the specifics of the developmental environment. If a trait plays a role in promoting survival and reproduction, surely its development cannot be left to chance? But this overlooks an alternative source of stability. Rather than making development independent of environmental parameters, evolution can make development reliable by stabilizing environmental parameters at the right value or by exploiting preexisting environmental regularities. Development occurs in an "ontogenetic niche" (Stotz 2008; West and King 1987). Birdsong researchers Meredith West, Andrew King, and collaborators have conducted a long-term study of the ontogenetic niche of the Brown-headed Cowbird *Molothrus ater* (West and King 1988; West, King, White, Gros-Louis, and Freed-Brown 2006). Cowbirds are obligate nest parasites (like cuckoos) and do not hear their parents sing. It was therefore assumed that they sing "innately." West and King showed that, instead, among other processes, male song is shaped by feedback from female cowbirds, whose wing stroking and gaping displays in response to the songs are strong reinforcers for males. Female song preferences are themselves socially transmitted. As a result of these and other processes, cowbirds reliably transmit not only species-typical song but also regional song dialects.

There is a strong evolutionary rationale for the existence of the "ontogenetic niche." Natural selection does not select for mechanisms which buffer traits against variation in the environment unless variation of that kind regularly occurs. Hence, organisms become "addicted" to innumerable aspects of their environment (Deacon 1997). For example, the human ascorbic acid synthesis pathway was disabled by mutation during the long period in which our fruit-eating ancestors had no chance of developing vitamin C deficiencies.

Typicality does not Imply Environmental Canalization

One of the functions of the ontogenetic niche is to ensure the reliable development of species-typical traits. The ontogenetic niche is a way of ensuring the transgenerational stability of adaptations without environmental canalization.

Another way to look at the concept of the ontogenetic niche is to consider the organism as a "developmental system" which includes both the traditional starting point of an organism—the egg—and those aspects of the environment which make up its ontogenetic niche (Griffiths and Gray 1994, 2004, 2005; Oyama 1985, 2000; Oyama, Griffiths, and Gray 2001). If we consider the whole developmental system, then many adaptations are, indeed, canalized against perturbations in developmental parameters—both genetic and environmental. The mechanisms that create this canalization extend far beyond the contents of the fertilized egg, let alone the genetic material which it contains: the reliability with which cowbirds develop species-typical song is ensured by mechanisms that include a flock of cowbirds. The burgeoning study of these distributed mechanisms for canalizing normal development has been termed "ecological developmental biology" (Gilbert and Epel 2009).

The concepts of ontogenetic niche and developmental system run counter to a widely shared intuition that the true nature of something is best revealed by removing exogenous influences and allowing it to develop under the influence of endogenous factors alone. When applied to organisms, that intuition is simply wrong (Oyama 1985, 2000). You cannot find out what an ant is "really like" by removing the influence of the rest of the ant colony. These "external influences" are an essential part of the biological nature of the ant. It is these factors that will determine whether it develops into a queen, a worker, or some other caste, to mention only the grossest aspects of its phenotype. It is equally absurd to suppose that the "biological" aspects of human beings can be revealed by removing the perturbing influence of society or culture.

Adaptation does not Imply Typicality

The process of evolution by natural selection has no intrinsic tendency to produce species-typical traits. Some adaptations are monomorphic—all human beings have lost the ability to synthesize vitamin C. Other adaptations are polymorphic—some people can metabolize lactose as adults, and others cannot and are lactose intolerant.

It is hard to see why the prejudice that adaptations will be species-typical persists, given the obvious evolutionary rationale for alternative outcomes. Natural selection will favor monomorphic traits when ecological factors are temporally stable and spatially uniform, and when there is a single winning strategy in evolutionary competition. When selection is frequency-dependent, it will often favor polymorphic outcomes. When ecological factors fluctuate over time, selection may favor polymorphism, so that a lineage can hedge its bets, or phenotypic plasticity. When ecological factors fluctuate across space, selection may favor the emergence of a range of "ecotypes" or the evolution of phenotypic plasticity.

The second function of the ontogenetic niche is to provide the input to mechanisms of phenotypic plasticity. The classical example of phenotypic plasticity is the water flea *Daphnia pulex*, which invests resources in growing a defensive "helmet" if it is exposed to chemical traces of predators, or if its mother has been exposed to those traces (Lüning 1992). The first of these triggers is an example of intragenerational phenotypic plasticity, and the second, of transgenerational phenotypic plasticity, or parental effects (Gluckman, Hanson, Spencer, and Bateson 2005; Mousseau and Fox 1998; Uller 2008).

Human Nature and Developmental Systems

The error at the heart of folkbiology is the idea that some traits are expressions of an internal factor—the nature of the animal—while others result from the action of the environment. If the best biological interpretation of an animal's inner nature is its genetic endowment, then this is false. Both "innate" and "acquired" traits result from the interaction of genes and environment—a truism sometimes known as the "interactionist consensus" (Kitcher 2001). The difference between innate and acquired traits is not a matter of the role of genes in development, but of the shape of the norm of reaction, which is a property of the developmental system. Moreover, "innate" and "acquired" are stereotypes based on a small selection of the many patterns of interaction between genes and environment.

At a still deeper level, the folkbiological conception of the organism is inadequate. It sees development as a goal-directed process toward the adult, and senescence as a departure from that end point. But "An animal is, in fact, a developmental system, and it is these systems, not the mere static adult forms which we conventionally take as typical of the species, which become modified in the course of evolution" (Waddington 1952:155). From an evolutionary point of view an animal is the implementation of a life-history strategy (Griffiths 2009). The real product of evolution is not an adult phenotype but a developmental system that gives rise to a life cycle.

The idea that an organism's genome is something like an inner nature is reinforced by the idea that genes are the only thing an animal inherits from its parents, or at least that the only alternative to genetic heredity is a special phenomenon called "cultural transmission." The environment may be essential, but it cannot be part of the nature of the organism because it is not inherited. This idea is refuted by the ubiquity and importance of mechanisms of "epigenetic inheritance" (Jablonka and Lamb 1995; Jablonka and Szathmáry 1995; Jablonka and Lamb 2005; Jablonka and Raz 2009; chapter 21 in this volume). If an animal's nature is what explains species-typical development, then its nature includes many of the environmental influences with which "nature" has traditionally been contrasted.

A Developmental Perspective on Human Diversity

Evolution designs developmental systems which reproduce themselves via multiple mechanisms, of which genetic heredity is only one. In a species like ours, the developmental system includes socialization and exposure to all those factors that make up a culture. The traditional idea that the "biological nature" of human beings is to be discovered by factoring out the interfering effects of culture is fundamentally misguided. Human cultural variation does not "mask" human nature any more than variation between castes "masks" the true nature of the ant. So the search for a shared human nature cannot be a search for human universals; it must instead be a way to interpret and make sense of human diversity.

Contemporary Evolutionary Psychology recognizes that humans are designed to develop in a social context, and that a phenotype may be both a (phylogenetic) product of evolution and an (ontogenetic) product of a rich context of socialization. The diversity this produces, like the outcome of other forms of phenotypic plasticity, is a legitimate target of evolutionary explanation. However, evolutionary psychologists have restricted themselves unnecessarily with the concept of a "genetic program" (e.g., Barkow, Cosmides, and Tooby 1992). The role of the developmental context is not restricted to activating alternative outcomes which have played a role in the evolution of the mechanism of plasticity (Griffiths and Stotz 2000). Developmental systems are often competent to produce viable phenotypes outside the specific parameter ranges in which they have historically operated. As a result, plastic organisms can generate novel functional phenotypes, as has been dramatically demonstrated in Alexander Badyaev's work on the recent evolution of the North American House Finch *Carpodacus mexicanus* (Badyaev, Hill, Beck, Dervan, and Duckworth 2002; Pennisi 2002).

To do justice to the role of evolution in producing human diversity, we need a principle of classification that embodies two insights. The first is that very different traits can be united by their common origins. The second is that traits can be genuine novelties, despite being built by evolved developmental mechanisms. Fortunately, biology already has just such a principle.

Classification by homology is universal in anatomy, physiology, and neuroanatomy, and the same fundamental principles underlie the classification of genes and of molecular-level structures in the cell (Brigandt and Griffiths 2007). The classification of behavior by homology poses special difficulties, but is a familiar part of behavioral biology. Yet, surprisingly, classification by homology has been neglected as an approach to human cultural diversity. Homology has just the features we need to conceptualize human diversity.

First, homology is a principle of identity through difference. Human facial expressions of emotion are homologous to certain facial expressions in chimpanzees, as was first pointed out by Darwin. But the homologous pairs of expression differ substantially in both form and function. Homology is not a matter of similarity, but of identifying the *corresponding* components of two systems. One important use of homology is as a principled way to use one system as a model for investigating another system. The best model for a human emotion in chimpanzees is the homologous emotion, and this conclusion does not depend on any particular degree of similarity between the two. Exactly the same approach can be applied to diverse phenotypes generated by the human developmental system in different cultural contexts. Hence, homology allows us to appreciate human diversity while seeking to illuminate that diversity in terms of common origins.

Second, recent developments in biology have reinterpreted the relationship between homology and genetics in a way that can act as a model for solving the problematic issue of the interaction of novel developmental environments with the evolved developmental system. The identity—homology—of parts at one level of biological organization is independent of the identity of their constituent parts at a lower level of organization. This realization came about primarily as a result of work in evolutionary developmental biology showing that evolution can conserve a trait while transforming the molecular mechanism that produces it and, conversely, evolution can redeploy existing mechanisms to underpin the development of a novel trait.

Human diversity results primarily from the interaction between the evolved developmental system and a wide range of environments, including novel environments. Exploring this microdiversity in the way that biologists explore the diversity of living forms would mean replacing the question "Is this the same?" with the question "Is this homologous, and at what level of analysis?" For example, the question of whether certain emotions are found in all human cultures has been addressed by asking how similar they are, or by identifying some aspect of similarity and arguing that it is (or is not) essential to the identity of that emotion (Griffiths 1997).

It would be both more meaningful and more tractable to seek to identify the corresponding (homologous) elements of the emotional repertoires of the two cultures, and to determine at what level of analysis claims of homology can be defended. It need not be whole emotional responses that are homologous. A behavioral expression might be homologous although it has now been recruited for a very different function. Or an early phase in the development of two very different emotions might be homologous, reflecting their diversification from some shared precursor.

Comparative biology provides sophisticated ways to think about the commonalities that underlie biological diversity. Bringing order to that diversity is not about identifying universal elements, but about finding order in the pattern of similarity and difference. This order reflects the fact that diverse organisms descend from a common ancestor and also the fact that many developmental mechanisms are shared among organisms and are reused in new contexts. Hence, corresponding parts can be identified in different organisms at various levels of analysis. Taking a biological perspective on human nature can mean treating human diversity in much the same way. It need not be restricted to demonstrating or refuting the existence of human universals.

Conclusions

In place of the understanding of human nature derived from folkbiology I have offered a vision of human nature in which development is at the fore. The primary sense which should be attached to the term "human nature" is simply what human beings are like. I have argued that this means the pattern of similarity and difference among human beings, not some special set of universal characteristics. Humans have a shared nature in the way that vertebrate skeletons have a shared nature. There is structure to their diversity, and a biological perspective, incorporating both development and evolution, can help us identify that structure. Human nature in the other sense—the cause of this pattern—is the human developmental system. This is distributed far beyond the traditional boundaries of the organism in the whole matrix of resources that previous generations bring into existence through multiple mechanisms of heredity. Our nature is plastic, but still entirely amenable to biological analysis.

Acknowledgment

This research was supported under the Australian Research Council's Discovery Projects funding scheme DP0878650. Portions of the text are based on my inaugural professorial lecture at the University of Sydney, published in the journal *Arts* (2009).

References

Atran S. *Cognitive Foundations of Natural History: Towards an Anthropology of Science.* Cambridge and New York: Cambridge University Press; 1990.

Atran S. Causal constraints on categories and categorical constraints on biological reasoning across cultures. In: *Causal Cognition: A Multidisciplinary Debate.*, Sperber D, Premack D, Premack AJ, eds. Oxford: Oxford University Press; 1995:263–265.

Atran S, Medin D, Lynch E, Vaparansky V, Edilberto UE, Sousa P. 2001. Folkbiology doesn't come from folkpsychology: Evidence from Yukatek Maya in cross-cultural perspective. J Cog Cult. 1,1: 3–42.

Badyaev AV, Hill GE, Beck ML, Dervan AA, Duckworth RA. 2002. Sex-biased hatching order and adaptive population divergence in a passerine bird. Science 295,5553: 316–318.

Barkow JH, Cosmides L, Tooby J, eds. *The Adapted Mind: Evolutionary Psychology and the Generation of Culture.* Oxford: Oxford University Press; 1992.

Berlin B. *Ethnobiological Classification: Principles of Classification of Plants and Animals in Traditional Societies.* Princeton, NJ: Princeton University Press; 1992.

Brigandt I, Griffiths PE. 2007. The importance of homology to biology and philosophy. Biol Philos. 22,5: 633–641.

Deacon TW. *The Symbolic Species: The Coevolution of Language and the Brain.* New York: W.W. Norton; 1997.

Gilbert SF, Epel D. *Ecological Developmental Biology: Integrating Epigenetics, Medicine and Evolution.* Sunderland, MA: Sinauer Associates; 2009.

Gluckman PD, Hanson MA, Spencer HG, Bateson PPG. 2005. Environmental influences during development and their later consequences for health and disease: Implications for the interpretation of empirical studies. Proc R Soc Lond. B272,1564: 671–677.

Griffiths PE. *What Emotions Really Are: The Problem of Psychological Categories.* Chicago: University of Chicago Press; 1997.

Griffiths PE. 2002. What is innateness? The Monist 85,1: 70–85.

Griffiths PE. 2009. Reconstructing human nature. Arts: The Journal of the Sydney University Arts Association 31: 30–57.

Griffiths PE. 2009. In what sense does "nothing in biology make sense except in the light of evolution"? Acta Biotheor. 57: 11–32.

Griffiths PE, Gray RD. 1994. Developmental systems and evolutionary explanation. J Philos. XCI,6: 277–304.

Griffiths PE, Gray RD. The developmental systems perspective: Organism-environment systems as units of development and evolution. In: *Phenotypic Integration: Studying the Ecology and Evolution of Complex Phenotypes*. Pigliucci M, Preston K, eds. Oxford and New York: Oxford University Press; 2004:409–431.

Griffiths PE, Gray RD. 2005. Three ways to misunderstand developmental systems theory. Biol Philos. 20: 417–425.

Griffiths PE, Machery E, Linquist S. 2009. The vernacular concept of innateness. Mind Lang. 24,5: 605–630.

Griffiths PE, Stotz K. 2000. How the mind grows: A developmental perspective on the biology of cognition. Synthèse 122: 29–51.

Jablonka E, Lamb MJ. *Epigenetic Inheritance and Evolution: The Lamarckian Dimension.* Oxford: Oxford University Press; 1995.

Jablonka E, Lamb MJ. *Evolution in Four Dimensions: Genetic, Epigenetic, Behavioral, and Symbolic Variation in the History of Life.* Cambridge, MA: MIT Press; 2005.

Jablonka E, Raz G. 2009. Transgenerational epigenetic inheritance: Prevalence, mechanisms, and implications for the study of heredity and evolution. Q Rev Biol. 84,2: 131–176.

Jablonka E, Szathmáry E. 1995. The evolution of information storage and heredity. Trends Ecol Evol. 10: 206–211.

Kitcher P. Battling the undead: How (and how not) to resist genetic determinism. In: *Thinking About Evolution: Historical, Philosophical and Political Perspectives (Festchrift for Richard Lewontin)*. Singh R, Krimbas K, Paul D, Beatty J, eds. Cambridge: Cambridge University Press; 2001:396–414.

Linquist S, Machery E, Griffiths PE, Stotz K. 2011. Exploring the folkbiological conception of human nature. Phil Trans R Soc Lond B. In press.

Lüning J. 1992. Phenotypic plasticity of *Daphnia pulex* in the presence of invertebrate predators: Morphological and life history responses. Oecologia 92: 383–390.

Mameli M, Bateson PPG. 2006. Innateness and the sciences. Biol Philos. 21,2: 155–188.

Medin D, Atran S. 2004. The native mind: Biological categorization and reasoning in development and across cultures. Psychol Rev. 111,4: 960–983.

Medin DL, Atran S, eds. *Folkbiology.* Cambridge, MA: MIT Press; 1999.

Mousseau TA, Fox CW, eds. *Maternal Effects as Adaptations.* New York: Oxford University Press; 1998.

Oyama S. *The Ontogeny of Information: Developmental Systems and Evolution.* Cambridge: Cambridge University Press; 1985.

Oyama S. *Evolution's Eye: A Systems View of the Biology-Culture Divide.* Durham, NC: Duke University Press; 2000.

Oyama S, Griffiths PE, Gray RD, eds. *Cycles of Contingency: Developmental Systems and Evolution.* Cambridge, MA: MIT Press; 2001.

Pennisi E. 2002. Finches adapt rapidly to new homes. Science 295,5553: 249–250.

Pittendrigh CS. Adaptation, natural selection and behavior. In: *Behavior and Evolution*. Roe A, Simpson GG, eds. New Haven, CT: Yale University Press; 1958:390–416.

Schlichting CD, Pigliucci M. *Phenotypic Evolution: A Reaction Norm Perspective.* Sunderland, MA: Sinauer Associates; 1998.

Stotz K. 2008. The ingredients for a postgenomic synthesis of nature and nurture. Philos Psychol. 21,3: 359–381.

Sultan S. 1992. What has survived of Darwin's theory? Phenotypic plasticity and the neo-Darwinian legacy. Evol Trends in Plants 6: 61–70.

Sultan SE. 2000. Phenotypic plasticity for plant development, function and life history. Trends Plant Sci. 5,12: 537–542.

Uller T. 2008. Developmental plasticity and the evolution of parental effects. Trends Ecol Evol. 23,8: 432–438.

Waddington CH. 1942. Canalization of development and the inheritance of acquired characters. Nature 150: 563–565.

Waddington CH. The evolution of developmental systems. In *Proceedings of the Twenty-eighth Meeting of the Australian and New Zealand Association for the Advancement of Science*, Herbert DA, ed. Brisbane, Australia; A.H Tucker Government Printer; 1952: 155–159.

West MJ, King AP. 1987. Settling nature and nurture into an ontogenetic niche. Devel Psychobiol. 20,5: 549–562.

West MJ, King AP. 1988. Female visual displays affect the development of male song in the cowbird. Nature 334,6179: 244–246.

West MJ, King AP, White DJ, Gros-Louis J, Freed-Brown J. 2006. The development of local song preferences in female cowbirds (Molothrus ater): Flock living stimulates learning. Ethology 112,11: 1095–1107.

West-Eberhard MJ. *Developmental Plasticity and Evolution.* Oxford: Oxford University Press; 2003.

31 The Relative Significance of Epigenetic Inheritance in Evolution: Some Philosophical Considerations

James Griesemer

Philosophy *for* Science

The role of epigenetic inheritance in evolution is the subject of a lively debate. Some claim that recently discovered epigenetic mechanisms of gene regulation constitute nongenetic inheritance systems pointing to a "Lamarckian dimension" of inheritance, and therefore of evolution. Others judge epigenetic inheritance to be relatively insignificant, even in principle, due to disanalogies with the genetic system that make epigenetic inheritance implausible as a platform for evolution: unstable states, high mutation rates, horizontal and within-generation-only transmission patterns, and environmental feedback mechanisms contribute to making epigenetically driven evolution appear non–Mendelian, Lamarckian, and probably rare—or, if common, weak. In this chapter, I explore the character and role of relative significance arguments and "heuristic reductionism" in strategies for investment in theoretical epigenetics research. I show that differing theoretical goals and commitments of distinct research specialties with overlapping domains help drive conflicting assessments of relative significance.

Relative significance claims in distinct research specialties about the role of epigenetic inheritance in evolution make presuppositions about the character of relative significance, often entangling epistemological presuppositions about the effective conduct of research with theoretical presuppositions about the causal contribution of various causes to empirical phenomena under study. Surprisingly, most of the relative significance claims in the twenty-year history of the modern controversy over epigenetic inheritance in evolution come from mechanistic molecular (MM) biology rather than from quantitative evolutionary (QE) specialties.[1]

MM and QE sciences identify and manage risks of investing in the study of epigenetic inheritance differently, in part due to very different conceptions of theory construction, theory structure, and formulation of hypotheses and explanations. An asymmetry between the relative significance claims of MM and QE sciences underlies some of the reasons for the resistance of evolutionary biologists to claims of molecular and cell biologists that epigenetic inheritance is, or might be, significant in evolutionary dynamics. I propose that some theoretical changes which look relatively straightforward—conservative "normal science"—from the point of view of MM science turn out to look potentially radical—transformative "revolutionary science"—from the point of view of QE theory.[2] In brief, the discovery of "Lamarckian" inheritance would seem to merely add a novel mechanism to the growing list in molecular biology, while it would take a major shift of conceptualization to incorporate it into neo-Mendelian, neo-Darwinian evolutionary theory. Indeed, to many, "incorporation" would seem nonsensical, a conjunction of contraries, so such a theoretical shift would indeed be revolutionary—a "paradigm shift" in Kuhn's sense.

The argument of this chapter is painted with a broad brush because the prospects are limited for philosophical contributions at the front line of rapidly advancing fields such as epigenetic inheritance. Here, I treat MM specialties as similar with respect to certain broad features of theory structure. I also treat QE specialties as similar in pursuing (or being amenable to) quantitative modeling of dynamical systems. These are caricatures, to be sure, but my goals are modest: to reposition discussion of the controversy amid questions of theoretical risk and research investment, in terms of a rough sketch of two kinds of theory structure in biology. I do not expect to get the details right because they will become different tomorrow anyway.

Philosophy can not only be *of* science, it also can be *for* science (Griesemer 2008). It can contribute to fast-moving scientific fields in three ways: (1) offering new organizing descriptions of phenomena to help articulate scientific research agendas, (2) describing theory and model structures in relevant specialties to clarify differences of measurement and explanatory strategies, and (3) articulating heuristic research strategies operating in the sciences—in the present context, reductionistic research strategies in molecular epigenetics and quantitative evolutionary biology. Here I focus on the second and third projects, as others have ably pursued the first in the case of epigenetic inheritance (e.g., Jablonka and Lamb 1989, 1995, 1998, 2005; Rakyan and Whitelaw 2003; cf. Griesemer 1998a, 1998b, 2000, 2002a, 2010).

In mechanistic molecular and quantitative evolutionary sciences, concepts of inheritance and evolution have different theoretical significance, and reductionism supports different explanatory styles. In MM sciences, reductionism generally adds consideration of lower levels in a hierarchy of material mechanisms to understand the operation of complex wholes, while in QE sciences reductionism generally adds consideration of lower levels to represent the engagement of forces, which usually requires simplification of the collection (and description) of forces used to explain the operation of mechanisms.[3]

A research agenda for epigenetic inheritance is taking shape in quantitative evolutionary theory. An example of how the agenda is beginning to consider just these sorts of issues is the list of questions from a recent workshop at the U.S. National Evolutionary Synthesis Center (2009):

1. What is epigenetics?
2. What methodologies are available to investigate epigenetic variation and inheritance in model systems?
3. How can we assess the frequency of heritable epigenetic processes in natural populations?
4. How do we go from studying epigenetic variation to assessing its ecological relevance?
5. How can we separate genetic from epigenetic effects in natural populations?
6. How do we evaluate the relative importance of epigenetic effects for phenotypic evolution?

These are the sorts of questions that would advance the state of the debate beyond critiques of neo-Darwinism and calls to revive Lamarckism and would begin to formulate quantitative models and testable hypotheses.

A Three-Stage Toy Model of Relative Significance Claims

Cross-specialty controversies about relative significance may be illuminated by a toy model of relative significance claims in three stages of research, even when the phenomena are rapidly moving targets. Mechanistic molecular research into epigenetic inheritance may claim relative significance corresponding to demonstration of (1) the existence of the phenomenon, (2) sufficient causal "power" or "force" to be worthy of note in specific cases, and (3) sufficiently widespread occurrence to be worth testing for in every case. Debates about the relative significance of a novel kind of cause with respect to currently understood causes may thus address the following questions in turn: (1) Is the novel potential cause worthy of research investment even though its *existence* is in doubt or if means of detection and measurement are lacking? (2) Does the novel cause, even if known to exist, produce effects of comparable magnitude to known causes or figure in "important" cases, even though the *competence* of the cause is in doubt? ("Important" cases might include: humans, model organisms, animals, the whole tree of life, the progenote, or other origin-of-life entity.) (3) Does the novel cause potentially contribute widely enough across the domain of interest to be worth tracking, even though the *responsibility* of the cause in nature is in doubt? These historical stages, in which the existence, competence, or responsibility of a potential cause is doubted, correspond to a classical view of the causal basis for all mechanical sciences going back to Newton.[4]

Diverse judgments of relative significance can be in play in each stage about the wisdom of investing in research activities. Of course, research investment judgments are made in risky contexts, else the research would not be worth doing. In concert with diversity of judgments, a plurality of investment decisions may be made in light of a plurality of risk management strategies among researchers and specialties. The process of science, according to the toy model, runs a certain course in order to resolve these different kinds of doubts, and in due course supports different modes of pluralism about causes in each stage. Beatty (1995, 1997) and Mitchell (2003) address problems that center on issues of causal responsibility and causal competence in their concentration on matters of laws in biology and the unity or disunity of science. However, pluralism in the first stage, when doubt about the existence of a phenomenon is in play, is new to the philosophical literature. It emerges from the variety of judgments among scientific specialties or lines of work about investment worthiness of a given *possible* cause or phenomenon.

Serious engagement of epigenetic inheritance with quantitative evolutionary theory is at an early stage, even though it has been heralded for at least twenty years (Jablonka and Lamb 1989). The existence of epigenetic inheritance is by now well established, and debate is moving to the second stage. MM literature on relative significance for *evolution* concerns competence to produce molecular epigenetic inheritance effects, and is not the same problem as competence to generate theoretically significant evolutionary forces. My diagnosis of the current state of epigenetic inheritance theory is that QE scientists have resisted the move from stage 1 to stage 2 implicit in MM scientists' proposal of evolutionary significance because the significance within MM theory of a competent *molecular* mechanism of epigenetic inheritance is substantially different from the significance in QE theory of a competent *evolutionary* mechanism of epigenetic inheritance. What looks like a simple domain extension of hereditary phenomena from an MM point of view, looks like a more radical theoretical transformation for QE.

Toy Models for Two Kinds of Theories

Neo-Darwinian quantitative evolutionary theory is open to a variety of possible mechanisms of heredity. As widely noted, this was a necessity when Darwin framed his theory because so little was known about heredity at the cellular and molecular levels, not to mention the ecological level. A multiplicity of possible mechanisms leads inevitably to relative significance disputes over which mechanisms occur, which are more relevant, more powerful, more widespread, and so on. But which kind of dispute, at which stage(s)?

The variety of modes of resistance to an evolutionary role for epigenetic inheritance (e.g., DNA methylation or histone acetylation) ranges over various doubts about the competence of epigenetic systems to cause stable, heritable, high-fidelity variations that are transmitted to offspring organism generations. Evolutionary theorists (Haig and Westoby 1989; Walsh 1996; Hall 1998; Pál and Hurst 2004) have suggested that epigenetic inheritance systems, in contrast to the genetic system, exhibit the following:

- Limited rather than unlimited heredity
- Transmission through mitosis but not meiosis
- Epigenetic mark resetting in embryogenesis and de novo reestablishment in offspring rather than true transgenerational transmission
- Limited phylogenetic distribution
- High, variable epimutation rates
- Limited distribution of methylation marks in the genome (mainly retrotransposons' silencing).

Moreover, the enzymes involved in molecular epigenetic inheritance systems are themselves gene-encoded. Thus, epigenetic inheritance depends in turn on genetic inheritance, so epigenetic inheritance "reduces to" genetic inheritance one generation back, just as maternal effects reduce to gene expression with transgeneration time lags.[5]

To elaborate on this chapter's main hypothesis: MM research, showing the existence of epigenetic inheritance and the competence to produce epigenetic heritability, appeals to a different notion of competence than does QE theory. Competence of a mechanism to produce cell-level heritability (a key role for epigenetic inheritance in multicellular development) is not the same as competence to produce stable, high heritability necessary for multigenerational response to cumulative selection.

QE science resisted significance claims for epigenetic inheritance in evolution by MM scientists because even if epigenetic inheritance mechanisms exist and are competent to produce the *developmental* phenomena of cell heritability (and other regulatory effects), it does not follow that these mechanisms are competent to support substantial organism-level *evolutionary* forces. The tension arises because, to MM sciences, the change in evolutionary theory looks conservative: Darwinians need only add another mechanism to their list of mechanisms that promote heritability. But in QE theory, the dynamics cannot be modeled just by adding a new parameter for epigenetic heritability.

Consider toy models for MM and QE theories. MM theories are lists of molecular mechanisms (maybe also lists of interactions among mechanisms). MM explanations are causal narratives that string mechanisms together in a particular structure and account for effects by describing conditions that trigger cascading operations of mechanisms. Reductionism means looking for the smallest list adequate to account for phenomena of interest. Mechanisms that exist, but are unnecessary for explanation, count as relatively insignificant.

QE theories are expressions of the time rate of change in a set of variables according to mathematical operations. They are dynamical theories if the variables represent causal forces. The mere list of variables and operations does not constitute the theory. The "stringing together" in a particular mathematical expression is part of the structure of the theory. Explanation can also be understood as a causal narrative, but by filling variables already embedded in a structure with values so as to solve the expression (e.g., for equilibria) under specified conditions. Reductionism here means looking for the most general expression adequate to phenomena of interest. Mechanisms that exist, but whose variable expressions do not alter dynamics very much over cases in the domain or relative to other mechanisms, count as relatively insignificant.

Conservative and Transformative Theoretical Change

A key difference lies in where the structure of causal narratives is located. For MM sciences, that structure is assembled to some extent ad hoc in the act of explaining a phenomenon while, for QE sciences, structure is embedded in the expression of the theory. The adaptive, self-organizing, "autopoietic" character of biomolecular mechanisms has inspired a different approach to theory structure in MM sciences. Particular explanations supply not only the conditions but also the relevant causal structure. Thus the relative significance debate for MM scientists should be about the extent of applicability or the causal contribution of epigenetic inheritance relative to genetic inheritance in evolution, and hence should raise the stakes only in particular causal explanations of particular empirical cases.

In QE science, the theoretical change called for by MM scientists investigating epigenetic inheritance looks like whole theory *transformation* because epigenetic heritability cannot be modeled by anything like the genetic theory of heritability, which depends fundamentally on a (roughly) Mendelian genotype–phenotype map. Merely expanding the list of evolutionary mechanisms producing heritability will not work because the phenotypic nature of heritability entails that inheritance systems interact to produce the phenotype. These interactions cannot be addressed only in particular explanations—they must be embedded as structures in the theory itself, or else the models as well as explanations will be ad hoc. This theoretical difference is key to differing judgments in MM and QE sciences about the relative significance that research investments into epigenetic inheritance would make.

There has been a lot of work in applied quantitative genetics on birth weight, longevity, and age at first reproduction (Falconer 1981). These are the very sorts of phenotypes that are now being shown to be under epigenetic regulatory control (see chapter 23 in this volume). Predictive failure of standard quantitative genetics models in cases where molecular mechanists can demonstrate that epigenetic inheritance operates for the phenotype might get the attention of quantitative evolutionary theorists, just as teratologies and disease conditions represented early candidates for Mendelian explanation, and especially since the mathematical theory of selection has advanced considerably beyond the classical theories of population and quantitative genetics, such as in the Price equation formalism (Price 1970, 1972). To make real progress in QE theory for such cases, the mapping from inheritance mechanisms into phenotypes would have to change, and that is a deep conceptual problem.

Epigenetic Inheritance in Evolution: Conservative or Transformative?

Transformative theory change in MM sciences occurs when additions or specifications to the theory list involve claims of interaction with other mechanisms already on the list, which necessitate revision or elimination of mechanisms previously understood from the list. In molecular epigenetics, working out the details (e.g., of how a methytransferase enzyme causes DNA methylation) is conservative "specification work" that merely provides detail to a mechanism already on the theory list. Adding histone acetylation to the list of epigenetic mechanisms for gene regulation (via chromatin remodeling rather than DNA alteration) is "addition work" (i.e., it further specifies the *theory* in the sense that it adds to it without further specifying other items on the list or altering other list elements, as opposed to specifying a *mechanism* already on the list).

Resolving the way in which histone acetylation interacts with DNA methylation, and the possibility that only the two acting in concert are competent to regulate genes, are potentially transformative for MM theory in the sense that all those causal explanations of how DNA methylation silences genes, and even what methylation is for, would have to be rethought. We can see the transformative effect of discovering new kinds of interactions among mechanisms, in that successful explanatory causal narratives (which string mechanisms together) would have to be "restrung" if the novel interactions dictated different "assembly rules" for narrative explanations. In classical genetics, discovering that DNA is the carrier of genetic information dictated that inheritance narratives assemble mechanisms following the paths described by Crick's central dogma of the flow of genetic information (DNA to RNA to protein).

Griesemer (2010) discusses molecular mechanisms of epigenetic inheritance and cases in which molecular discoveries support speculation about the evolutionary implications and theoretical significance in more detail. Here, I repeat one example of MM science to illustrate how a seemingly modest proposal of a molecular mechanism might have transformative consequences for QE theory.

Rakyan, Preis, Morgan, and Whitelaw (2001) offer a pair of mechanisms for the incomplete erasure and stochastic reestablishment of epigenetic marks in offspring which have the capacity to generate the variably expressive phenotypes observed in crosses of agouti- and yellow-colored mice. Such mechanisms would have evolutionary consequences insofar as the stability of epigenetic marks bears on questions of epigenetic heritability, and thus on the potential for and magnitude of evolutionary response to selection and drift of epigenetic variation.

One mechanism assumes that epigenetic marks are reestablished in the preimplantation embryo (rather than in the primordial germ cells of the parents). To justify their model of the erasure and reestablishment mechanism, the authors cite evidence that maternal and paternal genomes are demethylated at different times in zygotic DNA, "so parent-of-origin-specific erasure does exist" (Rakyan, Preis, Morgan, and Whitelaw 2001:7). The existence of the phenomenon of parent-of-origin-specific erasure licenses invoking it in a model of a mechanism which could generate the incomplete erasure pattern, provided the mechanism is "inefficient," leading to variable expressivity.

Genetic experiments revealing variable "expressivity" or "penetrance" of traits were puzzling to classical geneticists who expected Mendelian mechanisms to yield clean, mathematically explicable results. The fur colors of mice that are genetically identical and raised in similar environments should be the same, but it isn't always so. Studies through the 1970s and 1980s, using classical genetic crossing techniques (e.g., reciprocal crosses between heterozygotes and wild-type outbred mice) attributed the variability to "unlinked modifier genes" (see Rakyan, Preis, Morgan, and Whitelaw 2001:5). This work represented only conservative theory change for both MM and QE sciences in the sense that the ways modifier genes work had been investigated empirically and mathematically by population and evolutionary geneticists, so explanations in both fields merely applied a known kind of interaction, already on the MM list and already accounted for in QE models, to new cases, although their dynamical sufficiency is hard to test. Jablonka and Lamb (1995), in their argument for the significance of epigenetic inheritance in evolution, reviewed many cases of variable expression from classical genetics and argued that the conservative strategy swept the epigenetic phenomena under the rug rather than faced up to the need to transform evolutionary theory.

Emma Whitelaw and colleagues conducted experiments on these same classically understood traits in mice (agouti fur color and kinked tail shape), using inbred strains so that it would be unlikely that modifier gene differences could explain the variations among offspring (Rakyan, Preis, Morgan, and Whitelaw 2001). They found correlations in fur color between parents and offspring (though not as strong as would be expected if the traits were determined by Mendelian genes). They also did transplant experiments to rule out metabolic differences in intrauterine environments of mothers with different fur color as the cause of the variable offspring traits. They found that transplanting fertilized eggs of yellow females into black females did not alter the proportion of fur colors among offspring compared with those gestated by yellow mothers (Rakyan, Preis, Morgan, and Whitelaw 2001:5). Although environmental causes prior to transplantation could not be ruled out, they concluded the inheritance pattern "is likely to be due to an epigenetic modification" (Rakyan, Preis, Morgan, and Whitelaw 2001:6).

However, their interpretation of the results is conservative, because only genetic mechanisms are supposed to account for the trait's heritability. For example, they deny that their mechanisms are Lamarckian. They suppose that epigenetic marks in the parents are erased either in the formation of germ cells or in the preimplantation zygote, and are randomly reestablished at either gamete maturation or early embryogenesis. This occurs either because the *erasure* mechanism is not 100 percent effective in erasing all epigenetic marks or, for genes that are unmarked in the parent, the *reestablishment* mechanism is not 100 percent effective. This incomplete erasure or reestablishment explains the degree of heritability of the trait (Rakyan, Preis, Morgan, and Whitelaw 2001:7). The *randomness* of reestablishment is interpreted to mean that the environment does not direct the production of heritable variation, but rather that variation arises as a result of "random failure to completely erase marks at certain alleles during development" (Rakyan, Preis, Morgan, and Whitelaw 2001:8). The speculation that the mechanism is stochastic is plausible, given that biomolecular mechanisms are generally less than 100 percent efficient, but it introduces a measure of conservatism into considerations about what sort of investment in theoretical evolutionary research would be needed to accommodate epigenetic inheritance mechanisms. If the molecular epigenetic mechanism generates random variation, as does the genetic system, then there would seem to be no need to consider radical, nonrandom Lamarckian alternatives. Rakyan, Preis, Morgan, and Whitelaw (2001) don't close the door completely on Lamarckian epigenetic inheritance but leave the existence of the phenomenon in doubt, thus leaving the licensing of modeling efforts uncertain as well.

I have focused on this 2001 paper because it marks a pivotal moment in the history of molecular epigenetics. The basic phenomena of molecular epigenetics are discoveries of the late 1970s and 1980s; their incorporation into an MM theory is more or less a product of the 1990s, especially with the development of bisulfite sequencing methods and a broadening survey of taxa for methylation (and other epigenetic) mechanisms by application of bioinformatics tools to the growing list of sequenced taxa. In the early 2000s, it appears that epigenetic mechanisms for gene regulation, and especially for the maintenance of differentiated cell types in development, became widely accepted in the molecular community (i.e., reported in textbooks such as Turner 2001; cf. Gilbert and Epel 2009). The potential evolutionary significance of epigenetic inheritance dawned on many in the molecular epigenetics community as the evidence for transmission through meiosis mounted, particularly in *Drosophila*, yeast, mice, and humans (Klar 1998). In 2001, it made sense to evolutionists to remain cautious about evolutionary implications, particularly because the molecular mechanisms were still being specified, but from the molecular point of view, epigenetic inheritance phenomena were fully demonstrated, so evolutionary modeling was justified.

Conservative QE research specifies variables in established models by appealing to mechanisms whose molecular competence, as well as their existence, is demonstrated, in order to establish evolutionary competence and to guide empirical investigation into responsibility in nature. Transformative QE research recognizes new quantities and interactions or new kinds of potential mechanisms that require mathematical expressions with different structure to adequately describe their behavior. It is clear in the example from Rakyan, Preis, Morgan, and Whitelaw (2001) that the MM theory specification is conservative, that Whitelaw and colleagues are sensitive to the potentially transformative effects of their proposed mechanisms on QE theory, and that they describe features of the mechanisms that might avoid transformative QE implications. What is not clear is whether claims about the properties of their proposed mechanisms that assimilate them to genetic inheritance are sufficient to avoid full investigation of alternative QE theories, in order to determine whether the consequences are indeed relatively modest theoretically.

It is a hallmark of relative significance claims in QE theory that the only way to judge a causal factor as relatively insignificant is to model it and show that it cannot be significant (causally competent) even if it exists, and therefore cannot be responsible in nature, even if it sometimes occurs. This is a significant and key difference from MM science: QE theory has to embed a quantity in the theory to judge the relative significance of what it represents. Thus, claims from MM sciences that some mechanism exists and may have evolutionary implications nearly always imposes risky, potentially transformative research burdens on QE theorists. Hence the generally skeptical stance toward calls to move from stage 1 to stage 2.

Conclusion: Reductionism as a Heuristic Research Strategy

If the hypothesis that differing claims about relative significance of epigenetic inheritance in evolution result in part from mismatched judgments about the need for new theoretical work in evolutionary biology is correct, what can be done about it? One answer is to wait a week (such is the pace of this research), and the empirical landscape will be substantially different. This is effective but not theoretically satisfying. A second answer is to look for a general characterization of heritability and quantitative dynamical theory that can accommodate epigenetic inheritance as a possible mechanism for phenotypic heritability. If taken seriously, this answer is likely to be transformative, and that looks really hard. It would be nice, therefore, to have a conservative (not risky) strategy toward QE theory construction that could nevertheless lead to transformative solutions if they turn out to be necessary.

Our problem is one of theory *construction*. Here, a *heuristic* use of reductionism may help (see Griesemer 2000, 2002b). Heuristic reductionism for MM theories might look like this: add a model, m_j, of a novel mechanism to the theory $T = \{M\}$ (a set of models $m_i \in M$), making any necessary adjustments to the specification of existing mechanisms, in order to construct a new version of the theory, T*. Then check to see if the new theory supports causal narratives of familiar phenomena without the involvement of the new mechanism. If that is the case, the old theory reduces (in the classical sense of approximate derivation) to the new theory insofar as the old theory can now be seen as a simplification of the new one (i.e., lacking the new model, m_j).

A similar story applies to QE theories: add a quantity, Q_j, to T to get T*. Then check to see if T* explains (causally narrates) phenomena, holding Q_j constant (or using other ceteris paribus assumptions). If so, T reduces to T* as a simplification (e.g., by taking a limit).

These are heuristic strategies for theory construction. They are reductionistic via additivity assumptions. If such strategies succeed, they are conservative. If they fail, then a nonadditive, transformative change of theory may be adequate. These are risk- "tolerant" strategies which aim to be conservative but can take advantage of failures to facilitate transformation when necessary. Differently put, it's no great loss if it turns out that a new, more adequate theory is not reducible to an old, less adequate theory.

The discovery of epigenetic mechanisms of gene, cell, and organism regulation presents investment trade-offs for molecular biologists and for evolutionary biologists—but not necessarily the *same* trade-offs. Moreover, because the disagreement is built around phenomena of shared interest and interdependent lines of work, relative significance claims and investments in one line of work can have a bearing on the evaluation of relative significance in the other. What one line of work sees as risky another may not, so risk-averse conservatives and risk-tolerant heuristic reductionists may call for patterns of research investment in other lines of work based on differing relative significance assessments. Reconciliation of this kind of "entangled" disagreement between lines of work may be central to interdisciplinary scientific arenas such as evo-devo and epigenetic inheritance. Reconciliation is beyond the scope of this chapter, but by identifying sources of disagreement, perhaps more productive cooperation can ensue.

Acknowledgments

I thank Snait Gissis and Eva Jablonka, organizers of the Transformations of Lamarckism workshop, Tel Aviv University, and the Van Leer Jerusalem Institute, and the workshop participants for their stimulating talks and feedback on my paper (especially Yemima Ben-Menahem and Sam Schweber). I also thank Snait and Eva for their editorial advice. I also thank Elihu Gerson, Chris diTeresi, and Brian Hall for comments; Steve Tonsor, Sue Kalisz, and Rasmus Winther for helpful discussion; and the Center for Philosophy of Science, University of Pittsburgh, as well as the Bay Area Biosystematists, for providing critical audiences for an earlier version. Many of the ideas in this chapter are discussed in Griesemer (2010). I thank Benedikt Hallgrimsson, Brian Hall, and the University of California Press for their permission to use material from that essay here.

Notes

1. Quantitative theories such as those arising from population, quantitative, and evolutionary genetics models of selection, drift, and evolution are not, of course, all there is to the evolutionary sciences. My focus is on those specialties in evolutionary biology which quantify evolutionary change, whether kinetically in terms of rates or dynamically in terms of forces and causes, because these provide one important target of claims about the significance of epigenetic inheritance.

2. The "normal science"/"revolutionary science" distinction is due to Kuhn (1970). I do not intend the contrast between conservative and transformative to evoke all of Kuhn's ideas in his theory of scientific revolutions, especially not his claims of (meaning) "incommensurability."

3. Thus, molecular mechanists seek molecular explanations of cell behavior in their reductionist research strategies while reductionist quantitative evolutionary biologists explain dynamics of populations of organisms in terms of (simplified) descriptions of dynamics of populations of genes. On reductionistic research strategies in general, see Wimsatt (2007). On mechanistic explanation and reductionism, see Bechtel (2006), who argues that mechanistic science is different from the sort of science traditionally described by philosophy of science, which is more aptly applied to explain quantitative physical sciences as well as the population biological sciences.

4. Hodge (1977).

5. This is a species of molecular "Weismannism" which traces all causal determination in development to a specific molecular origin. See Griesemer and Wimsatt (1989) and Griesemer (2005).

References

Beatty J. The evolutionary contingency thesis. In: *Concepts, Theories and Rationality in the Biological Sciences*. Wolters G, Lennox J, eds. Pittsburgh, PA: University of Pittsburgh Press; 1995:45–81.

Beatty J. 1997. Why do biologists argue like they do? Phil Sci. 64,4 (Proceedings):S432–S443.

Bechtel W. *Discovering Cell Mechanisms: The Creation of Modern Cell Biology.* New York: Cambridge University Press; 2006.

Falconer DS. *Introduction to Quantitative Genetics.* 2nd ed. London: Longman; 1981.

Gilbert SF, Epel D. *Ecological Developmental Biology: Integrating Epigenetics, Medicine, and Evolution.* Sunderland MA: Sinauer Associates; 2009.

Griesemer J. 1998a. Commentary: The case for epigenetic inheritance in evolution. J Evol Biol. 11,2: 193–200.

Griesemer J. 1998b. Review: Turning back to go forward. A review of *Epigenetic Inheritance and Evolution, The Lamarckian Dimension.* Biol Philos. 13,1: 103–112.

Griesemer J. Reproduction and the reduction of genetics. In: *The Concept of the Gene in Development and Evolution, Historical and Epistemological Perspectives.* Beurton P, Falk R, Rheinberger H-J, eds. New York: Cambridge University Press; 2000:240–285.

Griesemer J. 2002a. What is "epi" about epigenetics? Ann NY Acad Sci. 981: 97–110.

Griesemer J. Limits of reproduction: A reductionistic research strategy in evolutionary biology. In: *Promises and Limits of Reductionism in the Biomedical Sciences.* Van Regenmortel MHV, Hull D, eds. Chichester, UK: John Wiley and Sons; 2002b:211–231.

Griesemer J. The informational gene and the substantial body: On the generalization of evolutionary theory by abstraction. In: *Idealization XII: Correcting the Model, Idealization and Abstraction in the Sciences.* Jones MR, Cartwright N, eds.Poznan Studies in the Philosophy of the Sciences and the Humanities 86. Amsterdam: Rodopi; 2005:59–115.

Griesemer J. 2008. Philosophy and tinkering. Biol Philos. Accessed December 17, 2009. doi:10.1007/s10539-008-9131-0.

Griesemer J. Reductionism and the relative significance of epigenetic inheritance in evolution. In: *Epigenetics: Linking Genotype and Phenotype in Development and Evolution.* Hallgrimsson B, Hall B, eds. Berkeley: University of California Press; 2010.

Griesemer J, Wimsatt WC. Picturing Weismannism: A case study of conceptual evolution. In: *What the Philosophy of Biology Is: Essays for David Hull.* Ruse M, ed. Dordrecht, Netherlands: Kluwer; 1989:75–137.

Haig D, Westoby M. 1989. Parent-specific gene expression and the triploid endosperm. Am Nat. 134,1: 147–155.

Hall B. 1998. Epigenetics: Regulation not replication. J Evol Biol. 11,2: 201–205.

Hodge MJS. 1977. The structure and strategy of Darwin's "long argument." Br J Hist Sci. 10: 237–246.

Jablonka E, Lamb M. 1989. The inheritance of acquired epigenetic variations. J Theor Biol. 139: 69–83.

Jablonka E, Lamb M. *Epigenetic Inheritance and Evolution.* New York: Oxford University Press; 1995.

Jablonka E, Lamb M. 1998. Epigenetic inheritance in evolution. J Evol Biol. 11: 159–183.

Jablonka E, Lamb M. *Evolution in Four Dimensions: Genetic, Epigenetic, Behavioral, and Symbolic Variations in the History of Life.* Cambridge, MA: MIT Press; 2005.

Klar AJS. 1998. Propagating epigenetic states through meiosis: Where Mendel's gene is more than a DNA moiety. Trends Genet. 14: 299–301.

Kuhn TS. *The Structure of Scientific Revolutions.* 2nd ed. Chicago: University of Chicago Press; 1970.

Mitchell SD. *Biological Complexity and Integrative Pluralism.* New York: Cambridge University Press; 2003.

National Evolutionary Synthesis Center. Catalysis meeting: What role, if any, does heritable epigenetic variation play in phenotypic evolution? 2009. http://www.nescent.org/cal/calendar_detail.php?id=282. Accessed December 17, 2009.

Pál C, Hurst L. Epigenetic inheritance and evolutionary adaptation. In: *Organelles, Genomes and Eukaryotic Phylogeny: An Evolutionary Synthesis in the Age of Genomics.* Hirt RP, Homer DS, eds. London: Taylor and Francis; 2004:353–370.

Price GR. 1970. Selection and covariance. Nature 227: 520–521.

Price GR. 1972. Extension of covariance selection mathematics. Ann Hum Genet. 35: 485–490.

Rakyan V, Preis J, Morgan H, Whitelaw E. 2001. The marks, mechanisms and memory of epigenetic states in mammals. Biochem J. 356,1: 1–10.

Rakyan V, Whitelaw E. 2003. Transgenerational epigenetic inheritance. Curr Biol. 13,1: R6.

Turner BM. *Chromatin and Gene Regulation: Molecular Mechanisms in Epigenetics.* Oxford: Blackwell; 2001.

Walsh JB. 1996. The emperor's new genes. Evolution 50,5: 2115–2118.

Wimsatt WC. *Re-Engineering Philosophy for Limited Beings: Piecewise Approximations to Reality.* Cambridge, MA: Harvard University Press; 2007.

32 The Metastable Genome: A Lamarckian Organ in a Darwinian World?

Ehud Lamm

In the context of a workshop celebrating the bicentennial of Lamarck's *Philosophie zoologique,* a paper focused on the genome may seem overly reductionistic and insensitive to the reciprocal interaction between the organism and its environment. My discussion of the dynamic nature of the genome will illustrate why overly zealous reduction fails even at the level of understanding the genome, and why the genome is a significant developmental unit.

This article is arranged around two general claims and a thought experiment.

I begin by suggesting that the genome should be studied as a developmental system, and that genes supervene on genomes (rather than the other way around). I move on to present a thought experiment that illustrates the implications a dynamic view of the genome has for central concepts in biology, in particular the information content of the genome, and the notion of responses to stress.

The Developing Genome

The genome, understood as the concrete physicochemical *system* carrying hereditary and developmental information—including the DNA, the non-DNA components of the chromosomes, and mechanisms related to them—develops during ontogeny, both as part of the regular cell cycle and through organized responses to various stimuli (Lamm and Jablonka 2008). Thus, the genome and the set of interrelated developmental processes immediately associated with it should be considered a developmental system, not because it encapsulates the genes, nor because it is a physically localized entity, but rather because genomes exhibit recurring developmental processes. Moreover, these processes and related mechanisms (specifically, genomic epigenetic mechanisms or GEMs) share an evolutionary history, and in all likelihood evolved at least partly for their participation in these developmental processes, resulting in a system that reacts in a consistent way to developmental cues of various kinds, genetic cues among them (cf. Griffiths and Gray 1994).

It might seem that by talking about the genome as a delineated developmental system nested inside the developmental system constituting the organism, we implicitly privilege the genetic system in relation to other developmental resources. Individuating the genome as a developmental system, however, does not have to rely on the assumption that genes have a unique role in determining the organism. Developmental systems may be embedded inside one another (Griffiths and Gray 1994: 294–295), and thus considering the genome as a developmental system does not rule out the organism of which the genome is part, as well as more encompassing systems, from being developmental systems as well. These developmental systems can interact and affect the genome.

Viewing the genome as an integrated developmental system is used here as an explanatory strategy rather than an ontological claim. The rest of this chapter attempts to highlight some of the implications of this approach.

Genomes as Prior to Genes

My next explanatory move is to argue that genomes (as active, responsive systems) should be considered as conceptually prior to genes. This claim can be loosely summarized by the claim that *genes supervene on genomes*. A convenient biological summary is that *genes are manifestations of the physiology of genomes.*

The problems associated with the gene concept have been recognized by many (Pearson 2006). As Keller (2000) argues, the gene concept has in all likelihood run its course, and the gene as an explanatory concept combining both biological structure and biological function will have to be replaced with a different explanatory framework. The discussion below will make it apparent that some of the difficulties with the gene concept Keller and others have identified (for example, that genes may be coded in pieces that need to be combined, that the same sequence can result in different proteins due to alternative splicing, and the different ways in which gene regulation can span multiple genes) are the result of mechanisms that depend on genomic context. Indeed, it can be argued that the source of the difficulties in defining the gene and individuating genes is that the definition must acknowledge developmental and genomic context and mechanisms.

The basic strategy for substantiating the claim that genomes should be considered as prior to genes is to show that properties associated with genes are in fact properties of genomes. The classic gene concept articulated the gene as a unit of structure, function, mutation, and recombination. These properties are problematized in different ways, depending on how the relation between genes and genomes is elucidated. I suggest that there are three fundamental ways of interpreting the gene/genome relation, which I call the Waddingtonian, the Mendelian, and the McClintock interpretations.

Dominance and recessivity are two generic properties often attributed to "genes." However, the degree of dominance of an allele is influenced, as Waddington emphasized, by the whole set of genes (Waddington 1961:62). Generic properties of genes are thus seen as the result of (*generic*) properties of the organism's developmental system. They cannot be properties of genes per se, but rather are properties of genes by virtue of genomic properties (indeed, properties of gene networks). It could be argued, however, that recessivity and dominance should indeed not be considered properties of genes per se but rather as properties of organismic development, and that the problems with attributing them to genes rather than to genomes do not generalize to other properties of genes, specifically nonfunctional ones.

The Mendelian interpretation of the relation between genes and genomes, in contrast, focuses on the developmental aspects of the genome, not those of the organism. If Mendelian inheritance is the result of chromosomal organization and developmental dynamics, then the generic properties of the genetic system are seen to be the result of the developmental behavior of the genomic system.

A third way of privileging genomes, which makes explicit the functional significance of their ontogenetic capabilities, is inspired by Barbara McClintock's arguments about genome responses to challenges (McClintock 1984). This suggests that the genome is a reactive system, and hence possibly homeostatic or even teleological.

The basic idea behind the claim that genomes should be considered as explanatorily prior to genes would be that properties of genes either originate from, or need to survive, genomic behavior. Genes, then, are not individuated entities apart from genome mechanics; they are passive, and do not control their own fate after they experience "random mutations." Gene networks and genomes, in contrast, can respond to stimuli, reorganize, and adapt. If we observe "Darwinian genes" that are individuated and for the most part suffer random mutations passively, this should be attributed to the reaction of the genome—or lack of it.

The Dynamic Genome

It is well known that the genome is highly dynamic over various timescales and across wide intra- and interchromosomal distances. Taken together, the wide variety of genomic mechanisms, the timing and timescales of their operation, the scope of the changes, as well as their relationship to gene expression, and hence to development, demand a radical reevaluation of the genome as a static container of genetic information.

The following is a cursory look at some of the types of genomic maintenance mechanisms. It will illustrate some of the activity that is going on in eukaryotic genomes simultaneously with gene expression (transcription), and in addition to DNA replication, the two prominent functions typically attributed to genomes (for further details see appendix B of this volume).

DNA is wrapped around histone proteins, forming the nucleosomes that make up the chromatin. There are several histone variants (e.g., histones appearing only in specific genomic regions), and histones may be in one out of several possible states. These affect the physical conformation of the chromatin, and hence gene expression. This is shown most clearly in the differences between the typically coding-gene rich, and not highly condensed, euchromatin, and gene-poor, highly condensed, heterochromatin. Some histone variants are deposited in a replication-independent manner, and may be evacuated and redeposited due to transcriptional events (Mito, Henikoff, and Henikoff 2007).

Histone modifications, and hence chromatin state, are maintained through cell division (Henikoff, Furuyama, and Ahmad 2004). DNA methylation, an additional epigenetic process which can lead to gene silencing, is also maintained when DNA is replicated. The supposedly gene-poor heterochromatic regions of the genome may have an important functional role enabling various genomic mechanisms (Grewal and Jia 2007; Zuckerkandl and Cavalli 2007). The state of the chromatin is correlated with recombination and DNA repair, and thus "mutation rate" (Prendergast, Campbell, Gilbert, Dunlop, Bickmore, and Semple 2007). More generally, there are mutational hotspots in the genome and regions of genomic instability, whose instability is the result of interaction with other, nonlocal, genomic elements (Aguilera and Gómez-González 2008).

Chromosomes may interact with neighboring chromosomes, and their location in the nucleus and the identity of their neighbors is not random (Fraser and Bickmore 2007). Centromeres and telomeres also turn out to be dynamic. Remarkably, the location of centromeres, which are pivotal for mitosis and meiosis, may be determined by non–DNA-sequence factors, and inherited via a chromatin-based inheritance mechanism, rather than being defined solely by DNA sequence motifs (Henikoff, Ahmad, and Malik 2001). The end regions of linear chromosomes, the telomeres, are actively maintained either by the telomerase ribozyme or by other mechanisms.

RNA-mediated processes are involved in gene regulation by transcriptional repression as well as posttranscriptional gene silencing. They are also involved in genome rearrangement events (Lamm and Jablonka 2008), and RNAi and small RNAs may play a role in the maintenance of heterochromatin (Grewal and Elgin 2007).

It is worth including, in this picture of the dynamic genome, developmentally regulated genome rearrangements that operate on a different timescale than the processes mentioned above. Processes of this type are often induced by environmental and genomic stress conditions (Zufall, Robinson, and Katz 2005; Lamm and Jablonka 2008).

Each of the processes alluded to may or may not be active in different organisms, and each mechanism is the result of the evolution of several enzymes and structures. As could be expected, homologous genomic mechanisms have different functions in different organisms. The existence of such a range of mechanisms operating in the genome, at times necessarily simultaneously, suggests that they may occasionally interact and interfere with one another, whether directly or indirectly. Moreover, an enzyme may play a role in several genomic processes, serving different functions. All this leads to a complicated picture, both ontogenetically and evolutionarily.

The Metastable Genome

The notion of *metastability* may help uncover the conceptual implications of genomic dynamism. A system is said to be in a metastable state when it is in a delicate equilibrium state, likely to switch to a different attractor as a result of even small perturbations. In the case of the genome, the attractors I will consider consist of the configuration of the genome, the operating genomic mechanisms, and possibly the expression pattern of relevant genes. It should be emphasized that the systems I am talking about consume and accumulate energy, and thus the attractors are *not* simply deduced from considerations of minimal energy.

By focusing on metastability and the role of genomic mechanisms, I do not mean to discount various other forms of stability. Indeed, these form the context in which genomic metastability has to be understood. Metastability, however, highlights the fact that the genome is constantly dynamic and in constant danger of unintentional switching from one metastable state to another, and hence that the distance between attractors and the mechanisms maintaining their stability are sources of variation and are subject to evolutionary pressure. Thus, as a thought experiment, I am going to consider genome functioning as being dependent solely on the genomic system occupying and transitioning between attractor states that are metastable. The purpose of the metastable genome (MSG) thought experiment is to highlight the possible evolutionary consequences that might ensue if genomes are indeed metastable, or when their behavior approaches this extreme case. This allows me to explore the conceptual implications of the suggestion that the stability of many genomic states does not stem from structural stability or even simple dynamic stability, but is rather metastability, actively maintained by GEMs, such as those responsible for histone replacement (Mito, Henikoff, and Henikoff 2007) and the RNAi GEMs involved in chromatin maintenance (Grewal and Elgin 2007).

The metastable states (or attractors) of the genome, which underlie the thought experiment, consist of the physical organization of the genome and the activation pattern (quantitative) of the GEMs. Attractors are not, I should stress, merely the expression level of genes in the transcriptome (cf. Bar-Yam, Harmon, and De Bivort 2009). They are, however, less than the complete cellular state.

To remain in the metastable states, in spite of ever present noise, active (energy-consuming) mechanisms are needed. These mechanisms, however, not only maintain stability; they may cause perturbations. This is one reason why the physical stability of DNA molecules is largely irrelevant to the question of genome stability. The MSG thought experiment helps elucidate the implications of current knowledge about the division of labor between the supposedly inert DNA molecules (but see Shapiro 2006) and their surrounding machinery.

The stage is now set to consider how this setup affects our understanding of key biological notions. I discuss two: responses to stress conditions and the notion of genomic information.

Stress Response

Various environmental stresses, such as temperature shocks and pathogen attacks, as well as genomic stress, such as hybridization and polyploidy, can cause anything from small-scale genomic changes to wide-ranging and repeatable genomic repatterning (Lamm and Jablonka 2008). These genomic reorganization events may involve a variety of molecular mechanisms (e.g., RNAi) and activate transposition of transposable elements. Whether the ensuing genomic reorganization is an adaptive response that was selected for remains an open question. How would the MSG perspective account for such phenomena?

There is disagreement about whether similar stress conditions are frequent enough to lead to selection for stress responses. When the genome is thought of as being a metastable system, these issues take on a new shape. Stress, according to this perspective, would be any situation or event that disturbs precariously stable states. If a system is relatively stable, stress would be relatively infrequent, and hence stress responses would be less amenable to selection. If, in contrast, the system occupies metastable states most of the time, and their stability and the transition between them have to be actively maintained by homeostatic mechanisms (such as histone replacement, maintenance of small RNAs, etc.), these mechanisms will experience strong selection pressure. Stress, and consequently the stress response, are according to this perspective a matter of degree; normal conditions and stress are not a priori distinct, nor does stress occur only when some threshold is crossed. Rather, homeostatic mechanisms are constantly under pressure to retain the stability of the attractors, which is precarious.

Since stress is a matter of degree, stress responses will, inter alia, be selected (although this will not, of course, necessarily result in mechanisms that always manage to maintain the stable state or responses that always have adaptive value in the environment). There is no principled distinction to be made between normally occurring selection of regulatory mechanisms and rare selection for stress response.

In the scenario we are considering, the very same mechanisms, and generally the same attractors, are involved in frequently and infrequently occurring conditions. Some states of the genome, which may exist in various external conditions, may, however, be more stable than others and exert less evolutionary influence on GEMs and genome organization. States of the genome whose stability is more likely to be affected by changes in the GEMs will exert a larger effect on the selection pressure of GEMs. Accordingly, the less stable functional states of the genome are likely to have a larger effect on the selection pressure of the GEMs than more stable states employing the same GEMs.

This perspective is in marked contrast to the standard account which focuses on frequency of "stress conditions" rather than stability of genomic states.

If stress is understood to be any situation or event that disturbs precariously stable states, it follows that stress may involve conditions that affect the functioning of the genomic mechanisms as well as conditions that affect the genomic substrate (i.e., chromatin) more directly. Additionally, selection may operate both on the dynamic homeostatic mechanisms *and* on the more stable genomic substrate. This is because changes to either can change the stable states, or attractors, of the system, and a change to either can alter the "distance" between attractors, a crucial factor in maintaining their developmental stability (homeorhesis) and the developmental trajectory of the system as whole. Exploring empirically whether selection can indeed achieve similar results when variation in only one of these is permitted, is an intriguing possibility.

The observation that selection may operate both on the dynamic homeostatic mechanisms and on the more stable substrate of the genome, in order to establish the genomic attractors and the developmental trajectories through them, is important when considering coordinated or large-scale genomic reorganization in response to various stresses (see, for example, chapters 24 and 25 in this volume). Hybridization and polyploidy involve a change in the relatively inert "substrate" upon which the homeostatic mechanisms operate (i.e., the chromatin). Hybridization and polyploidy thus create new attractors (e.g., by creating new target sites for mechanisms or sinks that cause depletion of various enzymes) and possibly change distances between stable states. This may create new stable states "between" two existing states, making the transition fraught with the possibility of accidentally ending up in the new state.

Similar types of genomic outcomes resulting from different types of disruptions become easier to explain with the aid of the MSG framework: The new attractor depends not only on the direction of disruption and homeostatic mechanisms stabilizing the new state, but also on the topography of the attractor space. It is thus probable that different types of disruptions will end up leading to the same attractor. In the metastable scenario, most perturbations cause the system to move between attractors, making this possibility more likely.

It is critical to note that the effect of the physical organization of the genome, a fundamental factor shaping the attractor space, depends on the functional GEMs in each particular genome. One significant reason for this is that various perturbations to the genomic substrate cause various repair mechanisms to be activated—and these affect the subsequent organization of the genome. For example, the distribution of homologous sequences in the genome is particularly important for genomes that repair double-stranded breaks using homologous recombination rather than nonhomologous end-joining (Argueso, Westmoreland, Mieczkowski, Gawel, Petes, and Resnick 2008; Scheifele and Boeke 2008). Naturally, these factors become more significant the more frequent the triggering events are (Zahradka, Slade, Bailone, Sommer, Averbeck, Petranovic, Lindner et al. 2006).

For understanding the response to stress it is therefore not enough merely to consider the frequency and repeatability of the stress conditions, since the organization of the system can affect the *effective repeatability* of stress (or conditions more generally); it is not the "external" conditions, but internal ones, that affect the response.

Genomic Information

Heredity, as well as developmental regulation, is often explained in terms of information. The MSG perspective has significant implications for what might be termed "genomic information." This can be thought of as analogous to the notion of genetic information, which is typically understood as referring to the coded information in the DNA sequence.

The functional effect of the genome depends on the attractor state which the genome occupies, since this determines which regions are expressible, which genomic processes can occur, where they can occur, and so on. The basic argument relating the MSG scenario and genetic information is straightforward: In an MSG, the functional information *is* the attractor, and if these attractors depend on GEMs, then genomic information is dependent on GEMs. Ergo, epigenetic mechanisms *underlie* genetic information. This argument undermines the traditional account of epigenetic phenomena as an information or inheritance system operating "on top of" the genetic system.

The genome, according to this picture, is not only, as traditionally understood, a *message*, but also a *receiver* involved in the interpretation of messages (cf. Jablonka

2002). Putting this idea more concretely, the possible interpretations of genetic messages—and hence the set of possible messages that can be encoded—are constrained by the fact that the incorporation of a message (i.e., sequence) into the genome potentially affects the functioning of the genomic system.

This analysis leads to the following conclusions about genomic information: (1) The physical organization, and not only the nucleotide sequence, is relevant for assessing the information content, as it affects the GEMs and, hence, the identity of stable states and their effective proximity to one another. (2) Information is holistic because attractors are typically influenced by nonlocal events in the genome (e.g., a change in conformation can affect multiple loci). The holism is both structural and dynamical. The appearance of individuated elements, such as genes, calls for an evolutionary explanation. (3) The genetic information system and genome-related epigenetic phenomena should be understood as a single information system, not as separate information channels (Lamm 2009; cf. Jablonka 2002). The present argument does not rest on an in-principle objection to the possibility of distinguishing inheritance channels in general, but rather provides a specific reason why the division between the genetic inheritance channel and genomic epigenetic inheritance should be rejected (cf. Griffiths and Knight 1998; Wimsatt and Griesemer 2007). This argument also means that the genomic inheritance system may inherit properties (or limitations) of the epigenetic inheritance system, leading, for example, to restrictions on the modularity of variation of genomic attractors (cf. Jablonka 2002). (4) Information content is relative to a specific environment (external or internal), and life strategy: Information depends on GEMs whose activation depends on environmental stimuli. For instance, the identity of the active GEMs may determine if noncoding DNA sequences have a functional role (Zahradka, Slade, Bailone, Sommer, Averbeck, Petranovic, Lindner et al. 2006).

Conclusions

The metastable genome suggests that the view of genomic homeostasis, inspired by McClintock's suggestions, is a productive way to conceptualize the genome. The genome, according to this picture, functions by virtue of being in (metastable) attractor states that consist of the configuration of genome, the operating genomic mechanisms, and the expression pattern of relevant genes. The attractors are maintained (and perturbed) by genomic mechanisms as well as by external forces. This has functional consequences, and affects and constrains genomic variation, resulting in what might be termed "genomic inherency" (cf. Müller and Newman 2005).

Understanding the transition between attractors (e.g., in development) requires attention to the topography of the attractor space, namely, to how close stable genomic states are to one another. The analysis may also need to appeal to the *ecology* of the genome, in the sense of considering the distribution of elements (e.g., transposable elements) rather than their specific location or exact copy number, since their distribution may affect the attractor space. Elements may be "useful without being indispensable" and "non-specific in their activity" (Darlington, quoted in Vanderlyn 1949). A set of stable genome states, nonidentical in sequence, epigenetic marks, or pattern of GEM activity, may nonetheless be functionally equivalent. These equivalence classes may be significant units of genomic analysis.

The systemic, reactive properties of genomes suggest that "neo-Darwinian genes" that are manifested in some, though not necessarily all, circumstances, and hereditary variation more generally, have to be explained by appealing to the reactive nature of the system and the propensities that result from its organization. Neo-Darwinian genetic elements, in this picture, are an outcome of the behavior of a reactive organized system. Mutation-inducing events should not be considered as random on the basis of a priori reasoning, since the type of randomness experienced by a system can be judged only with respect to its organization. Ironically, "Darwinian" systems biology may be "Lamarckian" in this sense, due to its attention to system-level properties.

Acknowledgments

The work presented in this chapter is part of the author's Ph.D. research, which was supported by the Cohn Institute for the History and Philosophy of Science and Ideas and the School of Philosophy at Tel Aviv University. I thank the participants of the Annual International Workshop on History and Philosophy of Science, Transformations of Lamarckism: 19th–21st Centuries, for their helpful comments, the editors of this volume, and Joel Velasco for comments on earlier drafts of this chapter.

References

Aguilera A, Gómez-González B. 2008. Genome instability: A mechanistic view of its causes and consequences. Nat Rev Genet. 9: 204–217.

Argueso JL, Westmoreland J, Mieczkowski PA, Gawel M, Petes TD, Resnick MA. 2008. Double-strand breaks associated with repetitive DNA can reshape the genome. Proc Natl Acad Sci USA. 105,33: 11845–11850.

Bar-Yam Y, Harmon D, De Bivort B. 2009. Systems biology: Attractors and democratic dynamics. Science 323,5917: 1016–1017.

Fraser P, Bickmore W. 2007. Nuclear organization of the genome and the potential for gene regulation. Nature 447: 413–417.

Grewal SIS, Elgin SCR. 2007. Transcription and RNA interference in the formation of heterochromatin. Nature 447: 399–406.

Grewal SIS, Jia S. 2007. Heterochromatin revisited. Nat Rev Genet. 8: 35–46.

Griffiths PE, Gray RD. 1994. Developmental systems and evolutionary explanation. J Philos. 91,6: 277–304.

Griffiths PE, Knight RD. 1998. What is the developmentalist challenge? Philos Sci. 65: 253–258.

Henikoff S, Ahmad K, Malik HS. 2001. The centromere paradox: Stable inheritance with rapidly evolving DNA. Science 293,5532: 1098–1102.

Henikoff S, Furuyama T, Ahmad K. 2004. Histone variants, nucleosome assembly and epigenetic inheritance. Trends Genet. 20,7: 320–326.

Jablonka E. 2002. Information: Its interpretation, its inheritance, and its sharing. Philos Sci. 69: 578–605.

Keller EF. *The Century of the Gene.* Cambridge, MA: Harvard University Press; 2000.

Lamm E. 2009. Conceptual and methodological biases in network models. Ann NY Acad Sci. 1178,4676: 291–304.

Lamm E, Jablonka E. 2008. The nurture of nature: Hereditary plasticity in evolution. Philos Psychol. 21,3: 305–319.

McClintock B. 1984. The significance of responses of the genome to challenge. Science 226: 792–801.

Mito Y, Henikoff JG, Henikoff S. 2007. Histone replacement marks the boundaries of cis-regulatory domains. Science 315,5817: 1408–1411.

Müller GB, Newman SA. 2005. The innovation triad: An EvoDevo agenda. J Exp Zoolog. B: Mol Dev Evol. 304B: 487–503.

Pearson H. 2006. What is a gene? Nature 441,7092: 399–401.

Prendergast JGD, Campbell H, Gilbert N, Dunlop MG, Bickmore WA, Semple CAM. 2007 Chromatin structure and evolution in the human genome. BMC Evol Biol. 7: 72.

Scheifele LZ, Boeke JD. 2008. From the shards of a shattered genome, diversity. Proc Natl Acad Sci USA. 105,33: 11593–11594.

Shapiro JA. 2006. Genome informatics: The role of DNA in cellular computations. Biol Theory 1,3: 288–301.

Vanderlyn L. 1949. The heterochromatin problem in cytogenetics as related to other branches of investigation. Bot Rev. 15,8: 507–582.

Waddington CH. *The Nature of Life.* London: Allen & Unwin; 1961.

Wimsatt WC, Griesemer JR. Reproducing entrenchments to scaffold culture: The central role of development in cultural evolution. In: *Integrating Evolution and Development: From Theory to Practice.* Sansom R, Brandon R, eds. Cambridge, MA: MIT Press; 2007:227–323.

Zahradka K, Slade D, Bailone A, Sommer S, Averbeck D, Petranovic M, Lindner AB et al. 2006. Reassembly of shattered chromosomes in *Deinococcus radiodurans.* Nature 443: 569–573.

Zuckerkandl E, Cavalli G. 2007. Combinatorial epigenetics, "junk DNA," and the evolution of complex organisms. Gene 390: 232–242.

Zufall RA, Robinson T, Katz LA. 2005. Evolution of developmentally regulated genome rearrangements in eukaryotes. J Exp Zoolog. B: Mol Dev Evol. 304B: 448–455.

33 Self-Organization, Self-Assembly, and the Inherent Activity of Matter

Evelyn F. Keller

For all Darwin's indisputable achievements, he left a sizable problem for future generations to solve: namely, the question of how the first "primordial form, into which life was first breathed," from which "all the organic beings which have ever lived on this earth have descended"—that "simple beginning" from which "endless forms most beautiful and most wonderful have been, and are being, evolved"—first came into existence.

In fact, it is surprising how often, or how easily, this conspicuous lacuna in Darwin's argument is overlooked—perhaps especially by advocates of natural selection as the universal solvent of life, capable of generating any biological property whatever. For example, let's look at the hallmark of biological systems that goes under the name of function—as in "the function of *X* is to do *Y*." Over the last decade or so, something of a consensus has emerged among philosophers of science about how to treat this problem. Proper function, as first argued by Ruth Millikan and as now widely asserted, should be understood solely in the context of natural selection—that is, the function of *X* is that "which caused the genotype, of which *X* is the phenotypic expression to be selected by natural selection" (Neander 1998:319).

In this way, it is often claimed that the problem of function has been solved. But I disagree. I think, rather, that it has been circumvented. It entirely avoids the problem of how function, understood as a property internal to a biological structure, might have first arisen, particularly in the light of recent arguments that it almost certainly emerged prior to the onset of natural selection. That is to say, natural selection, as conventionally understood since the neo-Darwinian synthesis (and I will adhere to the conventional understanding), requires the prior existence of stable, autonomous, and self-reproducing entities, such as single-cell organisms, or simply stable, autonomous cells capable of dividing. But these first cells were, of necessity, already endowed with numerous subcellular entities (or modules) endowing the primitive cell with the functions minimally required for the cell to sustain itself and reproduce. In other words, even if the first cells lacked many features of the modern cell, they had to have had primitive mechanisms to support metabolism, cell division, and so on; there needed to have come into being primitive embodiments of function that would work to keep the cell going and to protect it from insult.

But perhaps I should say what I mean by "function." Let me try to clarify my meaning by taking off from Michael Ruse's argument against the use of the term for inanimate systems, and more specifically against sufficiency of circular causality (Ruse 2003). Ruse offers the familiar example of the cyclical process by which rain falls on mountains, is carried by rivers to the sea, and is evaporated by the sun, whereby it forms new rain clouds, which in turn discharge their content as rain. The river is there because it produces or conveys water to form new rain clouds. The rain clouds are a result of the river's being there. But Ruse argues that we would not want to say the function of the river is to produce rain clouds, and he is right. What is missing, he claims, is the means by which "Things are judged useful." I won't follow Ruse in his deployment of such worrisomely adaptationist notions as "value" and "desire." Instead, I want to salvage his observation by redescribing what he calls "judgement" as a measurement of some parameter or, if you like, as an evaluation that is performed by a mechanical sensor and, when exceeding some preset limit, is fed back into a controller which is able to restore the proper range of parameters. In other words, I use the term "function" in the sense that the philosophers Ernst Nagel and Morton Beckner originally did, that is, in the sense of a simple feedback mechanism such as a thermostat. Once such a mechanism is added to the rain–cloud–river cycle (say, a mechanism that triggers a change in evaporation rates when the water level falls too low), we can, in this sense of the term, legitimately speak of function and say, for instance, that the function of such a mechanism is to maintain the water level within a certain range of parameters. But crucially, this device differs from the thermostat in that maintaining the room temperature at comfortable levels does not contribute to the persistence of the home, whereas maintaining the water level does contribute to the persistence of the entire cycle; hence it also contributes to the persistence of the mechanism that performs this function. In much the same vein, I suggest that we can refer to the many different cellular mechanisms (proofreading and repair, chaperones, cell-cycle regulation) that maintain the cellular dynamics necessary to the persistence of the cell (and its progenitors) as mechanisms that have functions. They survive not as a result of natural selection but as a consequence of the internal selection that follows automatically from their contribution to the persistence of the system of which they are part. It is a form of selection that does not depend on reproduction (which might be regarded as one way of ensuring persistence, rather like autocatalysis), but rather a more general kind of selection of which natural selection is a particular example. Indeed, their existence is what lends the cell the stability necessary for natural selection to operate.

The existence of such mechanisms is crucial to what makes a system qualify as biological, and the difficult task is to account for how they might have originally come into being. How might such devices—devices that bear all the marks of design—have arisen naturally, without a designer? In other words, we need to explain the emergence of those properties (and here I take function to be a stand-in for purpose and agency as well) that led Immanuel Kant to introduce the term "self-organization," that is, to attribute to the living organism a self with the capacity for its own organization.

This is the problem of accounting for the origin of entities capable of persisting long enough for Darwinian selection to operate, entities therefore capable of subsequently evolving into all those "endless forms most beautiful and most wonderful." That is, it is the problem of accounting for the origin of that system "into which life was first breathed"—the primordial cell.

The question boils down to this: By what processes (or dynamics) did these early machines come together and combine to constitute a primitive cell? Clearly, if natural selection is a product of this early evolution, that process cannot be evoked as an answer. What are the alternatives? What, other than intelligent design, can provide the requisite directionality to the random processes of change to which entities in the physical world are subject? Is it, in fact, possible to account for the emergence of natural design, of a "self" that can be said to organize—indeed, for the emergence of natural selection itself—from purely physical and chemical processes?

Although Darwin did not himself attempt to answer this question, Lamarck did. In fact, one might say that Lamarck saw it as the central problem of evolution. Firmly rejecting any evocation of extranatural causes, he sought a purely physical account of the "power of life," of its natural tendency to increased complexity, and of the origin of entities that could be said to self-organize. "Life," as he wrote in the *Philosophie zoologique*, "is an order and a state of things that permit organic movement there; and these movements, which constitute active life, result from the action of a stimulating cause that excites them" (quoted in Burkhardt [1977] 1995: 152). The stimulating or excitatory cause, which Lamarck likened to the spring of a watch, was to be found in subtle, imponderable, and invisible fluids (such as caloric and electricity) that came originally from outside, insinuating themselves into the interstices of the soft parts of the body, exciting movement, tension, and increasing organization of that body. Caloric he saw as the prime source of what he called orgasm, or irritability, electricity of animal motion. Initially the body's power of life depends on the external supply of these fluids, but with the increasing complexity of organization, he thought that "the productive force of movement" could eventually be internalized (see Burkhardt [1977] 1995:150–156).

Lamarck's terms and categories are alien to us, and his account hardly seems like an explanation at all. We are far more likely to be satisfied by explanations couched in our own contemporary terms and categories. Here is one such account that has become popular in recent years, and for many, the principal—perhaps even the only possible—scientific alternative to evolution by natural selection. I refer to the view of the origin of life as an example of the spontaneous emergence of order, of the kind associated with the process of self-organization as that process has come to be understood in the physical sciences over the last several decades—the kind of self-organization that can be seen in a nonlinear dynamical system that can "mold itself," as Paul Davies put it, "into thunderstorms, people and umbrellas." Many workers in this field have drawn inspiration from statistical mechanics, aiming to describe the emergence of organized structured systems out of blind, random physical processes. Self-organization becomes a kind of phase transition. Or, as Stuart Kauffman writes, "metabolic networks need not be built one component at a time; they can spring full-grown from a primordial soup. Order for free, I call it" (1995:45).

Kauffman is correct. Many complex structures—including networks—can and do arise spontaneously. Indeed, we can find examples of order-for-free all around us. The problem is that such structures do not yet have function, agency, or purpose. They are not yet alive. Self-organization, as mathematicians and physicists use the term, may indeed be necessary for the emergence of biological forms of organization, but as I have argued on a number of occasions, and as Stuart Kauffman now acknowledges, for understanding living processes it is not enough.

Despite all our efforts, the critical properties of function, agency, and purpose continue to mark organisms (even if not machines) apart from thunderstorms—indeed, apart from all the emergent phenomena of nonlinear dynamical systems, remaining conspicuously absent from the kinds of systems with which physics deals. An account of how properties of this sort might emerge from the dynamics of effectively homogeneous systems of simple elements, however complex the dynamics of their interaction might be, continues to elude us. Such properties seem clearly to require an order of complexity that goes beyond that which spontaneously emerges from complex interactions among simple elements—a form of complexity that control engineers have been struggling to characterize ever since the 1940s, and that Warren Weaver, Herbert Simon, and now John Mattick and John Doyle, have dubbed *organized complexity*.

For Weaver (1948), the domain of problems characterized by organized complexity lay in sharp contrast to the problems of statistical mechanics that made up the domain of disorganized complexity. For Simon, organized complexity was complexity with an architecture, and in particular the architecture of hierarchical composition (or modularity) whereby a system "is composed of interrelated subsystems, each of the latter being in turn hierarchic in structure until we reach some lowest level of elementary subsystem" (1962:468). For Mattick, the organization of complexity is mandated by the meaninglessness of the structures generated by sheer combinatorics of complex interactions: "[B]oth development and evolution," he writes, "have to navigate a course through these possibilities to find those that are sensible and competitive" (2004:317). Yet none of these authors quite grapples with the question of just what kind of organization would warrant the attribution of the properties of agency or function, would turn a structure or pattern into a self.

Cybernetics, and its emphasis on the relation between feedback and function characteristic of homeostatic devices, offered one clue; I believe that Herbert Simon offered us another. In fact, it is sobering to go back and read Simon's 1962 essay "The Architecture of Complexity." Here Simon introduces a crucial if much neglected argument for a form of evolution that is alternative both to natural selection and to emergent self-organization: evolution by composition. The idea is this: if stable heterogeneous systems, initially quite simple, merge into composite systems that are themselves (mechanically, thermodynamically, chemically) stable, such composite systems in turn can provide the building blocks for further construction. Through repetition, the process gives rise to a hierarchical and modular structure that Simon claims to be the signature of systems with organized complexity. "Direction," he explains, "is provided to the scheme by the stability of the complex forms, once these come into existence. But this is nothing more than survival of the fittest—that is, of the stable" (1962:471).

We need to be a bit careful here about what we mean by stability—we are not interested in the stability of rocks, and perhaps not even of the limit cycles of dynamical systems closed to informational or material input. Rather, we are interested in the stability of nonequilibrium systems that are by definition open to the outside world, not only thermodynamically but also materially. Perhaps a better word would be "robustness." The systems that endure are those that are robust with respect to the kinds of perturbations that are likely to be encountered. The critical questions then become (1) How do new ways of persisting—new stable modes of organization—come about? and (2) How are they integrated into existing forms?

In neo-Darwinian theory, novelty arises through chance mutations in the genetic material and is integrated into existing populations by selection for the increased relative fitness such mutations might provide. In the picture Simon evokes, novelty arises through composition (or combination), is further elaborated by the new interactions that the proximity of parts brings into play, and finally is integrated into the changing population by selection for increased relative stability. Of particular importance is the stability of the composite acquired with the passage of sufficient time, as Bill Wimsatt puts it, "to undergo a process of mutual co-adaptive changes under the optimizing forces of selection" (Wimsatt 1972: 76). Symbiosis provides what is probably the best example of all three aspects of the process, and perhaps especially of the ways in which the net effect is to bring into being entirely new kinds of entities that would persist by virtue of their enhanced robustness.

But over the long history preceding the arrival of the first cell, a different kind of composition was required—not composition of existing life forms but composition of complex molecular structures (such as proteins, or nucleic acids, or complexes of these macromolecules): molecular composition rather than symbiosis. (Or perhaps simply what Jean Marie Lehn refers to as supramolecular chemistry).

A crucial question is how such molecular or supramolecular composites come about. To be sure, random collisions are a big part of the picture. But molecules, and especially large molecules like proteins, are not billiard balls. They are sticky, they have binding sites, hooks that actively engage other molecules, that invite the formation of larger complexes through the formation of covalent and noncovalent bonds. There is a kind of inherent activity to such collections of molecules—perhaps bearing some resemblance to what Lamarck sought in his imponderable fluids, but here is activity that has already been internalized—long before the complexities of animal movement could evolve. In many macromolecular complexes the springs of activity are built into their very structure; they are already internalized. Drawing energy from their interactions with the environment, they are molecular motors. We might speak of chemical forces and free energies rather than of caloric and electricity, but the idea seems to me to have distinct echoes in Lamarck's earlier vision.

In any case, there is another important point about molecular composition that I need to emphasize, perhaps even the most important point. The formation of such covalent and noncovalent bonds can also change the components with which the process started, thereby creating the possibility for new interactions, new binding sites, new hooks. These new capacities are not simply the consequence of the new proximities molecular binding creates, but also of the changes that have been triggered in the component parts. Macromolecules such as proteins not only are not billiard balls, but also are not simply sticky balls. They are composite structures that are often—perhaps usually—capable of stabilizing in a variety of distinctive shapes or forms. The binding of other molecules can trigger a shift from one conformation to another, thereby exposing new binding sites, new possibilities for subsequent composition. As Eva Jablonka pointed out when she heard me talk about this on an earlier occasion, *prions* provide a good example of what I am talking about. Prions are proteins that become infectious agents as a result of a change in folding —a change in folding endows the molecules with the capacity to transmit their new state to other (normally folded) proteins when they come into contact. The general process of changes in conformation that induce new properties is referred to as *allosteric transition* in molecular biology, and I suggest that it adds a new dimension to evolution—under the pressure for increased stability. Prion infection results from a single allosteric change, but more generally, and especially with the possibility of cumulative changes, allostery provides a mechanism for exploring new evolutionary spaces and for accelerating the formation of ever more complex, and perhaps even functional, structures.

Such processes would seem to be especially pertinent to the evolution of cellularity. Biological cells are replete with devices for ensuring survival, stability, and robustness. Think, for example, of the structures (devices) that have arisen to regulate cell division, ensuring that cell division is not triggered too early (when the cell is too small) or does not wait too long (when the cell has gotten too big); or of the vastly complex kinds of machinery for guaranteeing fidelity in DNA replication, the accuracy of translation, or the proper folding of proteins. Each of these processes—or functions—could presumably have evolved by virtue of the enhanced stability/persistence that the structures on which they depend lend to the system of which they are part.

Because each such mechanism transforms the available options and pathways for subsequent evolution, their arrival might be said to mark off different evolutionary epochs. Nucleic acid molecules, for example, appearing on the scene long before the advent of anything like a primitive cell, introduced a significant advance over mechanisms of autocatalysis for making more such molecules, because it made possible the replication of molecules with arbitrary sequences. The subsequent formation of a translation mechanism between nucleic acid sequences and peptide chains required the combination of already existing nucleic acid molecules and already existing protein structures. But the innovation of a translation mechanism—in effect, the advent of genes—ushered in an entirely new order of evolutionary dynamics dominated, according to Carl Woese, by horizontal gene transfer. Woese argues that cellular evolution, precisely because it needed so much componentry, "can occur only in a context wherein a variety of other cell designs are simultaneously ... [and] globally disseminated." He writes, "The componentry of primitive cells needs to be cosmopolitan in nature, for only by passing through a number of diverse cellular environments can it be significantly altered and refined. Early cellular organization was necessarily modular and malleable" (2002:8742). Indeed, only with the sealing off of these composite structures and the maintenance of their identity through growth and replication—that is, after a few hundred million years of extremely rapid evolution—did individual lineages become possible. As Freeman Dyson puts it, "one evil day, a cell resembling a primitive bacterium happened to find itself one jump ahead of its neighbors in efficiency. That cell separated itself from the community and refused to share. Its offspring became the first species. With its superior efficiency, it continued to prosper and to evolve separately" (2005:4). The rest, as they say, is history —that is, the history of Darwinian evolution.

But long before the advent of that cell, long before anything like the biological cell became possible, another (perhaps almost equally important) transition occurred, and this was the advent of what I might call "smart matter," or rather, smart molecules. Smart molecules are molecules that can both register (sense) signals in their environment and respond by changing their rules of engagement—such as allosteric molecules. I suggest that such molecules came on the scene somewhere over the course of the evolution of macromolecules such as DNA and proteins, and further, that their appearance was crucial to the subsequent evolution of living systems. Ray Kurzweil (2005:364), undoubtedly employing a somewhat different notion of smart, has written that "once matter evolves into smart matter . . . it can manipulate matter and energy to do whatever it wants." I wouldn't go quite that far, but I would suggest that once matter evolves into smart matter, the range of what it can do becomes enormously expanded.

References

Burkhardt RW. *The Spirit of System: Lamarck and Evolutionary Biology*. Cambridge, MA: Harvard University Press; 1977 [1995].

Dyson F. 2007. Our biotech future. New York Review of Books 54,12:4-8 (July 19).

Kauffman S. *At Home in the Universe.* New York: Oxford University Press; 1995.

Kurzweil R. *The Singularity Is Near: When Humans Transcend Biology.* New York: Viking Penguin; 2005.

Mattick J. 2004. RNA regulation: A new genetics? Nat Rev Genet. 5: 316–323.

Neander K. Functions as selected effects. In: *Nature's Purposes: Analyses of Function and Design in Biology*. Allen C, Bekoff M, Lauder G, eds. Cambridge, MA: MIT Press; 1998:313–333.

Ruse M. *Darwin and Design: Does Evolution Have a Purpose?* Cambridge, MA: Harvard University Press; 2003.

Simon H. 1962. The architecture of complexity. Proc Am Philos Soc. 106,6: 467–482.

Weaver W. 1948. Science and complexity. Am Sci. 36: 536–544.

Wimsatt W. C. Complexity and organization. In: *PSA-1972 (Boston Studies in the Philosophy of Science,* K. F. Schaffner and R. S. Cohen, eds. Dordrecht: Reidel, 1972: 67–86.

Woese C. 2002. On the evolution of cells. Proc Natl Acad Sci USA. 99: 8742–8747.

V RAMIFICATIONS AND FUTURE DIRECTIONS

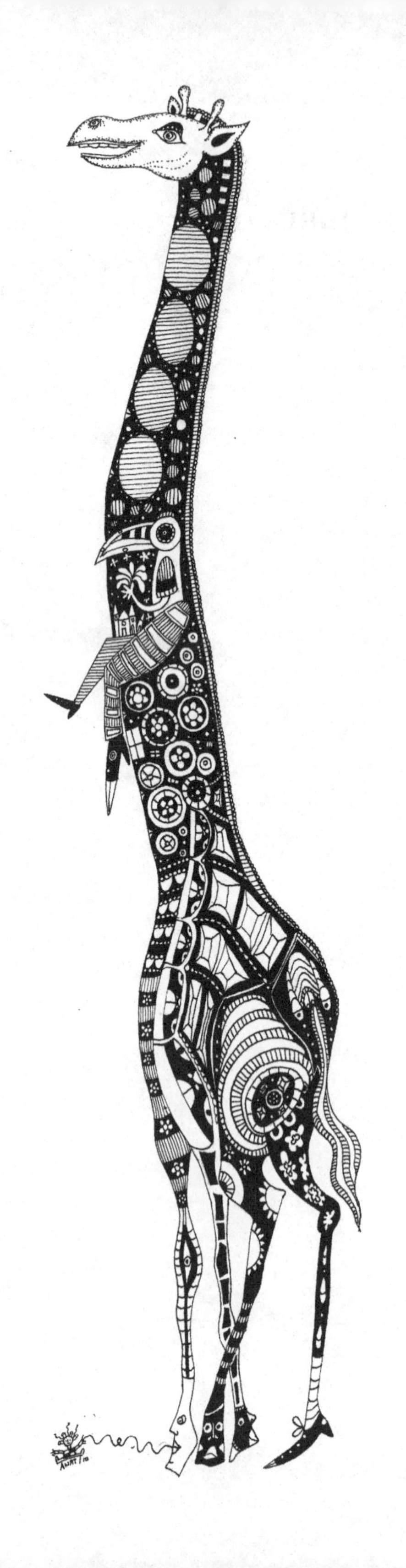

34 Introduction: Ramifications and Future Directions

Snait B. Gissis and Eva Jablonka

The short chapters in this part of the volume present points of view that add to or extend ideas that were presented in parts I–IV. We were not, and could not be, comprehensive, and are acutely aware of the many important aspects of Lamarckian problematics that are missing. As we noted in the Historical Introduction, some important facets of the rich and varied history of Lamarckism in the last two hundred years were not considered in the workshop, and consequently they are absent from this volume. On the biological front, among the important topics that are missing are the fascinating relationship between Lamarckism and immunology (although this was, in fact, discussed in the workshop); the role of learning in animal evolution; studies of adaptive mutations in bacteria; and the interactions between acquired, symbol-based knowledge and the genetic and epigenetic aspects of human evolution. These are only some of the most conspicuous lacunae. Important and directly relevant philosophical discussions pertaining to levels of selection, levels of variation, and levels of evolution are only alluded to, and the same is true for evolvability, modularity, and human evolutionary psychology.

We were, however, lucky that the participants in the workshop could highlight additional facets of Lamarckian problematics that add to the general picture we have been drawing, and point to future perspectives. Of the short chapters that follow, some present points of view that were important for Lamarck's thought and that may become important again in the twenty-first century. One such topic, presented by Simona Ginsburg, is Lamarck's original view on the evolution of psychology, and his take on the mind–body problem. This is a view that has been ignored during the last hundred years, but may be of relevance to current discussions. Another old Lamarckian theme, progressive evolution, is presented by Francis Dov Por, who emphasizes the materialistic elements in Lamarck's thought and their possible incarnation in modern biology. Luisa Hirschbein, on the other hand, discusses a very modern theme—epigenetic inheritance in bacteria—which she pioneered and which she believes may have played an important and as yet unappreciated role in bacterial evolution. The next two chapters, by Raphael Falk and Fred Tauber, look at the implications of the views presented in the volume for our conception of modern biology and its societal role, while Adam Wilkins, our devil's advocate, challenges some of the positions presented in the volume and points to the research directions which, he believes, can address his concerns. The challenges he presented and other important issues are summarized in the last chapter of the book, the Final Discussion.

35 Lamarck on the Nervous System: Past Insights and Future Perspectives

Simona Ginsburg

Lamarck's work on the evolution of the nervous system and the ability to feel and think is of interest today because the problematics he defined are still relevant, although, of course, in a very different modern incarnation.

First, as many have pointed out, Lamarck's approach was thoroughly materialistic. In his opinion, biological complexity—including psychological complexity—has to be explained by employing physical and chemical laws and their biological extensions. These extensions are the result of the dynamic organization of complex material compounds.

Second, Lamarck saw behavior as the engine of animal evolution. For him, the formation of habits was the initial phase in the process. The habits acquired led to altered organization of the nervous system, to altered behavior, morphology, and physiology. When the ontogenetic changes that become habits occur in a lineage for a long time, they lead eventually to instincts—behaviors which require a minimal amount of exposure to external stimuli and learning. This approach is recognized today as important for understanding animal behavior. It is of particular interest that Lamarck highlighted selective attention as a major element in the generation of thinking and acting in animals. Many current models in the cognitive sciences, such as the Global Workspace, focus on attention as a necessary element in explaining mental computations and consciousness. Lamarck referred to attention as "the first of the principal faculties of the intellect," "a special act of the inner feeling" (Lamarck [1809] 1914: 380).

Feeling, according to Lamarck, is a necessary condition for the formation of thought, which brings me to the main focus of my remarks. Lamarck confronted head-on one of the most important problems of biology—how are feelings generated from a mere ability to react adaptively? He used the eighteenth-century distinction between irritability and sensibility, and maintained that the ability to feel, to be sensible, is the result of a new organization in the nervous system, which leads to an overall sensory state. This idea seems crucial for developing a modern evolutionarily informed account of the ability to experience, of the basic ability to feel and perceive.

In book III of *Zoological Philosophy* Lamarck discussed the evolution of the nervous system and the kind of organization that such a system must exhibit in order to give rise to sensations, feelings, and thoughts. Necessary conditions for sensations, and for the precursor of feelings, which he called "the inner feeling," were electric conduction (a flow of what he called "electric fluid") and a nervous system organized so that it allows communication and integration of signals. He suggested the following sequence of events: a stimulus triggers electrical conduction through the nervous fluid along one of the nerves; the signal arrives at a "center of communication," probably a ganglion or the brain; from this center, signals travel throughout the entire nervous system to the whole body, and from the parts of the body signals return to the center of communication; only after integration at this center is a signal sent to a motor effector to produce action.

The systems of sensory and muscle-related nerves are constantly communicating. This is important for Lamarck, for it is the system as whole, the whole animal rather than a part of it, that has an inner feeling—an overall sensory state. This inner feeling becomes differentiated during evolution, giving rise to the various emotions seen clearly in vertebrates. The ability to reason, Lamarck argued, requires further evolution—the addition of new anatomical structures such as the neocortical hemispheres in mammals.

According to Lamarck, the inner feeling is, at its simplest, a feeling of existence. The first inner feeling has to do with the feeling of the body as a thing in space that can move, and it evolved in the context of internal and external motor responses:

> The feeling in question is now well recognized, and results from the confused assembly of inner sensations which are constantly arising throughout the animal's life, owing to the continual impressions which the movements of life cause on its sensitive internal parts.
> . . . the sum-total of these impressions, and the confused sensations resulting from them, constitute in all animals subject to them a very obscure but real inner feeling that has been called the feeling of existence. (Lamarck [1809] 1914:333–334)

Lamarck's inner feeling is somewhat similar to what Damasio (1999) called the "proto-self." Damasio restricted it to vertebrates, and defined it as "a coherent collection of neural patterns which map, moment by moment, the state of the physical structure of the organism in its many dimensions" (Damasio 1999:154). The inner feeling (the proto-self) is a precondition for what Damasio describes as "background feelings," which are feelings that are associated with overall neural and hormonal organismal states. Prominent background feelings are wellness/sickness, tension/relaxation, balance/imbalance. Damasio's background feelings are inner feelings with an intrinsic value attached—feelings that can stimulate reinforcing or neutralizing adaptive motor behaviors.

Lamarck made a conceptual leap—from integrated, global, whole-organism sensations to a feeling. Although this jump cannot be justified on logical grounds (it is not a deduction), it fits with the general intuition that there is an intimate connection between sensations and feelings: although sensation, that is, processing of sensory inputs, can occur without feelings, feelings cannot occur without some sensory processing. Lamarck's assumption that integrated, multiple sensations reverberating throughout the animal's interconnected nervous system constitute the basis for feelings is still used today, in the work of Damasio and others. Thus, for example, Ginsburg and Jablonka (2010) are trying to develop this idea, this leap from sensation—mere processing of signals—to full-fledged feeling and experiencing, like seeing red or smelling a rose. In order to bridge the gap between mere sensory processing and real feeling, to account for the qualitative difference, we chose a process that is both intuitively and philosophically connected to experiencing—the process of motivation; we have tried to explain this process from an evolutionary point of view, to give an account of the appearance of motivation, by following the transitions that occurred in early nervous systems. In a nutshell, we believe that neural trajectories formed ontogenetically by mechanisms of associative learning enable even very simple animals with a central nervous system to develop pathways that guide them to goals achieved in the past. These trajectories act as motivators, which are similar to Lamarck's ill-interpreted "need" (*besoin*) and "will."

References

Damasio A. *The Feeling of What Happens: Body and Emotion in the Making of Consciousness.* New York: Harcourt Brace; 1999.

Ginsburg S, Jablonka E. 2010. Experiencing: A Jamesian approach. J Cons Stud. 17: 102–124.

Lamarck J-B. *Philosophie zoologique.*.[1809]. *Zoological Philosophy.* Elliot H, trans. New York: Hafner; 1914.

36 Lamarck's "Pouvoir de la Nature" Demystified: A Thermodynamic Foundation to Lamarck's Concept of Progressive Evolution

Francis Dov Por

Evolutionism was born from the need to comprehend the diversity of the zoological world. This need was felt by Lamarck when, in his fifties, he turned zoologist. Strangely, two hundred years after his *Philosophie zoologique*, while evolutionism has spread to the most diverse branches of human knowledge, it is passing through a profound crisis in zoology, its very cradle.

Lamarck envisaged an evolutionary progression linking, through an immense period of time, a simple organism of the infusorian group with the complex orang-utan and thereafter with the "bi-mane," man. He saw the gradation from simple to complex as the result of a natural force, acting in all organisms, which was inherent in the organization of living matter; in his words, there was a "pouvoir de la nature" still to be discovered. The accusation that he invoked a mystical vital force is unjustified, as is that of a supposed anthropocentric teleology, or an Aristotelian *scala naturae*. In fact, Lamarck published a diverging evolutionary tree of the animal world decades before Darwin made the famous branching sketch in his notebook.

Modern neo-Darwinism views evolution as aimless and contingent, uniformitarian and occasionally catastrophistic. It is merely transformist, and basically anti-progressivist. Gould's position is symptomatic: "We are glorious accidents of an unpredictable process with no drive to complexity . . ." (Gould 1996:216).

For many, this reigning view is philosophically deeply unsatisfactory. For example, Wilson (1992) does not want to deny the subjectively felt truth of progressive evolution; Ruse (2003) concludes that, for many, the causes of evolution are still "here to stay" and Conway-Morris (2003), after concluding that zoological evolution was not accidental, admits to an interest in reunifying the scientific worldview with a religious instinct.

Russian and Soviet Darwinians of the first half of the twentieth century harmoniously assimilated Lamarckian progressivism within a materialistic framework. Severtsov spoke of increasing complexity and vital energy during the history of animal evolution (see Levit, Hössfeld, and Olsson 2004). Schmalhausen, sadly famous as the victim of Lysenko's pseudo–Lamarckian campaign, advocated Lamarck's view of "increasing levels of organization in evolution" (Schmalhausen 1949). Representing a more recent Russian school, Konoplev and his colleagues saw "energetic metabolism [as] a criterion for the direction of evolutionary progress of animals. Deviation from equilibrium increases in the process of evolution" (Konoplev, Sokolov, and Zotin 1975:756).

The concept that complex organisms are "anti-entropic" exceptions in a world which tends toward maximizing entropy still weighs heavily against evolutionary progressivism in modern mainstream biological thinking. Lotka (1922) first proposed that Darwinian natural selection acted not only to increase individual survival and reproductive success, but also to increase energy flow in ecological systems. It took Morowitz (1968) to extrapolate this principle of increased energy flow to the history of the biosphere as a whole.

Vermeij has emphasized the importance of animal predation, of increasing metabolism, and of increasing animal "power budgets" (Vermeij 1987, 2004). Animals have been considered as energy conveyers and boosters in the biosphere, and as such their complexity and functional efficiency are measures of their progressive evolution (Por 1980, 1994). Greater animal activity created feedback loops at many levels, resulting in the increased energy capture and plant biomass (Por 1994).

Chaisson's (2001) presentation of thermodynamic cosmic evolution has allowed me to situate zoological evolution within a new and comprehensive framework. A rise in complexity is seen in all inanimate and animate systems open to energy flow. Systems are selected according to their ability to command energy sources. Energy flow is quantified by Chaisson in units of $erg^{-1}\ s^{-1}$ of free energy rate density, or Φm. The tendency toward increased complexity and energy utilization which emerges in open systems can be interpreted as Lamarck's power of nature, and the selection of the more active thermodynamic systems is a broadening of Darwinian selection. In his fairly extended discussion of the zoological phase of evolution, Chaisson equated Φm with the metabolic rate.

A sketch of the progressive trend, as I envisage it, starts with slow-moving and small-particle-feeding organisms, with average Φm around 1000, not much higher than that of plants. Highly mobile and visually able consumers then develop, with doubled Φm values, expressed in higher food and oxygen needs. The seas are, however, relatively limited in nutrients and dissolved oxygen. Out of the more than thirty animal phyla, or structural types, only a handful, such as the arthropods, mollusks, and vertebrates, expanded to the estuarine and inland waters where food is plentiful. In order to do this, they had to be able to actively regulate the osmotic pressure of their body fluids against the fluctuating salinity of the surrounding aquatic medium. The metabolic cost of osmoregulation raised the Φm, for instance of fish, to values around 10,000. Natural selection further restricted the number of major phyla that could take advantage of the rich vegetation of the sunlit continents and the oxygen-rich air to two: the arthropods (insects and arachnids) and vertebrates, which developed costly drought-resistance features. But only two branches of terrestrial vertebrates, the mammals and the birds, acquired the capacity for maintaining stable individual body heat and extended activity despite the fluctuating diurnal and seasonal air temperatures, at a cost starting around 50,000 Φm. Warm-blooded animals obtained a stable internal medium that allowed functions leading to higher neural capacities. Songbirds require a Φm of 80,000 and so, probably, did the Neolithic *Homo sapiens*, but he had already trespassed the animal–human boundary.

Thermodynamic evolution in zoology is not, of course, limited to vertebrates. The insects, which because of their small size could not maintain stable temperatures individually, in some cases did so in a different way: by forming huge colonies, thermostat-like controlled structures developed, as well as a type of collective memory and learning.

In chapter 6 of his *Philosophie zoologique,* Lamarck wrote, "It is obvious that, if nature had given existence to none but aquatic animals...we should then no doubt have found a regular and even continuous gradation in the organization of these animals" (Lamarck [1809] 1914:69). Unbeknown to him, the modern marine Cephalopoda, the squids and octopus-like animals, independently reached a peak of evolution, having an anatomical and functional complexity with a high metabolism and Φm comparable with that of the vertebrates, and behavioral and learning capacities equaling those of mammals. They vindicate Lamarck's thesis of a "pouvoir de la nature" acting across the entire animal world. However, the cephalopods use up all their energy resources in the poor marine environment and die soon after their reproductive effort. Therefore, there is no passing on of their life experience to their progeny. No wise octopus could have possibly met us.

References

Chaisson E. *Cosmic Evolution: The Rise of Complexity.* Cambridge, MA: Harvard University Press; 2001.

Conway-Morris S. *Life's Solution: Inevitable Humans in a Lonely Universe.* Cambridge: Cambridge University Press; 2003.

Gould SJ. *Full House: The Spread of Excellence from Plato to Darwin.* New York: Harmony Books; 1996.

Konoplev VA, Sokolov VE, Zotin AI. 1975. Classification of animals by energetic metabolism. [English summary]. Zool Zh. 65: 756.

Levit GS, Hössfeld U, Olsson L. 2004. The integration of Darwinism and evolutionary morphology: Alexey Nikolajevich Sewertzoff (1866-1936), and the developmental basis of evolutionary change. J Exp Zool. (Mol Evol Dev.) 302B,4:343–354.

Lotka A. 1922. Contributions to the energetics of evolution. Proc Natl Acad Sci USA. 8,6: 147–155.

Morowitz HJ. *Energy Flow in Biology.* New York: Academic Press; 1968.

Por FD. 1980. An ecological theory of animal progress: A revival of the philosophical role of zoology. Perspect Biol Med. 23: 389–399.

Por FD. *Animal Achievement: A Unifying Theory of Zoology*. Rehovot, Israel: Balaban Press; 1994.

Ruse M. *Darwin and Design: Does Evolution Have a Purpose?*Cambridge, MA.: Harvard University Press; 2003.

Schmalhausen II. *Factors of Evolution. The Theory of Stabilizing Selection.* Philadelphia: Blakiston Press; 1949.

Vermeij GJ. *Evolution and Escalation: An Ecological History of Life.* Princeton, NJ: Princeton University Press; 1987.

Vermeij GJ. *Nature, an Economic History.* Princeton, NJ: Princeton University Press; 2004.

Wilson EO. *The Diversity of Life.* Cambridge, MA: Harvard University Press; 1992.

37 Prokaryotic Epigenetic Inheritance and Its Role in Evolutionary Genetics

Luisa Hirschbein

In prokaryotes such as bacteria, it is usually assumed that heritable variations that are maintained in the progeny through many generations are always associated with genetic inheritance and primary DNA alterations. However, examples from bacteriophages and bacteria have empirically proven the existence of alternative inherited chromosome variations, called epigenetic variations (see Jablonka and Raz 2009: table 1).

Epigenetic Regulation: Characterization

In eukaryotes, cell-heritable epigenetic inactivation was initially associated with a particular condensation of the chromatin, and was considered to be a regulation process specific to eukaryotic systems. This view changed when evidence of heritable phenotypic variations, which are not mediated by DNA mutation, was found in several prokaryotic systems.

In prokaryotic biology, an "epigenetic variation" is characterized by the following minimum properties: (a) the variation is inherited, that is, it is transmitted to the progeny through many cell divisions; (b) the modified phenotypic expression is *not* encoded by changed DNA sequence variation; and (c) the expression patterns can be reversed by environmental conditions (Holliday 2006).

Potential Mechanisms of Epigenetic Inheritance

Epigenetic inheritance mechanisms have been found in both gram-negative and gram-positive bacteria. These mechanisms involve (1) self-sustaining feedback loops, (2) the propagation of a nucleoprotein structure, and (3) DNA methylation, mainly but not exclusively at the cytosine base, which forms 5-methyl cytosine (m^5C). The importance of phenotypic variations due to m^5C, and the functional contribution of the modified DNA to epigenetic regulation in bacteria, are well reviewed elsewhere (Casadesús and Low 2006), so here we focus on self-sustaining feedback loops and on chromatin propagation mechanisms. Unlike eukaryotic systems, evidence for microRNA (miRNA) that can efficiently regulate prokaryotic epigenetic processes has so far proved elusive (Pichon and Felden 2008).

Feedback Loops

When a cell is infected by a temperate bacteriophage, such as the phage lambda in the bacterium *Escherichia coli*, two alternative patterns of gene expression may be generated: a lysogenic pattern and a lytic pattern. In lysogeny, the phage is integrated within the bacterial chromosome and remains silent during many cell divisions due to the synthesis of the repressor protein *cI*. This lysogenic mode entails permanent expression of *cI,* whose protein product binds to the operator site and represses the transcription of the other phage genes. Alternatively, the repressor protein *cI* can be degraded, leading to the expression of *cro* (known as the secondary repressor), thus allowing *both* the lytic mode (involving the independent multiplication of the phage and the eventual death of the cell) and the lysogenic mode of the cell to be activated. The genes that kill the cell can, however, be deleted, and by using a thermosensitive *cI* repressor protein, one can produce a system in which, at the appropriate temperature, *cro* is expressed and *cI* is not expressed, and vice versa. Such a system replicates for many cell divisions, generating two genetically identical cell lines in the same cellular medium (Herskowitz 1973).

Experiments involving the *lactose* operon have shown similar regulatory epigenetic mechanisms mediated by feedback loops (Novick and Weiner 1957). The operon contains the permease gene (*lacY*) and the gene for the enzyme β-galactosidase (*lac Z*). The operon is turned on by sugar, for example, by lactose, or by an analogue of lactose. In order to induce the expression of the β-galactosidase enzyme, lactose (the inducer) must enter the cell, and this requires the presence of a permease. However, at high concentrations of inducer, the inducer can enter by diffusion even without the permease. In these conditions, the operon is switched ON in the cells, the permease gene is activated, and the permease protein is synthesized. When induced and noninduced cells are mixed together and grown at low concentrations of inducer, both types of cells (induced and noninduced) breed true: in these conditions two genetically identical cell lines manifest and pass on different phenotypes for many generations. One cell line expresses the *lactose* operon genes (as a result of the expression of the permease gene which was induced many generations ago, when the cells were transiently grown at high inducer concentrations), and the other cell line has a silent operon, resulting from a lack of permease.

These two examples, which involve the different states that a positive feedback loop may assume, show that no genetic changes are required for the persistent coexistence of two states when the same population of the bacteria is kept under the same environmental conditions.

Propagation of a Nucleoprotein Complex: Noncomplementing Diploids

Epigenetic inheritance was found in gram-positive bacteria such as *Bacillus subtilis* when cells with different genotypes were experimentally fused, forming heterodiploids. Among the fused cells, a subpopulation was found in which a whole chromosome was silenced. In spite of the cells being diploid and having two chromosomes, these heterodiploids, which were called NCDs (noncomplementing diploids), were functionally haploid (Hotchkiss and Gabor 1980).

The nature of the biochemical modifications in the NCDs suggests that an alteration in chromosome structure plays a role in the silencing of the chromosome. It was found that the expression of the *spoOA* gene is required for the establishment of the cryptic state, and the SPOOA protein is known to condense the chromosome (Setlow, Magill, Febbroriello, Makhimovsky, Koppel, and Setlow 1991). In line with this, an antibiotic resistance gene resident in the silent chromosome was reactivated from the expressed chromosome when the *spoOA* gene was deleted (Grandjean, Le Hegarat, and Hirschbein 1996; Grandjean, Hauck, Beloin, Le Hegarat, and Hirschbein 1998). Additional studies showed that recombinant bacteria derived from unstable progeny exhibit a mosaic structure composed of DNA fragments characteristic of both the parental chromosomes, as well as the inactivation of a large fragment of coding DNA. The cryptic, inactivated state was maintained through cell divisions for twenty to forty generations. However, no more than 2 to 20 percent of cells carrying the *spoOA* deletion can be reactivated. These results show that although SPOOA is involved in the silencing, more than one compacting protein chromosome participates in the maintenance of the silent state (Grandjean, Hauck, Beloin, Le Hegarat, and Hirschbein 1998).

It is tempting to speculate that the transitory silencing of coding DNA can be stabilized and that silenced sequences can accumulate, thus leading to new epialleles at several loci of the bacterial genome. If so, epigenetic inheritance of the changed phenotype could have direct effects on adaptive bacterial evolution and the production of new species. Prokaryotic epigenetic inheritance may have been significant early in bacterial evolution, mediating Lamarckian–like processes. It could be that a consideration of epigenetic inheritance will lead to Lamarckism's emerging essentially intact, despite two centuries of controversy and debate.

References

Casadesús J, Low D. 2006. Epigenetic gene regulation in the bacterial world. Microbiol Mol Biol Rev. 70,3: 830–856

Grandjean V, Hauck Y, Beloin C, Le Hegarat F, Hirschbein L. 1998. Chromosomal inactivation of Bacillus subtilis exfusants: A prokaryotic model of epigenetic regulation. Biol Chem. 379: 553–557.

Grandjean V, Le Hegarat F, Hirschbein L. Prokaryotic model of epigenetic inactivation: Chromosomal silencing in Bacillus subtilis fusion products. In: Russo V, Martiensen RA, Riggs AD, eds. *Epigenetic Mechanisms of Gene Regulation.*: Cold Spring Harbor Laboratory Press; 1996.

Herskowitz I. 1973. Control of gene expression in bacteriophage lambda. Annu Rev Genet. 7: 289–324.

Holliday R. 2006. Epigenetics: A historical overview. Epigenetics 1: 76–80.

Hotchkiss RD, Gabor MH. 1980. Biparental products of bacterial protoplast fusion showing unequal parental chromosome expression. Proc Natl Acad Sci USA. 77: 3553–3557.

Jablonka E, Raz G. 2009. Transgenerational epigenetic inheritance: Prevalence, mechanisms, and implications for the study of heredity and evolution. Q Rev Biol. 84,2: 131–176.

Novick A, Weiner M. 1957. Enzyme induction as an all-or-none phenomenon. Proc Natl Acad Sci USA. 43,7: 553–566.

Pichon C, Felden B. 2008. Small RNA gene identification and mRNA target predictions in bacteria. Bioinformatics 24,24: 2807–2813.

Setlow B, Magill N, Febbroriello P, Nakhimovsy L, Koppel D, Setlow P. 1991. Condensation of the forespore nucleoid early in sporulation of Bacillus species. J Bacteriol. 173: 6270–6278.

38 Evolution as Progressing Complexity

Raphael Falk

An important insight of Lamarck, and I think also of Buffon, was their characterization of the difference between physics and biology: the crucial feature was that biological systems are physical systems constrained by their *history*, while physical systems—for the most part—have no history. Or, to be more contemporary, life as we know it today is the *uninterrupted*, ongoing sequel of improbable physical processes of complex molecular structures that draw on their environment for material, energy, and entropy.

Although Lamarck believed that the founding events leading to the emergence of living entities occurred repeatedly, he advanced the theory that the history of organic evolution is based on a *progressive increase of complexity* as the very consequence of the cumulative continuity of the process. Thus, contrary to chemical theory, which considers systems whose components are stably interacting, biological systems build on the emergence of reshuffling systems, which do not just replicate but also *reproduce*. Such a conception is based not on teleological principles of populations but on the reorganizational basis of existing systems of individuals which are, as Darwin evoked fifty years later, constrained within their environment. Consequently, Darwinians put the emphasis on the dynamics of populations, their properties, and their ecology, whereas the Lamarckians emphasized the dynamics of individual development.

Although the conception of complexity appeared early in the twentieth century in descriptions such as that of the "protoplasm" of cellular theories and the "colloids" of biochemistry, nonetheless by that time the goal of most biologists was to establish the life sciences on the first principles of the physical sciences. Niels Bohr went so far as to believe that it would be possible to extend the scope of the laws of physics by studying biological systems. In these efforts to reduce biology to physics, genetics took a leading role. By the 1940s it abstracted traditional physiology and embryology to the notion of "one gene–one enzyme," and that of evolution to the dynamics of population genetics. Max Delbrück (Bohr's disciple) and his colleagues' extraordinary accomplishments in molecular biology further drove genetics to the extremes of reductionism, culminating in the formulation of Crick's (1958) Sequence Hypothesis and the Central Dogma of genetics.

This reductionist concept assumed that discoveries based on model organisms, such as the bacterium *Escherichia coli*, applied directly to all living systems. However, the very reductionist methodologies used by geneticists soon made it clear that what is true for *E. coli* is not true for the elephant. This insight was forced on the molecular biologists when they extended their analyses to viruses infecting eukaryotic, especially primate, cells. The multifunctional interactions of the T-antigens of the simian SV-40 polyoma virus foreshadowed complications to come. It was, however, the refinements of reductionist methodologies, ranging from Britten and Kohne's (1968) in vitro DNA hybridization, which led to the unexpected discovery of redundant DNA, to Lewontin and Hubby's (1966) application of gel electrophoresis, which in turn led to the discovery of the enormous molecular polymorphism among individuals in natural populations, that drove the point home: reduction is an indispensable research method, but conceptually life cannot be reduced to the sum of the properties of its components.

During the last decades of the twentieth century, biology became very much the science of complex systems, and evolution again became the progressive history of such systems. More and more research in biology turned its attention to *epistemic* top-down organismic and systemic analyses of the inherent complexity of life, while it enormously refined both the *reductionist* methodologies of molecular biology and the development of techniques of computational physical sciences for the bottom-up analyses of these problems. This inherent paradox of applying bottom-up analyses for seeking answers to top-down problems was addressed at two levels, conceptually and empirically.

Conceptually, following one variable at a time, ceteris paribus, is acceptable in living systems only as long as scientific explanations make sense of nature *as we see it,* providing intervening variables rather than establishing hypothetical constructs. Following a multiplicity of variables simultaneously is not merely a multiplication or addition of many single effects: a better approximation to a top-down view of systems is achieved by looking at a multiplicity of factors which leads to a multiplicity of interactions, many of which are nonlinear. Empirically, modern methodologies that allow one to obtain data about a very large number of variables and computational methods incorporating nonlinear dynamics enable an integration of interacting processes and factors that generate emergent phenomena, which resolves the paradox (see, for example, Kaiser 2008).

It is important to realize that although a reemergence of "emergence" is a general cultural phenomenon of our time, the power of accounts that allocate to specific factors a dominant causal role in phenomena (ceteris paribus notwithstanding) has not diminished. Actually, it is even more prominent with every new, spectacular, localized achievement of research when this is transferred and interpreted in contexts of sociomedical and technological discourse. It is in the best tradition of genetic analysis that breeders developed QTL (quantitative trait locus) techniques for allocating genetic markers to identify chromosome segments affecting the inheritance of quantitatively interesting properties, whether milk production in cattle or psychiatric disorders in men. Reducing qualitative traits such as behavioral, cognitive, or cultural properties to specific genes, is, however, another thing altogether. Thus, for many geneticists, profound intellectual efforts were needed to accept claims (like those of Gilbert Gottlieb and Susan Oyama) that genes and environments are not separable: that information is not coded in genes but rather emerges from gene–environment interactions (see Looren de Jong 2000). It is an irony of history that genetics, the most explicitly reductionistic of the life sciences, not only lost its independence—*all* of biology is nowadays "genes" and "DNA"—but also that it was pivotal in reestablishing integrative evo-devo, the intermediary between studies of complex ontogenetic development and the dynamics of natural populations that are exposed both to their system's constraints and to natural selection.

I have never felt comfortable with making predictions, but I would like to return to an evolutionary question that is of concern to many biologists and laypeople: Are we really at the stage of overcoming the hurdle of the "animate" character of life? Are we on the verge of—within a few decades, or a few centuries—replicating *in silico* life as we have known it? I doubt this. But it is meaningful to ask, in this context, whether this unique life-as-we-know-it has reached a stage where technology, as an extension of the human phenotype, might not only have an evolution of its own—as is the case for human culture today—but also that as the technology impacts and interacts with physical life, it will make the mechanisms and processes of biological Darwinian evolution irrelevant compared with the possibilities of technology-extended evolution.

I would like to turn to some words of Peter Medawar: "This proxy evolution of human beings through exosomatic instruments has [indeed] contributed more to our biological success than the conventional evolution of our own, endosomatic organs" (Medawar 1984:185). However, instructions for making such exosomatic organs depend, in the final analysis, completely on our endogenic instructions. Exogenic heredity is Lamarckian or *instructional* in style, rather than Darwinian or *selective*. We can and do *learn* things in the course of life which are transmitted to succeeding generations, and admittedly, any computational-style extension of our endosomatic instructions must be considered a profound modification of the very life-as-we-know-it.

Trying to adopt a global perspective, I am not sure if the progressive or, more precisely, accelerative instructional evolution does not loosen or lose many of the checks and balances of selective-style evolution. Whether Darwinian selective evolution will regain its role as the invisible hand, and reestablish a new equilibrium rather than the unchecked, open-ended Lamarckian instructive evolution, is difficult to say. I, for one, hope so.

References

Britten RJ, Kohne DE. 1968. Repeated sequences in DNA Science 161,3841: 529–540.

Crick FH. 1958. On protein synthesis. Symp Soc Exp Biol. 12: 138–163.

Kaiser J. 2008. Cast of 1000 proteins shines in movies of cancer cells. Science 322,5905: 1176–1177.

Lewontin RC, Hubby JL. 1966. A molecular approach to the study of genic heterozygosity in natural populations. II. Amount of variation and degree of heterozygosity in natural populations of *Drosophila pseudoobscura.* Genetics 54,2: 595–609.

Looren de Jong H. 2000. Genetic determinism: How not to interpret behavioral genetics. Theory Psychol. 10,5: 615–637.

Medawar PB. Technology and evolution. In: *Pluto's Republic*. New York: Oxford University Press; 1984:184–190.

39 Epigenetics and the "New Biology": Enlisting in the Assault on Reductionism

Alfred I. Tauber

I regard this conference and the book based on it as part of a larger movement that has been called various names, and for my purposes, "new biology" suffices. This term captures a widely held view that the elucidation of complex biological functions requires a new kind of analysis. According to the proponents of the new program, explanatory models of the dynamic, emergent properties characteristic of biosystems demand a holistic approach, albeit one coupled to elemental analysis (reductionism). This purported novel orientation builds on the general intuition that evolution, development, metabolism, immune responsiveness, and neurological functions each require better explanations of the plasticity, emergent phenomena, and nonlinear, dynamic causation pathways characteristic of organic phenomena. Epigenetics attains membership in this new alignment because the reductive pathway from gene to protein fails to follow the singular linear sequence predicted by the central dogma. Further, what some see as a new evolutionary synthesis arising from revised disciplinary boundaries (developmental biology and ecology are melded with classical genetics and molecular biology [Jablonka and Lamb 2005; Gilbert and Epel 2009]) portends a reframing of the basic theoretical questions underlying current research in line with these other developments.

For those engaged in what they believe is a conceptual shift, the reductionism of the "old biology" does not adequately address the challenges of characterizing biology in these expansive terms. In seeking to comprehend the *complexity* of biological phenomena as *intact systems* (Woese 2004), the "new biology" is hardly new. Dissidents waged attacks against the triumphant march of reductionism throughout the twentieth century in both evolutionary theory (Levins and Lewontin 1987) and developmental biology (Beurton, Falk, and Rheinberger 2000). That history begins in the 1840s, when German physiologists (led by Hermann von Helmholtz) declared their intention to establish all biological phenomena on a common chemical–physical basis that would characterize the organic on the basis of forces, analogous to what was established in the physical sciences (Galaty 1974). These so-called reductionists did not argue against the unique character of life, only that all causes must have certain elements in common. They connected biology and physics by equating the ultimate basis of their respective explanations in physical laws, and proceeded to link the physical sciences to the biological. Theirs was a metaphysical move against vitalism, and in that challenge they pronounced a new biology, and justly so. Later in that stage, genetics joined biochemistry, and it is within this context that soft inheritance encountered a massive and unyielding resistance.

The earlier tensions between Old and New biologists were not so much over methodological reductionism, or even reductive epistemology, as over the underlying metaphysical issue of establishing *cause*. Traditionally (and typically) the biologist with reductionist commitments seeks linear causal linkages that are best described as "mechanical," and only in the late twentieth century did complexity theory coupled to systems-wide analyses suggest how multiple nonlinear causation might serve as a better depiction of physiological and genetic processes (Alon 2007). Evolutionary theory has also reflected such shifts (Depew and Weber 1995; Jablonka and Lamb 2005). In addition, as already mentioned, the borders of the various life sciences have been blurred of late, and this reflects the appreciation that to understand developmental biology—for instance, in an evolutionary context—the "boundary conditions" must be expanded and larger integrative efforts must be made (Gilbert and Epel 2009). A new biology must move in both directions—outward in expanding disciplinary borders and inward with continued reductive analyses.

Those who now envision how epigenetics might lead to a more comprehensive account of development and evolution share the basic insight—broadly acknowledged in the other life sciences—that regulation and organization cannot be explained by simple mechanical models, and thus more complex modes of causation must be elucidated. Accordingly, the "Modern Synthesis" of Fisher and Wright will eventually be dubbed the "Old Synthesis" as a new, epigenetic thought collective becomes the "Refreshed Synthesis" (or something like it) to capture an inflection in modeling biological processes—perhaps less than a revolution but something more than a modification. While stirrings of success have appeared, whether a different paradigm has emerged is not clear, because it is hardly apparent that the basic reductive strategy of the twentieth century has been eclipsed, notwithstanding current efforts to revise the basic questions that organize research in genetics, developmental biology, metabolism, neurosciences, ecology, and immunology (Tauber 2009). As new investigative strategies in each of the biological subdisciplines seem noteworthy and portend potential new vistas of research, the criteria for successfully establishing *new kinds* of analysis have not been well established, nor have the outlines of research programs directed at these challenges been unambiguously explained. So, despite the excitement, a clearly articulated new biology has not yet emerged, and we are left with promissory notes that must await collection.

So I remain agnostic about the more radical promises of New Biology in its contemporary incarnation. My caution is based on the assessment that neither reductive methodology nor reductive epistemology seems fundamentally altered by new technologies, and contemporary biology, while acknowledging more complex causation streams, has yet to reveal biological systems with *metaphysical* characteristics different from those envisioned by Helmholtz and his colleagues. Plainly put, elements still must be defined by reductive techniques, and while changing the questions being pursued will likely reveal more complex kinds of organization and regulation, predictably, these, too, will be explainable by physical laws and derivative understandings of causation originating in the nineteenth century. On this view, while epigenetics (highlighting the plasticity of individual development as well as the plasticity of the organism in its environment) demonstrates dynamical, dialectical processes, our understanding of the underlying character of biological causation has not fundamentally changed. Thus the reductionists' metaphysical agenda remains, albeit more complex than they might have envisioned, given the state of the material sciences of their own era.

References

Alon U. *An Introduction to Systems Biology. Design Principles of Biological Circuits.* Boca Raton, FL: Chapman and Hall; 2007.

Beurton PJ, Falk R, Rheinberger HJ, eds. *The Concept of the Gene in Development and Evolution: Historical and Epistemological Perspectives.* New York: Cambridge University Press; 2000.

Depew DJ, Weber BH. *Darwinism Evolving: Systems Dynamics and the Genealogy of Natural Selection.* Cambridge, MA: MIT Press; 1995.

Galaty DH. 1974. The philosophical basis for mid-nineteenth-century German reductionism. J Hist Med Allied Sci. 29,3: 295–316.

Gilbert S, Epel D. *Ecological Developmental Biology.* Sunderland, MA: Sinauer Associates; 2009.

Jablonka E, Lamb MJ. *Evolution in Four Dimensions: Genetic, Epigenetic, Behavioral, and Symbolic Variation in the History of Life.* Cambridge, MA: MIT Press; 2005.

Levins R, Lewontin R. *The Dialectical Biologist.* Cambridge, MA: Harvard University Press; 1987.

Tauber AI. The biological notion of self and nonself. In: *Stanford Encyclopedia of Science.* Rev. ed. 2009. http://plato.stanford.edu/entries/biology-self.

Woese CR. 2004. A new biology for a new century. Microbiol Mol Biol Rev. 68,2: 173–186.

40 Epigenetic Inheritance: Where Does the Field Stand Today? What Do We Still Need to Know?

Adam Wilkins

As the last speaker at the round table, I would like to provide a perspective on where the field of epigenetic inheritance stands today. That subject is only one of several that have been presented, but it is central to many of the scientific talks that we have heard.

It should be mentioned at the outset that I was invited to this conference to play the role of devil's advocate, namely, someone to inject caveats and objections, where appropriate, into a conference dedicated, in part, to commemorating the memory and achievements of Jean-Baptiste Lamarck. As a neo-Darwinian (in the twentieth-century sense) and as someone who believes that a genuine core of wisdom is embodied in the Modern Synthesis, I should be a suitable choice for this role. Nevertheless, I regard the Synthesis, as it appeared in the first half of the twentieth century, as a highly incomplete evolutionary theory. Such an admission is neither a surprising nor a heretical view for a neo-Darwinian. The original Synthesis was essentially complete by 1947 and was celebrated as a fait accompli at a major conference at Princeton that year, the proceedings of which were published the following year (Jepsen, Simpson, and Mayr 1948). But, of course, 1947 was six years before James Watson and Francis Crick developed their model of DNA. In effect, the Synthesis was regarded as an established fact half a decade before there was a good model of what the gene was or any ideas about gene expression. In light of that fact alone, how could the Evolutionary Synthesis of the mid-twentieth century be anything but highly incomplete?

Moreover, despite my appointed role as devil's advocate, I have to say that I agree with most of what has been said at this conference. In particular, with respect to history, it is abundantly clear that Lamarck suffered a major historical injustice. It is both ironic and sad that the man who coined the word "biology" (in 1802) should have his name incorporated into what is probably the most pejorative term in biology. Second, with respect to the core subject of the scientific talks here, namely, epigenetic inheritance, it is clear that this is a real and very important phenomenon, and that some acquired traits can be passed on to following generations. Altogether,

I have few disagreements with either the main themes or the particular talks at this conference, though I would point out that nothing said here really supports the idea of the preferential or frequent inheritance of acquired *adaptive* traits, a tenet of classic Lamarckian views.

I will now turn to epigenetic inheritance and try to sum up what we know today and sketch the areas where I think we need to know a lot more. For this topic, where do things stand today? We should remember that "epigenetics" was formulated as a term in the 1950s by Conrad Waddington (Waddington 1957) and, at least by present-day standards, it was a rather vague expression. It was meant to encompass everything that happened in development "above" the level of the gene. In effect, it was a synonym for "developmental biology" before that term had acquired wide currency. "Epigenetics" did not catch on, and was even considered silly or pretentious by many biologists. In part, this was a case of guilt by association, reflecting Waddington's reputation as a gadfly, a role that he evidently relished, at least to a degree (Wilkins 2002). His reputation in recent decades, however, has improved considerably.

By the 1980s, the idea of "epigenetics" had revived, but not as a synonym for biological development in general. It was now used almost exclusively in connection with the idea of "epigenetic inheritance," those modes of cellular inheritance that are not encoded directly in the DNA but involve mechanisms "above" the DNA level. This change occurred because by then there was beginning to be evidence of stable modification of chromatin states that were at least inheritable through mitotic divisions and, in a smaller number of cases, mostly involving plants, through meiotic divisions as well. There had also been some conceptual advances in dealing with how such states might occur. I am thinking, in particular, of the highly influential paper by Robin Holliday and John Pugh on how states of DNA methylation could be inherited through cell divisions (Holliday and Pugh 1975).

Today the field of epigenetic inheritance is well established. It has moved from being seen as a rather odd, perhaps even peripheral, subject to one that is central to understanding normal development in plants, fungi, and animals (see chapter 21 in this volume; Kim, Samaranayake, and Pradban 2009; Martiennsson, Kloc, Slotkin, and Tanurdzic 2009; Henderson and Jacobsen 2007) and in disease states (chapter 23 in this volume; Jones and Laird 1999; Karpinets and Foy 2005; Wadhwa, Buss, Etringer, and Swanson 2009). Currently, it is a huge research enterprise, ranging across such phenomena as imprinting in mammalian embryos, X-chromosome silencing, cosuppression in plants, transgenerational inheritance of epimutations in plants, meiotic silencing of unpaired DNA segments, and cancer biology. Furthermore, the ideas about what constitutes epigenetic inheritance are considerably more advanced and diverse than they were in the 1970s and 1980s. Today, we know that it involves far more than just DNA methylation. These mechanisms include stable association of inhibitor RNAs with chromatin, regulatory loops, and modification of histones (Márton and Zhang 2007; see appendix B in this volume). In short, epigenetics has gone from being a marginal and suspect phenomenon to a mainstream subject.

Recognizing this, however, immediately prompts a two-sided question: What are the sorts of things that we do *not* know about epigenetic inheritance, and how might we begin to fill in those gaps in our knowledge? The question seems straightforward, but the answer(s) promise to be complex. In the first place, the exact mechanisms involved in many cases of documented epigenetic inheritance are unknown. Furthermore, even when we have an approximate answer—for instance, that it seems to involve some form of chromatin modification—often we do not know exactly what the chromatin state change involves. The answers will be of great interest both to biologists and to biophysicists who work on the details of chromosome structure and activity.

More important, however, the frequency and ease of transmission through meiosis of particular epigenetic chromatin states is largely unknown. Many epigenetic states are reversed—for example, in the mouse, where there is demethylation of methylated cytosines in germ line development. So what kind of epigenetic modifications are passed through the germ line? Can any, or many, be passed through with near 100 percent efficiency? We do not know, and this is of the greatest importance generally to assessing the importance of epigenetic inheritance in evolution (see chapter 31 in this volume; Slatkin 2009). There are other important related questions in this regard. In particular, are some organisms more prone to meiotic transmissible epigenetic modifications? (For instance, are plants better than, say, platypuses at cross-generational transmission of epimutations?) Can external conditions influence the probabilities of such transmission? Can epimutations predispose toward conventional genetic mutations at the affected loci?

The key point is that if epigenetic states are important to evolution, they are important through stable changes in these states, namely, transmissible epimutations. And if epimutations are not transmitted with reasonable stability over generations, they cannot have any long-term evolutionary potential (Slatkin 2009). If an epimutation is to have evolutionary importance, it must persist. Either epigenetic states must be highly stable or they have to be converted to some form of hard inheritance or, quite probably, both. This matter is central to whether epimutations can be treated as equivalent to conventional mutations or whether, if they have some degree of stability, some new population genetic theory is needed (see chapter 31 in this volume).

The relevant comparison is with the stability of true mutations; and, at least, unique DNA sequences—the kinds associated with coding regions—are generally very stable. For point mutations, the mutation rate is generally estimated to be 10^{-10}/base pair/replication. This is almost certainly more stable than most epimutations. Furthermore, while changes in DNA sequence in the germ line are passed on faithfully, we know that many epimutations, particularly those involving methylation of DNA base pairs, are not. On the other hand, some epimutations seem to be transmitted transgenerationally with high stability in plants and animals (Roemer, Reik, Dean, and Klose 1997; Cubas, Vincent, and Coen 1999; Johannes, Porcher, Teixeira, Saliba-Colombani, Simon, Agier, Bulski et al. 2009). The questions concern the frequency of such stable epimutations (relative to the total set) and the factors that affect whether or not epimutations can have high heritability. The other possibility that might give epimutations evolutionary significance is that they might preferentially, and with relatively high frequency, be converted into mutations at the same or neighboring sites in the same genes (Karpinets and Foy 2005).

Indeed, unless they have similar transmissibility to true (DNA sequence-based) mutations, epimutations would presumably have to be replaced by DNA sequence mutations, or to prompt the selection of mutations of similar effect, and there is a phenomenon of relevance that we have discussed in this conference that could have this effect: genetic assimilation. Waddington worked on converting developmentally plastic states through experimental means to give states that had stable inheritance (Waddington 1957). How pervasive is this in nature? This is hard to know, because once an epimutation has been converted to a DNA sequence mutation, you lose all trace of its origin as an epimutation. To assess these matters, we need more information on genetic assimilation. Most of our knowledge comes from that small body of experiments that Waddington carried out in the 1950s, and they need to be repeated and extended. The most dramatic experiment, involving ether induction of *bithorax* phenocopies in *Drosophila* and its dependence on certain genetic constitutions, has, however, been repeated (Gibson and Hogness 1996). Nevertheless, the full state of genetic assimilation was not demonstrated and, in general, we need more experiments. Yeast is a good candidate organism for such tests because good epigenetic states can be induced and studied in yeast.

It would also be very helpful to have hybrid lab–field experiments involving engineered animals or plants with epimutations, releasing them into controlled field conditions, and following their transmission over several generations. This might seem an unrealistic aim in times of tight funding, but it would be a very helpful approach.

I would like to conclude by throwing this issue open to discussion: What do you think about this assessment of the importance of genetic assimilation for assessing the probable importance of epigenetic inheritance in evolution?

References

Cubas P, Vincent C., Coen E. 1999. An epigenetic mutation responsible for natural variation in floral symmetry. Nature 401: 157–161.

Gibson G, Hogness DS. 1996. Effect of polymorphism in the *Drosophila* regulatory gene *Ultrabithorax* on homeotic stability. Science 271,5246: 200–203.

Henderson IR, Jacobsen SE. 2007. Epigenetic inheritance in plants. Nature 447: 418–424.

Holliday R, Pugh JE. 1975. DNA modification mechanisms and gene activity during development. Science 187,4173: 226–232.

Jepsen G, Simpson GG, Mayr E, eds. *Genetics, Paleontology and Evolution.* Princeton, NJ: Princeton University Press; 1948.

Johannes F, Porcher E, Teixera FK, Saliba-Colombani V, Simon M, Agier N, Bulski, A. et al. 2009. Assessing the impact of transgenerational epigenetic variation on complex traits. PLoS Genet. 5,6: e1000530.

Jones PA, Laird PW. 1999. Cancer: Epigenetics comes of age. Nat Genet. 21: 163–167.

Karpinets TV, Foy BD. 2005. Tumorigenesis: The adaptation of mammalian cells to sustained stress environment by epigenetic alterations and succeeding matched mutations. Carcinogenesis 26,8: 1323–1334.

Kim JK, Samaranayake M, Pradban S. 2009. Epigenetic mechanisms in mammals. Cell Mol Life Sci. 66,4: 596–612.

Martiennssen RA, Kloc A, Slotkin RK, Tanurdzic M. 2008. Epigenetic inheritance and reprogramming in plants and fission yeast. Cold Spring Harb Symp Quant Biol. 73: 265–271.

Márton C, Zhang Y. 2007. Mechanisms of epigenetic inheritance. Curr Opin Cell Biol. 19: 266–272.

Roemer I, Reik W, Dean W, Klose J. 1997. Epigenetic inheritance in the mouse. Curr Biol. 7: 277–280.

Slatkin M. 2009. Epigenetic inheritance and the missing heritability problem. Genetics 182: 845–850.

Waddington CH. *The Strategy of the Genes.* London: Allen & Unwin; 1957.

Wadhwa PD, Buss C, Entringer S, Swanson JM. 2009. Developmental origins of health and disease: Brief history of the approach and current focus on epigenetic mechanisms. Semin Reprod Med. 27,6: 358–368.

Wilkins AS. *The Evolution of Developmental Pathways.* Sunderland, MA: Sinauer Associates; 2002.

41 Final Discussion

The workshop closed with a session, chaired by Everett Mendelsohn, in which participants were invited to look at both the present state of Lamarckism and its future. It was an animated debate, which continued long after the formal session had ended, as everyone wandered, glass in hand, in the garden of the Van Leer Institute. Here we can offer only a summary of the main topics discussed. We started by considering the kind of biological evidence and the types of models that would convince evolutionary biologists that soft inheritance in particular, and Lamarckian problematics in general, should be a central part of modern evolutionary thinking; we then moved on to the transfer of ideas between evolutionary biology and sociology, and finally considered the nature and significance of the changes that evolutionary biology is undergoing today—the question of whether we are observing a revolution or merely a reform.

Lamarckism and Evolutionary Biology Today

Adam Wilkins kick-started the discussion by asking whether a population-genetics-minded evolutionary biologist would regard soft inheritance as of any importance. He argued that the only way in which an unstable epigenetic variation can become evolutionarily important is if a genetic change occurs in genes affecting the same trait. This is quite possible, he pointed out, because there are a lot of genes underlying a trait; if there is selection for a particular phenotype that is initially caused by an epimutation, then there will be many targets for mutations that genetically simulate the epigenetically induced trait. In his opinion, the evolutionary significance of soft inheritance depends on how common and important genetic assimilation proves to be, and on the stability and fidelity of epigenetic inheritance.

The Prevalence and Evolutionary Significance of Genetic Assimilation

Scott Gilbert and Eva Jablonka responded to Adam Wilkins's query about the prevalence of genetic assimilation. Gilbert pointed out that there is quite a lot of evidence for genetic assimilation from laboratory experiments, and that although there is indeed a dearth of direct evidence from studies of natural populations, there is plenty of *indirect* evidence. He gave two examples, one old and one recent. The old example goes back to late nineteenth-century experiments on butterfly wing pattern and coloration. These showed that heat shocks given during the early development of butterflies that normally live in cool areas induce wing patterns and coloration that resemble those of closely related subspecies or species living in hotter regions. This suggests that the conditional responses of ancestral species that lived in a fluctuating environment and exhibited polyphenism (that is, they had the ability to produce alternative morphologies in response to the different environmental cues) may have become partially canalized in the monomorphic species that display just a single type of wing pattern and coloration.

Gilbert's recent example was a study by Sangster and his colleagues of heat-shock-induced variation in the plant *Arabidopsis thaliana.* They showed that heat shock interferes with the activity of Hsp90, a chaperone protein that in normal conditions takes care of the way proteins fold and "covers up" mutational changes in proteins. When the amount of Hsp90 that is available for this normal buffering activity is lowered by a heat shock, many of the genetic variations that were previously buffered are revealed, and the different phenotypes can become genetically assimilated.

Eva Jablonka described another example of recent work that supported the idea that genetic assimilation is a factor in evolution. This was Richard Palmer's work on the evolution of asymmetry. There are two basic types of asymmetry. The first is *fixed asymmetry*, which is usually genetically determined and nonplastic; a population consists of individuals with either left asymmetry or right asymmetry, but not both. The second is *random asymmetry*: both right and left types exist in a population in more or less equal proportions, and whether an individual is "left" or "right" is the result of local conditions or developmental noise (i.e., it is a plastic, condition-dependent trait). From the phylogeny of a species-rich group in which fixed asymmetry has evolved independently several times, it is possible to determine whether fixed asymmetry evolved directly from symmetry, or whether the evolutionary route was from symmetry to random plastic asymmetry, and only then to fixed asymmetry. Palmer carried out such an analysis for species of bottom-dwelling flatfish in which one of the eyes migrates to the other side of the head during development, so the adult ends up having both eyes on one side. There are species with fixed right asymmetry, fixed left asymmetry, and random asymmetry. Comparative analysis of living and fossil species strongly suggested that fixed asymmetry evolved from random asymmetry. In this case, and in several others analyzed by Palmer, the conditional phenotype preceded the genetically fixed one, which suggests that fixed asymmetry evolved by genetic assimilation.

Both Gilbert and Jablonka insisted that the basic requirements for genetic assimilation are not demanding, and this makes it likely that it is common. The first requirement is that environmental conditions are both inducing and selecting, and this is frequently so. The second is that a there is a sufficient heritable variation, and this is known to be true for many natural populations. The third requirement, that genetically internalizing all or part of a plastic response is often beneficial, seems likely because there are many circumstances in which a reduced dependence on environmental conditions, which may be unreliable, is an advantage.

Konstantin Anokhin maintained that genetic assimilation of learned behaviors is probably of great importance in the evolution of complex behavior. He argued that the present neo-Darwinian account of how these behaviors evolved is not satisfactory, and genetic assimilation may be the missing mechanism. Usually, the solution to a new challenge that an animal finds by learning is based on multiple and simultaneous changes in the nervous system. It is almost impossible to find such a solution immediately through mutation. But if learning enables adaptation, genetic assimilation leading to the internalization of some of the components of the previously learned behavior may occur, thus enabling the descendants to respond to the same challenge more easily. Such local canalization of a behavioral response does not need to impair the animal's general learning ability, so responses to various evolutionary stable challenges can be gradually assimilated and become part of the a maturation process which anticipates learning. Anokhin said that he and his colleagues are studying the neurobiological bases of behavioral adaptations by mapping learning-induced gene expression in neurons of whole brains by optical tomography, and they find millions of altered circuits at each learning episode. As he sees it, only gradual genetic assimilation can accomplish developmental internalization and systematization of the correlates of such complex, learning-based adaptations.

Beyond Genetic Assimilation: Selection for Polyphenism

Gilbert pointed out that when thinking of the phenotype-first principle, we can go beyond genetic assimilation. West-Eberhard had suggested the term *genetic accommodation* for the process through which, following developmental response to environmental change or to a new mutation, selection leads to genotypic change. Genetic accommodation may lead to canalization (which decreases the range of phenotypic responses, making them more buffered), to enhanced plasticity (which increases the range of responses, making them more condition-dependent), or to the amelioration of the deleterious side effects of a new environmental or mutational challenge. From this point of view, genetic assimilation is just a special case of genetic accommodation that occurs when selection at the genetic level leads to a more specialized and canalized response.

Gilbert illustrated selection for induced plasticity (rather than canalization) by describing Suzuki and Nijhout's experiments on the hornworm moth, *Manduca sexta*. The caterpillar of this moth is always green, even when heat-shocked, but there is a black mutant form that when heat-shocked can develop coloration ranging from black to green. Suzuki and Nijhout selected and bred from the individuals of this mutant strain that became green after the heat shock, and also from those that remained black. In the first line, they ended up, after thirteen generations, with individuals that had become polyphenic—black when it was cool and green when it was mildly hot. In the other line, where nonresponsiveness to the heat shock was selected, all individuals remained black, even in very high temperatures, so in this case black coloration became genetically assimilated. What these beautiful experiments show is that both enhanced canalization and enhanced plasticity can evolve through selection for the regulation of responsiveness.

Mutational Assimilation

The possibility that epigenetic variation can have a direct effect on mutability was suggested by two of the participants. Evelyn F. Keller pointed out that epigenetic variations can bias mutation rates: DNA methylation at CpG sites can have a role in this, because methylated cytosines are more readily converted to thymines than nonmethylated ones. The epigenetic state of chromatin also influences genetic change through its effect on transposition and recombination, with active chromatin being more prone to both. Such effects, Keller argued, could be of enormous importance in directing evolution, in terms of both pace and direction. She thought that it is misleading to present directed and random processes as the two alternatives for the origin of mutants. In fact, the occurrence of mutations is never strictly random, but always biased, and the evolutionary implications of conditional and site-specific mutation rates must be studied.

This point was reinforced by Peter Gluckman, who drew attention to the biased and higher rate of mutations in CpG sites in cancer. Many mutations in human cancers are at potentially methylated CpG sites. Increased methylation of suppressors of oncogenes can lead to cancers, and there are many examples of this. For example, CpG islands are hypermethylated in retinoblastoma tumor suppressor gene (*Rb*) and in *BRCA1*, a breast-cancer susceptibility gene. Changes in the pattern of DNA methylation are also very common in some colorectal cancers. An example is the *TP53* gene that normally protects against cancer, which is mutated in CpG sites in colorectal cancer. Interestingly, this cancer, which is one of the most common malignancies worldwide, is associated with dietary risk factors, so the relations among diet, changes in the methylation of CpG sites, and mutability are of particular interest.

Beyond Genetic Assimilation: Plasticity and Niche Construction in Evolution

Paul Griffiths, Sonia Sultan, and Ehud Lamm maintained that it was a mistake to focus only on genetic assimilation when thinking about the significance of developmental plasticity and epigenetic inheritance in evolution. Griffiths contended that the niche construction framework suggests that even if epigenetic effects are transient, they can have a very large evolutionary effect. The organisms' effects on their ecological niche can go on, as a separate heredity channel, exerting selection pressures on the genomes of the next generations. He highlighted two points. First, because of the selection pressure that is associated with ecological niche construction, the genes an individual gets will be different from the genes it would have gotten had there been no niche construction. In other words, the population dynamics changes if there is niche construction, even if the effects at the level of the epigenetic correlates of plasticity are transitory. The second point was that developmental plasticity leads to change in the *developmental niche*. This is a systemic change; here it is not the individual epigenetic elements (which can be transient in heredity) that are significant, but the system dynamics. The epigenetic system can be very robust even if elements change, as network models show. So, even when individual epigenetic variations are unstable in heredity, and even if genetic assimilation was found to be uninteresting and unimportant, epigenetic inheritance, which is part of developmental niche construction, can neveretheless be important for evolution. As Scott Gilbert had stressed in his talk, one must add developmental symbiosis to this picture, because it leads to the formation of developmental niches and new developmental interactions which can be selected. All this, Griffiths said, should have a transformative effect on how we think about evolution.

Sonia Sultan and Ehud Lamm added another facet to this discussion. Sultan discussed one of the implications of plasticity: the creation of an opportunity for what is being called "relaxed selection," which can lead to a greater possibility of evolutionary innovation than is found in less plastic systems. Consider two different environments, E_1 and E_2, and a plastic organism which responds to the environments by developing phenotypes P_1 and P_2, respectively. Now suppose that one of these environments, say E_2, becomes very rare, which means that P_2 will very rarely be expressed. As a consequence, this phenotype will be in effect selectively neutral. What happens to neutral traits is that they accumulate mutations. If, after many generations, the environmental distribution changes again and E_2 becomes common, it is very likely that the norm of reaction will have evolved through drift so that a very different phenotype will now be expressed. Even though it is likely that many such new phenotypes will be maladaptive, different mutations will accrue in different individuals, and given that when mutations accumulate, new epistatic combinations can be formed, the new norm of reaction may yield a new, adaptive trait. As a result of this process, Sultan argued, plasticity itself may be a source of evolutionary novelties. Ehud Lamm added that we should not think only about the selectively advantageous effects of plasticity and epigenetic variations. Epigenetic effects and mechanisms can be important, for example, during population bottlenecks when just by chance the genotypes of novel phenotypes can become fixed. Epimutations, and epigenetic mechanisms more generally, may be involved in development as well as in many selective and nonselective processes in evolution, and all these have to be considered and modeled.

Data, Models, and the Significance of Epigenetic Inheritance for Evolution

In their talks, several participants had pointed to the need for additional experimental data to assess the evolutionary importance of epigenetic inheritance, and also for models that would address the concerns of population and quantitative geneticists (chapters 21, 31, 40). Both issues were brought up in the discussion, and it was emphasized that experimental studies should include not only traits that are inherited in the absence of the original inducing stimulus, but also traits that, while still requiring a stimulus, need a significantly smaller one than in the previous generations.

An interesting and important type of transgenerational epigenetic effect that goes beyond the existing framework of most current epigenetic research was described by Inna Shamakina and Konstantin Anokhin. They are looking at the transgenerational effects of chronic intake of alcohol and opiates. In their studies they tested whether chronic exposure of male rats to alcohol or morphine affects the early somatic and behavioral postnatal development of the progeny—their sensitivity to acute administration of the drugs, whether or not they developed a tolerance to the drugs, their preference for ethanol when given a choice (free choice paradigm), and drug-induced immediate-early gene expression in the brain. They found that chronic alcohol consumption by the father resulted in the progeny having a pronounced induction of *c-fos* mRNA in their cerebral cortex following acute administration of ethanol, something that was not seen in control animals. Chronic administration of morphine to males before mating resulted in an increased analgesic effect of morphine in their offspring. These offspring also developed tolerance to morphine (dependence) more rapidly than control progeny. The data therefore show that chronic exposure of males to drugs of abuse can have transgenerational effects on their progeny, but some of these effects are opposite in direction to those seen in the fathers.

To explain these data, Shamakina and Anokhin proposed a biphasic model: (1) Chronic changes in the organism's milieu pose a physiological problem for it (problem I—increased levels of circulating opiates), which can be solved by corresponding biochemical adaptations (e.g., decreased sensitivity of opiate receptors). These adaptive adjustments can be transmitted to the progeny, thus preparing it for the changed milieu. (2) When the predictive transgenerational adjustments of the biochemical settings do not conform to the actual environment (e.g., the offspring *are not* exposed to the high level of circulating opiates that their fathers experienced), the offspring face a secondary adaptation problem (problem II—the absence of the high levels of opiates to which they are biochemically set). To deal with this discrepancy, the offspring need to develop a series of adaptations (e.g., increased sensitivity of opiate receptors) that are opposite in direction to the initial parental adaptation. These secondary adaptations might include not only biochemical, but also behavioral changes in the progeny (for example, drug-seeking strategies to meet the parentally transmitted set points of high circulating levels of opiates).

It was recognized that the observations and model go beyond traditional epigenetic studies, which focus on similarity between parents and offspring. They show that an external stimulus that induces phenotype P_1 in the parental generation can induce a phenotype P_2 in the offspring that may be the opposite of the parental one. The P_2 phenotype might then fade away, or be inherited for a number of generations (this has yet to be established experimentally). A more interesting and extreme case would be when the induction of P_1 leads to the development of phenotype P_2 in the offspring, which leads to P_3 in the grand-offspring, and so on, until a final stable state is reached (or the lineage dies out). All such possibilities need to be explored, because as many participants emphasized, transgenerational effects are of enormous importance in medicine as well as in evolutionary biology

The Stability of Epigenetic Variations

An important question raised by Adam Wilkins in his final "devil's advocate" lecture, and repeated in this discussion, was the stability of epigenetic variations. Wilkins argued that if epigenetic variations are lost at a rate of 1 percent or more with every meiotic (gametic) transmission to the next generation, it would probably be impossible for a population to maintain traits underlain by such variations. The direct effect of epigenetic inheritance on evolution—evolution on the epigenetic axis itself—depends on the stability of epimutations.

Several answers were given to this question. Eva Jablonka noted that, first, we do not have enough information about the stability of epigenetic variations, but what we do know suggests that there is a range of stabilities—some epigenetic variations are ephemeral, others very stable, and there seems to be everything in between. Second, even if epigenetic variations are not very stable—say as many as 10 percent are reset every generation—they can still have important effects on evolutionary dynamics. Studies of maternal effects, which consider a lag of just one generation, show that this lag can make a big difference to the dynamics of the population over time, and the same point was made in the model of Pearl, Oppenheim, and Balaban on the effects of bacterial persistence on population dynamics. Third, the effect of the reversibility of epigenetic variations can be balanced or overwhelmed by strong selection. Fourth, whereas a classical mutation usually appears in a single individual, epigenetic variation can be induced in many individuals simultaneously. Furthermore, the inducing conditions may occur again and again. The establishment of an epigenetic variation in a population can therefore be a lot faster and far more likely than that of a mutational change. Fifth, even if the stability of individual epigenetic variations is low, the stability of the trait can be very high. Stability depends on the regulatory architecture of the cellular–developmental network (networks can be very robust even if large parts are deleted) and on the strength of the attractors.

Jablonka agreed that it is essential to be able to measure the extent, stability, and heritability of epigenetic variations, for this would make it possible to evaluate the epigenetic component of disease risk, help to assess organisms' responses to ecological stresses, and, of course, improve our understanding of evolutionary dynamics. She told the participants about a recent model constructed by Omri Tal, Eva Kisdi, and herself which suggests a method for measuring epigenetic transmissibility. Using a classical quantitative genetics approach, but adding information about the number of opportunities for epigenetic reset between generations and assumptions about environmental induction, it is possible to estimate heritable epigenetic variance and epigenetic transmissibility for both asexual and sexual populations. This, she said, can be a first step toward identifying phenotypes and populations in which epigenetic transmission occurs; it enables a preliminary quantification of epigenetic transmissibility, which can then be followed by genome-wide association studies based on molecular data.

Social Dimensions and the Future

Two principal topics were considered in the second part of the discussion, both of which related to the sociological and historical dimensions of the Lamarckian problematics: (1) the relationship between evolutionary biological theories and sociological ones, and the possibility of any kind of transfer, and (2) the nature of the change that evolutionary biological theory has been undergoing.

Interactions and Transfer between Evolutionary Biology and the Social/Cultural Sciences

Snait Gissis discussed the problems that transfers between evolutionary biology and the social sciences and the cultural disciplines raise. Biologists and social scientists use each other's models and metaphors as a means of changing both general fields and their boundaries. From the 1970s onward, such transfer occurred through sociobiology and its offshoots in evolutionary psychology, and through the discussions of cultural evolution either as a selection-based process or as Lamarckian-epigenetic-developmental one. Those in favor of transfer from contemporary Darwinian evolutionary biology to the social sciences claim that variation, selection, retention, and transgenerational transmission/inheritance are principal elements in both the biological and the social/cultural fields. There are, however, disagreements about their biological characterization and relative significance. There have also been various attitudes among social scientists toward the prospect of such transfer, ranging from the hope of finally putting the social sciences on a firm, unified, and rigorous scientific footing, to a call for a renewed bifurcation between "nature" and "culture," as well as a claim that it is unlikely that any historical narrative can provide an adequate test of an evolutionary explanation. The question is whether transfer from the Lamarckian-developmental-epigenetic perspective can provide the sole basis for explaining the social-cultural field, particular cultures, and specific societies.

Gissis noted that historically, positions against transfer between evolutionary biology and the social sciences relied on two kinds of arguments: one was the inner logic of the transfer—its lack of empirical support and substantiation; the other was its possible political and social implications. Nevertheless, there has been a general tendency among both theoreticians and philosophers of evolutionary biology to think that one can make a simple, or at least a rather straightforward, transfer from biological models to explanations of culture and society, and that it should work. Two extreme modes of transfer have been used since the nineteenth century. One is a direct transfer, and the other is transfer by analogy. Most of those who at present either commend or practice transfer between the two fields use transfer by analogy, whether they acknowledge it or not.

Transfer assumes sufficient resemblance between the two fields, although one could argue that the important differences call for entirely new analytical tools. When one does work by analogy, one is faced with problematics which are central yet specific to social sciences, even though they may have some resonance in biology. For example, from the perspective of the social scientist, what constitutes the basic units in such a transfer is not clear, nor is how it is chosen. Would it be some kind of biological individual which would be transferred by analogy to the target field—sociology? Which level of biological or ecological entity would be chosen for that—the analogue of a gene, of an organism, of a species, or of a group? Or maybe a combination of them? What are the limits of contingency and determination? How can one demarcate the boundaries of the nonreproducing—yet evolving—entity "society," which is not a "natural unit," nor a "population," nor a simple aggregate of individuals? Another example is the problem of cultural transfer and biological reproduction. It is not clear that there is that much in common between biological transmission and cultural inheritance, certainly from the viewpoint of the Modern Synthesis. And not just there: in culture there is a blurring of the distinction between preserving and transforming. The meaning of "development" points to profound differences between cultural and biological frameworks, even when one chooses to relate to individual processes (and that is a controversial decision). Take Gluckman and his colleagues' contribution (chapter 23), in which the possible trajectory of the future—the developing organism's "anticipation" of the future environment—has a role in the shaping of the present, in how the organism responds to challenges. In social and cultural fields the trajectory of the future is far more complex and mediated through institutions, systems of symbolization, intentions, self reflectivity, modes of historicity, context dependence, etc. These are crucial for the analysis of memory, both collective and individual.

Moreover, in the contemporary resurgence of evolutionary thinking across the social sciences it is often forgotten that there is a group of terms—used metaphorically within evolutionary biology—such as tradition, culture, memory, and recall, that were originally taken over from the social/cultural sciences, and were then transported back analogically into them, carrying their new biological–evolutionary meanings as if they originated in biology.

Transgenerational Transmission: From the Social Sciences to Epigenetics?

A different slant on the interactions between biology and the social sciences was suggested by Ohad Parnes. He argued that the Modern Synthesis did not take over the social sciences. Contrary to our intuitions, the social sciences (sociology, anthropology, and psychology) were not dominated by Darwinism. Of course one has had sociobiology and its offshoots, but historically there was also an opposite move. In the 1920s and 1930s, 1940s, and 1950s, something very interesting happened in the social sciences. Practitioners of these disciplines realized that notions of biological inheritance were not applicable. One could not at the time use ideas from the Modern Synthesis to explain social phenomena, because these phenomena clearly did not belong to genetics. Transgenerational effects were therefore formulated in other ways, and one can see this already in Freud, who was not a Darwinian. Cultural anthropology, as emphasized by Ruth Benedict, was a science where dealing with transmission across generations was central. One of the principal topics of sociology and social anthropology at that time was the problem of what happens with immigrants in the third, rather than in the second, generation—those offspring of immigrants who were no longer exposed to the kind of conditions their parents and their grandparents experienced, but still showed some response to these conditions. In the 1950s and 1960s, psychology took over this idea of a third-generation experience—the idea that there is a transmission of something which is not genetic but still carries over to the third generation. Thus, in a way, social sciences predicted the discussions of epigenetics, the general idea of it, and developed methods to describe cross-generational effects. Some of these discussions are now reinterpreted biologically, in epigenetic terms, as discussed by various speakers.

The Nature of the Change in Evolutionary Biology

The character and significance of the change that is occurring in evolutionary biology as Lamarckian problematics not only becomes legitimate, but also assumes an important role, was the last topic discussed. Sander Gliboff gave a sharp formulation of three questions: What kind of historical changes are being observed, and how are they to be described by those who take part in bringing them about? Is a new synthesis coming into being? Can the change be considered a "paradigm shift" or, to put it bluntly, should it be viewed as adding a footnote to the evolutionary synthesis?

Sam Schweber argued that the shift in evolutionary biology has to be situated within a larger framework of change that is happening in all the sciences, a major transformation (an Ian Hacking kind of revolution) that also changes institutions and brings a different feel to the atmosphere. He pointed to "information" as the new site of change that has crept into every area from physics to chemistry to biology.

Two basic positions emerged among those discussing the nature of change within evolutionary biology: one group (Gilbert, Gluckman, Lamm, and Gissis) thought that the situation is still unstable and sought to characterize this instability; the other group (Sultan and Jablonka) affirmed that a radical, though not a total, shift has already taken place.

Scott Gilbert argued that at present we have in evolutionary biology what one might call an "unstable population." As he sees it, the question is: Is evolutionary biology as a whole evolving in a new direction, with a kind of a radical, new, anagenetic shift in a single lineage, or will there be a split, with the epigenetic view coexisting for a while in parallel with the Modern Synthesis view, somewhat like two species after a cladogenetic split. Eventually, one of the two theories will take primacy and become more widespread, depending on how much support and economic buttressing it gets.

Talking from the perspective of medical research, Peter Gluckman, too, maintained that the significance of epigenetics at the level of the organism is uncertain. He emphasized that its scientific importance has to be proven and established within larger frameworks that are biologically, medically, and therefore economically, significant. Conceptual change or experimental frameworks of limited scope will not suffice. The group of researchers, himself included, who believe that epigenetic inheritance is of biological importance, is relatively small in number and still at the fringe of the biomedical empire. They will remain there until there is unequivocal evidence that epigenetic processes are significant in the biology that matters in economic terms: for human health, for pharmaceutical and nutritional companies, for agricultural development. One needs access to experimental designs and challenging technologies in order to get the kind of empirical data that will shift mindsets. Instability and uncertainty will remain until such a stage is reached.

Ehud Lamm commented on that instability from a somewhat different point of view. He noted that biology is not one discipline, and its subdisciplines may be in different places in this historical change. In line with James Griesemer's arguments (chapter 31), he pointed out that people in the most reductionist and empirical parts of biology—those studying the molecular mechanisms of genetics—have the least difficulty in overthrowing the neo-Darwinian paradigm, because they are much less concerned with the evolutionary implications of their findings. This, he said, is very different from other subfields of biology such as population genetics and quantitative genetics. Thus, following Kuhn, one can say that different subdisciplines may retain their existing ways of thinking for a while, since paradigms often change when the older generation retires.

Snait Gissis suggested that rather than viewing the process of change as a sequential replacement of alternatives, one should regard it as a state of constant complicated dithering: there are "thought collectives" that hold the two systems of thought at once and use the resources of both, with a sense of loyalty to more than one system at the same time. It may take a long time to resolve, because it is generative to go back and forth from one to the other, and thus there is a holding on to both.

A Radical Shift—Probably a New Synthesis

Both Sonia Sultan and Eva Jablonka thought that a new synthesis was taking shape in evolutionary biology. Sultan argued that we are all presently in the midst of a paradigm shift. Relating her personal experience when discussing this question with colleagues, she said she was puzzled to encounter two incompatible reactions—people believed either that the paradigm has already shifted, or else that all is well with neo-Darwinism, apart from the need for a few tweaks and minor additions.

Jablonka argued that if one accepts that soft inheritance is of evolutionary significance, then the Modern Synthesis, which is in essence neo-Darwinian, is no longer valid. Of course, the insights that were gained and the important discoveries that were made must be preserved. Because of these contributions, many biologists who are very aware of the limitations of the neo-Darwinian approach and are actively engaged in evo-devo research, regard the present developments in biology as an extension of the Modern Synthesis rather than a replacement. However, the Synthesis defined itself partly by what it excluded; Lamarckian problematics was peripheral to it, and soft inheritance was anathema. If the variation-first view is returning to center stage in evolution, and if soft inheritance is found to be significant, then it seems that a new evolutionary synthesis—Darwinian–Lamarckian, but no longer neo-Darwinian—is dawning.

This final discussion ended with some observations and questions voiced by Everett Mendelsohn, and his words are an appropriate ending for this volume:

I have learned that the Lamarckian framework is broad: it is not a single theme, not even two or three; there are several important themes, and they are not competing, but seem to be interconnected. But I want to ask: Toward what end is this meeting going, or shall I even say, directed? What is it we want to see come out of it? And I don't just mean papers and transcripts, but what we want to see come out of these ideas.

The historians showed us that there are many Lamarcks: there was Lamarck the materialist, Lamarck the transformist, Lamarck the progressivist who saw a trajectory in nature, moving upward through time. Then there is the Lamarck that sometimes, but not always, talked of the inheritance of acquired characteristics. These Lamarcks have become the focus for different people who have used Lamarck as an emblem, an inspiration, a cover, or a dodge. Two hundred years after publication of the *Philosophie zoologique,* as historians we should bring clarity to the sense of what it is we want Lamarck to do for us, and what we should not be asking him to do, though some may think that we have to get on with arguments we have today.

What should *the sociological dimension* look like? Every figure we talked about operated in a real context, whether a university, a private institution, a Roman Catholic society, or whether you moved across boundaries from European to distant shores for experiments. Every figure we talked about and every participant in this room operates within institutional contexts. People tell me about work they hesitate doing because they are up for tenure, or meetings they did not want to become involved in because of the potential controversies, as they were building their own careers. There are more important sociological dimensions—the sociology of knowledge itself: On what are things predicated? Were they creating intellectual space? Were they revolutionary? Were you consciously setting yourself against something that exists? People have looked at the implications of attempts of marked shifts like moving lab to a new site, changing the nature of your research, the nature of the funding, the activities you are engaged in. These have social connotations.

Other questions concerned *the biologies* into which Lamarck and Lamarckism had been interpolated. Some biologists have been jousting with the Lamarckian tradition for years. At the moment they are in the strongest positions they have been in for a long time. There has been more progress at every level, not only on the theoretical and polemic levels, but also on the experimental level—beautiful experimental work that has been shared with us. I ask the experimentalists: What was it that generated the questions you asked? Where did they come from? Some of you can say they came from a person in the lab next door. But none of you were naive about the structure of the questions you were asking and the potential outcome of them. Almost everyone was aware that you were providing challenges to the central core of simple neo-Darwinism, an alternative explanation for things happening in nature, and you were demonstrating it through questions about nature and through the probes you were making.

Jim Griesemer's paper and the queries that it raised, as well as many other papers in the biology and philosophy sessions, make me ask: Are we witnessing a hardening of the program of soft inheritance? My sense is yes. Not only the intellectual strength of what's being presented and the nature of the empirical evidence being marshaled, but also the structure of the arguments brought forward in the theoretical presentations. They were not apologetic, they were challenging.

So toward what end are they heading? Toward a mode of finding shared space with neo-Darwinists, or are they heading for replacement? History is replete with examples of both. The genome rewiring paper and the papers about the epigenetic turn are suggesting that something more is happening than just adding elements to the existing view. For those who follow, the question is where is it heading? For those engaged, what is your strategy? Where do you want to be in five or ten years from now?

Further Reading

Genetic Assimilation and Accommodation

West-Eberhard MJ. Developmental Plasticity in Evolution. New York: Oxford University Press; 2003.

For more recent examples of selection for canalization and polyphenism, see Gilbert SF, Epel D. Ecological Developmental Biology. Integrating Epigenetics, Medicine, and Evolution. Sunderland, MA: Sinauer Associates; 2009.

Analysis of direction asymmetry can be found in Palmer AR. 2009. Animal asymmetry. Curr Biol. 19,12: R474–R477.

Niche construction, plasticity, and the origins of novelty

Moczek AP, West-Eberhard MJ, Sultan SE, Dworkin I, Abouheif E, Ledig C, Nijhout HF et al. The role of developmental plasticity in evolutionary innovation. Manuscript.

Odling-Smee FJ, Laland KN, Feldman MW. Niche Construction: The Neglected Process in Evolution. Princeton, NJ: Princeton University Press; 2003.

Snell-Rood EC, Van Dyken JD, Cruikshank T, Wade, MJ, Moczek AP. 2010. Toward a population genetic framework of developmental evolution: The costs, limits, and consequences of phenotypic plasticity. BioEssays 32,1: 71–81.

Stability of Epigenetic Inheritance and Its Significance

Jablonka E, Raz G. 2009. Transgenerational epigenetic inheritance: Prevalence, mechanisms, and implications for the study of heredity and evolution. Q Rev Biol. 84,1: 131–176.

The model for estimating epigenetic heritability is in Tal O, Kisdi E, Jablonka E. 2010. Epigenetic contribution to covariance between relatives. Genetics 184: 1037–1050.

Transfer between Disciplines

Ceccarelli L. Shaping Science with Rhetoric: The case of Dobzhansky, Schrödinger, and Wilson. Chicago: University of Chicago Press; 2001.

Maasen S, Mendelsohn E, Weingart P., eds. Biology as Society, Society as Biology: Metaphors. Dordrecht, Netherlands, and Boston: Kluwer Academic Publishing; 1995.

Maasen S, Weingart P. Metaphors and the Dynamics of Knowledge. London: Routledge; 2000.

Simons HW, ed. The Rhetorical Turn: Invention and Persuasion in the Conduct of Inquiry. Chicago: University of Chicago Press; 1990.

Wimmer A, Kössler R, eds. Understanding Change: Models, Methodologies and Metaphors. Houndsmill, UK: Palgrave Macmillan; 2006.

Recent Perspectives on the Present State of Evolutionary Biology Suggesting That the Evolutionary Synthesis Needs to be Extended

Pigliucci M, Müller GB, eds. Evolution: The Extended Synthesis. Cambridge, MA: MIT Press; 2010.

Appendix A: Mandelstam's Poem "Lamarck"

The great Russian poet Osip E. Mandelstam (1891–1938), who died in a prison camp during the Great Purges, wrote the poem "Lamarck" in 1932. Although the poem has several interpretations, many think that Mandelstam wrote it as an act of defiance against the overly materialistic atmosphere that surrounded him in the Soviet Union in the early 1930s.

According to this interpretation, the transformation from human to insect is exactly what happens when the teleology of Life is regarded merely as an accidental "trifle." By choosing Lamarck's vision, the poet emphasizes the importance of the mobile ladder that actually *leads* evolutionary development toward higher and nobler ends: Mozart's music, for example.

Mandelstam discovered the *Philosophie zoologique* through close and dear friends, the biologists Boris Kuzin and Nikolai Leonov, who later also suffered under the Stalinist purges. This acquaintance opened before him the "entry to a bright field of action," as he put it. "The lowest forms of organic life are the actual hell of men," wrote Mandelstam in his "Armenian Journey" (1933), and maybe that is why the idea of an evolutionary guiding hand leading mankind up the Lamarckian ladder, an idea so contradictory to the official strictly materialistic doctrine, was so dear to his heart.

The following is a new translation of the poem made by Yigal Liverant (2010).

Lamarck (1932)

Sage, that was as bashful as a child,
Awkward, clumsy, timid patriarch.
Who will fence, defending Nature's honor?
Well of course, the fiery Lamarck.

If our Life is nothing but a trifle
Of an empty, short and barren day,
On the mobile ladder of Lamarck then
On the lowest rung I have to stay.

I'll descend to barnacles and ragworms
Rustle past reptilia with fear.
Down the springy ramps and down the ravines
Like proteus I shall shrink and disappear.

I shall put on thick and scute-made mantle.
And proclaim my blood no longer warm.
I'll grow suckers overall and like a tendril
Swirl into depths of ocean foam.

Passing through insect classification
with their eyes that look like shots of wine.
He said: "Nature always lays in fragmentation,
There's no eyesight—you have seen for the last time!"

He said: "Now sonority must end,
And your love of Mozart is in vain,
Arachnoid deafness will descend,
For this lapse is stronger than our strength."

Nature has retreated, stepped away
As if we were of no use for Her
And She inserted the line of oblong brain
Into a dark scabbard, like a sword.

And the drawbridge—She forgot to lower.
Missed the chance of saving them instead—
Those who'll lie in green grave when it's over,
Those with supple laughter, those whose breath is red.

Appendix B: Mechanisms of Cell Heredity

Several chapters in the biological and philosophical sections refer to mechanisms of cell heredity (*epigenetic inheritance systems*, or *EISs*). We cannot give a comprehensive account of this immense and ever-growing topic, but in view of its importance for the subject of this book, we provide here a short overview for nonspecialists. In addition to the mechanisms of cell heredity discussed in this volume (those based on self-sustaining loops, chromatin-marking systems, and RNA-mediated systems), we include a very short description of structural inheritance. This is needed to give a more complete and balanced picture of the mechanisms of cell heredity.

Chromatin-Marking Systems

Chromatin is the material of which chromosomes are made. It consists of DNA, small chemical groups covalently attached to DNA (e.g., methyl groups), and DNA-associated proteins and RNA molecules. The organization of chromatin and the dynamic interactions of chromatin components within and between chromosomes determine when, where, and to what extent genes are expressed and repaired, and how chromosomes behave during the cell cycle. The organization of chromatin is different in different cells, and can change over time; it is reconstructed during cell division, and can be transmitted between generations.

DNA Methylation and the Inheritance of DNA Methylation Patterns

DNA methylation—the addition of methyl (CH3) groups to some of the bases in DNA—is an epigenetic modification found in all kingdoms of life. In bacteria, it is important for defense against viruses and for the coordination of DNA replication and repair. It is also associated with the regulation of gene activity, and variant patterns of DNA methylation and the associated states of gene activity can occasionally be transmitted between generations.

In eukaryotes (organisms with cells having a membrane-bound nucleus), DNA methylation is associated with the stabilization of specialized chromosomal regions such as centromeres and telomeres, and with DNA replication and repair. It also plays an important role in the long-term silencing of genomic regions such the inactive X-chromosome of female mammals, transposable elements, imprinted sequences in both animals and plants, and some tissue-specific genes in cells in which their activity is not required. Recently it has been found that when the coding regions of genes in invertebrates, mammals, and plants are highly methylated, transcriptional activity is enhanced (reviewed in Suzuki and Bird 2008). Thus, a high level of methylation in noncoding regulatory regions is associated with gene silencing, whereas methylation in the body of the genes has an opposite, activating, effect. The different patterns of methylation found on a particular DNA sequence in cells of different types are known as alternative *chromatin marks*.

Replication of DNA methylation patterns is the best-understood system of chromatin inheritance. In eukaryotes, methylation usually occurs at the carbon 5 position of cytosines in CpG doublets (and CpNpG triplets in plants). CpG and CpNpG sequences are palindromic: the two strands of the DNA duplex are mirror images of one another and have symmetrical patterns of methylation. Consequently, replication of methylation patterns is semiconservative (figure B.1). Following DNA replication, there are two hemimethylated duplexes, but the hemimethylated sites are recognized by an enzyme, maintenance DNA methylase, which preferentially adds methyls to Cs in the new DNA strand. In this way DNA methylation patterns are reconstructed and transmitted from one cell generation to the next. The high fidelity with which methylation patterns can be reconstructed in somatic cells is the result of the activities of additional proteins that interact with hemimethylated sites and the maintenance methylase (reviewed in Ooi and Bestor 2008).

DNA methylation is involved not only in within-organism cell heredity but also in intergenerational heredity. There are now many examples of altered DNA methylation patterns that are transmitted for several generations.

The Inheritance of Histone and Nonhistone Protein Marks

Nucleosomes are the basic components of chromatin in eukaryotes. The nucleosome core consists of 147 bp of DNA wrapped around an octamer consisting of pairs of four histone proteins: H2A, HB, H3 and H4. Although the nucleosomes are fundamental to DNA packaging and higher-order chromatin conformation, they are dynamic structures, constantly in flux, readily changing their chemical nature and position during the cell cycle and as the cell responds to changes in its environment. Post-translational modifications (PTMs) of histones, which alter their interactions with DNA, affect chromosome structure and cellular activities, including the regula-

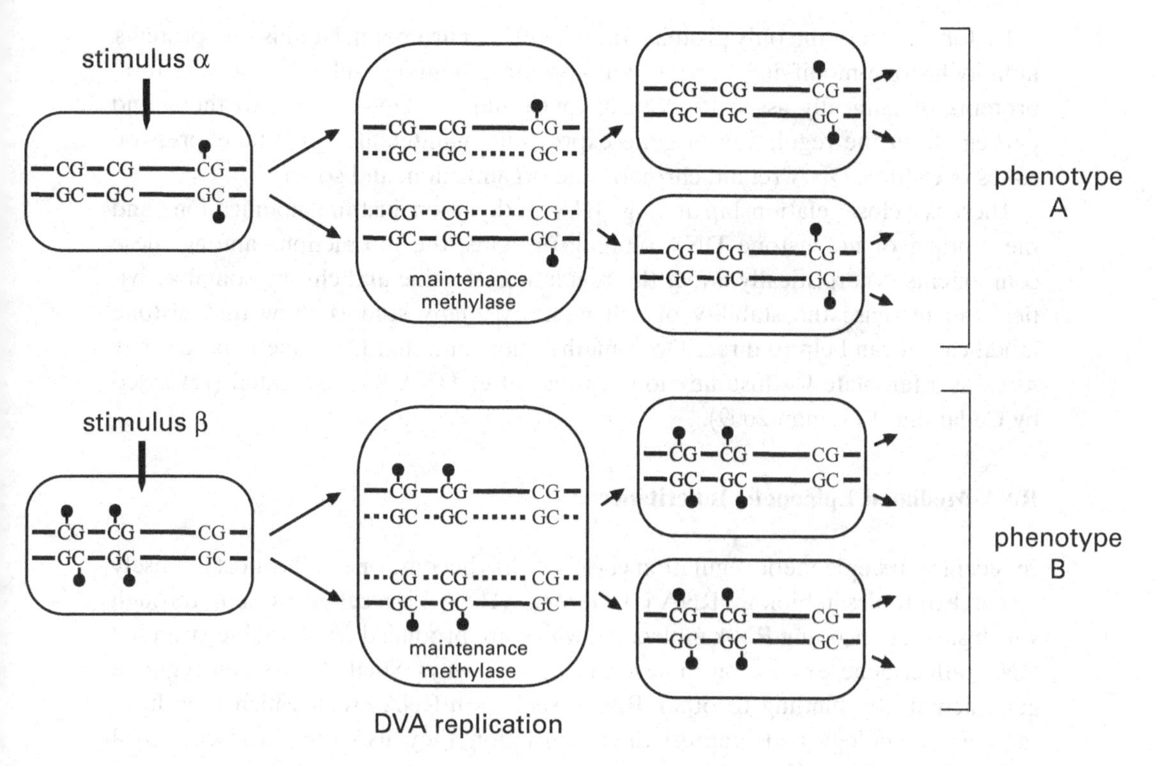

Figure B.1
The inheritance of methylation marks. As a result of receiving different stimuli, CG sites in the same locus of two genetically identical cells become differentially methylated (•). As a consequence, they have different phenotypes. Following replication, the old DNA strands remain methylated, but the new strands are initially unmethylated; a maintenance methylase recognizes CG sites that are asymmetrically methylated, and methylates the Cs of the new strand; the daughter cells thus inherit the mother cells' pattern of methylation (the marks). Based on Jablonka and Lamb 2010.

tion of gene expression, and DNA repair and replication. There are several DNA-encoded variants of particular histones within cells, and the replacement of one variant by another can remodel chromatin and affect gene activity. The histone variants and PTMs associated with a DNA sequence are chromatin marks that may be transmitted to daughter cells.

Several models for the inheritance of nucleosome configurations have been suggested. In all, the specific nucleosome variants and the PTMs of the parental nucleosomes are used as the blueprints for restoring nucleosome configuration in daughter cells. Other chromatin components, such as the methyl groups attached to DNA, RNAs, and the location of the chromosomal domains within the nucleus, may assist in restoring nucleosome configuration.

Histones are not the only protein components of chromatin. Nonhistone proteins, notably histone-modifying enzymes but also other bound regulatory and structural proteins, dynamically associate with histones and the DNA bound to them, and participate in the regulation of gene expression, maintenance of gene-expression states over time, DNA repair, chromosome organization, and so on.

There is a close relationship among DNA methylation, histone modifications, and the binding of nonhistone DNA-binding proteins; the interactions among these components synergistically affect the regulation of genic and chromosomal activities, and increase the stability of cell memory. Many studies show that histone modification can help to direct DNA methylation, and that DNA methylation may serve as a template for histone modifications after DNA has replicated (reviewed by Cedar and Bergman 2009).

RNA-Mediated Epigenetic Inheritance

In recent years, epigenetic regulation via RNAs has become one of the most intensely researched topics in biology. RNA interference (RNAi) is a cellular system through which small, noncoding RNA molecules, which are produced from double-stranded RNA, affect gene expression in several different ways. Small RNAs can regulate gene activity by binding to other RNAs, such as mRNAs with which they have sequence homology, and suppress their translation; they also affect transcriptional activity by modifying DNA.

Figure B.2 shows how small RNAs affect cell and organism heredity. They can do so in three different (nonexclusive) ways: (1) by guiding and targeting changes in DNA base sequences, which are then replicated; (2) by guiding, targeting, and assisting in transmitting variations in chromatin structure (DNA methylation, histone modification, recruitment of nonhistone DNA-binding proteins); (3) by a replication process that makes the small RNAs double-stranded, with the result that their supply is replenished and they may be transmitted to daughter cells during cell division. In some organisms, small RNAs can also move between cells, and hence act as systemic signals.

Self-Sustaining Loops

A self-sustaining loop is a regulatory circuit that dynamically maintains a particular pattern of cellular activity. A cell with such a regulatory circuit can have at least two stable states, one active (with the circuit operating), the other inactive. Complex circuits can have more than one persistent active state. A common and important type of positive feedback loop is one in which an inducible gene product acts as an

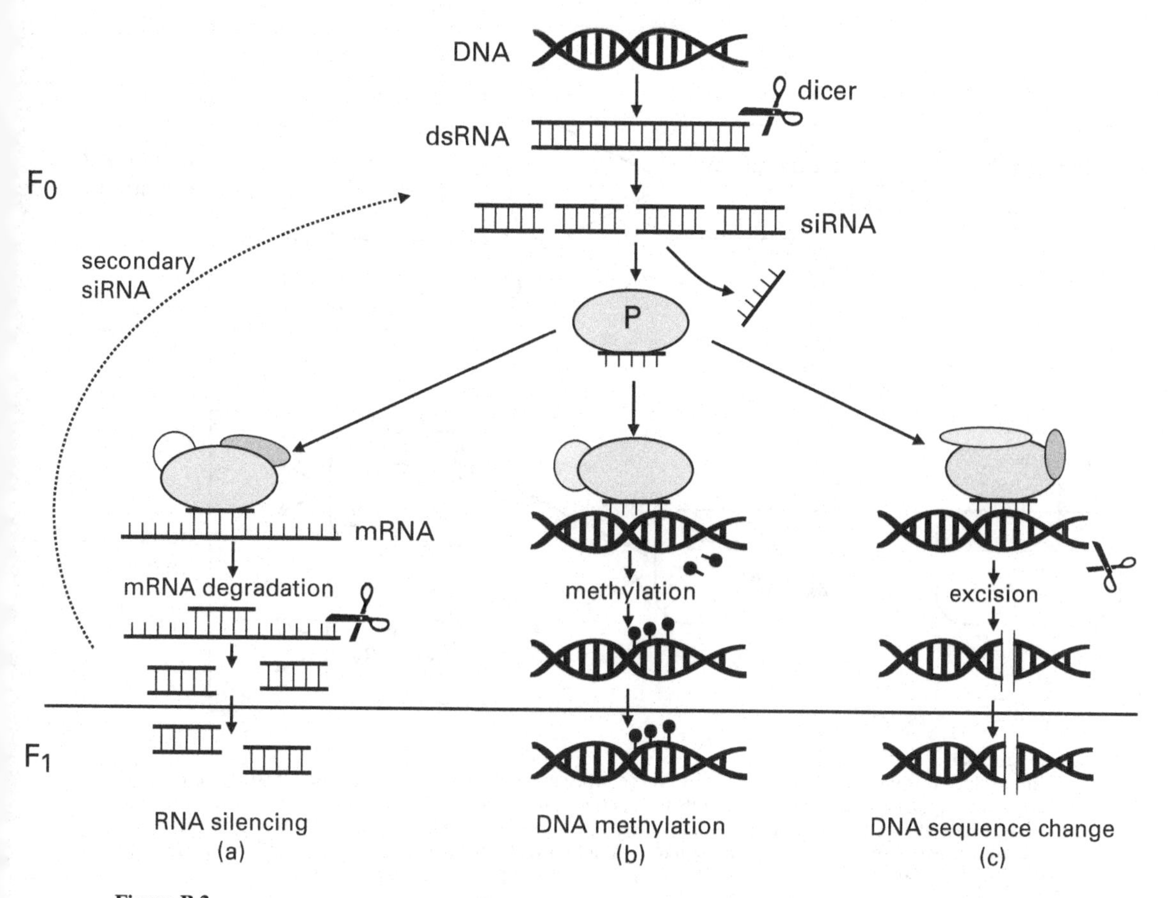

Figure B.2
Inheritance through RNA-mediated gene regulation. By associating with various protein complexes (P), small RNAs cut from double-stranded RNA can silence genes (a) by causing degradation of the target mRNA with which they have sequence homology (this silencing can be inherited when the small RNAs are replicated by RNA polymerase and transmitted to daughter, and sometimes nondaughter, cells; (b) by pairing with homologous DNA sequences and methylating them (the chromatin marks are then inherited); (c) by pairing with DNA and causing sequences to be excised (the changed DNA sequences are then inherited). Based on Jablonka and Lamb 2010.

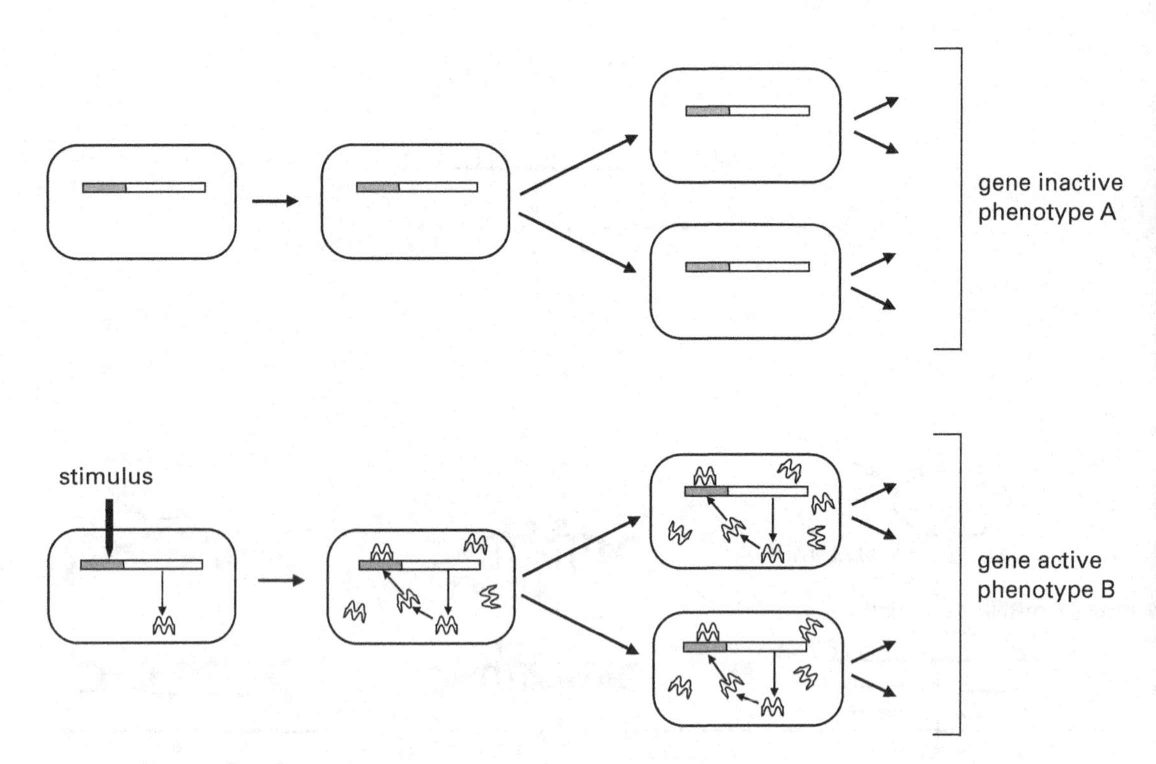

Figure B.3
Inheritance through self-sustaining loops. At the top, a gene having a control region (shaded) and coding sequence (open) is inactive and transmits its inactive state. Below, the same gene is transiently activated by an external stimulus, and produces a product which then associates with its control region and keeps it active; because the product is transmitted to daughter cells, the lineage retains the active state even in the absence of the external stimulus. Based on Jablonka and Lamb 2010.

activator of its own transcription (figure B.3). The components of the circuit (proteins, RNAs, and metabolites) can be transmitted to daughter cells during cell division, so the same patterns of gene activity can be reconstructed in daughter cells. Feedback-based, alternative cellular circuits that lead to alternative heritable phenotypes seem to be common in fungi and bacteria.

Structural Inheritance: Spatial Templating

Structural inheritance refers to the inheritance of alternative three-dimensional structures through spatial templating: a variant structure in a mother cell guides the formation of a similar structure in a daughter cell, leading to the transmission of the architectural variant. An early example was the inheritance of variations in the cortical structures of ciliates (Beisson and Sonneborn 1965). The molecular basis of cortical inheritance is not entirely clear, but it seems to be based on self-templating of complex molecular structures.

Structural continuity based on spatial templating is probably the basis of many cellular reconstruction processes. The production of new centrioles is an old example: a new centriole is normally formed in association with a preexisting centriole (Feldman, Geimer, and Marshall 2007). Another important type of structural continuity is found with membrane reproduction: Cavalier-Smith (2004) identified eighteen types of what he termed *genetic membranes*, and argued that the information embedded in the three-dimensional structure of the *membranome* is as essential for the reconstruction of a cell as the information in the genome. He suggested that many of the major events in the evolution of cells were associated with heritable changes in membranomes.

The discovery of *prions*, the infectious proteins that are the causative agents of mammalian neurodegenerative diseases such as bovine spongiform encephalopathy (BSE, or "mad cow disease"), has brought structural heredity into the limelight. Prions are misfolded proteins that propagate because the variant conformation converts the normal protein into its own form; the variant form can then be transmitted to daughter cells (figure B.4). Animal prions are usually transmitted to new individuals by being ingested, and once they reach the central nervous system, they spread from cell to cell, affecting the conformation of the normal protein. One of the most interesting finding about prions is that a single protein can often misfold into several different conformations, which results in differences in the incubation times, pathologies, and cross-species infectivity of the diseases they cause.

Prions have also been identified in yeast and other fungi, where they are not associated with disease. Some seem to have important biological functions. Their recognition has provided an explanation for the idiosyncratic biochemical and hereditary characteristics of a variety of traits that were difficult to account for in Mendelian terms. In yeast, prions are normally transmitted when the cell divides, and in filamentous fungi they can be transmitted horizontally during hyphal fusion.

The different EISs are obviously not mutually exclusive, and are often interrelated. For example, RNA-mediated heredity can involve chromatin inheritance, and self-sustaining loops may involve chromatin-modifying components and prions. However, in some cases and under some (unusual or laboratory) conditions, it is possible to dissociate the different mechanisms, and this has been crucial for their elucidation. The stability of epigenetic inheritance is variable, and the same system can lead to long-term, middle-term, or short-term persistence, depending on the organism, the specifics of the epigenetic mechanism, and the inducing conditions.

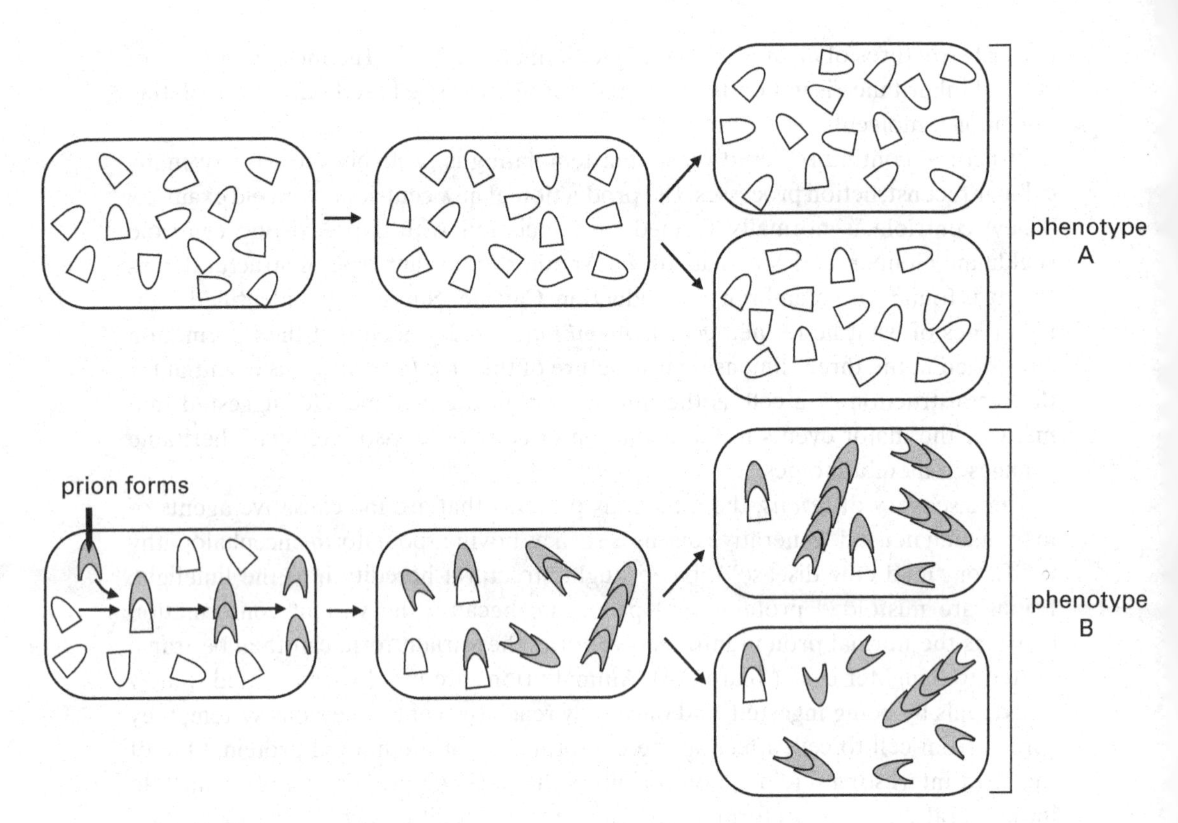

Figure B.4
Inheritance through three-dimensional templating. At the top, a cell with the normal form of a protein has a phenotype which is inherited by daughter cells, resulting in phenotype A. Below, a molecule with the same amino acid sequence adopts a prion conformation. The prion interacts with normal protein molecules and converts them to its own structure. Prions are transmitted through cell divisions, and phenotype B is formed when all or many of the newly formed proteins have assumed the prion conformation. Based on Jablonka and Lamb 2010.

Further Reading

Textbooks and Reviews Including Information on Epigenetic Inheritance

Allis CD, Jenuwein T, Reinberg D, Caparros M-L, eds. Epigenetics. Cold Spring Harbor, NY: Cold Spring Harbor Laboratory Press; 2007.

Gilbert SF, Epel D. Ecological Developmental Biology. Integrating Epigenetics, Medicine, and Evolution. Sunderland, MA: Sinauer Associates; 2009.

Jablonka E, Lamb MJ. Epigenetic Inheritance and Evolution: The Lamarckian Dimension. Oxford: Oxford University Press; 1995.

Jablonka E, Lamb MJ. Transgenerational epigenetic inheritance. In: Pigliucci M, Müller G, eds. Evolution: The Extended Synthesis. Cambridge, MA: MIT Press; 2010.

Jablonka E, Raz G. 2009. Transgenerational epigenetic inheritance: Prevalence, mechanisms, and implications for the study of heredity and evolution. Q Rev Biol. 84,2:131–176.

Chromatin Marking

Cedar H, Bergman Y. 2009. Linking DNA methylation and histone modification: Patterns and paradigms. Nat Rev Genet. 10:295–304.

Ooi STK, Bestor TH. 2008. Cytosine methylation: Remaining faithful. Cur Biol. 18,4:R174–R176.

Suzuki MM, Bird A. 2009. DNA methylation landscapes: Provocative insights from epigenomics. Nat Rev Genet. 9,6:465–476.

RNA-Mediated Inheritance

RNA interference. Wikipedia. http://en.wikipedia.org/wiki/RNA_interference.

Self-Sustaining Loops

Smits WK, Kuipers OP, Veening JW. 2006. Phenotypic variation in bacteria: The role of feedback regulation. Nat Rev Microbiol. 4:259–271.

Structural Inheritance

Beisson J, Sonneborn T. 1965. Cytoplasmic inheritance of the organization of the cell cortex in Paramecium aurelia. Proc Natl Acad Sci USA. 53:275–282.

Cavalier-Smith T. Genomes and eukaryotic phylogeny. In: Organelles, Genomes and Eukaryote Phylogeny: An Evolutionary Synthesis in the Age of Genomics. Hirt RP, Horner DS, eds. London: Taylor & Francis; 2004: 335–351.

Feldman JL, Geimer S, Marshall WF. 2007. The mother centriole plays an instructive role in defining cell geometry. PloS Biol. 5,6: e149.

Grimes GW, Aufderheide KJ. Cellular aspects of pattern formation: The problem of assembly. In: Monographs in Developmental Biology 22. Basel, Switzerland: S. Karger; 1991:1–94.

Prusiner SB. 1998. Prions. Proc Natl Acad Sci USA. 95,23:13363–13383.

Glossary

adaptability The ability of an individual organism to respond in an adaptive way to the variations of its environment (biotic and abiotic).

allopolyploidy Interspecific or intergeneric hybridization followed by chromosome doubling.

allostery The phenomenon in which the binding of a substrate, product, or other effector to a site other than the functional site, alters the enzyme's conformation and functional properties.

analytic philosophy A school of philosophy that turned to language and linguistic analysis, and later to the philosophy of language, as its subject matter. Its main contemporary topics are philosophy of mind, folk understanding, and folk psychology.

apoptosis Programmed cell death, which is part pf normal development in multiclellular organisms.

attractor The set of properties which a system tends to reach (the state it tends eventually to settle in), regardless of the conditions from which it started and the trajectory taken.

autopoiesis The process by which a system regenerates itself through the self-reproduction of its own elements and the network of interactions that characterizes them.

autopolyploidy Chromosome doubling that can result in an organism with twice the number of chromosomes of the parent.

bacteriophage A virus whose host is a bacterium; commonly called "phage."

base pairs (bp) Hydrogen-bonded bases (A-T and G-C) in double-stranded DNA.

biophors A Weismannian term for the vital subunits of **determinants**; biophores are responsible for elementary cell qualities.

canalization The adjustment of developmental pathways so as to bring about a uniform result in spite of genetic and environmental variations.

cell heredity The transmission of functional and structural cellular states in a lineage of dividing cells.

cell memory The retention of functional or structural states in dividing or nondividing cells in the absence of the conditions that originally induced these states. Cell memory mechanisms are necessary but not sufficient for cell heredity.

chaperones (chaperone proteins) Proteins that ensure that other proteins are correctly folded and thus able to fulfill their function.

chemostat An apparatus (a bioreactor) to which fresh culture medium is continuously added, while culture liquid is continuously removed, in order to keep a culture of microorganisms in a steady state. By changing the rate at which medium is added, the growth rate of the microorganisms can be controlled.

chromatin The complex of DNA, proteins, RNA, and other molecules that make up chromosomes.

chromatin marks (also **epigenetic marks**) The non-DNA part of a chromosomal locus that affects the nature and stability of gene expression. Marks can be inherited, and have a range of stabilities from transient to very persistent.

chromatin-marking EIS An epigenetic inheritance system based on mechanisms that reconstruct **chromatin marks** following cell division (see appendix B).

cis-trans interaction Cis arrangement—two genes are on the same chromosome strand; trans arrangement—two genes are on opposite chromosome strands or on different chromosomes; cis-trans interaction—interaction between proteins that are synthesized by cis-regulatory elements and trans-genetic elements that are involved in the regulation of the activity of a specific gene.

codevelopment The concept that the development of an organism depends upon its interactions with other organisms. It encompasses both symbiogenesis and the phenotypic plasticity that results from nonsymbiotic relations with other organisms.

comparative genomics The study of the relationships of genome structures and functions across different biological species or strains.

cytological diploidization The process in which the related chromosomes from the different genomes of an allopolyploid diverge and become unable to pair during meiosis.

cytoplasmic inheritance Inheritance that is based on the transmission of nonnuclear cellular factors. It includes the transmission of cytoplasmic genes (e.g., mitochondrial and chloroplast genes), and of cytoplasmic and cortical non–DNA elements.

cytosine methylation Addition of a methyl group to the 5′ position of cytosine (a pyrimidine base of DNA), which usually occurs in CpG (cytosine followed by guanine) dinucleotides. Cytosine methylation stably alters gene expression patterns, can be inherited through cell division, and is a crucial part of normal organismal development and cellular differentiation in higher organisms.

cytotaxis A term, suggested by Sonneborn, that is equivalent to **guided assembly**.

destabilizing selection A term suggested by Belyaev to describe a process of selection that destabilizes development. As well as uncovering preexisting heritable variation, it also leads to the generation of new variations (e.g., by increasing the rate of recombination).

determinant A Weismannian term referring to a primary constituent of the germ plasm, which controls a cell or group of cells when it decomposes into its constituents, the biophores.

developmental niche The environment that is constructed through reciprocal exchanges between the developing organism and its niche.

developmental noise Stochastic variation in phenotype caused by random events during development.

developmental symbiosis Symbiosis in which one organism changes the trajectory of development of another organism.

developmental system theory (DST) An approach to the philosophy of biology, developed by Susan Oyama, Paul Griffiths, and Russel Gray, which regards the unit of development as the (dynamic) system encompassing all the developmental resources (including interactions with all the components of the niche) that contribute to the life cycle of the focal individual. From this viewpoint, the unit of evolution is the whole life cycle of the system.

diploid The state of having two complete basic sets of chromosomes, which is characteristic of the zygote and most body cells of animals and plants.

drift (genetic) Changes in the allele frequencies in a population that are the result of chance events.

dynamical patterning module Association of one or more gene products with a physical force or process capable of acting on a multicellular aggregate so as to change its form or the arrangement of the constituent cells.

eco-devo (ecological developmental biology) The study of the mechanisms and outcomes of development in the context of the organism's environmental conditions.

eco-evo-devo (ecological evolutionary developmental biology) The integration of ecological developmental biology into evolutionary biology. Also, the integration of **evo-devo** studies within ecological frameworks.

ecological inheritance A process of organism–environment ecological interaction with developmental consequences, which recurs across different population generations.

endoreduplication Replication of chromosomes without intervening mitosis or cytokinesis, leading to an increase in the ploidy level (number of complete chromosome sets).

endosymbiosis A symbiotic relation in which one organism lives within the body of another organism (e.g., single-cell algae living inside reef-building corals).

epiallele One of the alternative epigenetic forms of a gene with an unchanged DNA sequence.

epigenetic inheritance 1. *Epigenetic inheritance* in the broad sense is the inheritance of phenotypic, developmental variations that do not stem from differences in DNA sequence or persistent inducing signals in the present environment. It includes cellular inheritance and the soma-to-soma transfer of information, which bypasses the germ line. 2. *Cellular epigenetic inheritance* as previously, but with the cell as previously, but the cell is the unit of transmission (see appendix B).

epigenetic inheritance system (EIS) The factors and mechanisms underlying the inheritance of functional and structural variations in cell lineages which do not involve DNA sequence changes. Cellular EISs include *self-sustaining loops*, *structural inheritance*, *chromatin marking*, and *RNA-mediated inheritance* (see appendix B).

epigenetic landscape A visual model, suggested by C.H. Waddington, which describes the dynamics of embryological development and its underlying genetic network of interactions.

epigenetic marks See **chromatin marks.**

epigenetics The study of the factors and mechanisms that result in persistent developmental effects, and underlie developmental plasticity and canalization.

epigenotype The total developmental system of interrelated developmental pathways through which the adult form of an organism is realized. A more restricted definition is: the actual pattern of gene activity in a specialized cell type.

epimutation A hereditary variation in gene expression that is not the result of a change in DNA base sequence.

euchromatin The relatively noncondensed chromatin conformation that is associated with potential or actual gene activity.

evo-devo (evolutionary developmental biology) The study of the control of evolution by development and of development by evolution. It includes the study of the role of the developmental constraints and affordance in evolution, developmental genetics and phylogenetics, evolutionary epigenetics, and the evolution of ontogeny.

experimental philosophy Philosophical investigations that rely on the gathering and analyzing of empirical data on how non-philosophers understand key philosophical concepts (for example, truth).

experimental philosophy of science Philosophical investigation based on gathering empirical data on how key scientific concepts are understood and put into practice by particular scientific communities (for example, the concept of the gene).

exploration and selective stabilization The generation of a large set of local variations from which only a small subset is eventually stabilized. Selective stabilization occurs when the system reaches an *attractor*.

expressivity The extent to which a particular genotype is expressed in the phenotype.

folkbiology The assumption that there is an everyday understanding of the living world which is implicit in everyday talk and behavior. It underpins the non-professionals' capacity to explain and predict the behavior of others and of themselves.

free energy flow Energy flow from energy source to sink in open, dissipative thermodynamic systems, quantified by Eric Chaisson as free energy rate density, Φm.

GEMs (genomic epigenetic mechanisms) The processes and related mechanisms that underlie changes in the genome that occur during development.

gene regulatory network A set of transcription factors and the genes that they regulate (including genes for the transcription factors themselves) through *cis-regulatory sequences*.

genetic accommodation The outcome of the selection for changes in developmental responsiveness that follows when altered conditions expose variation in responsiveness. Genetic accommodation can lead to *genetic assimilation* (increased canalization), to enhanced plasticity (more condition-dependent responses), and to the amelioration of the deleterious side effects of a new environmental or mutational challenge.

genetic assimilation The replacement through natural selection of a phenotypic response that originally depended on environmental induction or learning by a response that is independent of induction or learning.

genetic diploidization Divergence of the duplicated genes in the genomes of an allopolyploid, or the silencing of genes of one genome.

genetic membranes A term suggested by Cavalier-Smith for membranes that reproduce through *guided assembly* mechanisms.

genome The complete coding and noncoding DNA of an organism. A broader definition, encompassing function, is the physicochemical chromosomal system carrying hereditary and developmental information.

genomic imprinting The process whereby the expression of genetic information depends on the sex of the transmitting parent.

genomic plasticity The ability of the genome to assume more than one configuration in response to internal and external factors; genomic plasticity may, but need not, lead to transgenerational effects.

genotype The complement of genes present in a particular organism which interact with factors in the internal and external environments to give rise to the *phenotype*. (This is the common view at present. Wilhelm Johannsen, who introduced the terms phenotype, genotype, and gene in 1909, continued to view the genotype as a whole, not to be fully explained in terms of individual genes.)

Genotype-by-environment interaction Variation produced when genotypes express different patterns of response to a shared environmental gradient or environmental change.

germline The cell lineage that gives rise to the gametes (in sexually reproducing organisms, the sperm and eggs) that form the link between generations.

germplasm The cellular material constituting the hereditary information that in the *germ line* gives rise to gametes. Weismann identified the germplasm with the nuclear material, and assumed that it is built up of a hierarchy of heterogeneous vital units that control development.

guided assembly The assembly of new cell structures under the influence of preexisting cell structures.

heterochromatin The condensed, dark-staining *chromatin* of chromosomes. Usually it is poor in coding genes and rich in repetitive DNA sequences.

heterodiploid A bacterial cell that carries, in the same cellular milieu, two parental chromosomes, each bearing different genetic markers.

heterosis The superiority of the value of a particular character in F_1 hybrids over the midparental value of that character.

heuristic reductionism The claim that theories constructed at a higher level of generality should be using theories of lower levels as constraints (e.g., a general theory of inheritance would be used to explain epigenetic inheritance). See **reductionism**.

holobiont An organism composed of hosts and symbionts.

hologenome The DNA sequences of both a host organism and its symbionts. The genome of the *holobiont*.

homeoalleles Related gene loci that occur in the *homeologous chromosomes* of an allopolyploid.

homeologous chromosomes The nonidentical, yet related, chromosomes of different genomes that are derived from a common ancestor and that, despite some evolutionary divergence, show partial homology.

homeostasis The maintenance of relatively steady states in an organism through internal regulatory mechanisms which allows it to function effectively despite variations in internal and external conditions.

homologous chromosomes Chromosomes in individuals of the same species that contain the same gene loci and form bivalents in meiosis.

homology Similarity between characteristics of different organisms that is due to the inheritance of these characteristics (often with some modifications) from a common ancestor.

horizontal gene transfer The incorporation by one organism of genetic material from another organism which is not its parent, and may belong to a different species.

imprint A *chromatin mark* determined by the sex of the transmitting parent. (See **genomic imprinting.**)

infective heredity Inheritance of traits acquired by *horizontal gene transfer*, for example, through viruses.

inheritance of acquired characters For Lamarck, the idea that traits acquired in one generation as the result of new habits are passed on to the next generation. These traits could be behavioral (in the form of habits turned into instincts) or structural (where organs became stronger or weaker, depending on their use or disuse, respectively). The idea of the inheritance of acquired characters was commonly accepted in Lamarck's day, and Lamarck did not claim credit for it. He did claim credit, however, for the idea that the inheritance of acquired characters could lead to the transformation of species.

inner feeling A term used by Lamarck to describe the feeling of existence that animals with a centralized and interconnected nervous system have. According to Lamarck, psychological processes evolved from the inner feeling.

internal selection Selection resulting from differences in fitness of variant genomes within an organism, which derives from the lack of coherence of mutations or permutations with preexisting functional networks.

isoenzymes (also **isozymes)** Multiple, distinguishable forms of enzymes that catalyze the same reaction.

isogenic line A line of intercrossing, genetically identical (except for sex) individuals.

isoschizomers Pairs of restriction enzymes that are specific to the same recognition sequence but have different cutting specificities.

Lamarckian problematics A perspective on evolutionary problems that emphasizes developmental processes and phenotypic variation. Evolutionary analysis starts by studying the origins and causes of phenotypic variations, their developmental stabilization and alteration, and the modes of their hereditary transmission. Natural selection of heritable variations is seen as crucial for an account of evolution, so the Lamarckian approach is seen as complementary to the Darwinian, selection-oriented one.

Lamarckism A large spectrum of doctrines, having in common the recognition that life had a history and was subject to change, that focus on the evolutionary role of individual variations that emerge during the life of an organism in response to environmental stress. Concepts such as the power of life, or the tendency for inexorable, predetermined growth, also became associated with Lamarckism, and were appropriated by philosophical, teleological, and at times theologically oriented views of evolution.

limited heredity A term coined by Maynard Smith and Szathmáry to describe a system in which the number of possible heritable types is commensurate with, or much less than, the number of individuals in the population. In contradistinction, in unlimited heredity, the number of possible heritable types is much larger than the number of individuals in any realistic system.

lysogeny The phenomenon in which the genetic material of a virus and its bacterial host are integrated; the integrated state of the virus does not lead to the lysis of the host cell.

lytic cycle A viral life cycle in which the intracellular multiplication of the virus leads to the lysis of the host cell.

mechanistic materialism The concept that all natural phenomena, including those involving life and the mind, can be explained in terms of the interactions of forces and the simple or complex arrangement of material particles that obey the laws of mechanics/physics.

Mendelian genetics The principles of inheritance, worked out by Gregor Mendel in the nineteenth century, that describe the behavior of hereditary factors (later identified as nuclear genes) in sexually reproducing organisms. The principles include Mendel's "first law"—segregation of homologous hereditary factors, and "the second law"—independent assortment of different factors.

meristem An undifferentiated plant tissue from which new cells are formed.

metabolism The flow of chemical energy through the organism, which maintains its complexity far from the thermodynamic equilibrium. The basal metabolic rate is a specific measurable property of an organism and is calculated by its oxygen requirement when at rest.

metastability The state of a system when it is likely to switch to a different **attractor** as a result of even small perturbations.

mismatch hypothesis The idea that when the conditions of growth do not match the conditions experienced during fetal development (e.g., the fetus grew in poor nutritional conditions, but postnatal conditions were plentiful), the mismatch results in an exaggerated and sometimes pathological response (e.g., the development of obesity).

mneme A term coined by Richard Semon (1904), within the framework of analogy between memory and heredity, to describe a general conserving principle that operates during psychological processes of learning, during embryonic development, during regeneration, and during heredity. The mechanisms are different in each case, but in all cases the effects of a stimulation are conserved, and the subsequent interaction with the environment occurs on the basis of the conserved experience.

Modern Synthesis A wide-ranging consensus about the nature and dynamics of evolutionary change that emerged among biologists between the 1920s and 1950s, and was based on neo-Darwinian assumptions and population genetics.

monism Any view which holds that there is one overriding unity in the world, either of the kind of substance or of the ordering principle.

morphogenetic field A collection of cells with well-defined boundaries and characterizing properties (that make them members of the ensemble), which is established at each generation and whose interactions lead to the formation of a particular organ.

morphotype A phenotype defined by anatomical or structural characteristics.

neo-Darwinism A term originally used by Romanes (1896) to describe Weismann's version of Darwinism, which excluded the inheritance of acquired characters. It is now commonly used to describe the *Modern Synthesis* version of Darwinism.

neo-Lamarckism (neo-Lamarckianism) A term coined by Packard in 1885; used loosely to describe evolutionary theories that emphasize various ideas that, rightly or wrongly, are attributed to Lamarck, especially the idea that the inheritance of acquired characters is important in evolution.

niche construction The process by which organisms, through their metabolism, physiology, behavior, and dispersal, causally affect abiotic and biotic features of their environment. By doing so, organisms generate feedback from their environment that may operate on both ecological and evolutionary timescales.

non–Mendelian changes Genetic and epigenetic changes that cannot be explained by Mendelian principles, especially the first law.

norm of reaction The set of phenotypes expressed by a given genotype in response to a specified range of environmental conditions.

nucleosome The basic components of chromatin in eukaryotes. The nucleosome core consists of 147 bp of DNA wrapped around an octamer made up of pairs of four histone proteins: H2A, HB, H3, and H4.

old-school Darwinism A term coined by Sander Gliboff to describe the views of Haeckel, Plate, Semon, Kammerer, and a few other neo-Lamarckians, who adhered to Darwin's pluralistic approach to evolution, which included the inheritance of acquired characters; in contradistinction to *neo-Darwinism.*

operon A unit consisting of adjacent genes that functions coordinately under the control of a regulatory chromosomal region, the operator.

orthogenesis The assumption that evolutionary change goes along single or limited paths, and is driven mainly by internal factors, not by the impact of the environment or through interaction with it.

orthologous chromosomes Chromosomes in different species (or different genomes in allopolyploids) that derive from a common ancestor.

osmoregulation A set of energy-requiring processes which guarantee the stability of salinity value and ionic composition of the internal fluids of the organism, independent of the respective values and composition in the surrounding liquid environment.

pangenesis A hereditary mechanism, suggested by Darwin, in which all parts of the organism shed small particles (gemmules), which collect in the reproductive organs and are transmitted to the next generation, where they reconstruct the same parts and characteristics as those of the parents.

pangenome The ensemble of genomes of individuals of the same species.

panmixia Weismann's term for the relaxation of natural selection. Conditions in which panmixia occur allow the accumulation of new variations in traits that are currently not under selection.

paramutation An epigenetic interaction between alleles in a heterozygote that leads to directed, heritable change at one of the loci; it can occur with high frequency, and the changes can be transmitted across generations. Also refers to the product of this process.

penetrance The proportion of individuals with a specific genotype that express the genotype at the phenotypic level.

persistence (bacterial) An epigenetic state (e.g., a dormant state) that provides a bacterium with protection from adverse conditions. Also refers to the process conferring the persistent state.

phenotype The observable physical or behavioral traits of an organism, determined by the organism's *genotype* interacting with factors in the internal and external environments.

phenotypic plasticity The capacity of organisms to assume more than one morphological or physiological state in a fashion that may depend on both internal and external factors; usually used to describe the ability of a single **genotype** to respond to environmental inputs by changing aspects of its **phenotype** (form, movement, or rate of activity).

Piwi interacting RNA (piRNA) The largest class of small RNA molecules that are expressed in animal cells; piRNA forms RNA–protein complexes, and these are associated with the transcriptional gene silencing of some mobile elements in germ line cells.

plasmagenes Genes present in the cytoplasm; also used for other types of cytoplasmic factors.

plasmon A term coined by Fritz von Wettstein (1926) for the hereditary elements of the cytoplasm, including cytoplasmic genes (such as those of chloroplasts and mitochondria) and other hereditary factors.

plasticity See **phenotypic plasticity**; **genomic plasticity**.

polyphenism The occurrence of several distinct phenotypes arising from a single genotype as a result of differing environmental conditions or of developmental noise.

polyploid A species with two or more genomes, each containing two basic sets of chromosomes.

polyteny Formation of giant multistrand chromosomes by repeated replication of the chromosomes without separation of the daughter chromatids.

pouvoir de la nature (also puissance de la nature) The power or might of nature. According to Lamarck, the natural processes that bring about natural phenomena.

predator–prey cycles Correspondence between temporal fluctuations of a predator population and the population of its prey.

predictive adaptive response An irreversible developmental response to embryonic conditions that is made by a fetus and has long-term effects; it can be interpreted as anticipating that the same conditions will occur during postnatal development. The *thrifty phenotype* may be interpreted in these terms.

prion A proteinaceous infectious agent that propagates through spatial templating mechanisms. Prions can be transmitted vertically, as is the case with yeast prions, or horizontally, as is the case with the prions causing mad cow disease.

progenote A hypothetical ancestor of eubacteria, archaebacteria, and eukaryotes that has a primitive translation apparatus, and frequently exchanges genetic material with members of different lineages.

progressive evolution The view that evolution leads to an increase in morphological and functional complexity through the action of natural processes and without the intervention of nonnatural factors.

psycho-Lamarckism A variety of neo-Lamarckian theories, advocated by the German botanist–philosophers August Pauly and Raoul Francé, which assumed that an innate "striving for progress" characterizes all animals, regardless of the complexity of their organization. Wrongly attributed to Lamarck. Another version of it, attributable primarily to Edward Drinker Cope, argued for a consciously chosen reaction of the animal to its environment.

ramet An individual member of a plant clone.

recapitulation theory The theory that the ontogeny of the individual organism, mostly understood as the history of the embryo, follows and reconstructs the evolutionary path of its lineage (its phylogeny).

reductionism A claim that complex entities can be shown to be completely constituted by their component parts, can be completely explained by them (including how these entities function), and will be more fruitfully investigated at the level of their component parts.

retrotransposons Transposons that move by synthesizing an RNA intermediate that is converted back to DNA by reverse transcription and integrated into the chromosomes.

RNA-mediated epigenetic inheritance Inheritance mediated by RNA molecules. When transmitted to the next generation, RNAs that act as regulators of gene expression cause the expression patterns they influence to be inherited (see appendix B).

RNAi A set of related cellular mechanisms through which double-stranded RNA is degraded into short RNAs (siRNAs) that cause epigenetic or genetic changes in gene expression. The siRNAs can sometimes be inherited between generations.

saltational evolution The abrupt appearance, with few or no transitional forms, of a novel type of organism or a new species. It is attributed either to a mutational change in a single regulatory gene with large effects, or to genomic repatterning, or to the effects of heritable interactions between the organism and its environment, which drastically alters its phenotype.

self-assembly A process in which a disordered system of preexisting components forms an organized structure or pattern as a consequence of specific, local interactions among the components themselves, without external direction.

self-organization In biology, this refers to the capacity of matter to undergo changes as a result of its inherent properties and interaction with its environment, which result in the emergence of novel, specific, organized structures.

self-sustaining loop A regulatory circuit that dynamically maintains a particular pattern of cellular activity. The activity pattern can be inherited when some of the constituents of the circuit become part of daughter cells (see appendix B).

social imaginary The set of values, institutions, laws, and symbols common to a particular social group and the corresponding society. It is a historical construct which represents the system of meanings, defined by the interactions of the subjects that govern a given social structure.

soft inheritance A term originally coined by Darlington, but later used by Ernst Mayr to describe a gradual change of the hereditary material (which he assumed to be DNA) brought about by use or disuse, or by some internal progressive tendencies, or through the direct effect of the environment.

soma All the cells of the body of an organism, excluding the *germ line*.

somaclonal variation Variation between individual in vitro cultured plant cells and the plants regenerated from them.

stabilizing selection Selection for developmental robustness—for responses that are buffered from variations in the external and internal environments.

stress A state or a process that requires an organism to make significant adjustments (physiological, morphological, psychological, and/or behavioral) to changed conditions, in an attempt to restore *homeostasis*.

structural inheritance Refers to the reproduction of alternative three-dimensional cellular structures through spatial templating, and their subsequent transmission to daughter cells (see appendix B).

symbiopoiesis The development of the entire organism, including the reciprocal interactions between host and symbiont genomes.

symbiosis Close and often long-term interactions between different biological species; the symbiotic interactions may be mutually beneficial, beneficial to only one partner, or parasitic. The term is usually used more narrowly for describing mutually beneficial interactions.

systems biology An interdisciplinary field whose focus is complex interactions in biological systems. It assumes that the form and function of organisms are to be investigated and explained through the plurality of components and their interactions as parts of a system.

teratology The study of abnormalities in physiological development.

thrifty phenotype A phenotype exhibiting reduced growth and relative insulin resistance which is the result of irreversible developmental changes made in response to maternal undernutrition when it was a fetus.

toy model A simplified or reduced set of objects or fields and their relations which can be used to understand a mechanism in a scientific theory.

transfer Refers to the passage and matching of tools from one conceptual field/domain (the source) to another (the target). These tools—concepts, models, metaphors, analogies, ordering of hierarchies, and so on—are then used in the target to study phenomena or to solve problems. Certain features or properties of the source domain may disappear or become central in the target domain. Thus, transfer can bring about unforeseen restructuring, remodeling, and novelties.

transformism The doctrine that living organisms have evolved, and that existing species are the results of transformation of ancestral, different, species; evolution.

transposons Sequences of DNA that can move from and to different positions within the genome of a single cell, through a process known astransposition.

uniformitarianism The doctrine of the uniformity of nature (i.e., that the same laws apply in all of nature), so that whatever is true of any one case under a description of a process is true of all other cases under that description: past, present, and future (was originally applied in geology).

vernalization Treatment of germinating seed or more fully developed plants by periods of low temperature to facilitate flowering and setting seed within the same season (a latinization of the Russian term *iarovizatsiia*, "spring crop making" and *ver*, "spring" in Latin).

X-inactivation A process occurring during the embryogenesis of a female mammal that results in the transcriptional inactivation of one of her two X-chromosomes.

Contributors

Nathalie Q. Balaban Racah Institute of Physics, Center for Nanoscience and Nanotechnology, and Sudarsky Center for Computational Biology, Hebrew University, Jerusalem, Israel

Ramray Bhat Department of Cell Biology and Anatomy, New York Medical College, Valhalla, NY; present address: Life Sciences Division, Lawrence Berkeley National Laboratory, Berkeley, CA

Erez Braun Department of Physics and Network Biology Research Laboratories, Technion, Israel Institute of Technology, Haifa, Israel

Marcello Buiatti Department of Animal Biology and Genetics, University of Florence, Florence, Italy

Tatjana Buklijas Liggins Institute and the National Research Centre for Growth and Development, The University of Auckland, Auckland, New Zealand

Richard W. Burkhardt, Jr. Department of History, University of Illinois at Urbana-Champaign, Urbana, IL

Pietro Corsi Department of History, University of Oxford, Oxford, UK

Lior David Department of Animal Sciences, R.H. Smith Faculty of Agriculture, Food and Environment, The Hebrew University of Jerusalem, Rehovot, Israel

Raphael Falk Department of Genetics, Hebrew University, Jerusalem, Israel

Moshe Feldman Department of Plant Sciences, Weizmann Institute of Science, Rehovot, Israel

Evelyn Fox Keller Program of Science, Technology, and Society, MIT, Cambridge, MA

Scott Gilbert Department of Biology, Swathmore College, Swathmore, PA

Simona Ginsburg Department of Natural Sciences, Open University, Ra'anana, Israel

Snait B. Gissis Cohn Institute for the History and Philosophy of Science and Ideas, Tel-Aviv University, Tel-Aviv, Israel

Sander Gliboff Department of History and Philosophy of Science, Indiana University, Bloomington, IN

Peter D. Gluckman Liggins Institute and National Research Centre for Growth and Development, The University of Auckland, Auckland, New Zealand, and, Singapore Institute for Clinical Sciences, A STAR (Agency for Science, Technology and Research), Singapore

James Griesemer Department of Philosophy, University of California-Davis, Davis, CA

Paul E. Griffiths Department of Philosophy and Sydney Centre for the Foundations of Science, University of Sydney, Sydney, Australia, and, ESRC Centre for Genomics in Society, University of Exeter, Exeter, UK

Mark A. Hanson Developmental Origins of Health and Disease Division, Institute of Developmental Sciences, University of Southampton, Southampton, UK

Luisa Hirschbein Institute Curie (honorary), UMR 168 CNRS, Paris, France

Eva Jablonka Cohn Institute for the History and Philosophy of Science and Ideas, Tel-Aviv University, Tel-Aviv, Israel

Marion J. Lamb Independent Scholar, London, UK

Ehud Lamm Cohn Institute for the History and Philosophy of Science and Ideas, Tel-Aviv University, Tel-Aviv, Israel

Avraham A. Levy Department of Plant Sciences, Weizmann Institute of Science, Rehovot, Israel

Yigal Liverant Cohn Institute for the History and Philosophy of Science and Ideas, Tel-Aviv University, Tel Aviv, Israel

Laurent Loison Department of History of Science, Centre François Viète, Nantes University, Nantes, France

Arkady L. Markel Institute of Cytology and Genetics, Russian Academy of Sciences, Siberian Department, Novosibirsk State University, Novosibirsk, Russia

Everett Mendelsohn Department of History of Science, Harvard University, Cambridge, MA

Gabriel Motzkin Van Leer Jerusalem Institute, Jerusalem, Israel

Stuart A. Newman Department of Cell Biology and Anatomy, New York Medical College, Valhalla, NY

Amos Oppenheim Department of Molecular Genetics and Biotechnology, Hebrew University, Jerusalem, Israel

Sivan Pearl Racah Institute of Physics and Department of Molecular Genetics and Biotechnology, Hebrew University, Hadassah Medical School, Jerusalem, Israel

Francis Dov Por Department of Evolution, Ecology and Behavior, Hebrew University, Jerusalem, Israel

Minoo Rassoulzadegan Inserm U636, Laboratoire de Génétique du Développement Normal et Pathologique, University of Nice, Sophia Antipolis, Nice, France

Nils Roll-Hansen Department of Philosophy, Classics, History of Art and Ideas, University of Oslo, Oslo, Norway

Jan Sapp Department of Biology, York University, Toronto, Canada

Ayelet Shavit Department of Interdisciplinary Studies, Tel Hai Academic College, Tel Hai, Israel

Sonia E. Sultan Department of Biology, Wesleyan University, Middletown, CT

Alfred I. Tauber Center for the Philosophy and History of Science, Boston University, Boston, MA

Lyudmila N. Trut Institute of Cytology and Genetics, Russian Academy of Sciences, Siberian Department, Novosibirsk State University, Novosibirsk, Russia

Charlotte Weissman Cohn Institute for the History and Philosophy of Science and Ideas, Tel-Aviv University, Tel-Aviv, Israel

Adam Wilkins Wissenschaftskolleg zu Berlin, Berlin, Germany

Index

Note: page numbers ending with *g* indicate that the entry is defined in the Glossary

Accommodation
genetic (*see* Genetic accommodation)
physiological, 190
Acquired characters, 105, 130. *See also* Inheritance of acquired characters; Soft inheritance
induction of (*see* Direct induction; Parallel induction; Somatic induction)
Acrythosiphon pisum (pea aphid), 283, 287, 289, 290
Activation pattern, 303, 349–350
Adaptability, 71, 147, 269, 423g
Adaptation, xii, 4, 29, 48, 50, 51, 58, 63, 69, 70, 72, 74, 112, 115, 128, 130, 146, 149, 152, 161, 177–178, 227, 251, 258, 297, 301, 323–325, 401. *See also* Acquired characters; Bacteria, adaptations of; Bacterial persistence, as an adaptation; Evolution, adaptive; Evolution, adaptive and cumulative; Lamarckian inheritance
of allopolyploids, 269
behavioral, 397
to changing environments, 205, 212
and co-functionality, 189
in Darwin, 42
in Durkheim, 93, 94, 95
in French neo-Lamarckism, 69, 70, 72, 73, 74
in Haeckel, 47
physiological, 161, 167, 195
social, 97
in Spencer, 91–92
in Weismann, 58, 63
Adaptationist, 166, 358
Adaptive variation, 154, 286
Aegilops, 263, 264, 265, 266
Agassiz, Louis, 26
Agency, 22, 27, 47, 48, 93, 157, 178, 359, 360, 361
Agriculture, 81, 83
Agrobacterium rhizogenes, 255, 257, 259
Allison, Anthony, 140
Allopatric species, 201
Allopolyploidization, 152, 254, 261, 263
and cytological diploidization, 264–265, 268
and DNA elimination, 268
and DNA methylation, 267
and genetic diploidization, 266–267
and speciation, 261
and stress, 220
Allopolyploids, 152, 265, 266, 267, 268, 269, 428g
Allopolyploidy, 253, 261, 423g
Allostery, 362, 363, 423g
Amoeba, 117
Analogy, 14, 27, 28, 39, 90, 94, 403, 428
Analytic philosophy, 5, 6, 423g
Anatomy, 37, 58, 69, 272, 327
Anokhin, Konstantin, 397, 400, 401
Anthropology, 21, 90, 97, 319, 404
Apoptosis, 245, 285, 423g
Arabidopsis thaliana, 221, 222, 396
Archaea, 271, 273, 274, 429
Artificial selection, 171, 177, 252, 258
Asobara tabida, 285
Asparagopsis armata, 252
Aspergillus, 115
Assimilation. *See* Genetic assimilation
Asymmetry
directional, 158
fixed, 396
random and plastic, 396
Atomic Energy Commission, 124, 138
Attention, selective, 369–370
Attractor, 303, 349, 350, 351, 352, 353, 354, 402, 423g, 425, 428
Autopoiesis, 166, 291, 423g
Autopolyploidy, 423g
Ayala, Francisco, 297

Bacteria, 68, 116, 151, 227. *See also* Bacterial persistence; Blue-green algae; Evolution, coevolution; Horizontal gene transfer; Mutation, adaptive; Stress, role in bacterial evolution; Symbiosis, with bacteria; Symbiosis, co-evolution between host and symbiont
acquisition of genes by, 117, 153, 274
adaptations of, 70, 115, 222
biofilms, 158, 286
classification of, 271–272
definition of, 272–273
dormant, 207
epigenetic inheritance in bacteria, 151, 219, 367, 368, 377–379, 413, 418
evolution of and in, xiii, 153, 273–275, 368, 379
extremophile, 274
lysogenic, 276, 277
mutations in, 133, 138–139, 152, 177, 367
persistence, 183, 206–213
plasticity, 147
role in speciation, 277, 279
switching in, 147, 183
Bacterial persistence, 183, 205–213, 402, 429g
as an adaptation, 210–213
discovery of, 205
effect on growth rate, 206
high persistence, *hipA7* gene, 206, 208, 209
and predator-prey oscillations, 148, 210–212, 402
response to antibiotic, 205, 206, 207
types, 207, 213
Bacteroides fragilis, 285, 286, 290, 291
Bacteriophage. *See* Virus, bacterial
Badyaev, Alexander, 158, 177, 326
Baer, Karl Ernest von, 91
Balaban, Nathalie, 147, 219, 402
Baldwin, James Mark, 148
Baldwin effect, 148
Ballantyne, John William, 239
Barker, David, 240
Base pairs (bp), 392, 423g
Bateson, William, 117, 123, 160, 276
Beagle, The, 24, 40, 41
Beatty, John, 298, 334
Behavior, 22, 33, 52, 95, 104, 112, 207, 303. *See also* Habit; Plasticity, behavioral; Domestication; Genetics, behavioral; Instinct; Niche construction
classification of, 327, 328
for Darwin, 40–41
and epigenetics, 218, 220–221, 223, 244, 400, 401
and folk psychology, 299, 300, 320, 425g
for Frédéric Cuvier, 37–39
and genetic assimilation, 148, 397
in human societies, 92
induced by symbionts, 279
for Lamarck, 12, 23, 24, 33, 36, 369
social, 118, 298
for Spencer, 92
Beisson, Janine, 137, 279
Belyaev, Dimitri, 171, 172, 174, 175, 178, 424
Benedict, Ruth, 404
Ben Menahem, Yemima, xv, 99, 342
Bernard, Claude, 68, 69, 71
Bhat, Ramry, 146, 148, 224, 304
Biochemical pathway, 182
Biodiversity, 147, 181, 200, 300, 309, 314, 315
Biology, xi, xiii, xiv, 5, 48, 224, 246, 324, 327, 328, 332, 345, 406, 416, 424
developmental (*see* Developmental biology)
evolutionary, xi, xiv, 25, 31, 33, 116, 118, 131, 145, 246, 278, 291, 332, 395, 405, 406
folk, 319, 320, 321, 322, 325, 328, 425g
and French transformist microbiology, 67, 68, 70
and Lamarck, 11, 45, 57, 381, 389
and mathematics, 11, 138
and the Modern Synthesis, 104, 105, 106, 112, 115, 122, 124, 128, 138
molecular, 115, 118, 137, 141, 175, 332, 382, 385, 424, 425
"new biology," 385–387
and physics, 5, 80, 115, 138, 381
prokaryotic, 377
and sociology, 89, 90, 92–98, 395, 402–404
Soviet, 80, 81, 82, 86
and teleology, 47, 160
Biometry, 109, 115
Biophors, 58, 60, 424g
Biston betularia (peppered moth), 137, 138, 139
Blue-green algae, 272, 273, 276, 278
Body plan, 148, 158, 159, 164
Bohr, Niels, 80, 381
Bonnier, Gaston, 68–69
Bory de Saint Vincent, Jean Baptiste, 15, 17
Bowler, Peter, 64, 89
Buffon, George-Louis Leclerc Comte de, 9, 36, 39, 42, 160, 381
Buklijas, Tatjana, 151
Braun, Erez, 147, 152, 220, 304
Bread wheat, 263, 264, 265, 266
Breeding, 77, 79, 80, 83, 84, 85, 115, 172. *See also* Domestication; Hybridization; Inbreeding; Lysenkoism
Brink, Alexander, 216, 256
Buchnera amphidicola, 287, 289, 290
Buiatti, Marcello, 152, 303
Burbank, Luther, 83
Burkhardt, Richard W., Jr., xvi, 24
Burian, Richard, 68
Butler, Samuel, 28

Cabanis, Pierre J. G., 24
Caenorharbditis elegans, 230

Canalization, 105, 113, 145, 324, 397, 398, 408, 423g. *See also* Genetic assimilation; Genetic accommodation; Plasticity, evolution of
and cell determination, 150
definition of, 149
as Fixity, 323
in metazoan evolution, 148
and norms of reaction, 201
and plasticity, 323, 324
Cancer, 216, 218, 390, 398
Caullery, Maurice, 73, 74
Causality, xii, 4, 5, 6, 7, 93
circular, 358 (*see also* Causation, reciprocal)
Cause(s)
of evolution, 39, 50, 69, 178, 277, 308, 309, 373 (*see also The Causes of Evolution*)
of health in society, 239
of human nature, 319
"Lamarckian," 50, 162, 359
physiochemical, 9, 47, 48, 91, 385, 386
primary, 35, 36
relative significance of, 333–334
tension between, 74
of variation, 26, 30, 38, 39, 64, 91, 176, 234, 279, 285, 339
Causation
disease, 239
nonlinear, complex, 385, 386, 387
reciprocal, 307, 309
types of, 153
Causes of Evolution, The, Haldane, J. B. S., 110, 111
Cavalier-Smith, T., 419, 426
Cell, 96, 115, 116, 117, 130,133, 335
cortex of, 137
differentiation, 118, 136–138, 150, 162,163, 215, 216, 424
heredity, 119,126, 136, 151, 215, 216, 217, 413–420, 423g (*see also* Epigenetic inheritance, cellular)
inheritance, 215, 216–222, 377–379
learning, 224
lineages, 118, 151, 275, 425
lines, 58, 216, 378
memory, 145, 216, 416, 428 (*see also* Epigenetics)
networks, 145, 147, 259, 402
plasticity, 182, 190, 252
Central Dogma, 80, 106, 118, 337, 382, 385
Century of Darwin, A, Barnett, S. A., 116
Chaisson, E., 374, 425
Chambers, Robert, 17
Chaos infusoria, 271–272
Chaperones (chaperone proteins), 183, 184, 358, 396, 423g
Chemistry, 10, 11, 27, 70, 123, 161, 297, 362, 405
Chemostat, 184, 186, 187, 188, 189, 423g
Child, C. M., 122
Chlamydomonas, 177
Chloroplasts, 106, 134, 136, 273, 276, 278, 288, 424, 429. *See also* Cytoplasmic inheritance
Chromatin, 135, 222, 230, 257, 263, 348, 377, 378, 390, 391, 398, 413–416, 417, 423g, 425, 426
Chromatin marks, 218, 414, 415, 417, 423g. *See also* Epigenetic Inheritance Systems, chromatin marking; Genomic imprinting
Chromatin modifications, 229, 284, 337, 391. *See also* Histones, modifications of
Chromosomes, 52, 58, 79, 80, 111, 121, 122, 134, 135, 383. *See also* Chromatin; Mendelian inheritance; Genome, plasticity of
aberrations and changes in structure, 112, 255, 258, 430
bacterial, 207, 208, 377, 378, 379, 426
duplications, 112, 253, 263, 424, 429
homeologous, 261, 263, 264, 266, 268, 426g
homologous, 263, 265, 426g
interactions between, 348, 424 (*see also* Paramutation)
nuclear, 104
orthologous, 428
polytene, 253
recombination, 221, 261
structure of, 348, 391, 413, 414, 416, 423, 426
theory, 52, 79, 80, 84, 104, 133, 277, 303
X, 215, 431 (*see also* X-inactivation)
in yeast, 182, 186
Churchill, Frederick, 64, 65
Churchland, P. M., 6
Churchland, P. S., 6
Ciliates, 115, 116, 136, 215, 216, 219, 279, 419
Cis-regulatory elements, 163, 181, 284, 424, 425
Cis-trans interaction, 267, 424g
Class, 211, 230, 254, 271, 272, 354, 429
of animal, 35
social, 90, 97, 115
Classification, 9, 13, 263, 271, 273, 274, 320, 326, 412
of bacteria, 271–272
of DNA sequences in hybrids, 263–264
folk, 300, 320
by homology, 327
Climate, 49, 50, 59, 62, 63, 219, 310
Climate change, 289, 312, 314, 315
Clone, 116, 187, 252, 258
Cloning, 193, 228, 289, 430
Codevelopment, 283, 284, 286–287. *See also* Developmental symbiosis
Co-evolution. *See* Evolution, co-evolution; Symbiosis, host-symbiont co-evolution
Cognitive evolution, 93
Cognitive science, 6, 369
Cold Spring Harbor Laboratory, 138
Cold War, 31, 85, 86, 117, 124, 135, 138, 216, 277, 278

Cold War in Biology, The, Lindegren, C. C., 135, 216
Collectivity, 31, 89, 92, 94, 96
Colot, Vincent, 221
Community, 68, 286, 287, 308, 363. *See also* Holobiont
scientific, 134, 301, 302, 340
Comparative genomics, 261, 424
Competition, 22, 52, 118, 209, 258, 284, 325. *See also* Selection, levels of
between theories, 46
intracellular, 57, 61, 64
intra-organismal, 258
in society, 99
Complexity, 146, 181,230, 304–305, 373–374, 381–382, 385, 386, 428, 429
increase, 21, 25, 73, 238
in Lamarck's theory, 12, 13, 14, 34, 35, 165–166, 359, 369
organized and disorganized, 360–361
social, 95, 96
tendency toward, 12
theory of, 386
Composition, 70, 115, 124, 308, 360–362, 428
evolution by, 361 (*see also* Symbiosis)
of germ plasm according to Weismann, 57, 62, 64
Compositionality, 304–305
Constantin, Julien, 68
Contingency, 4, 96, 298, 403
Cooperation, xvi, 52, 81, 82, 92, 153, 257, 342. *See also* Evolution, co-evolution; Symbiosis
in evolution, 52
in Spencer's theory, 91, 92
Cope, Edward Drinker, 25–26, 28, 29, 50, 429
Corsi, Pietro, xii, xv, 9, 21, 24
CpG islands, 241
CpG sites, 398, 414, 415, 424
Creationism, 140
Crick, Francis, 106, 118, 134, 337, 382, 389
Cullis, Christopher A., 256
Culture
bacterial, 68, 139
biological, 183, 184, 188, 252, 255, 423, 430
cell, 119, 184, 185, 187, 258
changes in, 24, 57, 62, 110
cumulative, 63, 104, 127, 305, 335, 381
effects of, 14, 37, 109, 130, 245
human-social, 5, 31, 68, 95, 326–328, 383, 403–404
Cuvier, Frédéric, 24, 33, 41
on animal behavior, 37, 39
on heredity, 37, 39, 40, 42
on species, 37, 39, 40
Cuvier, Georges, 11, 15, 16, 17, 24, 37, 41, 46
Cytogenes, 136
Cytoplasmic inheritance, 80,106, 111, 115, 117, 129, 130, 134, 135, 136, 227, 288, 424g
Cytosine methylation. *See* Methylation of DNA
Cytotaxis, 279. *See also* Epigenetic inheritance systems, structural

Damasio, Antonio, 370–371
Daphnia, 123, 199, 200, 223, 325
Dareste, Camille, 68
Darlington, Cyril, 110, 112, 133–134, 135, 136, 430
Darwin, Charles, 3, 4, 17, 90, 103, 124, 148, 157, 159–163, 165, 271–272, 283
on animal behavior, 24, 40–41
on classification, 272
on evolution, 28, 75, 78, 79, 98, 110, 383
on Lamarck, 41, 42
on the origin of variation, 78
on species, 25, 40–42
on the struggle for existence, 284
theory of heredity (*see* Pangenesis)
Darwin, Erasmus, 24
Darwinian problematics, 154
Darwinism, xii, xiii, 97, 140, 149, 150, 271, 404, 428. *See also* Neo-Darwinism
characterization of, 159–160
classical, 57
components/doctrine, 9, 17, 30, 90
"creative Darwinism," 80, 82, 84
in Germany, 45, 46, 47, 51, 52, 53
and relation with Lamarckism, 21–23, 25, 26, 28, 29, 45–46, 48, 50, 78, 132
Das Keimplasma: Eine Theorie der Vererbung, Weismann, A., 58
David, Lior, 147, 152, 220
Dawkins, Richard, 298
De Bary, Anton, 275, 276
Degeneration, 98, 239
during evolution, 61, 62
Delage, Yves, xii
Delbrück, Max, 133, 138, 139, 381
Dennett, Daniel, 298
Descent of Man, and Selection in Relation to Sex, The, Darwin, C., 40
Descent with modification, 23. *See also* Darwinism
Destabilizing selection, 175, 178, 424g. *See also* Domestication
Determinism, 71
Development, 7, 12, 13, 22, 28, 77, 83, 119, 145, 154, 174, 228, 307, 390, 423, 424. *See also* Canalization; Epigenesis; Epigenetics; Genome, as developmental system; Norm of reaction; Plasticity; Polyphenism; Self-organization; Symbiosis, developmental; Symbiopoesis
adjustments, 195 (*see also* Adaptation, physiological)
as an aspect of heredity, 105, 107, 121, 145, 154
of bird song, 321
cell-type centered view of, 163

destabilization of, 175 (*see also* Domestication)
DNA changes during, 254–255, 303, 348
Durkheim's view of, 93–95
dynamic, physics-oriented view of, 163, 164, 166, 224, 347
embryological, animal, 13, 14, 26, 58, 68, 80, 91, 161, 162, 215, 425 (*see also* Morphogenesis)
epigenetically controlled (*see* Epigenetics; Epigenetic inheritance)
evolution of, 224 (*see also* Eco-Evo-Devo; Evo-Devo)
of flat fish, 396
gene controlled, 112, 114, 137
of germ cells, 49, 130, 228, 230, 233, 251, 252, 339, 391, 429
laws of, 42, 50, 121 (*see also* Lamarck, laws of)
pathological, 217, 219, 229, 238–240 (*see also* Teratology)
of plants, 83, 86, 251, 253 (*see also* Plants as open system)
and progress, 89, 90, 381–383, 411 (*see also* Progress)
as recall, 28
sensitive period during, 245
split with heredity, 104, 122–123, 124, 130
Weismann's view of, 58–65
Developmental biology, 104, 297, 298, 385, 386, 390. *See also* Developmental genetics; Evo-Devo
Developmental clocks, 216
Developmental genetics, 141, 162, 425
Developmental niche, 399, 424g. *See also* Developmental symbiosis
Developmental noise, 159, 219, 220, 221, 396, 424g, 429
Developmental plasticity, 145, 146, 241, 284. *See also* Development; Epigenetics
Developmental Plasticity and Evolution, West-Eberhard, M-J., 146
Developmental psychology, 298
Developmental resource, 299, 346, 424
Developmental symbiosis, 154, 284, 290, 399, 424g. *See also* Development, codevelopment
Developmental system, xiii, 147, 153, 154, 163, 166, 299, 300, 302, 303, 319, 324, 325, 326, 327, 328, 345, 346, 347, 425. *See also* Genome, as a developmental system
Developmental Systems Theory (DST), 299, 300, 424g
Developmental view of evolution, 15, 70, 84, 141, 146. *See also* Lamarckian problematics
d'Hérelle, Félix, 276
Diabetes, 239
Differentiation, 118, 136, 137, 138, 162, 163, 215, 216, 228, 234, 424. *See also* Development, of germ cells
of homeologous chromosomes, 268
morphogenetic, 136
sexual, 158
Diploid, 112, 187, 188, 253, 254, 263, 264, 265, 266, 268, 269, 379, 424g. *See also* Allopolyploidy; Polyploidy
Direct induction, 221, 223
Disease, 237. *See also* Epigenetics and disease; Epigenetics, of cancer; Teratology
bacteriological model of, 239
diabetes, 239
environmental, 218
familial, 231
fetal origins of, 240–241, 243 (*see also* Disease, prenatal effects on)
humoral model of, 237
infectious bacterial, 205, 243, 277
lifestyle, 237
neurodegenerative, 419, 429
noncommunicable chronic, 237, 243
predictive adaptive responses (PARs) in, 241–242
prenatal effects on, 238, 239, 240, 241, 246 (*see also* Thrifty phenotype)
risk, 237, 238, 242, 402
sickle-cell anemia, 140 (*see also* Globin paradigm)
transgenerational transmission of, 237, 239, 241, 243–246
Diversity. *See also* Biodiversity; Variability
of determinants, 60
ecological, 199
evolutionary, 200
genetic, 174, 193, 196, 198, 201, 271, 287
human, 4, 326–328
of organisms, 37, 39, 175, 301, 373
phenotypic, 35, 193, 228, 233
Division of labor
in biology, 94, 276, 350
social, 95, 98
DNA, 3, 5, 80, 118, 150, 345, 383. *See also* Base pairs; Chromatin; Development, DNA changes during; Methylation of DNA; Stress, genomic response; Transposable elements
acquisition of, 151, 153
amplification, 255, 256
coding, 379, 426
damage to, 152
elimination of sequences, 265, 266, 267, 268
junk, 255, 382
microarrays of cDNA, 189
in mitochondria and chloroplasts, 106, 278
model of, 389
modification of (*see* Methylation of DNA)
mutation in, 131, 175, 187, 377, 392
non-coding, 268, 353, 426
repair, 348, 413, 415, 416
repetitive, 426 (*see also* Heterochromatin)

DNA (cont.)
replication, 5, 135, 216, 218, 261, 348, 363, 413, 414, 415, 416
ribosomal (r-DNA), 255, 256
selfish, 255
sequencing, 140
transfer through conjugation, 274
transfer through transduction of, 207, 208, 274
transfer through transformation, 274
Dobzhansky, Theodosius, 3, 104, 116, 123, 124, 128, 138, 224, 297
D'Omalius d' Halloy, Jean-Baptiste-Julien, 17
Domestication, 37, 147, 152, 171–178. *See also* Darwin, on animal behavior
of dogs, 172
and epigenetic inheritance, 175
of silver foxes, 171–174, 175
variability in, 174
Dominance, genetic, 162, 303, 347
evolution of, 110
Dörner, Günther, 239
Drift, genetic, 4, 7, 104, 118, 127, 140, 166, 194, 198, 258, 338, 342, 399, 424g
Drosophila, 110, 117, 118, 119, 122, 137, 149, 178, 215, 230, 255, 272, 277, 340, 392
Durkheim, Émile, 31, 90
as Lamarckist, 94–95, 97
on the organism, 94–95
on social plasticity, 96–97
on transfer to the social, 93–94
Durrant, Alan, 215, 256
Dynamical patterning modules (DPNs), 163, 165

East, E. M., 277
Eco-Devo (ecological developmental biology), 193, 194, 246, 324, 424g
Ecological inheritance, 299, 307, 308, 424g. *See also* Niche construction
Economic(s), 23, 90, 96
considerations, 86, 90, 405, 406
of nature, 272
socioeconomic, 239, 243
Economy of the organism, 91, 94
Ecotypes, 198, 325
Eimer, Theodor, 29, 50, 59
Eldrege, Niles, 298
Embryology, 26, 78, 79, 112, 122, 123, 124, 238, 272, 381. *See also* Development, embryological
Emergence
of disease, 238
of genetics, 29
of life, 381
of new species, 25
of novelties, 35, 165, 178, 186, 189
and self-organization, 359–360, 382, 430
of sociology, 89
Endocrinology, 175
Endopolyploidy, 253
Endoreduplication, 253, 424g
Endosperm, 253
Endosymbiosis, 278, 288, 425g
Entropy, 7, 73, 374, 381
Environment. *See also* Norm of reaction; Plasticity; Vernalization
causal role of, 22, 24, 26, 30, 59, 74, 162, 238, 290–291, 423 (*see also* Darwin; Durkheim; Haeckel; Lamarck; Spencer; Symbiosis; Use and disuse)
concept of, xii, 3, 91
developmental, 240, 323, 327, 424
future, 241–242
in French neo-Lamarckism, 68, 69, 72, 73, 74
in German biologists' thought, 47, 48, 49, 50, 51
internal, 49, 69, 115, 430g
in Lysenkoist USSR, 77, 79, 81
maternal, 237–239, 246
and the Modern Synthesis, 104–105, 110, 112, 115–116, 123, 127
and organism interaction, 23, 166, 298–299, 308–309, 319, 424, 427 (*see also* Developmental Systems Theory; Niche construction)
social, 31,79, 81, 95, 97, 174 (*see also* Domestication)
and soft inheritance, 105, 223 (*see also* Soft inheritance)
and stress, 184, 212 (*see also* Stress)
and temperature changes, 59, 61, 62, 63, 152, 159, 177, 195,197, 288, 289, 398
Weismann's view of, 57, 59, 61, 64, 65
Environmental
change, 59, 60, 68, 71–73, 146–147, 200–201, 209, 251
circumstances in Lamarck, 12, 14, 35
cues, 158, 201, 242, 396
effects, 46, 48–50, 52, 53, 115, 195, 198, 223
factors, 46, 63, 221, 258, 299 (*see also* Use and disuse)
influence, 48, 64, 46, 238, 320, 326
responses to, 194, 195, 200
stress, 18, 152, 199, 219, 275, 286, 350 (*see also* Domestication; Stress)
variation, 194, 195, 423 (*see also* Epigenetic variability)
Ephrussi, Boris, 215
Epiallele, 124, 222, 284, 329, 425g
Epigenesis, 121
Epigenetic control mechanisms, 174, 175, 216, 229, 244, 245, 251, 302, 303, 377. *See also* GEMs
Epigenetic heritability, 222, 335, 336, 338, 402
Epigenetic inheritance, xiii, 75, 119, 135, 150, 159, 175, 213, 215, 230, 251, 279, 326, 389, 390, 391, 413–420, 424g. *See also* Epigenetic Inheritance Systems; Heredity and memory; Non-Mendelian inheritance; Paramutation; Bacterial persistence; Soft inheritance

in animals, 151, 220, 221, 230–234, 244–246, 338–339 (*see also* Domestication)
in bacteria, 151, 219, 367, 368, 377–379, 413, 418
in the broad sense, 151
cellular, 215, 216–222, 377–379
defined, 150–151, 215, 424g
through developmental reconstruction, 199, 218 (*see also* Maternal effects)
in ecology, 219
evolutionary significance of, 145, 219, 234, 301–302, 331–342, 379, 389, 391–392, 395, 399–402, 405–407 (*see also* Evolutionary theory, extended; Genetic assimilation)
evolution of, 188, 200, 205, 213 (*see also* Norms of reaction, evolution of)
frequency of, 219, 220, 221
through gametes, 151, 218, 220, 221, 244
history of, 150, 215–217
inducibility of, 219, 220, 221
interactions with genetic inheritance, 152, 187, 222, 223, 224, 255, 256, 257, 303, 345 (*see also* Information, epigenetic/genomic)
mechanisms of, 151, 154, 188, 190, 223, 251 (*see also* Epigenetic control mechanisms; Epigenetic Inheritance Systems)
in medicine, 151, 216 (*see also* Disease, transgenerational transmission)
in mitotic cell lineages, 151, 215
in the narrow sense, 151
in plants, 59, 152, 220, 221–222, 231, 254, 256, 267, 390, 392 (*see also* Allopolyloidization)
in protists, 151, 219–220
stability of, 219, 220, 221, 392, 395, 401–402
and stress, 152, 175
in yeast, 188, 189
Epigenetic inheritance systems (EISs), 159, 217–221, 413–420, 424g
chromatin-marking, 218, 219, 220, 245, 253, 379, 413–416, 419, 424g
interactions among, 416, 419
RNA-mediated, 151, 218, 219, 220, 221, 223, 230–234, 245, 253, 413, 416, 417, 430g
self-sustaining loops, 217, 218, 219, 220, 378–379, 413, 416, 418, 430g
structural, 137, 218, 219, 220, 279, 413, 418–420, 430g
Epigenetic landscape, 113, 114
Epigenetic learning, 224
Epigenetic marks. *See* Chromatin marks
Epigenetic memory trace, 224
Epigenetic reprogramming (remodeling), 228, 229, 253. *See also* Development of germ cells; Cloning
Epigenetic variability, 175, 205, 207, 213, 228, 229, 251, 252, 253, 284, 291
adaptive, 222, 253, 263, 267, 268
in fluctuating environments, 188, 205, 213
generated stochastically, 219–221 (*see also* Developmental noise)
global, 189, 190, 254
induction of, 219, 220, 221, 223 (*see also* Soft Inheritance; Inheritance of acquired characters)
segregation of, 221–222
stress-associated, 179, 189, 190, 224 (*see also* Stress)
Epigenetic variations, xi, 190, 205, 217, 228, 377
Epigenetics, 7, 45, 75, 150, 385, 386, 390, 424g
of cancer, 216, 218, 390, 398
correlates in disease, 218, 241, 243–246 (*see also* Disease, transgenerational transmission of)
defined, 150
and psychobiology, 218
and social sciences, 403–405
Epigenotype, 425g
Epimutation, 217, 251, 255, 335, 390, 391, 392, 395, 400, 401, 425g
Epi-RILs (epigenetic recombinant inbred lines), 221, 222
Escherichia coli (*E. coli*), 138, 152, 177, 206, 207, 208, 210, 211, 212, 219, 275, 378, 382
Ethics, 4, 97, 298
Euchromatin, 222, 348, 425g
Eugenics, 51, 52, 82, 110, 134, 239
Eukaryote, 150, 182, 256, 273, 274, 280, 377, 414, 428, 429
Euprymna scolopes, 284, 285
Evo-Devo (evolutionary developmental biology), 45, 80, 86, 160, 161, 283, 246, 327, 341, 383, 406, 424, 425g
Evolution, xi, 3, 4. *See also* Darwinism; Epigenetic inheritance, evolutionary significance of; Evolutionary theory; Gradualism; Modern Synthesis; Natural selection; Neo-Darwinism; Phylogeny
adaptive, 59, 104, 112, 127, 150, 151, 152, 224
in bacteria, xiii, 153, 273–275, 368, 379 (*see also* Mutations, adaptive; Stress, role in bacterial evolution; Symbiosis, co-evolution between host and symbiont)
co-evolution, 154, 257, 258, 286, 287–290 (*see also* Holobiont; Hologenome; Symbiosis, co-evolution between host and symbiont)
Darwinian (*see* Darwinism)
dynamics of, 103, 104, 148, 181, 182, 184, 200, 210–212, 302, 332, 335, 342, 363, 381, 383, 401, 402, 428
of evolvability, 148 (*see also* Evolvability)
of genetic systems, 112
gradual, 12, 23, 45, 59, 62, 91, 92, 104, 105, 109, 110, 112, 115, 154, 165, 227, 275, 276, 278, 298, 430
human, 3, 40, 147, 178, 367
Lamarckian (*see* Lamarckism; Neo-Lamarckism)

Evolution (cont.)
levels of, 4, 5, 134, 367
macroevolution, 80, 104, 109, 115, 157, 160, 164
of metazoans, 148, 149, 159 (*see also* Morphogenesis; Body plan)
microevolution, 104, 128, 164, 165 (*see also* Evolution, gradual; Gradualism)
molecular, 273
neo-Darwinian (*see* Neo-Darwinism; Modern Synthesis)
neo-Lamarckian (*see* Neo-Lamarckism; Lysenkoism)
of plants, 104, 251 (*see also* Allopolyploidization; Polyploidy)
progressive, 21, 23, 45, 48, 59, 60, 67, 72, 73, 74, 367, 373, 374, 381, 383, 429g
rate of, 28, 227 (*see also* Evolution, gradual; Evolution, saltational)
role of stress in (*see* Domestication; Genome rewiring; Genome, shock and stress response; Stress, genomic response to)
saltational, 73, 104, 109, 146, 154, 164, 165, 224, 271, 275, 276, 298, 430g (*see also* Hybridization; Polyploidization)
theologically oriented views of, 18, 427
transitions in, 153, 298, 371
trends in, 57, 59, 60, 115, 157
Evolution: The Modern Synthesis, Huxley, J., 112
Evolutionary biology, 6, 25, 31, 33, 47, 89, 93, 94, 96, 97, 98, 104, 105, 106, 111, 112, 115, 116, 118, 124, 135, 145, 146, 246, 278, 297, 298, 301, 305, 332, 340, 395, 401–, 406
Evolutionary synthesis. *See* Modern Synthesis
Evolutionary theory, 3, 9, 23, 25, 29, 30, 31, 52, 78, 301, 332, 334, 335, 336, 338, 385, 386, 389. *See also* Darwinism; Evolution; Gradualism; Lamarckism; Neo-Darwinism; Neo-Lamarckism; Orthogenesis; Modern Synthesis
extended, 338, 386, 405–407, 408
history of (*see* Darwinism; Lamarckism; Lysenkoism; Neo-Darwinism; Neo-Lamarckism; Modern Synthesis)
transfers from, 31, 89, 95–98, 395, 402, 403
Evolvability, 148, 153, 303, 367
Experimental philosophy, 298, 425g
of biology, 300
Experimentation, 27, 29, 30, 49, 51, 69, 233, 276
Expression of the Emotions in Animals and Man, The, Darwin, C., 40
Expressivity, 233, 338, 425g

Feedback, 304, 305, 358, 374
circular, 134
from the environment, 307, 323, 331, 428 (*see also* Niche construction)
and function, 361
novel regulatory, 186, 189 (*see also* Genome, rewiring of)
positive, 61, 416
self-sustaining, 305, 377, 378, 416, 418 (*see also* Epigenetic inheritance systems, self-sustaining loops)
Feldman, Moshe, 152, 224, 254
Fieldwork, 307, 309, 313, 314, 315
Fisher, Ronald A., 110, 111, 112, 113, 118, 127, 131, 134, 140, 193, 386
Fitness, 4, 75, 209, 243, 257, 261, 297, 298, 361
across environments, 199, 209, 243
constant, 297, 298, 308, 361
of heterozygotes, 141
of symbiont and host, 257, 261, 287, 290
reduction in, 171
variation in, 195, 197, 198
Fixity
as lack of plasticity, 300, 321, 322, 323
of species, 16
Flax, 117, 215, 216, 256
Ford, E. B., 111, 134
Fossil, 13, 39, 73, 396. *See also* Paleontology
Foster, Michael, 27
Foster, Patricia, 152
Founder effects, 198
Free energy-rate density, 374, 425
Frege, Gottlieb, 5
Freinkel, Norbert, 239
Function, 297, 303, 304, 322, 357–358. *See also* Adaptation; Self-organization
and cofunctionality, 189
and feedback, 361
and the genome, 347, 348, 349, 350–351
and Newtonianism, 160
specialization of, 5, 7
and structure and form, 14, 24, 248, 161, 181, 253, 346, 360, 363, 423, 424, 430

GAL system, 182, 183, 186, 188
Gametogenesis resetting, 244, 253
Gayon, Jean, 42, 68
Gemmules, 70, 162, 429. *See also* Pangenesis
GEMs (genomic epigenetic mechanisms), 303, 345, 349–353, 425g
Gene. *See also* Genetics, Mendelian
action, 111, 266
composition of, 115, 116 (*see also* DNA)
core, 189
definition of, 78
flow, 128, 197
frequencies of, 107, 124, 299
horizontal transfer of (*see* Horizontal gene transfer)
locus (loci), 165, 187, 219, 220, 222, 224, 230, 245, 256, 263, 265, 266, 267, 353, 379, 383, 391, 415, 423, 426, 429

master, 227
materiality of, 123
Mendelian, 104, 105, 106, 127, 131, 135, 136, 276, 339 (*see also* Genetics, classical; Genetics, Mendelian)
Mobile, 117 (*see also* Transposable elements)
mutation in (*see* Mutation)
networks, 163, 347
promoter, 181, 183, 186, 187, 241, 254, 289
recessive, 174, 276
recombination among (*see* Recombination)
regulatory, 149, 181, 284
replication of (*see* Replication)
selfish, 118, 279
segregating, 117
as unit of ontogeny, 122
Gene expression, 209, 218. *See also* Differentiation, cellular; Epigenetics
changes in expression in, 123, 124, 220, 229, 234, 261, 267, 284, 348, 397, 400
induction by symbiont, 286, 290 (*see also* Symbiosis)
networks of, 228
regulation of, 152, 269, 415, 416, 430
transmission of (*see* Epigenetic inheritance)
Gene receptor, 135
Generative systems, 159
Generelle Morphologie, 47, 48
Genetic accommodation, 397, 426g
in *Manduca sexta*, 398
Genetical Theory of Natural Selection, The, Fisher, R. A., 110
Genetic assimilation, 113, 148, 392, 397, 426g
and animal behavior, 150, 397
conditions of, 397
defined, 149, 392, 396, 397
and evolution of fixed asymmetry, 396
experiments, 149–150, 392
mutational, 398
pervasiveness of, 392, 395, 396, 399
regulation of, 175
Genetic membrane, 419, 426g
Genetic polymorphism, 118, 122, 325, 382
Genetic variance, 176, 402
Genetics. *See also* Gene; Inheritance, hard
behavioral, 141
classical, 77, 79, 103, 115, 116, 127, 135, 137, 138, 327, 337, 338, 385, 402
developmental (*see* Developmental genetics)
ecological, 111
evolutionary, 141, 377 (*see also* Evolutionary biology)
human, 141
Laboratory of Evolutionary Genetics, 171
and mathematics, 111
Mendelian, xii, 103, 115, 123, 127, 128, 130, 427g
microbial, 130
molecular, 80, 85, 118, 119, 140, 273, 278
and neo-Lamarckism, 52, 53, 79, 84
in *Neurosopora*, 135
non-Mendelian, 115, 117, 123, 135 (*see also* Epigenetic inheritance; Hybridization, graft; Inheritance, cytoplasmic; Paramutation; Soft inheritance)
of populations, 118, 124, 127, 128, 130, 131, 140, 222, 283, 301, 308, 381, 395, 406, 428
quantitative, 130, 196, 301, 336, 337, 402
rise of, 3, 29, 51, 74, 78, 122
split with embryology, 122, 123
stages in development of, 116–117
suppression of, 77, 82, 84, 277 (*see also* Lysenkoism)
Genetics and the Origin of Species, Dobzhansky, T., 128
Genome, xiii, 130, 426g. *See also* GEMs; Gene; Genotype; Hologenome
acquisition and sharing of, 153, 207, 208, 275, 279, 280 (*see also* Horizontal gene transfer; Hologenome)
as a developmental system, 141, 302, 303, 345–346
as a dynamic system, 108, 145, 153, 178, 190, 220, 345–346, 347–349 (*see also* Transposable elements; Allopolyploidization)
doubling, 261 (*see also* Polyploidy; Allopolyploidy)
ensemble of (pangenome), 254
of humans, 141
induced changes in, 145, 220
as inner nature, 326
McClintock's view of, 152, 177, 216, 347
metastable (MSG), 303, 352, 349–350
methylation of, 222, 335, 338 (*see also* Genomic imprinting; Methylation of DNA)
origin of term, 135, 136
plasticity of, 145, 152, 426 (*see also* Plasticity, genomic)
as program, 158, 190
as a receiver, 352, 353
relation with genes, 346–347
rewiring of, 147, 152, 181, 182–189, 267, 408
shock and stress response, 152, 153, 177–178, 220, 223, 253, 254, 275, 289, 351 (*see also* Allopolyploidization; Hybridization; Polyploidization; Stress, genomic)
size reduction of, 263, 265
Genome-wide association studies, 237, 402
Genomic imprinting, 217, 338, 390, 426g
Genotype, 426g
concept of, 78–79, 84
discordance with phenotype, 157–158
as distinguished from phenotype, 78, 127, 400, 425
evolution and selection of, 131, 201, 257

Genotype (cont.)
"generalist," 198
information transfer to, 154
and norm of reaction, 194–195, 196, 198, 201 (*see also* Norm of reaction; Plasticity)
and symbiosis (*see* Hologenome; Symbiont)
Genotype-by-environment interaction (g × e), 426g. *See also* Norm of reaction
Geology, 15, 16, 26, 41. *See also* Uniformitarianism
Germ line, 426g. *See also* Germ plasm
epigenetic variations in, 151, 217, 218, 221, 223, 391 (*see also* Epigenetic inheritance)
segregation from soma, 117, 118, 128–130, 161, 251, 252, 253, 279
Weismann on, 58, 65, 70, 111
Germ plasm, 29, 117, 426g. *See also* Pangenesis
in bacteria, 272 (*see also* Holobiont)
French neo-Lamarckism view of, 70–71
Germinal selection, 29, 57, 60–65
Weismann's theory of, 58–65, 70, 71, 78, 116
Gilbert, Scott, xvi, 103, 105, 106, 138, 140, 141, 153, 397, 396, 397, 398, 399, 405
Giraffe, 33, 36
Gissis, Snait B., 103, 402, 403, 405, 406
Gliboff, Sander, 28, 29, 45, 405, 428
Globin paradigm, 133, 138, 140–141
Gluckman, Peter, 151, 398, 404, 405
Glushchenko, I. E., 85
Goethe, Johann Wolfgang von, 46, 160
Goldschmidt, Richard, 80
Goldsmith, Mary, xii
Gottlieb, Gilbert, 383
Gould, Steven J., 109, 297, 298, 373
Gradualism, 30, 154. *See also* Evolution, gradual
in Darwinism, 110, 164, 276, 278
Lamarck's, 113
Graham, Loren, 77
Gray, Russel, 299, 424
Grenier, Anne-Marie, 278
Griesemer, James, xiv, 300, 301, 302, 406, 408
Griffiths, Paul, 300, 301, 399, 424
Grinnell, Joseph, 310, 311, 312, 313, 314
Guided assembly, 424, 426g. *See also* Cytotaxis; Epigenetic inheritance systems, structural

Habit, 27, 31, 37, 50, 92, 94, 95, 97, 272, 427. *See also* Inheritance of acquired characters
Darwin on, 24, 40–41
Frédéric Cuvier, on, 24, 37–39
Lamarck on, 12, 24, 33–37, 238, 369
Hacking, Ian, 405
Hadorn, Ernst, 215
Haeckel, Ernst, 25, 28, 29, 49, 50, 238, 428
and Darwin, 25, 46–47
on environment, 48
followers of (Haeckelian school), 29, 51–53
and Lamarck, 46, 48
Haldane, J. B. S., 85, 110, 111, 112, 113, 115, 116, 118, 127, 131, 134, 140, 196
Hanson, Mark, 151
Haplodiploid, 112
Haploid, 182, 187, 188, 233, 263, 379
Haplopappus gracilis, 258
Hard inheritance, 78, 107, 135, 391
Harman, Oren, 133, 134
Heidegger, Martin, 5
Helicobacter hepaticus, 285
Helmholtz, Hermann von, 27, 385, 387
Hereditary symbiosis, 277, 278, 279, 280
Heredity, xii. *See also* Cell heredity; Cell memory; Cytoplasmic inheritance
infective, 277, 427g
levels of, 152
and memory, 27, 28, 52, 224
split with embryology, 122–124
Hertwig, Oskar, 59
Heslop Harrison, John S., 139
Heterochromatin, 222, 348, 426g
Heterochrony, 284
Heterodiploid, 379, 426g
Heterometry, 284
Heterosis, 266, 426g
Heterotopy, 284
Heterozygote superiority (heterosis), 141
Heuristic, 301, 307, 313, 315, 332, 426
reductionism, 331, 341
Hierarchical organization, 58, 158, 275, 360, 361. *See also* Classification; Selection, levels of
in Darwin, 272
in Lamarck, 12
Hinshelwood, C. N., 115
Hirschbein, Luisa, 218, 219, 367
Histoire naturelle des animaux sans vertèbres, Lamarck, J-B., 14, 15, 25
Histones, 135, 218, 348, 428
modifications of, 244, 256, 334, 337, 348, 391, 414, 415, 416
replacement of, 348, 349, 350
History of Biology: A Survey, The, Nordenskiöld, E., 25
Hitler, Adolf, 3–4
Holistic approach, 385
Holliday, Robin, 216, 217, 390
Holobiont, 154, 284, 286, 287, 289, 290, 426g
Hologenome, 284, 286, 287, 288, 290, 426g
Homeostasis, 69, 174–175, 177, 190, 241, 290, 291, 304, 353, 426g, 430
Homo sapiens, 147, 178
Homeoalleles, 267, 426
Homology, 264, 301, 327, 328, 427g. *See also* Chromosomes, homeologous; Chromosomes, homologous
Hooker, J. D., 41, 162
Horizontal gene transfer, 153, 255, 271, 274, 275, 279, 280, 285, 286, 331, 363, 419, 427g, 429. *See*

also Holobiont; Hologenome; Heredity, infective; Symbiosis
through conjugation, 274
through transduction, 274, 277
through transformation, 274, 278
Hormones, 174, 219, 253
glucocorticoid, 175
Houssay, Frédéric, 68
Hubby, J. L., 382
Hull, David, 298
Human universals, 319, 326, 328
Human evolution, 40, 147, 178, 367
Husserl, Edmond, 5
Huxley, Julian, 103, 112, 116, 272
Huxley, Thomas H., 51, 121
Hyatt, Alpehus, 25–26, 28, 50
Hybridization
between species, 112, 152, 261, 262–265, 268
DNA-RNA, 273, 274, 350, 351, 382
graft, 80, 83–86
in Lamarck's theory, 50
Mendelian rules of, 79
Hydrogeology/Hydrogéologie, Lamarck, J-B., 11

Imprint. *See* Genomic imprinting
Inbreeding, 112
Individuality, xiii, 22, 92, 145, 153–154, 252, 257. *See also* Holobiont; Hologenome; Levels of selection
and collectivity, 92, 94, 320
questioned/expanded, 277, 278
Infectious heredity, 277, 427g
Information, 30, 118, 159, 361, 383, 405. *See also* Central Dogma
epigenetic, 151, 222, 303, 345, 419, 425, 426
genetic/genomic, 303, 337, 345, 347, 350–353
Informational genes, 275
Inheritance. *See also* Epigenetic inheritance; Inheritance of acquired characters; Non-Mendelian inheritance; Soft inheritance; Infectious heredity
blending, 162 (*see also* Pangenesis)
cytoplasmic, 80, 106, 111, 115, 117, 129, 130, 134, 135, 136, 227, 288, 424g
ecological, 299, 307, 308, 424g (*see also* Niche construction)
hard, 78, 107, 135, 391
levels of governances, 134
Mendelian, 103–105, 117 (*see also* Genetics, Mendelian)
physicalist view of, 159
soft (*see* Soft inheritance)
Inheritance of acquired characters, xiii, 4, 23, 26, 59, 427g. *See also* Cytoplasmic inheritance; Direct induction; Ecological inheritance; Habit; Instinct; Lamarckism; Lysenkoism; Neo-Lamarckism; Parallel induction; Plasticity; Soft inheritance; Somatic induction
adaptive, 390
Darwin's view of, 162, 165
DST view of, 300, 321, 322
Frédéric Cuvier's view of, 36, 37, 39, 40
in French neo-Lamarckism, 70–74
and genetic assimilation, 113, 149
and horizontal gene transfer, 278, 279 (*see also* Symbiosis, hereditary)
Lamarck's view of, 3, 12, 31, 33, 36, 37, 40, 42, 407, 427
in Lysenkoist USSR, 77, 80–86
and the Modern Synthesis, 106, 111, 113, 116, 132, 57, 277, 278, 279
objections to, 71–72, 82 (*see also* Soft selection, objection to; Weismann, August, challenging Lamarckism; Modern Synthesis, exclusion of Lamarckism)
and social thought, 31, 89, 97
Waddington's take on, 113, 149
Instinct, 27, 150. *See also* Habit; Heredity, and memory
Darwin's view on, 24, 40–42
Frédéric Cuvier's view on, 24, 37–39
Lamarck's view on, 34, 35, 369, 427
Internal selection, 60, 61, 63, 65, 254, 358 (*see also* Germinal selection)
Invasive species, 166, 219
Isoschizomers, 267, 427g
Isoenzymes (isozymes), 266, 427
Isogenic line, 427

Jablonka, Eva, 7, 103, 159, 284, 362, 371, 395, 396, 397, 401, 402, 406
Jacob, François, 137, 227
Johannsen, Wilhelm, 78, 79, 82, 84, 194, 426
Joravsky, David, 77
Just, E. E., 161

Kammerer, Paul, 29, 51, 52, 53, 428
Kant, Immanuel, 160, 161, 359
Kappa elements, 137
Kaufman, Stuart, 304
Keller, Evelyn F., 303, 304, 305, 346, 398
Kettlewell, H. B. D., 138, 139, 140
Kingsley, Charles, 17
Kisdi, Eva, 402
Komarov, Vladimir, 82
Kosslyn, Stephen M., 6
Krementsov, Nikolai, 77

Lactose operon, 378
Lamarck, Jean-Baptiste de. *See also* Inheritance of acquired characters; Lamarckian inheritance; Lamarckian problematics; Soft inheritance
on classification, 271
contemporary views on his physicalist approach, 163, 164–165, 228, 389

Lamarck, Jean-Baptiste de (cont.)
"dangerous idea" of, 157, 166
Darwin's view of, 42
on earth, 9, 10, 11, 13, 16, 33, 40, 42
on fluid dynamics, 11–14, 24, 370
on individuality in plants, 252
laws of, 36, 72
on life and spontaneous generation, 9–13, 36
Mandelstam's poem, 411–412
as a materialist, 9–10, 35, 166, 369
on mind and inner feeling, 34, 157, 369–371, 427g
on Nature, 9, 33, 35, 36, 42
on organisms, xi, 11, 12, 18, 23, 157, 162–163, 373
on "power of life," 13–14, 35, 161, 166, 359, 427g
reception of his ideas, 9, 11, 14–18 (*see also* Neo-Darwinism; Neo-Lamarckism; Weismann, August)
on sensation and sensibility, 34, 369–371
on species, 10, 13–16, 33, 35–37
on subtle fluids, 35, 45, 46, 163, 167, 223
Lamarckian evolution (Lamarckian evolutionism), 26, 31, 73, 89, 94, 95, 98, 132
Lamarckian inheritance, xii, 64, 65, 80, 103, 105, 128, 130, 131, 132, 332. *See also* Inheritance of acquired characters; Use and disuse; Weismann, August
and Central Dogma, 106, 118
Fisher's view of, 110
future of, 395, 407, 408
Haldane's view of, 110–111
as heresy, 132 (*see also* Modern Synthesis)
historical definition, 17–18
in USSR, 77, 82–83, 85, 86 (*see also* Lysenkoism; Neo-Lamarckism)
Lamarckian problematics, xi, 57, 146, 367, 395, 402, 403, 405, 407–408, 427g
in biology, 145–154
characterization, xii
in historical perspective, 21–31
in philosophy of biology, 297–305
Lamarckians, 45, 50, 53, 57, 60, 64, 65, 81, 148, 154, 381 (*see also* Neo-Lamarckism)
American, 50
Lamarckism. *See* Inheritance of acquired characters; Lamarck; Lamarckian inheritance; Lamarckian evolution; Soft Inheritance; Use and disuse
and psycho-Lamarckism, 50–51, 429g
Lamb, Marion J. xvi, 7, 103, 104, 106, 133, 159, 217
Lamm, Ehud, 139, 153, 302, 303, 399, 400, 406
Lankester, Edward Ray, 72
Laplace, Pierre Simon, 11, 71, 130
Laurent, Goulven, xvi
Lavoisier, Antoine de, 9
Le Dantec, Félix, 70, 73, 74
Learning, 3, 27, 147, 148, 149,151, 367, 369, 371, 375, 397, 426, 428. *See also* Epigenetic, learning
Lederberg, Esther, 139
Lederberg, Joshua, 139, 277
Leuckart, Rudolph, 58
Levels of selection, 3, 4, 6, 7, 9, 29, 59, 61–63, 65, 153, 257, 287, 297, 298, 299, 358, 367
Levy, Avraham, 152, 224, 254
Lewontin, Richard, 165, 166, 297, 298, 382
L'Héritier, Charles Louis, 137
Life. *See* Lamarck on "power of life"
Liliaceae, 254
Limited heredity, 335, 427g
Lindegren, Carl, 135, 136, 139
Linnaeus, Carl, 272
Listeria, 286
Liverant, Yigal, xvi, 411
Logic, 4, 5, 6, 67, 403
Loison, Laurent, 26, 30
Lotka, A., 211, 212, 374
Liubimenko, Vladimir, 82
Luria-Delbrück experiments, 133, 138–139
Lwoff, André, 273
Lycopersicum chilense, 254
Lyell, Charles, 3, 16, 42, 91
Lyon, Mary, 215
Lysenko, Trofim, 3, 30, 77–79, 82–86, 103, 117, 373. *See also* Lysenkosim
Lysenkoism, 31, 77–79, 81–86, 103, 124. *See also* Neo-Lamarckism

Macroevolution, 80, 104, 109, 115, 157, 160, 164
Maize, 177, 215, 254, 256
Malthus, Thomas R., 41
Mandelstam, Osip E., xvi, 31, 411–412
Manduca sexta, 398
Margulis, Lynn, 279
Markel, Arkady, 146, 147, 152
Marxist, 30, 77, 78, 81, 82
Materialism, 21
dialectical, 77
mechanistic, 160, 427g
Materialist. *See* Lamarck, as a materialist; Materialism
Darwin as, 160, 161
in Lamarckism, 16, 26, 28, 367, 373, 411
and reductive physiology, 27
Maternal effects, 111, 231, 335, 401
Maternal environment, 199, 237, 238–241, 246
Mating patterns, 104, 127
Maynard Smith, John, 118, 119, 131, 132, 153, 217, 279, 427
Mayr, Ernst, 45, 53, 116, 123, 127, 128, 161, 297. *See also* Modern Synthesis
on germinal selection, 64
on inheritance, 104, 105, 106, 110, 124, 127, 131, 132, 133, 162, 223, 430

on relations between disciplines in biology, 109
on speciation, 104, 128
McClintock, Barbara, 117, 152, 177, 216, 254, 346, 347, 353
Mechanism of Mendelian Heredity, The, Morgan, T. H., 79
Mechanistic world view, 26, 46, 427. *See also* Lamarck as a mechanist; Materialism, mechanistic
in Haeckel's biology, 48, 50
Meckel, Johann Friedrich, 14, 15, 238
Medawar, Peter, 383
Meiosis, 167, 265, 335, 340, 348, 391, 424, 426
pairing in, 261, 264, 266
tetrad analysis, 187, 188
Meister, G. K., 82
Melanism, 111, 133, 139–140
Mendel, Gregor, 111, 161, 162
Mendelian gene. *See* Gene, Mendelian
Mendelian inheritance. *See* Gene, Mendelian; Genetics, Mendelian; Hybridization, Mendelian rules of; Inheritance, Mendelian
Mendelism and Evolution, Ford, E. B., 111
Mendelsohn, Everett, 141, 145, 395, 407
Meristem, 251, 252, 427g
Metaphor, 6, 31, 89, 90, 91, 93, 94, 96, 165, 178, 403, 404, 431
Metaphysical conception, 71, 72, 123, 386
agenda, 387
dogma, 30, 74
Metastable genome (MSG), 349. *See* Genome, metastable
Metazoa. *See* Evolution of metazoans
Methylation of DNA, 119, 131, 216, 217, 334, 335, 337, 339, 424g
and aging, 216
and alloployplidization, 267, 268 (*see also* Allopolyploidization)
as cell memory system, 217, 224
in different taxa, 339
in disease, 216 (*see also* Disease, transgenerational effects)
effects on gene expression, 124, 217, 229, 267
and evolution, 391, 392, 398, 401–402 (*see also* Epigenetic inheritance, evolutionary significance of)
functions of, 413, 414
and genomic imprinting, 217 (*see also* Genomic imprinting)
inheritance of patterns of, 124, 217, 221–222, 348, 377, 390
methylation-sensitive amplified polymorphism (MSAP), 267
regulation of, 175, 241, 258
replication of patterns, 413–415
and transposable elements, 230, 245, 267
Methylation of histones, 244, 256. *See also* Histones, modifications of
Michie, Donald, 116–117
Michurin, Ivan, 83, 85
Microbial world. *See* Bacteria; Bacteria, evolution of and in; Archaea
Microevolution. *See* Evolution, microevolution; Population, genetics
Migration, 104, 110, 127, 140, 210, 212
Millikan, Ruth, 357
Milne-Edwards, Henri, 91
Mind, 47
biology of, 77, 367
and brain, 6
Lamarck on, 369–371 (*see also* Lamarck, on mind and inner feeling)
philosophy of mind, 298, 423
and psycho-Lamarckism, 50, 51, 427
Mneme, 52, 224, 428g. *See also* Heredity and memory
Model
Bateson's, 160
Darlington's, 134
Darwin's, 162 (*see also* Darwinism; Pangenesis)
of disease, 237, 239, 241, 243, 245 (*see also* Disease, bacteriological model of; Disease, humoral model of; Disease, noncommunicable chronic; Disease, predictive adaptive responses in; Thrifty phenotype)
Eco-Devo, 193–194
ecological vs. statistical models, 314
of epigenetic inheritance, 413–420
epigenetic landscape, 113, 114, 425g
of evolutionary epigenetics, 301–302, 322–333, 338–342, 395, 400–401, 402
folkbiology, 321, 322
of gene regulation, 163, 182
of individual, 277 (*see also* Individuality)
Lamarckist-transformist, 24
Lamarck's hydraulic, 24
of modern society, 90, 95, 98
Neo-Darwinian/Mendelian, 3, 116, 157, 162, 194 (*see also* Neo-Darwinism)
network, 399
niche construction, 307, 308–315
operon, 138
organisms, 92, 99, 121, 229, 245, 263, 272, 382
population genetics, 127, 138, 210, 308
population genetics vs. quantitative genetics, 301–302, 332
predator-prey interaction, 210–213, 401, 402 (*see also* Bacteria, persistence)
recapitulation, 25, 26, 28, 29, 89
rheostat, 233
system, 242, 272, 277, 327, 333, 382
toy, 333, 334, 335, 431g

Model (cont.)
transfer of, 93–94, 96, 403, 431 (*see also* Transfer)
use of, 5, 6, 90, 298–299
Weismann's germ plasm, 57, 62 (*see also* Selection, germinal; Weismann August)
Modern society, 90, 95, 98
Modern Synthesis, xii, 103–142, 160, 162, 164, 165, 166, 271, 272, 386, 389, 405, 406, 428g. *See also* Gradualism; Lamarckism; Lysenkoism; Neo-Lamarckism; Mayr, Ernst; Neo-Darwinism; Soft inheritance
and Central Dogma, 106, 118
characterization of, 103, 105–106, 127, 131, 157, 159–160 (*see also* Inheritance, hard)
embryology's place in, 121–124
establishment of, 31, 45, 106, 109–110
exclusion of Lamarckism, 28, 45, 127–131, 162 (*see also* Neo-Darwinism; Soft inheritance, rejection of)
exclusion of plasmagene theories, 134–137
exclusion of saltationism, 164–166
extended, 386, 405–407, 408
Gould's view of, 109
institutionalization of, 106, 137–138
key supporting studies, 133, 138–141 (*see also* Luria-Delbrück experiment; Melanism; Globin paradigm)
and mathematics, 111
and microbial evolution, 272–274
and plant biology, 128
and politics, 133, 137–138, 157
and tree of life, 275
Monera. *See* Bacteria
Monism, 51, 428g
Monod, Jaques, 137
Morgan, Conway Lloyd, 148
Morgan, Thomas Hunt, 79, 81, 84, 122, 123, 124, 135, 148, 277
Morowitz, H. J., 374
Morphogenesis, 136, 137, 158, 161
Morphogenetic field, 105, 121, 122, 123, 428g
Morphology, 28, 47, 68, 69, 122, 146, 199, 223, 224, 369. *See also* Development, embryological; *Generelle Morphologie*; Plasticity
inducing variations in, 174, 175, 199, 200, 223, 276, 396
innovations in, 146, 165, 174, 175, 177
rational, 160, 162
revolt against, 51
and variations in cilate cortex, 219
Morphotype, 158, 428g
Motzkin, Gabriel, xv
mRNA. *See* Transcription; RNA
Müller, Hermann, 81, 84
Muralov, Aleksander, 82
Museum of Vertebrate Zoology (MVZ), 309, 310–315
Mutations, 48, 63, 78, 83, 110, 112, 122, 124, 131, 140, 161, 216
adaptive, 184, 185, 367
in bacteria (*see* Bacteria, mutation in)
biased, 223, 348, 398
deleterious, 234
fixation of, 185, 197, 246, 290
hotspots, 348
induced, 81, 110, 117, 124, 139, 140, 175, 177, 354
in population genetics, 113, 115, 131, 182, 186
pressure of, 104, 127
rate of (stability of), 117, 177, 184, 254, 258, 331, 348, 392
regulatory, 181, 227
spontaneous or random, 138, 173, 184, 185, 227, 272, 286, 347, 348, 361, 392
Mutator, 227, 254
Mycobacterium tuberculosis, 206

Nägeli, Karl Wilhelm von, 29, 49, 50
Nanney, David, 216, 217
Napoleon, 11
Nardon, Paul, 278
Narrative, 21, 22, 23, 28, 29, 30, 90, 92, 98, 123–124, 403
causal, 335, 336, 337, 341
National Museum of Natural History of Paris (Muséum d'Histoire Naturelle), 11, 24, 36, 238
Natural selection. *See also* Adaptation; Darwinism; Evolution, adaptive; Fitness; Function; Gradualism; Selection
Darwin's view of, 23, 24, 40, 41, 42, 128, 159–161, 164, 227 (*see also* Darwin)
and development, 113, 121
eliminative, 22, 93
evolution of, 357, 358
and function, 304, 305, 357, 358 (*see also* Function)
and plasticity, 146, 197 (*see also* Norm of reaction, evolution of; Plasticity)
as survival of the fittest, 65, 93, 227, 361 (*see also* Fitness)
Need, 33, 34, 45, 46, 50, 51, 69, 371
Neo-Darwinism, xii, 29, 79, 80,161–162, 428g
and bacterial evolution, 275
challenges to, 333, 406–408 (*see also* Epigenetic inheritance, evolutionary implications of)
definition of term, 105, 428g
and Lysenkoism, 79, 80, 82, 84, 86
and niche construction, 307
and symbiosis, 278
as theory of evolution, 78, 79, 140, 161, 162, 181–182, 187, 397
Neo-Lamarckians, 59, 62, 65, 71–73, 95, 105
Neo-Lamarckism/Lamarckianism, 4, 6, 7, 26, 103, 278, 428g. *See also* Lamarckism; Transformism
American, 25–27, 50
definition of, 428g

French, 30, 70–71, 73–74
golden age in Germany, 45–53
and inheritance of acquired characters, 30, 67, 70–71, 73–74
and its theoretical impasse, 72–73
Old School Darwinism, 51, 428g
and plasticity, 67, 68, 69, 71–73
Nervous system and behavioral evolution, 178, 252, 397
Networks, 6, 257, 258, 360, 399
cellular, 145, 147, 259, 402
neuroendocrine, 174
regulatory genetic, 149, 163, 188, 228, 230, 254, 347, 425g
self-sustaining (*see* Epigenetic inheritance systems, self-sustaining loops)
social, 95
Neuroendocrine stress, 147, 171, 174, 178
Neurospora, 135
New Biology, 385–387
Newman, Stuart, xvi, 146, 148, 224, 304
Newton, Isaac, 160
Newtonian mechanics, 160
Niche
adaptive, 131,
developmental/ontogenetic, 323, 324, 325, 399, 424g
ecological, 146, 207, 158, 177, 207, 311
intracellular, 289
Niche construction, 166, 244, 286, 299, 399, 428g
concept of, 300, 306–308, 428g
ecological, 177, 207, 399
and localization, 309–314, 315
modeling evolutionary effects of, 308, 309
Nicotiana glauca, 255, 257, 259
Noncomplementary diploids, 379
Nonlinearity
causation, 385, 386
dynamical system, 360
dynamics, 304, 382
process, 161
Non-Mendelian inheritance, 106, 117, 118, 135, 136, 215, 277, 331, 428g. *See also* Cyotoplasmic inheritance; Epigenetic inheritance; Plasmagenes
in animals, 231, 245, 246
in fungi, 136
in plants, 136, 268 (*see also* Allopolyploidization)
in protists, 136
Nordenskiöld, Eric, 25
Norm of reaction, 193–197, 201, 325, 428g
evolution of, 194, 197–198, 399
genetic diversity in, 195–196
Novak, J., 64
Novelty, 93, 110, 132, 158, 227, 267, 430
emergence in evolution, 93, 181, 319, 360, 361, 165–166
regulatory, 181 (*see also* Genome, rewiring of; Genome, shock and stress response)
Novosibirsk, 147, 172, 178
Laboratory of evolutionary genetics, 171, 176
Nucleosome, 348, 414, 415, 428g
Nutrition, 12, 49, 59, 239
of mother and fetus, 239, 240, 241, 243–244, 431 (*see also* Thrifty phenotype)
poor, 12, 152, 431
transition in, 243
in Weismann's theory, 60–62, 64, 65

Oken, Lorenz, 38, 46, 160
On the Origin of Species by Means of Natural Selection, or the Preservation of Favoured Races in the Struggle for Life, Darwin, C., 25, 40, 41, 42, 50, 67, 89, 93, 128, 167, 272, 278
Ontogeny/ontogenesis, xii, 28, 78, 26, 28, 122, 301, 303
Operon, 137, 138, 378, 428g
Oppenheim, Amos, 147, 402
Organism, 4, 21–22, 26, 104, 116,146, 159, 164–166,174, 347. *See also* Environment, and organism interaction; Holobiont; Individuality; Lamarck, on organisms; Model, organism; Self-organization
and agency/activity, xiii, 22, 91, 146, 166, 298, 299, 360
French Neo-Lamarckists's view of, 68, 69, 70–71, 73, 74 (*see also* Durkheim Émile)
Haeckel's view of, 46–48
superorganism, xiii, 92, 94 (*see also* Holobiont; Symbiosis)
Organization, biological, 29, 147, 160, 385
levels of, xii, 7, 12, 29, 147, 150, 174, 251, 257, 327, 374
Origin of life, 11, 298, 360
Orr, Henry P., 28
Orthologous chromosomes, 263, 428g
Orthogenesis, 26, 27, 50, 104, 105, 109, 113, 157, 160, 428g
Osborn, Henry Fairfield, 50, 148
Oyama, Susan, 298, 299, 383, 424

Packard, Alpheus S., 25, 26, 428
Paleontology, 21, 26, 80, 104, 113, 122, 128
Palmer, Richard A., 396
Pangenesis, 78, 84, 85, 116, 161, 162, 163, 429g
Pangenome, 254, 429g
Panmixia, 61, 429g
Parallel induction, 62, 64, 65, 221, 223
Parallelism, 238
Paramecium, 115, 136, 137
Paramutation, 119, 216, 231–234, 256, 429g
Parnes, Ohad, 404
Pasteur, Louis, 68
Paszkowski, Jerzy, 221
Pattern formation, 158. *See also* Morphogenesis

Pauly, August, 29, 51, 429
Pearl, Sivan, 147, 402
Penetrance, 338, 429g
Persistence in bacteria. *See* Bacterial persistence
Phage. *See* Virus, bacterial
Phenotype, 78, 114. *See also* Epigenetic landscape; Holobiont
continuity, 70, 146
heritable, 187, 222, 341, 418 (*see also* Epigenetic inheritance; Soft inheritance)
plasticity, 74, 123, 161, 166, 190, 193, 251, 429g
thrifty, 241, 429, 431g
variation in, 150, 154, 159, 183, 233, 377 (*see also* Variability)
Philosophie zoologique, Lamarck J-B., xi, 11,13, 14, 25, 34, 36, 41, 42, 67, 75, 162, 238, 252, 271, 345, 359, 370, 373, 375, 407, 411
Phylogenetics, 273, 278, 425
Phylogeny, 22, 84, 152, 257, 396, 430
Physics, 4, 5, 6, 7, 27, 45, 90, 111, 137, 381, 386. *See also* Materialism, mechanistic; Self-organization
of development, 146, 161, 163–166
terrestrial, 11
Physiology, 68, 84, 146, 174, 223, 246, 307, 327, 369, 381, 428
developmental, 83, 86
ecological, 82
of genomes, 303, 346
of plants, 69, 77, 82, 256
Pigliucci, Massimo, 4, 7
Planaria, 121, 122
Plasmagenes, 79, 134, 136, 137, 278, 429g
theories of, 136, 137, 278
Plasmid, 182, 274, 278
Plasmon, 135, 136, 429g
Plasticity, xiii, 26, 31, 301, 326, 385, 387
adaptive, xii, 195, 198, 199, 200, 252, 319
and disease, 241–246
bacterial, 147 (*see also* Bacterial persistence)
behavioral, 252, 253
cellular, 182, 190, 252
conditional, 148
definitions of, 145, 146, 193
developmental, 145, 146, 241, 284
effect on evolution, 148, 319, 399–400
evolution of, 145–146, 148, 193–201, 325, 397–398, 399 (*see also* Genetic accommodation; Genetic assimilation; Norm of reaction, evolution of)
and Fixity, 300, 321, 322, 323
among French neo-Lamarckians, 67–75
generic, 146, 157, 159 (*see also* Physics, of development)
genomic, 145, 152, 303, 426g (*see also* Genome, shock and stress response; Stress, genomic)
inevitable, 194, 195, 199
in Lamarck's theory, 72
in neo-Lamarckism, 26, 27, 30, 67–74
and "new biology," 385, 387
open-ended, 27, 148
phenotypic, 74, 123, 161, 166, 190, 193, 251, 429g
in plants, 158, 251, 252–257
programmed, 148
in rewired yeast, 181–190
and seasonal dimorphism, 62, 64
and sexual dimorphism, 158
social, 31, 94, 95, 96–97
transgenerational, 199–200, 284, 301 (*see also* Epigenetic inheritance)
under stress, 175
unprogrammed (nonadaptive), 158
Plastid/s, 130, 136, 273
Plastogenes, 134. *See also* Chloroplasts
Plate, Ludwig, 29, 51, 52, 53
Polygonum, 147, 198, 199, 200
Polymerase chain reaction, 124, 264
Polyommatus plaeas, 62
Polyphenism, 62, 64, 396, 397–398, 429g. *See also* Plasticity, and sexual dimorphism
Polyploidization, 112, 152, 220, 223, 253, 254. *See also* Allopolyploidization
Polyploidy, 112, 261, 351, 350
Polyteny, 253, 429g
Population, 22, 57, 104, 121
dynamics, 104, 181, 210, 211, 212, 399, 402
ecology, 308
genetics, 111, 124, 127, 128, 130, 131, 136, 140, 222, 283, 301, 308, 381, 391, 406, 428 (*see also* Modern Synthesis)
natural, 59, 61, 63, 104, 113, 118, 128, 138, 194, 198, 333, 382–383, 396, 397
of symbionts, 284, 287, 291 (*see also* Holobiont; Hologenome)
thinking, 127, 128
wild, 171, 174
Portier, Paul, 276
Position effects, 119
Power of nature ("pouvoir de la nature"), 373, 375, 429g. *See also* Lamarck, on "power of life"
Predator-prey interactions, 148, 210–213, 429g. *See also* Bacterial persistence; Model, of predator prey interactions
Predictive adaptive response (PAR), 429g
epigenetic correlates of, 241
hypothesis, 242, 243
and menarche, 242–243
Primitive cell, 357, 359, 363, 429. *See also* Progenote
Primitive life, 304, 305, 357, 429
Primitive plasticity, 159. *See also* Plasticity, generic
Princeton conference, 105
Principles of Psychology, The, Spencer, H., 91, 92

Prion, 218, 220, 304, 362, 419, 420, 429g. *See also* Epigenetic inheritance, structural
Principles of Geology, Lyell, C., 16
Problems of Genetics, Bateson, W., 276
Progenote, 273, 274, 429g
Progress, 25, 28, 53, 73, 89, 96, 97, 98, 374, 429
Progressive evolution, 21, 23, 45, 48, 59, 60, 67, 72, 73, 74, 367, 373, 374, 381, 383, 429g
Prokaryote, 150, 273, 377
Promoter, 181, 182, 183, 186, 187, 241, 254, 289
Protein chaperones, 183
Protists (protozoa), 116, 136, 151, 219, 220, 224, 276, 279
Protoplasm, 70, 371
Protoplasmic continuity, 70
Provine, William, xvi, 16, 31, 103, 106, 133
Pseudomonas aeruginosa, 206
Pugh, John, 216, 390
Psychological essentialism, 320
Psychology, 21, 90, 91, 92, 97, 298, 404, 423
 evolutionary, 326, 367, 403
Purpose, 47, 48, 320, 359, 360. *See also* Teleology

Rabaud, Étienne, 68
Race, 3–4, 36, 38, 39, 97, 99
Racism, 3–4
Random drift, 104, 194, 258. *See also* Drift, genetic
Random mutation, 131, 138, 182, 184, 186, 227, 286, 347
Random variation, 339
Rassoulzadegan, Minoo, xvi, 151
Rational morphology, 160, 162
Raz, Gal, 219, 221
Réaumur, Réne-Antoine Ferchault de, 36
Recapitulation, 22, 23, 25, 26, 28, 29, 89, 91, 238, 430g
Recessivity, 303, 347
Recherches sur l'organisation des corps vivans, Lamarck, J-B., 11, 13
Reciprocal causation, 307, 309
Reciprocal induction, 286. *See also* Codevelopment
Reciprocal transformation, 304, 305
Recombination, 78, 131, 227, 228, 261, 268, 272, 286, 346, 348, 352, 398, 424
 frequency changes in, 175
Reductionism, 4, 5, 239, 332, 342, 385, 386, 387
 concept of, 336, 430g
 genetic, 298, 381–382
 heuristic, 331, 340–341, 426g
 methodological, 386
 theoretical, 297
Regeneration, 121–122, 251, 428
Regulation, 137, 341, 385, 386, 387, 398. *See also* Cis-trans interactions; Plasticity, adaptive
 and co-expression of genes, 189
 epigenetic, 377 (*see also* Epigenetic control mechanisms)
 and gene activity (*see* Gene Expression, regulation of)
 global, 186, 188, 190 (*see also* Genome, rewiring; Genome, plasticity of changes under stress)
Replication, 58. *See also* Allopolyploidization; Endoreduplication; Polyploidy
 of DNA, 5, 135, 216, 218, 261, 348, 386, 363, 392, 413, 414, 415, 416
 of molecules, 363
 of phage, 207, 208, 209, 210, 273
 records and trappings, 312
 of RNA, 218, 417
Reproducer, 159
Reproduction, 5, 57, 70, 72, 78, 121, 151, 243, 257, 305, 323, 358, 404, 419
 asexual, 70, 116, 151, 252, 290, 303 (*see also* Clone)
 self-, 116, 357, 423 (*see also* Self-organization)
 sexual, 12, 78, 116, 151, 221, 255, 303
Resistance
 to antibiotics, 184, 205, 219, 221, 222, 274, 275, 379
 to drought, 375
 to leptin, 241
 to malaria, 141
 to phages (*see* bacterial persistence)
Responsiveness. *See* Genetic accommodation; Genetic assimilation; Norm of reaction; Plasticity
Restriction enzymes, 124, 427
Rheostat model, 233
Richards, O. W., 112, 113
Riggs, Arthur, 216
Risk
 medical and disease, 237, 238, 239, 240, 244, 246, 398, 402
 theoretical, 331, 332, 334, 340, 341
RNA. *See also* Central Dogma; Chromatin; Gene expression; Transcription; Transposable elements, retrotransposons
 amplification of, 230 (*see also* Replication, of RNA)
 analysis of, 125
 inheritance mediated by (*see* Epigenetic Inheritance System, RNA-mediated)
 interference (RNAi), 223, 253, 257, 348, 349, 350, 416, 430g
 microRNA (miRNA), 229, 245, 378
 non-coding, 229, 230, 416
 and paramutation, 231–233
 piwi (piRNA), 230, 429g
 regulatory role in bacteria, 153
 regulatory roles in eukaryotes, 186, 229, 230, 233, 234, 245, 348, 391, 417 (*see also* Rheostat model)

RNA (cont.)
ribosomal (rRNA), 256, 267, 273, 275, 278
role in germ cells, 230, 234
silencing, 229, 230, 234
small interfering (siRNA), 151, 218, 229, 348, 350, 416
in speciation, 151, 234
theoretical precursor of, 136
transfer (tRNA), 267, 274
RNAse III family, 229
Robson, G. C., 112, 113
Robustness, 361, 362, 362, 399, 402. *See also* Canalization
Rockefeller Foundation, 137–138
Roll-Hansen, Nils, 30, 31
Romanes, George, 59, 105, 428
Roux, Wilhelm, 60
Ruse, Michael, 298, 358, 373
Ryder, John, 26

Saccharomyces cerevisiae, 182, 220
Sachs, Julius von, 29, 49
Sagan, Dorion, 279
Saint Hilaire, Étienne Geoffroy, 13, 15, 16, 46, 64, 160, 164, 238
Saint Hilaire, Isidore, 17
Salmonella, 205, 275
Sapp, Jan, 105, 135, 136, 137, 138, 153
Schmalhausen, Ivan, 80, 105, 149, 175, 373
Schweber, Sam, 99, 137, 342, 405
Selection, 22, 25, 38, 51, 59, 113, 145, 305
artificial, 171, 177, 252, 258
canalizing/stablizing, 113, 149, 166, 195, 197, 395 (*see also* Genetic assimilation; Norm of reaction, evolution of; Plasticity, evolution of)
for coordination among levels, 152, 251, 257, 259
cumulative, 109, 110, 127, 130, 335
Darwin on, 23, 24, 40, 41, 42, 128, 159–161, 164, 227
destabilizing, 175, 178, 424g (*see also* Domestication; Stress, genomic response to)
for epigenetic inheritance, 188, 200, 205, 213
of epigenetic variations, 222, 228, 229, 233, 234, 253, 259, 263, 267, 268, 402 (*see also* Rheostat model)
germinal, 57, 60–65
group, 287
lab experiments on, 118
levels of, 3, 4, 6, 7, 9, 29, 59, 61–63, 65, 153, 257, 287, 297, 298, 299, 358, 367
in neo-Lamarckism, 28, 29, 47, 48, 49, 50, 59, 65, 70, 71, 73, 84, 89
and niche construction (*see* Niche construction)
for plasticity, 147, 150, 397, 398
pressure of, 171, 258, 307, 308, 309, 350, 351, 399
of regulatory mechanisms, 351
relaxed (panmixia), 61, 399, 429g
sexual, 3, 40, 112, 116
somatic (internal), 254, 258, 358, 427
for symbiotic interactions, 287
for tameness, 147, 171–174 (*see also* Domestication)
units of, 251, 297
Selectionism, xii, 53, 137
Selective stabilization, 98, 147, 164, 304–305, 414, 425g
Self-assembly, 163–164, 357, 430g
Self-organization, xi, 5, 22,146, 161, 304, 430. *See also* Development; Plasticity,
Kant on, 160–161, 359
Lamarck on, 146, 163, 304
in metazoan development, 164, 166 (*see also* Plasticity, generic)
and the origin of function, 304, 359–361
Semon, Richard, 28, 29, 51, 52, 53, 224, 428
Serebrovski, Aleksander, 81, 84
Serratia symbiotica, 289
Severtsov, A. N., 373
Sexual dimorphism, 158
Shamakina, Inna, 400
Shrew, 177
Sickle-cell anemia, 133, 138, 140–141
Silver foxes, 147
domestication of, 171–174
piebaldness in, 173
Simon, Herbert, 304, 360, 361
Simpson, George Gaylord, 116, 128, 297
Smart matter, 305, 364. *See also* Self-organization
Sober, Elliott, 298
Social imaginary, 90, 430g
Social sciences, 21, 52, 90, 92, 98, 402, 403, 404, 405. *See also* Sociology; Transfer
Sociobiology, 298, 403, 404
Sociology, xi, 21, 31, 89–98, 395, 403–404, 407
Soft inheritance, xii, 30, 86, 103, 150, 154, 219, 223, 224, 395, 430g. *See also* Darwinism; Direct induction; Epigenetic inheritance; Heredity, as memory; Inheritance of acquired characters; Lamarckian inheritance; Parallel induction; Somatic induction
in Britain, 109–119
and Central Dogma, 106
characterization of, xiii, 105, 110, 159, 223, 430g
and extended synthesis, 406–408
and French neo-Lamarckism, 67, 70
and genetic assimilation, 390–392, 396–399
molecular interpretation of, 223–224 (*see also* Epigenetic inheritance; Mutation, adaptive; Genome, as response system)
origin of term, 107, 135
reductionist methodology, 125
rejection of, 103, 105, 106, 117–119, 121–125, 127–131, 135–137, 386 (*see also* Modern Synthesis, exclusion of soft inheritance)
in the United States, 25–27, 50

Somaclonal variation, 252, 430g
Somatic induction, 221, 223
Sonneborn, Tracy, 79, 136, 137, 215, 279, 419, 424
Soyfer, Valerii, 77
Speciation, 4, 25, 46, 48, 50, 62, 63, 68, 86, 92, 104, 109, 116, 151, 152, 169, 227, 230, 379, 396. *See also* Epigenetic inheritance, evolutionary significance of; Evolution, macroevolution; Hybridization; Lamarckism; Modern Synthesis; Neo-Darwinism; Stress
by allopolyploidization (*see* Allopolyplodization)
by geographic isolation, 128, 201
Mayr's view of, 128
through symbiosis (*see* Symbiosis, evolutionary role of; Transformism)
Species, xii, 39, 82, 113, 424, 426, 427, 428, 429, 430, 431
concept of, 7, 13, 275, 298 (*see also* Holobiont; Progenote)
domesticated (*see* Domestication)
folk classification of, 320–326
geographic distribution of, 147
invasive, 166, 219
locality of, 300, 307, 308–315
Spencer, Herbert, 29, 31, 59, 65, 90–98
on organism and environment, 91–92
on psychology, 91–93, 97
on superorganism, 92
Spontaneous generation, 10–13, 163. *See also* Emergence; Self-organization
Stability
developmental, 303, 323, 351 (*see also* Canalization)
dynamic and the genome, 349–350, 351
and epigenetic inheritance, 219, 221, 222, 301, 338, 395, 401–402, 419, 423
and epimutation, 391, 392
and genome metastability, 349, 350, 351, 428 (*see also* Genome, metastable)
of heredity, 27, 30, 73, 79
as molecular persistence, 304, 361, 363
of social order, 90
Stalin, Joseph V., 82
Stalinist, 81, 82, 411
Staphylococcus aureus, 206
Stebbins, George Ledyard, 116, 128, 134
Sterelny, Kim, 298
Stress, 73, 147, 195, 198, 200, 201, 206, 207, 208, 209, 212, 216, 220, 223, 224, 239, 240, 287, 288, 289, 290, 427, 430
and domestication, 171, 174, 175
emotional, 176
environmental, 18, 152, 198, 199, 200, 219, 223, 275, 286, 350, 427
genomic response to, 150, 152, 174, 183–184, 188, 216, 263, 303, 345, 349, 350–352, 402
in human evolution, 178
inducing variability, 171–178, 183–184, 189, 190, 208, 219–222, 223, 224, 255 (*see also* Variability, stress-induced; Selection, destabilizing)
neuroendocrine, 147, 171, 174, 178
role in bacterial evolution, 152, 206, 208, 254, 289 (*see also* Evolution, in bacteria; Bacterial persistence)
role in plant evolution, 152, 198, 254–256, 263, 268 (*see also* Allopolyploidization; Hybridization; Polyploidy)
Structural inheritance. *See* Epigenetic Inheritance Systems, structural
Structure. *See also* Morphology; Self-organization
of chromatin, 229, 263, 377, 416
composite molecular, 262, 272, 327, 381, 419
conceptual, 72, 319, 331, 336, 340, 407, 408
Darwin on, 41–42, 272
of genome, 152, 223, 255, 261, 263, 269, 424 (*see also* Chromosomes)
of germ plasm (*see* Weismann August, germ plasm theory)
Lamarck on, 10, 12, 24, 35, 36 (*see also* Habit)
Neo-Lamarckism on, 59
supramolecular, 279
Subtle fluids, 35, 45, 46, 163, 167, 223
Sultan, Sonia, 147, 148, 399, 400, 405, 406
Superorganism, 92, 94, 154, 275. *See also* Holobiont
Supervening relations, 7, 303, 345, 346
Symbiodinium, 287, 288, 289
Symbiogenesis, 124, 151, 276, 424. *See also* Codevelopment; Symbiopoeisis
Symbiont, xiii, 124, 283. *See also* Symbiosis; Codevelopment; Symbiopoiesis
chloroplasts as, 276, 278
mitochondria as, 278
shuffling, 288, 289
switching, 288
transfer between hosts, 286
variations in, 286, 289, 290
Symbionticism and the Origin of Species, Wallin, I. E., 276
Symbiogenesis, 276
Symbiopoiesis, 284, 291, 430g
Symbiosis, 153, 154, 275, 276, 278, 279, 284, 361, 430g. *See also* Symbiogenesis; Holobiont; Hologenome; Horizontal gene transfer
with bacteria, 279, 284, 285, 291
between coral and algae, 287–289
developmental, 154, 284, 399, 424g (*see also* Development, codevelopment)
endosymbiosis, 278, 288, 425g
evolutionary dynamics, 286–287, 288, 289, 290
evolutionary role of, 257, 275–280, 285, 286, 287, 290
hereditary, 277, 278, 279, 280
in mammalian gut, 285, 286
molecular, 304, 362

Symbiosis (cont.)
neo-Darwinist reaction to, 277
between pea aphid and bacterium, 289–290
selection for, 286–291
between squid and bacterium, 284–285
with viruses, 153, 277, 278 (*see also* Bacteria lysogenic)
with *Wolbachia*, 279, 285
Synchronization (of oscillations), 164
Systema Naturae, Linneus, C., 272
Systematics and the Origin of Species from the Viewpoint of a Zoologist, Mayr, E., 128
Système des animaux sans vertèbres, Lamarck, J-B., 11
Systems biology, 354, 430g
Szathmáry, Eörs, 153, 427

Tal, Omri, 402
Tauber, Alfred, 103, 368
Teleology, 30, 47, 51, 160, 297, 300, 321, 322, 323, 373, 411. *See also* Function
Teleomechanists, 160, 162
Tempo and Mode in Evolution, Simpson, G. G., 128
Teratology, 68, 238, 239, 431g
Tetrad analysis, 187–188
Theory of Evolution, The, Maynard Smith, J., 118
Thought experiment, 345, 349, 350
Thrifty phenotype, 241, 429, 431g. *See also* Disease, prenatal effects on
Tiedemann, Friedrich, 15
Toy model, 333, 334, 335, 431g
Transcription
factors, 163, 164, 254, 274, 425
global effect on mRNA, 186, 188–189
Transfer, 31, 89, 90, 95–98, 395, 402, 403, 431g
Transformism, 16, 17, 67, 68, 73, 74, 93, 431g. *See also* Lamarck; Lamarckism; Neo-Lamarckism
experimental, 67, 73, 74
Transformist, 15, 24, 39, 68, 89, 373, 407
Transgenerational effects, 219–222, 400–401. *See also* Epigenetic inheritance; Soft inheritance
of alcohol and morphine, 400–401
of malnutrition (*see also* Disease, fetal origins of; Maternal effects)
of psychological stress, 217, 244
Transgenes, 217
Transmutation, Lamarck's view on, 23, 41, 89, 161
Transposable elements, 177, 222, 228, 254, 256, 354, 414
activated by stress, 254
epigenetically controlled, 230, 267, 350
helitrons, 254
retrotransposons, 267, 268
Transposons. *See* Transposable elements
Triticum, 263, 264, 265, 266
Trut, Lyudmila, 146, 147, 152
Typicality, 300, 321–324
Typological thinking, 124

Uniformitarianism, 158, 373, 431g
Units of selection, 251, 297
Use and disuse, 23, 24, 37, 40, 46, 50, 104, 148, 223. *See also* Lamarckism; Soft inheritance

Variability, 25, 159, 234, 251, 258, 338 (*see also* Epimutation)
bursts of, 174
capacitated, 183
cryptic, 176
epigenetic (*see* Epigenetic variability)
genetic (*see* Mutation)
social, 95
somatic and somatoclonal, 251, 252, 253, 290
stress-induced, 173, 174, 175–178, 255 (*see also* Allopolyploidization; Domestication; Genome, shock and response; Polyploidization; Selection, destabilizing; Stress)
tissue-specific, 233
Variation. *See* Variability
developmental, xii, 145, 153, 154, 284, 425
Variation and Evolution in Plants, Stebbins, G. G., 128
Variation of Animals and Plants under Domestication, The, Darwin, C., 162, 283
Variation of Animals in Nature, The, Robson, G. C. and Richards, O. W., 112
Vavilov, Nikolai, 81–84
Vermeij, J. G., 374
Vernalization (Russian *iarovizatsiia*), 83, 84, 103, 431g
Vibrio fischeri, 284, 285
Viruses, 153, 230, 239, 273, 382. *See also* Infectious heredity
and bacterial persistence, 207–213 (*see also* Bacterial persistence)
bacterial (phages), 138, 147, 208, 209, 274
λ, 207, 219, 378
and lysogenic phase, 207–208
and lytic phase, 207–208
Vulpes vulpes (silver fox), 147, 171–174

Waddington, Conrad, 80, 105, 113, 114, 115, 118, 149, 150, 323, 347, 390, 392, 425
Wallace, Alfred Russel, 93, 160
Wallin, Ivan, 276, 277
Watson, James, 134, 389
Weaver, Warren, 137, 138, 360
Weismann, August, 22, 26, 52, 57, 58, 70, 105, 129, 428, 429. *See also* Inheritance of acquired characters; Neo-Darwinism; Neo-Lamarckism; Parallel induction
on cell, 29, 58–61

challenging Lamarckism, 26, 29, 49, 70, 71, 78, 93, 98
on germinal selection, 57, 60–65
germ line segregation, 118, 128, 129, 251
germ plasm theory, 29, 57–59, 60, 63, 64–65, 78, 111, 136, 423, 424, 426g
impact of, 27, 29–30, 48, 50, 51, 71, 93, 98, 111, 116, 118, 128, 161
notion of organism, 57, 58, 59–61, 65
on seasonal dimorphism, 62, 63, 64
Weissman, Charlotte, 29, 49, 57
West-Eberhard, Mary Jane, 146, 284, 397
Wilkins, Adam, xvi, 103, 106, 368, 395, 401
Wilson, David Sloan, 298
Wilson, Edward O., 298, 373
Wimsatt, William, 298, 361
Woese, Carl, 273, 274, 363
Wolbachia, 279, 285, 290, 291
Wolterek, Richard, 123, 194
World War I, 51, 52, 121
World War II, 31, 110, 111, 113, 131, 136, 138, 239, 243, 272, 274
Wright, Sewall, 110, 118, 127, 131, 136, 386
Wyville, Thomson, 40

Xenopus, 230
X-inactivation, 119, 215, 216, 229, 390

Yeast, 139, 147, 150, 152, 182, 185, 188, 215, 220, 230, 340, 392, 419, 429
epigenetic inheritance in, 188
genome rewiring in, 182–189, 184, 185
tetrad analysis, 187–188
Yongsheng, Liu, 85

Zebra fish, 230
Zooxanthella algae, 287, 288

Printed in the United States
By Bookmasters